Asterisk™: The Definitive Guide

THIRD EDITION

Asterisk™: The Definitive Guide

Leif Madsen, Jim Van Meggelen, and Russell Bryant

O'REILLY®

Beijing · Cambridge · Farnham · Köln · Sebastopol · Tokyo

Asterisk™: The Definitive Guide, Third Edition

by Leif Madsen, Jim Van Meggelen, and Russell Bryant

Published by O'Reilly Media, Inc., 1005 Gravenstein Highway North, Sebastopol, CA 95472.

O'Reilly books may be purchased for educational, business, or sales promotional use. Online editions are also available for most titles (*http://my.safaribooksonline.com*). For more information, contact our corporate/institutional sales department: (800) 998-9938 or *corporate@oreilly.com*.

Editor: Mike Loukides
Production Editor: Teresa Elsey
Copyeditor: Rachel Head
Proofreader: Andrea Fox
Production Services: Molly Sharp

Indexer: Fred Brown
Cover Designer: Karen Montgomery
Interior Designer: David Futato
Illustrator: Robert Romano

Printing History:

June 2005:	First Edition.
August 2007:	Second Edition.
April 2011:	Third Edition.

ISBN: 978-0-596-51734-2

[LSI]

1302020201

Table of Contents

Foreword

"There's more than one way to do it." I've been working with Asterisk for nine years, and this motto becomes more true with each release, each added feature, and each clever person who attacks a telecommunications problem with this incredibly flexible toolkit. I had the fantastic opportunity to work as the community manager for the Asterisk project at Digium for two years, which gave me one of the best vantage points for seeing the scope and imagination of the worldwide development effort pushing Asterisk forward. The depth and breadth of Asterisk is staggering—installations with hundreds of thousands of users are now commonplace. I see Asterisk making deep inroads into the financial, military, hospital, Fortune 100 enterprise, service provider, calling card, and mobile environments. In fact, there really aren't any areas that I can think of where Asterisk isn't now entrenched as the default choice when there is a need for a generalized voice tool to do "stuff."

Asterisk has been emblematic of the way that open source software has changed business—and changed the world. My favorite part of any Asterisk project overview or conference talk is answering questions from someone new to Asterisk. As I continue to answer "Yes, it can do that," I watch as the person's eyes grow wide. The person starts to smile when he *really* starts to think about new things to do that his old phone or communication system couldn't possibly have done. Radio integration? Sure. Streaming MP3s into or out of phone calls? OK. Emailing recorded conference calls to the participants? No problem. Integration of voice services into existing Java apps? Easy. Fax? Instant messages? IVRs? Video? Yes, yes, yes, yes.

The affirmative answers just keep flowing, and at that point, the best thing to do is to sit the person down and start showing him quick demonstrations of how Asterisk can be quickly deployed and developed. Then, I typically point the person toward the first edition of this book, *Asterisk: The Future of Telephony*, and set him loose. In just a few hours of development (or longer, of course), companies can change the way they deliver products to customers, nonprofits can overhaul how their users interact with the services they offer, and individuals can learn to build a perfectly customized call-handling system for their mobile and home phones. Asterisk scales up and down from individual lines to vast multiserver installations across multiple continents, but the way to start is to install the package, open up some of the configuration files, and start looking at examples.

From the basic beginnings of a PBX that Mark Spencer coded in 1999, the Asterisk project, with the help of thousands of developers, has moved from simply connecting phone calls and has matured into a platform that can handle voice, video, and text across dozens of virtual and physical interface types. The creation and growth of Asterisk were the inescapable results of the convergence of the four horsemen of the proprietary hardware apocalypse: open source development ideas, the Internet, Moore's Law, and the plummeting costs of telecommunications. Even hardware vendors who may be frightened of Asterisk from a competitive standpoint are using it in their labs and core networks: almost all devices in the Voice-over-IP world are tested with Asterisk, making it the most compatible system across vendors.

At a recent communications conference I attended, the question "Who uses Asterisk?" was posed to the 1000-plus crowd. Nearly 75 percent raised their hands. Asterisk is a mature, robust software platform that permeates nearly every area of the telecommunications industry and has firmly cemented itself as one of the basic elements in any open source service delivery system. I tell people that it's reasonable for anyone delivering services both via phone and web to want to add an "A" for Asterisk to the LAMP (Linux, Apache, MySQL, [Perl/Python/PHP]) acronym, making it LAAMP. (LAMA-P was another option, but for some reason nobody seems to like that version...I don't know why.)

The expansion of this book to include more examples is something I've been looking forward to for some time. Asterisk is accessible because of the ease with which a novice can understand basic concepts. Then it continues to succeed as the novice becomes a pro and starts tapping the "other ways to do it" with more sophisticated implementations, using AGI with Java, Perl, or Python (or one of the other dozen or so supported languages), or even writing her own custom apps that work as compile-time options in Asterisk. But the first step for anyone, no matter what his or her skill level, is to look at examples of basic apps others have written. Leif, Jim, and Russell have not only put together a fantastic compendium of Asterisk methods, but they have also provided an excellent list of examples that will let the novice or expert quickly learn new techniques and "more than one way to do it."

Asterisk 1.x is fantastically powerful and can solve nearly any voice problem you might have. For those of you building the most complex installations, there is even more interesting work—which will be realized quite soon—in development. The currently-in-development Asterisk SCF (Scalable Communications Framework) is being built as an adjunct open source project to allow Asterisk 1.x systems to scale in even more powerful ways—stay tuned, or better yet, get involved with the project as a developer.

If you're an experienced Asterisk developer or integrator, I'm sure this book will have a few "Hey, that's a neat way to do it!" moments for you, which is one of the joys of Asterisk. If this is your first project with Asterisk, I'd like to welcome you to the huge community of users and developers dedicated to making Asterisk better. This book will take you from a vague idea of doing something with computers and voice communication to the point where you're able to stun everyone you know with your phone system's sophisticated tricks.

You're encouraged to participate in the online mailing lists, IRC chatrooms, and yearly AstriCon conference that provide up-to-the-second news and discussion surrounding the project. Without your interest, input, and code, Asterisk wouldn't exist. Open source projects are hungry for new ideas and excellent contributions: I encourage you to be a participant in the Asterisk community, and I look forward to seeing your questions and examples in the next edition of this book.

—John Todd

Preface

This is a book for anyone who uses Asterisk.

Asterisk is an open source, converged telephony platform, which is designed primarily to run on Linux. Asterisk combines more than 100 years of telephony knowledge into a robust suite of tightly integrated telecommunications applications. The power of Asterisk lies in its customizable nature, complemented by unmatched standards compliance. No other PBX can be deployed in so many creative ways.

Applications such as voicemail, hosted conferencing, call queuing and agents, music on hold, and call parking are all standard features built right into the software. Moreover, Asterisk can integrate with other business technologies in ways that closed, proprietary PBXs can scarcely dream of.

Asterisk can appear quite daunting and complex to a new user, which is why documentation is so important to its growth. Documentation lowers the barrier to entry and helps people contemplate the possibilities.

Produced with the generous support of O'Reilly Media, *Asterisk: The Definitive Guide* is the third edition of what was formerly called *Asterisk: The Future of Telephony*. We decided to change the name because Asterisk has been so wildly successful that it is no longer an up-and-coming technology. Asterisk has arrived.

This book was written for, and by, members of the Asterisk community.

Audience

This book is intended to be gentle toward those new to Asterisk, but we assume that you're familiar with basic Linux administration, networking, and other IT disciplines. If not, we encourage you to explore the vast and wonderful library of books that O'Reilly publishes on these subjects. We also assume you're fairly new to telecommunications (both traditional switched telephony and the new world of Voice over IP).

However, this book will also be useful for the more experienced Asterisk administrator. We ourselves use the book as a reference for features that we haven't used for a while.

Organization

The book is organized into these chapters:

Chapter 1, A Telephony Revolution
This is where we chop up the kindling and light the fire. Welcome to Asterisk!

Chapter 2, Asterisk Architecture
Discusses the file structure of an Asterisk system.

Chapter 3, Installing Asterisk
Covers obtaining, compiling, and installing Asterisk.

Chapter 4, Initial Configuration Tasks
Describes some initial configuration tasks for your new Asterisk system. This chapter goes over some of the configuration files required for all Asterisk installations.

Chapter 5, User Device Configuration
Provides guidance on configuring Asterisk to allow devices such as telephones to connect and make calls.

Chapter 6, Dialplan Basics
Introduces the heart of Asterisk, the dialplan.

Chapter 7, Outside Connectivity
Discusses how to configure Asterisk to connect to other systems, such as other Asterisk servers, Internet telephony service providers, or the plain old telephone network.

Chapter 8, Voicemail
Covers the usage of one of the most popular applications included with Asterisk, the voicemail system.

Chapter 9, Internationalization
Focuses on issues that an Asterisk administrator should be aware of when deploying a system outside of North America.

Chapter 10, Deeper into the Dialplan
Goes over some more advanced dialplan concepts.

Chapter 11, Parking and Paging
Describes the usage of two popular telephony features included with Asterisk, call parking and paging.

Chapter 12, Internet Call Routing
Covers techniques for routing calls between different administrative domains on the Internet.

Chapter 13, Automatic Call Distribution (ACD) Queues
Discusses how to build call queues in Asterisk.

Chapter 14, Device States
> Introduces the concept of device states and how they can be used as presence indicators.

Chapter 15, The Automated Attendant
> Covers how to build a menuing system using the Asterisk dialplan.

Chapter 16, Relational Database Integration
> Discusses various ways that Asterisk can be integrated with a database.

Chapter 17, Interactive Voice Response
> Goes over how Asterisk can be used to build applications that act on input provided by a caller.

Chapter 18, External Services
> Provides instructions on how to connect to external services including LDAP, calendars, IMAP for voicemail, XMPP, Skype, and text-to-speech.

Chapter 19, Fax
> Discusses the various options for integrating sending and receiving faxes with an Asterisk system.

Chapter 20, Asterisk Manager Interface (AMI)
> Introduces a network API for monitoring and controlling an Asterisk system.

Chapter 21, Asterisk Gateway Interface (AGI)
> Introduces the Asterisk API that allows call control to be implemented in any programming language.

Chapter 22, Clustering
> Discusses a number of approaches for clustering multiple Asterisk servers together once the demands of a deployment exceed the capabilities of a single server.

Chapter 23, Distributed Universal Number Discovery (DUNDi)
> Covers a peer-to-peer protocol native to Asterisk that can be used for call routing.

Chapter 24, System Monitoring and Logging
> Introduces some of the interfaces available for logging and monitoring an Asterisk system.

Chapter 25, Web Interfaces
> A survey of some of the web interfaces that complement an Asterisk installation.

Chapter 26, Security
> Discusses some common security issues that Asterisk administrators should be aware of.

Chapter 27, Asterisk: A Future for Telephony
> In conclusion, we discuss some of the things we expect to see from open source telephony in the near future.

Appendix A, Understanding Telephony
> Explores the technologies in use in traditional telecom networks. This used to be a chapter in old versions of this book. Although not directly relevant to Asterisk

we felt that it might still be useful to some readers, so we've left it in the book as an appendix.

Appendix B, Protocols for VoIP
Delves into all the particularities of Voice over IP. This was also a chapter in old versions of this book.

Appendix C, Preparing a System for Asterisk
Contains information you should be aware of and take into consideration when planning an Asterisk deployment.

Software

This book is focused on documenting Asterisk version 1.8; however, many of the conventions and much of the information in this book is version-agnostic. Linux is the operating system we have run and tested Asterisk on, and we have documented installation instructions for both CentOS (Red Hat Enterprise Linux–based) and Ubuntu (Debian-based) where they differ from each other.

Conventions Used in This Book

The following typographical conventions are used in this book:

Italic
Indicates new terms, URLs, email addresses, filenames, file extensions, pathnames, directories, and package names, as well as Unix utilities, commands, options, parameters, and arguments.

`Constant width`
Used to display code samples, file contents, command-line interactions, library names, and database commands.

`Constant width bold`
Indicates commands or other text that should be typed literally by the user. Also used for emphasis in code.

`Constant width italic`
Shows text that should be replaced with user-supplied values.

`[ Keywords and other stuff ]`
Indicates optional keywords and arguments.

`{ choice-1 | choice-2 }`
Signifies either `choice-1` or `choice-2`.

 This icon signifies a tip, suggestion, or general note.

This icon indicates a warning or caution.

Using Code Examples

This book is here to help you get your job done. In general, you may use the code in this book in your programs and documentation. You do not need to contact us for permission unless you're reproducing a significant portion of the code. For example, writing a program that uses several chunks of code from this book does not require permission. Selling or distributing a CD-ROM of examples from O'Reilly books does require permission. Answering a question by citing this book and quoting example code does not require permission. Incorporating a significant amount of example code from this book into your product's documentation does require permission.

We appreciate, but do not require, attribution. An attribution usually includes the title, author, publisher, and ISBN. For example: "*Asterisk: The Definitive Guide*, Third Edition, by Leif Madsen, Jim Van Meggelen, and Russell Bryant (O'Reilly). Copyright 2011 Leif Madsen, Jim Van Meggelen, and Russell Bryant, 978-0-596-51734-2."

If you feel your use of code examples falls outside fair use or the permission given above, feel free to contact us at *permissions@oreilly.com*.

Safari Books Online

 When you see a Safari Books Online icon on the cover of your favorite technology book, that means the book is available online through the O'Reilly Network Safari Bookshelf.

Safari offers a solution that's better than ebooks. It's a virtual library that lets you easily search thousands of top tech books, cut and paste code samples, download chapters, and find quick answers when you need the most accurate, current information. Try it for free at *http://safari.oreilly.com*.

How to Contact Us

Please address comments and questions concerning this book to the publisher:

O'Reilly Media, Inc.
1005 Gravenstein Highway North
Sebastopol, CA 95472
(800) 998-9938 (in the United States or Canada)
(707) 829-0515 (international or local)
(707) 829-0104 (fax)

We have a web page for this book, where we list errata, examples, and any additional information. You can access this page at:

http://oreilly.com/catalog/9780596517342

To comment or ask technical questions about this book, send email to:

bookquestions@oreilly.com

For more information about our books, conferences, Resource Centers, and the O'Reilly Network, see our website at:

http://www.oreilly.com

Find us on Facebook: *http://facebook.com/oreilly*

Follow us on Twitter: *http://twitter.com/oreillymedia*

Watch us on YouTube: *http://www.youtube.com/oreillymedia*

Acknowledgments

To David Duffett, thanks for the excellent chapter on internationalization, which would not have been served well by being written by us North Americans.

Next, we want to thank our fantastic editor, Michael Loukides, for your patience with this third edition, which took too long to get off the ground, and many long months to finally get written. Mike offered invaluable feedback and found incredibly tactful ways to tell us to rewrite a section (or chapter) when it was needed, and make us think it was our idea. Mike built us up when we were down, and brought us back to earth when we got uppity. You are a master, Mike, and seeing how many books have received your editorial oversight contributes to an understanding of why O'Reilly Media is the success that it is.

Thanks also to Rachel Head (nee Rachel Wheeler), our copyeditor, who fixes all our silly grammar, spelling, and style mistakes (and the many Canadianisms that Leif and Jim feel compelled to include), and somehow leaves the result reading as if it was what we wrote in the first place. Copyeditors are the unsung heroes of publishing, and Rachel is one of the very best.

Also thanks to Teresa Elsey, our production editor, and the rest of the unsung heroes in O'Reilly's production department.

These are the folks that take our book and make it an *O'Reilly book*.

During the course of writing this book, we had the pleasure of being able to consult with many people with specific experience in various areas. Their generous contributions of time and expertise were instrumental in our research. Thanks to Randy Resnick, organizer of the VoIP User Group; Kevin Fleming of Digium; Lee Howard, author of iaxmodem and hylafax; Joshua Colp of Digium; Phillip Mullis of the Toronto Asterisk Users Group; Allison Smith, the Voice of Asterisk; Flavio E. Goncalves, author of books on Asterisk, OpenSER, and OpenSIPS; J. Oquendo, Security Guru; Tzafrir Cohen, font of knowledge about security and lots of other stuff; Jeff Gehlbach, for SNMP; Ovidiu Sas, for your encyclopedic knowlege of SIP; Tomo Takebe, for some SMDI help; Steve Underwood, for help with fax and spandsp; and Richard Genthner and John Covert, for helping with LDAP.

A special thanks should also go to John Todd for being one of the first to write comprehensive Asterisk how-tos, all those years ago, and for all the many other things you do (and have done) for the Asterisk community.

Open Feedback Publishing System (OFPS)

While we were writing this book, O'Reilly introduced its Open Feedback Publishing System (OFPS), which allowed our book to appear on the Web as we were writing it. Community members were able to submit feedback and comments, which was of enormous help to us. The following is a list of their names or handles[*]:

> Matthew McAughan, Matt Pusateri, David Van Ginneken, Asterisk Mania, Giovanni Vallesi, Mark Petersen, thp4, David Row, tvc123, Frederic Jean, John Todd, Steven Sokol, Laurent Steffan, Robert Dailey, Howard Harper, Joseph Rensin, Howard White, Jay Eames, Vincent Thomasset, Dave Barnow, Sebastien Dionne, Igor Nikolaev, Arend van der Kolk, Anwar Hossain, craigesmith, nkabir, anest, Nicholas Barnes, Alex Neuman, Justin Korkiner, Stefan Schmidt, pabelanger, jfinstrom, roderickmontgomery, Shae Erisson, Gaston Draque, Richard Genthner, Michael S Collins, and Jeff Peeler

Thanks to all of you for your valuable contribution to this book.

Thanks to Sean Bright, Ed Guy, Simon Ditner, and Paul Belanger for assisting us with clarifying best practices for user and group policies for Asterisk installation. In the past it was common to just install Asterisk with *root* permissions, but we have elected to describe an installation process that is more in keeping with Linux best practices,[†] and these fine gents contributed to our discussions on that.

Kudos to all the folks working on the FreeSWITCH, YATE, SER, Kamailio, OpenSIPS, SER, sipXecs, Woomera, and any other open source telecom projects, for stimulating new thoughts, and for pushing the envelope.

Everyone in the Asterisk community also needs to thank Jim Dixon for creating the first open source telephony hardware interfaces, starting the revolution, and giving his creations to the community at large.

---

[*] We tried wherever possible to include the contributors' names, but in some cases could not, and therefore included their handles instead.

[†] Without starting a holy war!

Finally, and most importantly, thanks go to Mark Spencer, the original author of Asterisk and founder of Digium, for Asterisk, for Pidgin (*http://www.pidgin.im*), and for contributing his creations to the open source community. Asterisk is your legacy!

Leif Madsen

It sort of amazes me where I started with Asterisk, and where I've gone with Asterisk. In 2002, while attending school, a bunch of my friends and myself were experimenting with voice over the Internet using Microsoft's MSN product. It worked quite well, and allowed us to play video games while conversing with each other—at least, until we wanted to add a third participant. So, I went out searching for some software that could handle multiple voices (the word was conferencing, but I didn't even know that at the time, having had little exposure to PBX platforms). I searched the Internet but didn't find anything in particular I liked (or that was free). I turned to IRC and explained what I was looking for. Someone (I wish I knew who) mentioned that I should check out some software called Asterisk (he presumably must have thought I was looking for MeetMe(), which I was).

Having the name, I grabbed the software and started looking at what it could do. Incredibly, the functionality I was looking for, which I thought would be the entirety of the software, was only one component in a sea of functionality. And having run a BBS for years prior to going to college, the fact that I could install a PCI card and connect it to the phone network was not lost on me. After a couple of hours of looking at the software and getting it compiled, I started telling one of my teachers about the PCI cards and how maybe we could get some for the classroom for labs and such (our classroom had 30 computers at 10 tables of 3). He liked the idea and started talking to the program coordinator, and within about 30 minutes an order had been placed for 20 cards. Pretty amazing considering they were TDM400Ps decked out with four daughter cards, and they had only heard about them an hour prior to that.

Then the obsession began. I spent every extra moment of that semester with a couple of computers dedicated to Asterisk use. In those two months, I learned a lot. Then we had a co-op break. I didn't find any work immediately, so I moved home and continued working on Asterisk, spending time on IRC, reading through examples posted by John Todd, and just trying to wrap my head around how the software worked. Luckily I had a lot of help on IRC (for these were the days prior to any documentation on Asterisk), and I learned a lot more during that semester.

Seeing that the people who took a great interest in Asterisk at the time had a strong sense of community and wanted to contribute back, I wanted to do the same. Having no practical level of coding knowledge, I decided documentation would be something useful to start doing. Besides, I had been writing a lot of papers at school, so I was getting better at it. One night I put up a website called The Asterisk Documentation Assigned (TADA) and started writing down any documentation I could. A couple of weeks later Jared Smith and I started talking, and started the Asterisk Documentation

Project (*http://www.asteriskdocs.org*), with the goal of writing an Asterisk book for the community. That project became the basis of the first edition of this book, *Asterisk: The Future of Telephony*.

Nine years later, I'm still writing Asterisk documentation and have become the primary bug marshal and release manager for the Asterisk project, spoken at every single AstriCon since 2004 (at which Jared and I spoke about the Asterisk Documentation Project; I still have the AsteriskDocs magnet his wife made), and become a consultant specializing in database integration (thanks Tilghman for func_odbc) and clustering (thanks Mark Spencer for DUNDi). I really love Asterisk, and all that it's allowed me to do.

First, thanks to my parents Rick and Carol, for the understanding and support in everything I've done in my life. From the first computer they purchased for far too much money when I was in grade 6 (I started taking an interest in computers in grade 2 using a Commodore 64, and they got me a computer after a parent-teacher interview a few years later) to letting me use the home phone line for my BBS endeavors (and eventually getting me my own phone line), and everything else they have ever done for me, I can never thank them enough. I love you both more than you'll ever imagine.

Thanks to my Grandma T for letting me use her 286 during the years when I didn't have a computer at home, and for taking me shopping every year on my birthday for 15 years. Love lots!

To my beautiful wife, Danielle, for setting the alarm every morning before she left for work, letting me sleep those extra 10 minutes before starting on this book, and understanding when I had to work late because I went past my 9 A.M. stop-writing time, thank you and I love you so much.

There are so many people who help me and teach me new things every day, but the most influential on my life in Asterisk are the following: Mark Spencer for writing software that has given me a fantastic career, John Todd for his early examples, Brian K. West for his early help and enthusiasm on IRC, Steve Sokol and Olle Johansson for flying me to my first AstriCon (and subsequent ones!) and letting me be part of the first Asterisk training classes, Jared Smith for helping start the documentation project and doing all the infrastructure that I could never have done, Jim Van Meggelen for joining in early on the project and teaching me new ways to look at life, and Russell Bryant for being an amazing project leader and easy to work with every day, and for not holding a grudge about the bush.

Jim Van Meggelen

When we set out to write the very first edition of this book over five years ago, we were confident that Asterisk was going to be a huge success. Now, a half-decade later, we've written this third edition of what the worldwide Asterisk community calls "The Asterisk Book," and we've matured from revolutionaries into Asterisk professionals.

Asterisk has proven that open source telecom is a lasting idea, and the open source telecom landscape is nowadays complemented by more than just Asterisk. Projects like Freeswitch, sipXecs (from SipFoundry), OpenSER/Kamailio/OpenSIPS, and many, many more (and more to come) help to round out the ecosystem.

I want to take this opportunity to thank my very good friend Leif Madsen, who has been with me through all three editions. In our daily lives we don't always have many opportunities to work with each other (or even grab a pint, these days!), and it's always a delight to work with you. I also want to thank Russell Bryant, who joined us for this edition, and whose dedication to this project and the Asterisk project in general is an inspiration to me. You're a Renaissance man, Russell. To Jared Smith, who helped found the Asterisk Documentation Project and coauthored the first two editions with Leif and me (but has since moved on to the Fedora project), I can only say: Asterisk's loss is Fedora's gain.

I would like to thank my business partners at Core Telecom Innovations and iConverged LLC, without whom I could not do all the cool things I get to do in my professional career.

I would like to thank all my friends in the improv community, for helping me to keep laughing at all the challenges that life presents.

Thanks to all my family, who bring love into my life.

Finally, thanks to you, the Asterisk community. This book is our gift to you. We hope you enjoy reading it as much as we've enjoyed writing it.

Russell Bryant

I started working on Asterisk in 2004. I was a student at Clemson University and was working as a co-op engineer at ADTRAN in Huntsville, Alabama. My first job at ADTRAN was working in the Product Qualification department. I remember working with Keith Morgan to use Asterisk as a VoIP traffic generator for testing QoS across a router test network. Meanwhile, a fellow co-op and friend, Adam Schreiber, introduced me to Mark Spencer. Over the next six months, I immersed myself in Asterisk. I learned as much as I could about Asterisk, telephony, and C programming. When Asterisk 1.0 was released in the fall of 2004, I was named the release maintainer.

At the beginning of 2005, I was hired by Digium to continue my work on Asterisk professionally. I have spent the past six amazing years working with Digium to improve Asterisk. I have worked as a software developer, a software team lead, and now as the engineering manager of the Asterisk development team. I am extremely grateful for the opportunity to contribute to so many areas of the Asterisk project. There are many people that deserve thanks for the support they have provided along the way.

To my wife, Julie, I cannot thank you enough for all the love and support you have given me. Thank you for keeping my life balanced and happy. You are the best. I love you!

To my parents, thank you for giving me so many great opportunities in my life to explore different things and find what I really enjoy. You taught me to work hard and never give up.

To Leif and Jim, thank you for your invitation to contribute to this book. It has been a fun project, largely due to the pleasure of working with the two of you. Thanks for the laughs and for your dedication to this book as a team effort.

I have learned a lot from many people at Digium. There are three people who stand out the most as my mentors: Mark Spencer, Kevin P. Fleming, and David Deaton. Thank you all for going the extra mile to teach me along the way. I am extremely grateful.

To the software development team at Digium, thank you for being such an amazing team to work with. Your dedication and brilliance play a huge part in the success of Asterisk and make Digium a great place to work.

To Travis Axtell, thank you for your help in my early days of learning about Linux and for being a good friend.

To my dogs, Chloe and Baxter, thanks for keeping me company while I worked on the book every morning.

To all of my friends and family, thank you for your love, support, and fun times.

To the entire Asterisk community, thank you for using, enjoying, and contributing to Asterisk. We hope you enjoy the book!

## CHAPTER 1

A Telephony Revolution

First they ignore you, then they laugh at you,
then they fight you, then you win.

—Mahatma Gandhi

When we first set out—nearly five years ago—to write a book about Asterisk, we confidently predicted that Asterisk would fundamentally change the telecommunications industry. Today, the revolution we predicted is all but complete. Asterisk is now the most successful private branch exchange (PBX) in the world, and is an accepted (albeit perhaps not always loved) technology in the telecom industry.

Unfortunately, over the past five years the telecom industry has continued to lose its way. The methods by which we communicate have changed. Whereas 20 years ago phone calls were the preferred way to converse across distances, the current trend is to message via text (email, IM, etc.). The phone call is seen as a bit of a dead thing, especially by up-and-coming generations.

Asterisk remains pretty awesome technology, and we believe it is still one of the best hopes for any sort of sensible integration between telecom and all the other technologies businesses might want to interconnect with.

With Asterisk, no one is telling you how your phone system should work, or what technologies you are limited to. If you want it, you can have it. Asterisk lovingly embraces the concept of standards compliance, while also enjoying the freedom to develop its own innovations. What you choose to implement is up to you—Asterisk imposes no limits.

Naturally, this incredible flexibility comes with a price: Asterisk is not a simple system to configure. This is not because it's illogical, confusing, or cryptic; on the contrary, it is very sensible and practical. People's eyes light up when they first see an Asterisk dialplan and begin to contemplate the possibilities. But when there are literally thousands of ways to achieve a result, the process naturally requires extra effort. Perhaps it can be compared to building a house: the components are relatively easy to understand, but a person contemplating such a task must either a) enlist competent help or

b) develop the required skills through instruction, practice, and a good book on the subject.

Asterisk and VoIP: Bridging the Gap Between Traditional and Network Telephony

Voice over IP (VoIP) is often thought of as little more than a method of obtaining free long-distance calling. The real value (and—let's be honest—challenge as well) of VoIP is that it allows voice to become nothing more than another application in the data network.

It sometimes seems that we've forgotten that the purpose of the telephone is to allow people to communicate. It is a simple goal, really, and it should be possible for us to make it happen in far more flexible and creative ways than are currently available to us. Technologies such as Asterisk lower the barriers to entry.

The Zapata Telephony Project

When the Asterisk project was started (in 1999), there were other open-source telephony projects in existence. However, Asterisk, in combination with the Zapata Telephony Project, was able to provide public switched telephone interface (PSTN) interfaces, which represented an important milestone in transitioning the software from something purely network-based to something more practical in the world of telecom at that time, which was PSTN-centric.

The Zapata Telephony Project was conceived of by Jim Dixon, a telecommunications consulting engineer who was inspired by the incredible advances in CPU speeds that the computer industry has now come to take for granted. Dixon's belief was that far more economical telephony systems could be created if a card existed that had nothing more on it than the basic electronic components required to interface with a telephone circuit. Rather than having expensive components on the card, digital signal processing (DSP)[*] would be handled in the CPU by software. While this would impose a tremendous load on the CPU, Dixon was certain that the low cost of CPUs relative to their performance made them far more attractive than expensive DSPs, and, more importantly, that this price/performance ratio would continue to improve as CPUs continued to increase in power.

Like so many visionaries, Dixon believed that many others would see this opportunity, and that he merely had to wait for someone else to create what to him was an obvious improvement. After a few years, he noticed that not only had no one created these cards,

---

[*] The term DSP also means digital signal processor, which is a device (usually a chip) that is capable of interpreting and modifying signals of various sorts. In a voice network, DSPs are primarily responsible for encoding, decoding, and transcoding audio information. This can require a lot of computational effort.

but it seemed unlikely that anyone was ever going to. At that point it was clear that if he wanted a revolution, he was going to have to start it himself. And so the Zapata Telephony Project was born:

> Since this concept was so revolutionary, and was certain to make a lot of waves in the industry, I decided on the Mexican revolutionary motif, and named the technology and organization after the famous Mexican revolutionary Emiliano Zapata. I decided to call the card the "tormenta" which, in Spanish, means "storm," but contextually is usually used to imply a big storm, like a hurricane or such.[†]

Perhaps we should be calling ourselves Asteristas. Regardless, we owe Jim Dixon a debt of thanks, partly for thinking this up and partly for seeing it through, but mostly for giving the results of his efforts to the open source community. As a result of Jim's contribution, Asterisk's PSTN engine came to be.

Over the years, the Zapata Telephony interface in Asterisk has been modified and improved. The Digium Asterisk Hardware Device Interface (DAHDI) Telephony interface in use today is the offspring of Jim Dixon's contribution.

Massive Change Requires Flexible Technology

Every PBX in existence suffers from shortcomings. No matter how fully featured it is, something will always be left out, because even the most feature-rich PBX will always fail to anticipate the creativity of the customer. A small group of users will desire an odd little feature that the design team either did not think of or could not justify the cost of building, and, since the system is closed, the users will not be able to build it themselves.

If the Internet had been thusly hampered by regulation and commercial interests, it is doubtful that it would have developed the wide acceptance it currently enjoys. The openness of the Internet meant that anyone could afford to get involved. So, everyone did. The tens of thousands of minds that collaborated on the creation of the Internet delivered something that no corporation ever could have.[‡]

As with many other open source projects, such as Linux and so much of the critical software running the Internet, the development of Asterisk was fueled by the dreams of folks who knew that there had to be something more than what traditional industries were producing. These people knew that if one could take the best parts of various PBXs and separate them into interconnecting components—akin to a boxful of LEGO bricks—one could begin to conceive of things that would not survive a traditional

---

† Jim Dixon, "The History of Zapata Telephony and How It Relates to the Asterisk PBX" (*http://www .asteriskdocs.org/modules/tinycontent/index.php?id=10*).

‡ We realize that the technology of the Internet formed out of government and academic institutions, but what we're talking about here is not the technology of the Internet so much as the cultural phenomenon of it, which exploded in the early '90s.

corporate risk-analysis process. While no one can seriously claim to have a complete picture of what this thing should look like, there is no shortage of opinions and ideas.§

Many people new to Asterisk see it as unfinished. Perhaps these people can be likened to visitors to an art studio, looking to obtain a signed, numbered print. They often leave disappointed, because they discover that Asterisk is the blank canvas, the tubes of paint, the unused brushes waiting.‖

Even at this early stage in its success, Asterisk is nurtured by a greater number of artists than any other PBX. Most manufacturers dedicate no more than a few developers to any one product; Asterisk has scores. Most proprietary PBXs have a worldwide support team comprising a few dozen real experts; Asterisk has hundreds.

The depth and breadth of the expertise that surrounds this product is unmatched in the telecom industry. Asterisk enjoys the loving attention of old telco guys who remember when rotary dial mattered, enterprise telecom people who recall when voicemail was the hottest new technology, and data communications geeks and coders who helped build the Internet. These people all share a common belief—that the telecommunications industry needs a *proper* revolution.#

Asterisk is the catalyst.

Asterisk: The Hacker's PBX

Telecommunications companies that choose to ignore Asterisk do so at their peril. The flexibility it delivers creates possibilities that the best proprietary systems can scarcely dream of. This is because Asterisk is the ultimate hacker's PBX.

The term *hacker* has, of course, been twisted by the mass media into meaning "malicious cracker." This is unfortunate, because the term actually existed long before the media corrupted its meaning. Hackers built the networking engine that is the Internet. Hackers built the Apple Macintosh and the Unix operating system. Hackers are also building your next telecom system. Do not fear; these are the good guys, and they'll be able to build a system that's far more secure than anything that exists today. Rather than being constricted by the dubious and easily cracked security of closed systems,

---

§ Between the releases of Asterisk 1.2 and Asterisk 1.4, over 4,000 updates were made to the code in the SVN repository. Between the releases of Asterisk 1.4 and 1.8, over 10,000 updates were made.

‖ It should be noted that these folks need not leave disappointed. Several projects have arisen to lower the barriers to entry for Asterisk. By far the most popular and well known is the FreePBX interface (and the multitude of projects based on it). These interfaces (check out *http://www.voip-info.org/wiki/view/Asterisk +GUI* for an idea of how many there are) do not make it easier to learn Asterisk, because they separate you from the platform or dialplan configuration, but many of them will deliver a working PBX to you much faster than the more hands-on approach we employ in this book.

The telecom industry has been predicting a revolution since before the crash; time will tell how well it responds to the *open source* revolution.

---

the hackers will be able to quickly respond to changing trends in security and fine-tune the telephone system in response to both corporate policy and industry best practices.

Like other open source systems, Asterisk will be able to evolve into a far more secure platform than any proprietary system, not in spite of its hacker roots, but rather because of them.

Asterisk: The Professional's PBX

Never in the history of telecommunications has a system so suited to the needs of business been available, at any price. Asterisk is an enabling technology, and as with Linux, it will become increasingly rare to find an enterprise that is not running some version of Asterisk, in some capacity, somewhere in the network, solving a problem as only Asterisk can.

This acceptance is likely to happen much faster than it did with Linux, though, for several reasons:

- Linux has already blazed the trail that led to open source acceptance. Asterisk is following that lead.
- The telecom industry is crippled, with no leadership being provided by the giant industry players. Asterisk has a compelling, realistic, and exciting vision.
- End users are fed up with incompatible and limited functionality, and horrible support. Asterisk solves the first two problems; entrepreneurs and the community are addressing the latter.

The Asterisk Community

One of the compelling strengths of Asterisk is the passionate community that developed and supports it. This community, led by the fine folks at Digium, is keenly aware of the cultural significance of Asterisk and has an optimistic view of the future.

One of the more powerful side effects of the Asterisk community's energy is the cooperation it has spawned among telecommunications, networking, and information technology professionals who share a love for this phenomenon. While these cadres have traditionally been at odds with each other, in the Asterisk community they delight in each others' skills. The significance of this cooperation cannot be underestimated.

If the dream of Asterisk is to be realized, the community must continue to grow—yet one of the key challenges that the community currently faces is a rapid influx of new users. The members of the existing community, having birthed this thing called Asterisk, are generally welcoming of new users, but they've grown impatient with being asked the kinds of questions whose answers can often be obtained independently, if one is willing to devote some time to research and experimentation.

Obviously, new users do not fit any particular kind of mold. While some will happily spend hours experimenting and reading various blogs describing the trials and tribulations of others, many people who have become enthusiastic about this technology are completely uninterested in such pursuits. They want a simple, straightforward, step-by-step guide that'll get them up and running, followed by some sensible examples describing the best methods of implementing common functionality (such as voicemail, auto attendants, and the like).

To the members of the expert community, who (correctly) perceive that Asterisk is like a web development language, this approach doesn't make any sense. To them, it's clear that you have to immerse yourself in Asterisk to appreciate its subtleties. Would one ask for a step-by-step guide to programming and expect to learn from it all that a language has to offer?

Clearly, there's no one approach that's right for everyone. Asterisk is a different animal altogether, and it requires a totally different mind-set. As you explore the community, though, be aware that it includes people with many different skill sets and attitudes. Some of these folks do not display much patience with new users, but that's often due to their passion for the subject, not because they don't welcome your participation.

The Asterisk Mailing Lists

As with any community, there are places where members of the Asterisk community meet to discuss matters of mutual interest. Of the mailing lists you will find at *http://lists.digium.com*, these three are currently the most important:

Asterisk-Biz
> Anything commercial with respect to Asterisk belongs in this list. If you're selling something Asterisk-related, sell it here. If you want to buy an Asterisk service or product, post here.

Asterisk-Dev
> The Asterisk developers hang out here. The purpose of this list is the discussion of the development of the software that is Asterisk, and its participants vigorously defend that purpose. Expect a lot of heat if you post anything to this list not specifically relating to programming or development of the Asterisk code base. General coding questions (such as queries on interfacing with AGI or AMI) should be directed to the *Asterisk-Users* list.

> The Asterisk-Dev list is not second-level support! If you scroll through the mailing list archives, you'll see this is a strict rule. The Asterisk-Dev mailing list is about discussion of core Asterisk development, and questions about interfacing your external programs via AGI or AMI should be posted on the *Asterisk-Users* list.

Asterisk-Users

> This is where most Asterisk users hang out. This list generates several hundred messages per day and has over ten thousand subscribers. While you can go here for help, you are expected to have done some reading on your own before you post a query.

Asterisk Wiki Sites

The Asterisk Wiki (which exists in large part due to the tireless efforts of James Thompson—*thanks James!*) is a source of much enlightenment and confusion. Another important resource is the community-maintained repository of VoIP knowledge at *http://www.voip-info.org*, which contains a truly inspiring cornucopia of fascinating, informative, and frequently contradictory information about many subjects, just one of which is Asterisk. Since Asterisk documentation forms by far the bulk of the information on this website,* and it probably contains more Asterisk knowledge than all other sources put together (with the exception of the mailing list archives), it is a popular place to go for Asterisk knowledge.

An important new wiki project is the official Asterisk Wiki, found at *http://wiki.asterisk.org*. While not yet as full of content as *voip-info.org*, this wiki will be more formally supported and is therefore more likely to contain information that is kept current and accurate.

The IRC Channels

The Asterisk community maintains Internet Relay Chat (IRC) channels on *irc.freenode.net*. The two most active channels are *#asterisk* and *#asterisk-dev*.† To cut down on spam-bot intrusions, both of these channels now require registration to join.‡

Asterisk User Groups

Over the past decade, in many cites around the world, lonely Asterisk users began to realize that there were other like-minded people in their towns. Asterisk User Groups (AUGs) began to spring up all over the place. While these groups don't have any official affiliation with each other, they generally link to one anothers' websites and welcome members from anywhere. Type "Asterisk User Group" into Google to track down one in your area.

---

* More than 30%, at last count.

† The *#asterisk-dev* channel is for the discussion of changes to the underlying code base of Asterisk and is also not second-tier support. Discussions related to programming external applications that interface with Asterisk via AGI or AMI are meant to be in *#asterisk*.

‡ To register, run */msg nickserv help* when you connect to the service via your favorite IRC client.

The Asterisk Documentation Project

The Asterisk Documentation Project was started by Leif Madsen and Jared Smith, but several people in the community have contributed.

The goal of the documentation project is to provide a structured repository of written work on Asterisk. In contrast with the flexible and ad hoc nature of the Wiki, the Docs project is passionate about building a more focused approach to various Asterisk-related subjects.

As part of the efforts of the Asterisk Docs project to make documentation available online, this book is available at the *http://www.asteriskdocs.org* website, under a Creative Commons license.

The Business Case

It is very rare to find businesses these days that do not have to reinvent themselves every few years. It is equally rare to find a business that can afford to replace its communications infrastructure each time it goes in a new direction. Today's businesses need extreme flexibility in all of their technology, including telecom.

In his book *Crossing the Chasm* (HarperBusiness), Geoffrey Moore opines, "The idea that the value of the system will be discovered rather than known at the time of installation implies, in turn, that product flexibility and adaptability, as well as ongoing account service, should be critical components of any buyer's evaluation checklist." What this means, in part, is that the true value of a technology is often not known until it has been deployed.

How compelling, then, to have a system that holds at its very heart the concept of openness and the value of continuous innovation.

Conclusion

So where to begin? Well, when it comes to Asterisk, there is far more to talk about than we can fit into one book. This book can only lay down the basics, but from this foundation you will be able to come to an understanding of the concept of Asterisk—and from that, who knows what you will build?

## CHAPTER 2

Asterisk Architecture

First things first, but not necessarily in that order.

—Doctor Who

Asterisk is very different from other, more traditional PBXs, in that the dialplan in Asterisk treats all incoming channels in essentially the same manner.

In a traditional PBX, there is a logical difference between stations (telephone sets) and trunks (resources that connect to the outside world). This means, for example, that you can't install an external gateway on a station port and route external calls to it without requiring your users to dial the extension number first. Also, the concept of an off-site resource (such as a reception desk) is much more difficult to implement on a traditional PBX, because the system will not allow external resources any access to internal features.*

Asterisk, on the other hand, does not have an internal concept of trunks or stations. In Asterisk, everything that comes into or goes out of the system passes through a channel of some sort. There are many different kinds of channels; however, the Asterisk dialplan handles all channels in a similar manner, which means that, for example, an internal user can exist on the end of an external trunk (e.g., a cell phone) and be treated by the dialplan in exactly the same manner as that user would be if she were on an internal extension. Unless you have worked with a traditional PBX, it may not be immediately obvious how powerful and liberating this is. Figure 2-1 illustrates the differences between the two architectures.

---

* To be fair, many traditional PBXs do offer this sort of functionality. However, it is generally kludgy, limited in features, and requires complex, proprietary software to be installed in the PBX (such as vendor-specific protocol extensions).

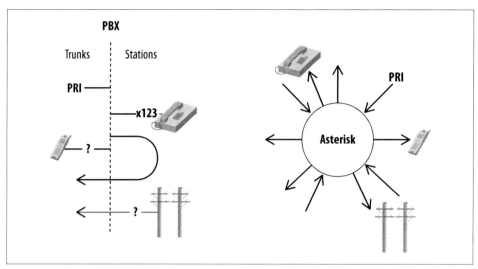

Figure 2-1. Asterisk vs. PBX architecture

Modules

Asterisk is built on *modules*. A module is a loadable component that provides a specific functionality, such as a channel driver (for example, *chan_sip.so*), or a resource that allows connection to an external technology (such as *func_odbc.so*). Asterisk modules are loaded based on the */etc/asterisk/modules.conf* file. We will discuss the use of many modules in this book. At this point we just want to introduce the concept of modules, and give you a feel for the types of modules that are available.

It is actually possible to start Asterisk without any modules at all, although in this state it will not be capable of doing anything. It is useful to understand the modular nature of Asterisk in order to appreciate the architecture.

> You can start Asterisk with no modules loaded by default and load each desired module manually from the console, but this is not something that you'd want to put into production; it would only be useful if you were performance-tuning a system where you wanted to eliminate everything not required by your specific application of Asterisk.

The types of modules in Asterisk include the following:

- Applications
- Bridging modules
- Call detail recording (CDR) modules
- Channel event logging (CEL) modules

- Channel drivers
- Codec translators
- Format interpreters
- Dialplan functions
- PBX modules
- Resource modules
- Addons modules
- Test modules

In the following sections we will list each module available within these categories, briefly identify its purpose, and give our opinion on its relative popularity and/or importance (while some modules are proven and deservedly popular, others are quite old, are barely ever used anymore, and are only maintained for the purpose of backward-compatibility). The details of how specific modules work will be covered in various chapters throughout the book, depending on what the module is and what it does. Some modules will be covered thoroughly; others may not be covered at all.

Regarding the Popularity/Status column in the tables that follow, the following list contains our opinions with respect to the meanings we have chosen (your mileage may vary):

Insignificant
> This module is ancient history. If you use it, be aware that you are mostly on your own when it comes to any sort of community support.

Unreliable
> This module is new or experimental, and is not suitable for production.

Useful
> This module is current, maintained, popular, and recommended.

Usable
> This module works but may be incomplete or unpopular, and/or is not recommended by the authors.

New
> This module is quite new, and its completeness and popularity are difficult to gauge at this time.

Deprecated
> This module has been replaced by something that is considered superior.

Limited
> This module has limitations that may make it unsuitable to your requirements.

Essential
> This module is one you'll never want to be without.

And now, without further ado, let's take a look at the modules, grouped by module type.

Applications

Dialplan applications are used in *extensions.conf* to define the various actions that can be applied to a call. The `Dial()` application, for example, is responsible for making outgoing connections to external resources and is arguably the most important dialplan application. The available applications are listed in Table 2-1.

Table 2-1. Dialplan applications

Name	Purpose	Popularity/Status
app_adsiprog	Loads Analog Display Services Interface (ADSI) scripts into compatible analog phones	Insignificant
app_alarmreceiver	Supports receipt of reports from alarm equipment	Insignificant
app_amd	Detects answering machines	Unreliable
app_authenticate	Compares dual-tone multi-frequency (DTMF) input against a provided string (password)	Useful
app_cdr	Writes ad hoc record to CDR	Useful
app_celgenuserevent	Generates user-defined events for CEL	New
app_chanisavail	Checks the status of a channel	Unreliable
app_channelredirect	Forces another channel into a different part of the dialplan	Useful
app_chanspy	Allows a channel to listen to audio on another channel	Useful
app_confbridge	Provides conferencing (new version)	New—not fully featured yet
app_controlplayback	Plays back a prompt and offers fast forward and rewind functions	Useful
app_dahdibarge	Allows barging in on a DAHDI channel	Deprecated—see app_chanspy
app_dahdiras	Creates a RAS server over a DAHDI channel (no modem emulation)	Insignificant
app_db	Used to add/change/delete records in Asterisk's built-in Berkeley database	Deprecated—see func_db
app_dial	Used to connect channels together (i.e., make phone calls)	Essential
app_dictate	Plays back a recording and offers start/stop functions	Useful
app_directed_pickup	Answers a call for another extension	Useful
app_directory	Presents the list of names from *voicemail.conf*	Useful
app_disa	Provides dialtone and accepts DTMF input	Useful[a]
app_dumpchan	Dumps channel variables to Asterisk command-line interface (CLI)	Useful

Name	Purpose	Popularity/Status
app_echo	Loops received audio back to source channel	Useful
app_exec	Contains Exec(), TryExec(), and ExecIf(); executes a dialplan application based on conditions	Useful
app_externalivr	Controls Asterisk as with an AGI, only asynchronously	Useful
app_fax	Provides SendFax() and ReceiveFax()	Useful[b]
app_festival	Enables basic text to speech using Festival TTS engine	Usable
app_flash	Performs a hook-switch flash on channels (primarily analog)	Useful
app_followme	Performs find me/follow me functionality based on *followme.conf*	Useful
app_forkcdr	Starts new CDR record on current call	Usable
app_getcpeid	Gets the ADSI CPE ID	Insignificant
app_ices	Sends audio to an Icecast server	Usable
app_image	Transmits an image to supported devices	Limited
app_ivrdemo	Sample application for developers	Insignificant
app_jack	Works with JACK Audio Connection Kit to share audio between compatible applications	Useful
app_macro	Triggers dialplan macros	Deprecated—see GoSub()
app_meetme	Provides multiparty conferencing	Useful—fully featured
app_milliwatt	Generates 1004-Hz tone for testing loss on analog circuits	Useful
app_minivm	Provides primitive functions to allow you to build your own voicemail application in dialplan	Usable
app_mixmonitor	Records both sides of a call and mixes them together	Useful
app_morsecode	Generates Morse code	Usable
app_mp3	Uses *mpg123* to play an MP3	Insignificant
app_nbscat	Streams audio from Network Broadcast Stream (NBS)	Insignificant
app_originate	Allows origination of a call	Useful
app_osplookup	Performs Open Settlement Protocol (OSP) lookup	Usable
app_page	Creates multiple audio connections to specified devices for public address (paging)	Useful
app_parkandannounce	Enables automated announcing of parked calls	Usable
app_playback	Plays a file to the channel (does not accept input)	Useful
app_playtones	Plays pairs of tones of specified frequencies	Useful
app_privacy	Requests input of caller's phone number if no CallerID is received	Insignificant
app_queue	Provides Automatic Call Distribution (ACD)	Useful

Name	Purpose	Popularity/Status
app_read	Requests input of digits from callers and assigns input to a variable	Useful
app_readexten	Requests input of digits from callers and passes call to a designated extension and context	Usable
app_readfile	Loads contents of a text file into a channel variable	Deprecated—see the FILE() function in func_env
app_record	Records received audio to a file	Useful
app_rpt	Provides a method to interface with an audio board for the app_rpt project	Limited
app_sayunixtime	Plays back time in specified format	Useful
app_senddtmf	Transmits DTMF to calling party	Useful
app_sendtext	Sends a text string to compatible channels	Insignificant
app_setcallerid	Sets CallerID on a channel	Deprecated—see func_call erid
app_skel	Sample application for developers	Useful[c]
app_sms	Sends SMS message in compatible countries	Limited
app_softhangup	Requests hangup of channel	Useful
app_speech_utils	Provides utilities relating to speech recognition	Useful[d]
app_stack	Provides Gosub(), GoSubIf(), Return(), Stack Pop(), LOCAL(), and LOCAL_PEEK()	Essential
app_system	Executes commands in a Linux shell	Useful
app_talkdetect	Similar to app_background, but allows for any received audio to interrupt playback	Useful
app_test	Client/server testing application	Usable
app_transfer	Performs a transfer on the current channel	Useful
app_url	Passes a URI to the called channel	Limited
app_userevent	Generates a custom event in the Asterisk Manager Interface (AMI)	Useful
app_verbose	Generates a custom event in the Asterisk CLI	Useful
app_voicemail	Provides voicemail	Essential
app_waitforring	Waits for a RING signaling event (not to be confused with RINGING); most likely unnecessary, as only chan_dahdi with analog channels where ringing is received (such as an FXO port) generates the RING signaling event	Insignificant
app_waitforsilence	Includes WaitForSilence() and WaitForNoise(); listens to the incoming channel for a specified number of milliseconds of noise/silence	Useful
app_waituntil	Waits until current Linux epoch matches specified epoch	Useful

Name	Purpose	Popularity/Status
`app_while`	Includes `While()`, `EndWhile()`, `ExitWhile()`, and `ContinueWhile()`; provides while-loop functionality in the dialplan	Useful
`app_zapateller`	Generates SIT tone to discourage telemarketers	Usable

[a] The use of (DISA) is considered to be a security risk.

[b] Requires a suitable DSP engine to handle encoding/decoding of fax signaling (see Chapter 19).

[c] If you are a developer.

[d] Requires an external speech recognition application.

Bridging Modules

Bridging modules are new in Asterisk 1.8: they perform the actual bridging of channels in the new bridging API. Each provides different features, which get used in different situations depending on what a bridge needs. These modules, listed in Table 2-2, are currently only used for (and are essential to) `app_confbridge`.

Table 2-2. Bridging modules

Name	Purpose	Popularity/Status
`bridge_builtin_features`	Performs bridging when utilizing built-in user features (such as those found in *features.conf*).	New
`bridge_multiplexed`	Performs complex multiplexing, as would be required in a large conference room (multiple participants). Currently only used by `app_confbridge`.	New
`bridge_simple`	Performs simple channel-to-channel bridging.	New
`bridge_softmix`	Performs simple multiplexing, as would be required in a large conference room (multiple participants). Currently only used by `app_confbridge`.	New

Call Detail Recording Modules

The CDR modules, listed in Table 2-3, are designed to facilitate as many methods of storing call detail records as possible. You can store CDRs to a file (default), a database, RADIUS, or *syslog*.

Call detail records are not intended to be used in complex billing applications. If you require more control over billing and call reporting, you will want to look at channel event logging, discussed next. The advantage of CDR is that it just works.

Table 2-3. Call detail recording modules

Name	Purpose	Popularity/Status
cdr_adaptive_odbc	Allows writing of CDRs through ODBC framework with ability to add custom fields	Useful
cdr_csv	Writes CDRs to disk as a comma-separated values file	Usable
cdr_custom	As above, but allows for the addition of custom fields	Useful
cdr_manager	Outputs CDRs to Asterisk Manager Interface (AMI)	Useful
cdr_odbc	Writes CDRs through ODBC framework	Usable
cdr_pgsql	Writes CDRs to PostgreSQL	Useful
cdr_radius	Writes CDRs to RADIUS	Usable—does not support custom fields
cdr_sqlite	Writes CDRs to SQLite2 database	Deprecated—use sqlite3_custom
cdr_sqlite3_custom	Writes CDRs to SQLite3 with custom fields	Useful
cdr_syslog	Writes CDRs to *syslog*	Useful
cdr_tds	Writes CDRs to Microsoft SQL or Sybase	Usable—requires an old version of libtds

We will discuss some reporting packages that you may wish to use with CDR in Chapter 25.

Channel Event Logging Modules

Channel event logging provides much more powerful control over reporting of call activity. By the same token, it requires more careful planning of your dialplan, and by no means will it work automatically. Asterisk's CEL modules are listed in Table 2-4.

Table 2-4. Channel event logging modules

Name	Purpose	Popularity/Status
cel_custom	CEL to disk/file	Useful
cel_manager	CEL to AMI	Useful
cel_odbc	CEL to ODBC	Useful
cel_pgsql	CEL to PostgreSQL	Useful
cel_radius	CEL to RADIUS	Usable—does not support custom fields
cel_sqlite3_custom	CEL to Sqlite3	Useful
cel_tds	CEL to Microsoft SQL or Sybase	Usable—requires an old version of libtds

Channel Drivers

Without channel drivers, Asterisk would have no way to make calls. Each channel driver is specific to the protocol or channel type it supports (SIP, ISDN, etc.). The channel module acts as a gateway to the Asterisk core. Asterisk's channel drivers are listed in Table 2-5.

Table 2-5. Channel drivers

Name	Purpose	Popularity/Status
chan_agent	Provides agent channel for Queue()	Useful
|chan_alsa	Provides connection to Advanced Linux Sound Architecture	Useful
chan_bridge	Used internally by the ConfBridge() application; should not be used directly	Essential [a]
chan_console	Provides connection to portaudio	New
chan_dahdi	Provides connection to PSTN cards that use DAHDI channel drivers	Useful
chan_gtalk	Provides connection to Google Talk	Usable
chan_h323	Provides connection to H.323 endpoints	Deprecated—see chan_ooh323 in Table 2-11
chan_iax2	Provides connection to IAX2 endpoints	Useful
chan_jingle	Provides connection to Jingle-enabled endpoints	Usable
chan_local	Provides a mechanism to treat a portion of the dialplan as a channel	Useful
chan_mgcp	Media Gateway Control Protocol channel driver	Usable
chan_misdn	Provides connection to mISDN supported ISDN cards	Limited
chan_multicast_rtp	Provides connection to multicast RTP streams	Useful
chan_nbs	Network Broadcast Sound channel driver	Insignificant
chan_oss	Open Sound System driver	Useful
chan_phone	Linux telephony interface driver (very old)	Insignificant
chan_sip	Session Initiation Protocol channel driver	Essential
chan_skinny	Cisco Skinny Client Control Protocol (SCCP) channel driver	Usable
chan_unistim	Nortel Unistim protocol channel driver	Usable
chan_usbradio	Channel driver for CM108 USB cards with radio interface	Usable
chan_vpb	Voicetronix channel driver	Insignificant [b]

[a] If you are using the ConfBridge() application.

[b] Some Voicetronix hardware is supported by Zaptel using an addon Zaptel module distributed by Voicetronix. However, Zaptel is no longer supported by Asterisk and this driver has not been ported to DAHDI.

Codec Translators

The codec translators (Table 2-6) allow Asterisk to convert audio stream formats between calls. So if a call comes in on a PRI circuit (using G.711) and needs to be passed out a compressed SIP channel (e.g., using G.729, one of many codecs that SIP can handle), the relevant codec translator would perform the conversion.[†]

 If a codec (such as G.729) uses a complex encoding algorithm, heavy use of transcoding can place a massive burden on the CPU. Specialized hardware for the decoding/encoding of G.729 is available from hardware manufacturers such as Sangoma and Digium (and likely others).

Table 2-6. Codec translators

Name	Purpose	Popularity/Status
codec_adpcm	Adaptive Differential Pulse Coded Modulation codec	Insignificant
codec_alaw	A-law PCM codec used all over the world (except Canada/USA) on the PSTN	Essential
codec_a_mu	A-law to mu-law direct converter	Useful
codec_dahdi	Utilizes proprietary Digium hardware transcoding card	Essential[a]
codec_g722	Wideband audio codec	Useful
codec_g726	Flavor of ADPCM	Insignificant
codec_gsm	Global System for Mobile Communications (GSM) codec	Useful
codec_ilbc	Internet Low Bitrate Codec	Insignificant
codec_lpc10	Linear Predictive Coding vocoder (extremely low bandwidth)	Insignificant
codec_resample	Resamples between 8-bit and 16-bit signed linear	Usable
codec_speex	Speex codec	Usable
codec_ulaw	Mu-law PCM codec used in Canada/USA on PSTN	Essential

[a] If you are using a Digium codec transcoder card.

Format Interpreters

Format interpreters (Table 2-7) perform the function of codec translators, but they do their work on files rather than channels. If you have a recording on a menu that has been stored as GSM, a format interpreter would need to be used to play that recording to any channels not using the GSM codec.[‡]

---

† More information about what codecs are and how they work is available in "Codecs" on page 625.

‡ It is partly for this reason that we do not recommend the default GSM format for system recordings. WAV recordings will sound better and use less CPU.

If you store a recording in several formats (such as WAV, GSM, etc.), Asterisk will determine the least costly format[§] to use when a channel requires that recording.

Table 2-7. Format interpreters

Name	Plays files stored in	Popularity/Status
format_g723	G.723 *.g723*	Insignificant
format_g726	G.726 *.g726*	Insignificant
format_g729	G.729 *.g729*	Useful
format_gsm	RPE-LTP (original GSM codec) *.gsm*	Usable
format_h263	H.263—video *.h263*	Usable
format_h264	H.264—video *.h264*	Usable
format_ilbc	Internet Low Bitrate Codec *.ilbc*	Insignificant
format_jpeg	Graphic file *.jpeg .jpg*	Insignificant
format_ogg_vorbis	Ogg container *.ogg*	Usable
format_pcm	Various Pulse-Coded Modulation formats: *.alaw, .al, .alw, .pcm, .ulaw, .ul, .mu, .ulw, .g722, .au*	Useful
format_siren14	G.722.1 Annex C (14 kHz) *.siren14*	New
format_siren7	G.722.1 (7 kHz) *.siren7*	New
format_sln16	16-bit signed linear *.sln16*	New
format_sln	8-bit signed linear *.sln .raw*	Useful
format_vox	*.vox*	Insignificant
format_wav	*.wav*	Useful
format_wav_gsm	GSM audio in a WAV container *.WAV, .wav49*	Usable

Dialplan Functions

Dialplan functions, listed in Table 2-8, complement the dialplan applications (see "Applications" on page 12). They provide many useful enhancements to things like string handling, time and date wrangling, and ODBC connectivity.

Table 2-8. Dialplan functions

Name	Purpose	Popularity/Status
func_aes	Encrypts/decrypts an AES string	Useful
func_audiohookinherit	Allows calls to be recorded after transfer	Useful
func_base64	Encodes/decodes a base-64 string	Usable
func_blacklist	Writes/reads blacklist in *astdb*	Useful

§ Some codecs can impose a significant load on the CPU, such that a system that could support several hundred channels without transcoding might only be able to handle a few dozen when transcoding is in use.

Name	Purpose	Popularity/Status
func_callcompletion	Gets/sets call completion configuration parameters for the channel	New
func_callerid	Gets/sets CallerID	Useful
func_cdr	Gets/sets CDR variable	Useful
func_channel	Gets/sets channel information	Useful
func_config	Includes AST_CONFIG(); reads variables from config file	Usable
func_connectedline	Changes connected line information on supported handsets	New
func_curl	Uses cURL to obtain data from a URI	Useful
func_cut	Slices and dices strings	Useful
func_db	Provides *astdb* functions	Useful
func_devstate	Gets state of device	Useful
func_dialgroup	Creates a group for simultaneous dialing	Useful
func_dialplan	Validates that designated target exists in dialplan	Useful
func_enum	Performs ENUM lookup	Useful
func_env	Includes FILE(), STAT(), and ENV(); performs operating system actions	Useful
func_extstate	Returns status of a hinted extension	Useful
func_global	Gets/sets global variables	Useful
func_groupcount	Gets/sets channel count for members of a group	Useful
func_iconv	Converts between character sets	Usable
func_lock	Includes LOCK(), UNLOCK(), and TRYLOCK(); sets a lock that can be used to avoid race conditions in the dialplan	Useful
func_logic	Includes ISNULL(), SET(), EXISTS(), IF(), IFTIME(), and IMPORT(); performs various logical functions	Useful
func_math	Includes MATH(), INC(), and DEC(); performs mathematical functions	Useful
func_md5	Converts supplied string to an MD5 hash	Useful
func_module	Checks to see if supplied module is loaded into memory	Usable
func_odbc	Allows dialplan integration with ODBC resources	Useful
func_pitchshift	Shifts the pitch of an audio stream	Useful
func_rand	Returns a random number within a given range	Useful
func_realtime	Performs lookups within the Asterisk Realtime Architecture (ARA)	Useful
func_redirecting	Provides access to information about where this call was redirected from	Useful
func_sha1	Converts supplied string to an SHA1 hash	Useful
func_shell	Performs Linux shell operations and returns results	Useful
func_speex	Reduces noise and performs dB gain/loss on an audio stream	Useful

Name	Purpose	Popularity/Status
func_sprintf	Performs string format functions similar to C function of same name	Useful
func_srv	Perform SRV lookups in the dialplan	Useful
func_strings	Includes over a dozen string manipulation functions	Useful
func_sysinfo	Gets system information such as RAM, swap, load average, etc.	Useful
func_timeout	Gets/sets timeouts on channel	Useful
func_uri	Converts strings to URI-safe encoding	Useful
func_version	Returns Asterisk version information	Usable
func_vmcount	Returns count of messages in a voicemail folder for a particular user	Useful
func_volume	Sets volume on a channel	Useful

PBX Modules

The PBX modules are peripheral modules that provide enhanced control and configuration mechanisms. For example, pbx_config is the module that loads the traditional Asterisk dialplan. The currently available PBX modules are listed in Table 2-9.

Table 2-9. PBX modules

Name	Purpose	Popularity/Status
pbx_ael	Asterisk Extension Logic (AEL) offers a dialplan scripting language that looks like a modern programming language.	Usable[a]
pbx_config	This is the traditional, and most popular, dialplan language for Asterisk. Without this module, Asterisk cannot read *extensions.conf*.	Useful
pbx_dundi	Performs data lookups on remote Asterisk systems.	Useful
pbx_loopback	Performs something similar to a dialplan include, but in a deprecated manner.	Insignificant[b]
pbx_lua	Allows creation of a dialplan using the Lua scripting language.	Useful
pbx_realtime	Provides functionality related to the Asterisk Realtime Architecture.	Useful
pbx_spool	Provides outgoing spool support relating to Asterisk call files.	Useful

[a] We have not found too many people using AEL. We suspect this is because most developers will tend to use AGI/AMI if they do not want to use traditional dialplans.

[b] We've never heard of this being used in production.

Resource Modules

Resource modules integrate Asterisk with external resources. For example, res_odbc allows Asterisk to interoperate with ODBC database connections. The currently available resource modules are listed in Table 2-10.

Table 2-10. Resource modules

Name	Purpose	Popularity/Status
res_adsi	Provides ADSI	Essential[a]
res_ael_share	Provides shared routines for use with pbx_ael	Essential if you're using AEL
res_agi	Provides Asterisk Gateway Interface	Useful
res_ais	Provides distributed message waiting indication (MWI) and device state notifications via an implementation of the AIS standard, such as OpenAIS	Useful
res_calendar	Enables base integration to calendaring systems	Useful
res_calendar_caldav	Provides CalDAV-specific capabilities	Useful
res_calendar_exchange	Provides MS Exchange capabilities	Useful
res_calendar_icalendar	Provides Apple/Google iCalendar capabilities	Useful
res_clialiases	Creates CLI aliases	Useful
res_clioriginate	Originates a call from the CLI	Usable
res_config_curl	Pulls configuration information using cURL	Useful
res_config_ldap	Pulls configuration information using LDAP	Usable
res_config_odbc	Pulls configuration information using ODBC	Useful
res_config_pgsql	Pulls configuration information using PostgreSQL	Usable
res_config_sqlite	Pulls configuration information using SQLite	Usable
res_convert	Uses the CLI to perform file conversions	Usable
res_crypto	Provides cryptographic capabilities	Useful
res_curl	Provides common subroutines for other cURL modules	Useful
res_fax	Provides common subroutines for other fax modules	Useful
res_fax_spandsp	Plug-in for fax using the spandsp package	Useful
res_http_post	Provides POST upload support for the Asterisk HTTP server	Usable
res_jabber	Provides Jabber/XMPP resources	Useful
res_limit	Enables adjusting of system limits on the Asterisk process	Usable
res_monitor	Provides call recording resources	Useful
res_musiconhold	Provides music on hold (MOH) resources	Essential
res_mutestream	Allows muting/unmuting of audio streams	New
res_odbc	Provides common subroutines for other ODBC modules	Useful
res_phoneprov	Provisions phones from Asterisk HTTP server	New
res_pktccops	Provides PacketCable COPS resources	New
res_realtime	Provides CLI commands for the Asterisk Realtime Architecture (ARA)	Useful
res_rtp_asterisk	Provides RTP	Essential
res_rtp_multicast	Provides multicast-RTP	New

Name	Purpose	Popularity/Status
`res_security_log`	Enables security logging	New
`res_smdi`	Provides voicemail notification using the SMDI protocol	Limited
`res_snmp`	Provides system status information to an SNMP-managed network	Usable
`res_speech`	Generic speech recognition API	Limited[b]
`res_timing_dahdi`	Provides timing using the DAHDI kernel interface	Useful
`res_timing_kqueue`	Provides timing using a kernel feature in some BSDs, including Mac OS X	New
`res_timing_pthread`	Provides timing using only parts of the standard `pthread` API; less efficient but more portable than other timing modules.	Useful
`res_timing_timerfd`	Provides timing using the `timerfd` API provided by newer versions of the Linux kernel	Useful

[a] While most of the ADSI functionality in Asterisk is never used, the voicemail application uses this resource.

[b] Requires a separately licensed product in order to be used.

Addon Modules

Addon modules are community-developed modules with different usage or distribution rights from those of the main code. They are kept in a separate directory and are not compiled and installed by default. To enable these modules, use the *menuselect* build configuration utility. Currently available addon modules are listed in Table 2-11.

Table 2-11. Addon modules

Name	Purpose	Popularity/Status
`app_mysql`	Executes MySQL queries with a dialplan application	Deprecated—see `func_odbc`
`app_saycountpl`	Says Polish counting words	Deprecated—now integrated in *say.conf*
`cdr_mysql`	Logs call detail records to a MySQL database	Usable—we recommend `cdr_adaptive_odbc` instead
`chan_mobile`	Enables making and receiving phone calls using cell phones over Bluetooth	Limited[a]
`chan_ooh323`	Enables making and receiving VoIP calls using the H.323 protocol	Usable
`format_mp3`	Allows Asterisk to play MP3 files	Usable
`res_config_mysql`	Uses a MySQL database as a real-time configuration backend	Useful

[a] While `chan_mobile` works great with many phones, problems have been reported with some models. When a problem does occur, it is very difficult for developers to solve unless they have a phone of the same model to test with.

Test Modules

Test modules are used by the Asterisk development team to validate new code. They are constantly changing and being added to, and are not useful unless you are developing Asterisk software.

If you are an Asterisk developer, however, the Asterisk Test Suite may be of interest to you as you can build automated tests for Asterisk and submit those back to the project, which runs on several different operating systems and types of machines. By expanding the number of tests constantly, the Asterisk project avoids the creation of regressions in code. By submitting your own tests to the project, you can feel more confident in future upgrades.

More information about installing the Asterisk Test Suite is available in this blog post: *http://blogs.asterisk.org/2010/04/29/installing-the-asterisk-test-suite/*. More information about building tests is available in this document: *http://svn.asterisk.org/svn/test suite/asterisk/trunk/README.txt* or you can join the *#asterisk-testing* channel on the Freenode IRC network.

File Structure

Asterisk is a complex system, composed of many resources. These resources make use of the filesystem in several ways. Since Linux is so flexible in this regard, it is helpful to understand what data is being stored, so that you can understand where you are likely to find a particular bit of stored data (such as voicemail messages or log files).

Configuration Files

The Asterisk configuration files include *extensions.conf*, *sip.conf*, *modules.conf*, and dozens of other files that define parameters for the various channels, resources, modules, and functions that may be in use.

These files will be found in */etc/asterisk*. You will be working in this folder a lot as you configure and administer your Asterisk system.

Modules

Asterisk modules are usually installed to the */usr/lib/asterisk/modules* folder. You will not normally have to interact with this folder; however, it will be occasionally useful to know where the modules are located. For example, if you upgrade Asterisk and select different modules during the *menuselect* phase of the install, the old (incompatible) modules from the previous Asterisk version will not be deleted, and you will get a warning from the install script. Those old files will need to be deleted from the *modules* folder. This can be done either manually or with the "uninstall" make (*make uninstall*) target.

The Resource Library

There are several resources that require external data sources. For example, music on hold (MOH) can't happen unless you have some music to play. System prompts also need to be stored somewhere on the hard drive. The */var/lib/asterisk* folder is where system prompts, AGI scripts, music on hold, and other resource files are stored.

The Spool

The *spool* is where Linux stores files that are going to change frequently, or will be processed by other processes at a later time. For example, under Linux print jobs and pending emails are normally written to the spool until they are processed.

For Asterisk, the spool is used to store transient items such as voice messages, call recordings,[||] call files, and so forth.

The Asterisk spool will be found under the */var/spool/asterisk* directory.

Logging

Asterisk is capable of generating several different kinds of log files. The */var/log/asterisk* folder is where things such as call detail records (CDRs), channel events from CEL, debug logs, queue logs, messages, errors, and other output are written.

This folder will be extremely important for any troubleshooting efforts you undertake. We will talk more about how to make use of Asterisk logs in Chapter 24.

The Dialplan

The dialplan is the heart of Asterisk. All channels that arrive in the system will be passed through the dialplan, which contains the call-flow script that determines how the incoming calls are handled.

A dialplan can be written in one of three ways:

- Using traditional Asterisk dialplan syntax in */etc/asterisk/extensions.conf*
- Using asterisk Extension Logic (AEL) in */etc/asterisk/extensions.ael*
- Using LUA in */etc/asterisk/extensions.lua*

Later in this book, we'll have devoted several chapters to the subject of how to write a dialplan using traditional dialplan syntax (by far the most popular choice). Once you learn this language, it should be fairly easy to transition to AEL or LUA, should you desire.

[||] Not call detail records (CDRs), but rather audio recordings of calls generated by the `MixMonitor()` and related applications.

Hardware

Asterisk is capable of communicating with a vast number of different technologies. In general, these connections are made across a network connection; however, connections to more traditional telecom technologies, such as the PSTN, require specific hardware.

Many companies produce this hardware, such as Digium (the sponsor, owner, and primary developer of Asterisk), Sangoma, Rhino, OpenVox, Pika, Voicetronix, Junghanns, Dialogic, Xorcom, beroNet, and many others. The authors prefer cards from Digium and Sangoma; however, the products offered by other Asterisk hardware manufacturers may be more suitable to your requirements.

The most popular hardware for Asterisk is generally designed to work through the Digium Asterisk Hardware Device Interface (known as DAHDI). These cards will all have different installation requirements and different file locations.

In Chapter 7, we will discuss DAHDI in more detail; however, we will limit our discussion to DAHDI only. You will need to refer to the specific documentation provided by the manufacturers of any cards you install for details on those cards.

Asterisk Versioning

The Asterisk release methodology has gone through a couple of iterations over the last few years, and this section is designed to help you understand what the version numbers mean. Of particular relevance is the change in versioning that happened with the 1.6.x series of releases, which followed a different numbering logic than all other Asterisk releases (1.0 to 1.8 and onward for the foreseeable future).

Previous Release Methodologies

When we had just Asterisk 1.2 and 1.4, all new development was carried out in trunk (it still is), and only bug fixes went into the 1.2 and 1.4 branches. The Asterisk 1.2 branch has been marked as EOL (End of Life), and is no longer receiving bug fixes or security updates. Prior to the 1.6.x branches, bug fixes were carried out only in trunk and in the 1.4 branch.

Because all new development was done in trunk, until the 1.6 branch was created people were unable to get access to the new features and functionality. This isn't to say the new functionality wasn't available, but with all the changes that can happen in trunk, running a production server based on it requires a very Asterisk-savvy (and C code–savvy) administrator.

To try to relieve the pressure on administrators, and to enable faster access to new features (in the time frame of months, and not years), a new methodology was created. Branches in 1.6 would actually be marked as 1.6.0, 1.6.1, 1.6.2, etc., with the third number increasing by one each time a new feature release was created. The goal was to provide new feature releases every 3–4 months (which would be branched from trunk), providing a shorter and clearer upgrade path for administrators. If you needed a new feature, you'd only have to wait a few months and could then upgrade to the next branch.

Tags from these branches look like this:

- 1.6.0.1 -- 1.6.0.2 -- 1.6.0.3 -- 1.6.0.4 -- etc.
- 1.6.1.1 -- 1.6.1.2 -- 1.6.1.3 -- 1.6.1.4 -- etc.
- 1.6.2.1 -- 1.6.2.2 -- 1.6.2.3 -- 1.6.2.4 -- etc.

Figure 2-2 gives a visual representation of the branching and tagging process in relation to Asterisk trunk.

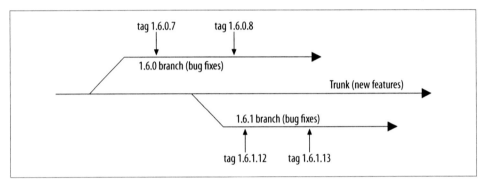

Figure 2-2. The Asterisk 1.6.x release process

So, so far we have branches, which are 1.2, 1.4, 1.6.0, 1.6.1, and 1.6.2 (there is no 1.6 branch). Within each of those branches, we create *tags* (releases), which look like 1.2.14, 1.4.30, 1.6.0.12, 1.6.1.12, and 1.6.2.15.

Unfortunately, it ended up not working out that 1.6.x branches were created from trunk every 3–4 months: the development process has led to a minimum release time of 6–8 months. Not only that, but the 1.6.x numbering methodology adds problems of its own. People got confused as to what version to run, and that the 1.6.0, 1.6.1, and 1.6.2 branches were all separate major version upgrades. When you increase the number from 1.2 to 1.4, and then to 1.8, it is obvious that those are distinct branches and major version changes. With 1.6.0, 1.6.1, and 1.6.2, it is less obvious.

The New Release Methodology

The development team learned a lot of things during the 1.6.x releases. The idea surrounding the releases was noble, but the implementation ended up being flawed when put into real use. So, with Asterisk 1.8, the methodology has changed again, and will look a lot more like that used in the 1.2 and 1.4 releases.

While the development team still wants to provide access to new features and core changes on a more regular basis (every 12 months being the goal), there is recognition that it is also good to provide long-term support to a stable, popular version of Asterisk. You can think of the Asterisk 1.4 branch as being a long-term support (LTS) version. The 1.6.0, 1.6.1, and 1.6.2 branches can be thought of as feature releases that continue to receive bug fixes after release, but are supported for a shorter period of time (about a year). The new LTS version is Asterisk 1.8 (what this book is based on); it will receive bug fixes for four years and an additional year of security releases after that, providing five years of support from the Digium development team.

During the long-term support phase of Asterisk 1.8, additional branches will be created on a semi-regular basis as feature releases. These will be tagged as versions 1.10, 1.12, and 1.14, respectively. Each of these branches will receive bug fixes for a period of one year, and security releases will continue to be made for an additional year before the branches are marked as EOL.

The current statuses of all Asterisk branches, their release dates, when they will go into security release–only mode, and when they will reach EOL status are all documented on the Asterisk wiki at *https://wiki.asterisk.org/wiki/display/AST/Asterisk+Versions*.

Conclusion

Asterisk is composed of many different technologies, most of which are complicated in their own right. As such, the understanding of Asterisk architecture can be overwhelming. Still, the reality is that Asterisk is well-designed for what it does and, in our opinion, has achieved remarkable balance between flexibility and complexity.

## CHAPTER 3

Installing Asterisk

*I long to accomplish great and noble tasks, but it is my
chief duty to accomplish humble tasks as though they
were great and noble. The world is moved along, not
only by the mighty shoves of its heroes, but also by the
aggregate of the tiny pushes of each honest worker.*

—Helen Keller

In this chapter we're going to walk through the installation of Asterisk from the source code. Many people shy away from this method, claiming that it is too difficult and time-consuming. Our goal here is to demonstrate that installing Asterisk from source is not actually that difficult to do. More importantly, we want to provide you with the best Asterisk platform on which to learn.

In this book we will be helping you build a functioning Asterisk system from scratch. In this chapter you will build a base platform for your Asterisk system. Given that we are installing from source, there is potentially a lot of variation in how you can do this. The process we discuss here is one that we've used for many years, and following it will provide you with a suitable foundation for Asterisk.

As part of this process we will also explain installation of some of the software dependencies on the Linux platform that will be needed for topics covered later in this book (such as database integration). We will show instructions for installing Asterisk on both CentOS (*http://www.centos.org*) (a Red Hat–based distribution) and Ubuntu (*http://www.ubuntu.com*) (a Debian-based distribution), which we believe covers the vast majority of Linux distributions being installed today. We'll try to keep the instructions general enough that they should be useful on any distribution of your choice.*

---

* If you are using another distribution, we're willing to bet you are quite comfortable with Linux and should have no trouble installing Asterisk.

We have chosen to install on CentOS and Ubuntu because they are the most popular options, but Asterisk is generally distribution-agnostic. Asterisk will even install on Solaris, BSD, or OS X† if you like. We won't be covering them in this book, though, as Asterisk is most heavily developed on the Linux platform.

Asterisk Packages

There are also packages that exist for Asterisk that can be installed using popular package-management programs such as *yum* or *apt-get*. You are welcome to experiment with them. Prebuilt packages may not always be kept up-to-date, though, so for the latest version we always recommend installing from source.

You can find package instructions at *http://www.asterisk.org/downloads/yum*.

Some commands you see in this chapter will be split into separate rows, each labeled for the distribution on which the command should be performed. Commands for which distributions are not specified are for common commands that can be run on both distributions.

Asterisk-Based Projects

Many projects have been created that use Asterisk as their underlying platform. Some of these, such as Trixbox, have become so popular that many people mistake them for the Asterisk product itself. These projects generally will take the base Asterisk product and add a web-based administration interface, a complex database, and a rigid set of constraints on how changes can be made to the configuration.

We have chosen not to cover these projects in this book, for several reasons:

1. This book tries, as much as possible, to focus on Asterisk and only Asterisk.

2. Books have already been written about some of these Asterisk-based projects.

3. We believe that if you learn Asterisk in the way that we will teach you, the knowledge will serve you well, regardless of whether you eventually choose to use one of these prepackaged versions of Asterisk.

4. For us, the power of Asterisk is that it does not attempt to solve your problems for you. These projects are an excellent example of what can be built with Asterisk. They are truly amazing. However, if you are looking to build your own Asterisk application (which is really what Asterisk is all about), these projects will impose limitations on you, because they are focused on simplifying the process of building a business PBX, not on making it easier to access the full potential of the Asterisk platform.

---

† Leif calls this "Oh-Eh-Sex," but Jim thinks it should be pronounced "OS Ten." We wasted several precious minutes arguing about this.

Some of the most popular Asterisk-based projects include:

AsteriskNOW	*http://www.asterisk.org/asterisknow*
Trixbox	*http://www.trixbox.org*
Elastix	*http://www.elastix.org*
PBX in a Flash	*http://www.pbxinaflash.net*

We recommend that you check them out.[‡]

Installation Cheat Sheet

If you just want the nitty-gritty on how to get Asterisk up and running quickly, perform the following at the shell prompt. We encourage you to read through the entire chapter at least once, though, in order to better understand the full process.[§]

The instructions provided here assume you've already installed either CentOS or Ubuntu using the steps outlined in "Distribution Installation" on page 35.

 Remember that Ubuntu requires commands to be prefixed with *sudo*.

1. Perform a system update and reboot:

CentOS	`yum update -y && reboot`
CentOS 64-bit	`yum remove *.i386 && yum update -y && reboot`
Ubuntu	`sudo apt-get update && sudo apt-get upgrade && sudo reboot`

2. Synchronize time and install the NTP (Network Time Protocol) daemon:

CentOS	`yum install -y ntp && ntpdate pool.ntp.org && chkconfig ntpd \` `on && service ntpd start`
CentOS 64-bit	`yum install -y ntp && ntpdate pool.ntp.org && chkconfig ntpd \` `on && service ntpd start`
Ubuntu	`sudo apt-get install ntp`

[‡] After you read our book, of course.

[§] Once you have experience with several Asterisk installations, you'll agree that it's a quick and painless process. Nevertheless, this chapter may make the process look complex. This is simply because we have an obligation to ensure you are provided with all the information you need to accomplish a successful install.

Some additional configuration of text files is required on Ubuntu. See "Enable NTP for accurate system time" on page 43.

3. On CentOS, add a new system user:

CentOS (32 and 64 bit)
```
adduser asteriskpbx && passwd asteriskpbx && yum install \
sudo && visudo
```

See "Adding a system user" on page 39 for specific information.

For an Ubuntu install, we are assuming that the user created during the installation process is *asteriskpbx*.

4. Install software dependencies:

CentOS
```
sudo yum install gcc gcc-c++ make wget subversion \
libxml2-devel ncurses-devel openssl-devel \
vim-enhanced
```

CentOS 64-bit
```
sudo yum install gcc.x86_64 gcc-c++.x86_64 \
make.x86_64 wget.x86_64 subversion.x86_64 \
libxml2-devel.x86_64 ncurses-devel.x86_64 \
openssl-devel.x86_64 vim-enhanced.x86_64
```

Ubuntu
```
sudo apt-get install build-essential subversion \
libncurses5-dev libssl-dev libxml2-dev vim-nox
```

5. Create your directory structure:
```
$ mkdir -p ~/src/asterisk-complete/asterisk
$ cd ~/src/asterisk-complete/asterisk
```

6. Get the latest source code via Subversion:
```
$ svn co http://svn.asterisk.org/svn/asterisk/branches/1.8
```

Or alternatively, you could check out a specific tag:
```
$ svn co http://svn.asterisk.org/svn/asterisk/tags/1.8.1
```

7. Build and install the software:
```
$ cd ~/src/asterisk-complete/asterisk/1.8/
$ ./configure
```

```
$ make
$ sudo make install
$ sudo make config
```

8. Install additional sound prompts from *menuselect*:

```
$ cd ~/src/asterisk-complete/asterisk/1.8/
$ make menuselect
$ sudo make install
```

9. Modify the file permissions of the directories Asterisk was installed to:

```
$ sudo chown -R asteriskpbx:asteriskpbx /usr/lib/asterisk/
$ sudo chown -R asteriskpbx:asteriskpbx /var/lib/asterisk/
$ sudo chown -R asteriskpbx:asteriskpbx /var/spool/asterisk/
$ sudo chown -R asteriskpbx:asteriskpbx /var/log/asterisk/
$ sudo chown -R asteriskpbx:asteriskpbx /var/run/asterisk/
$ sudo chown asteriskpbx:asteriskpbx /usr/sbin/asterisk
```

10. On CentOS, disable SELinux:

```
$ sudo vim /etc/selinux/config
```

Change the value of SELINUX from enforcing to disabled, then *reboot*.

11. Create the */etc/asterisk/* directory and copy the *indications.conf* sample file into it:

```
$ sudo mkdir -p /etc/asterisk
$ sudo chown asteriskpbx:asteriskpbx /etc/asterisk
$ cd /etc/asterisk/
$ cp ~/src/asterisk-complete/asterisk/1.8/configs/indications.conf.sample \
./indications.conf
```

12. Copy the sample *asterisk.conf* file into */etc/asterisk* and change runuser and run group to have values of *asteriskpbx*:

```
$ cp ~/src/asterisk-complete/asterisk/1.8/configs/asterisk.conf.sample \
/etc/asterisk/asterisk.conf
```

```
$ vim /etc/asterisk/asterisk.conf
```

See "indications.conf and asterisk.conf" on page 53 for more information.

13. Create the *modules.conf* file. Enable loading of modules automatically, and disable extra modules:

```
$ cat >> /etc/asterisk/modules.conf

; The modules.conf file, used to define which modules Asterisk should load (or
; not load).
;
[modules]
autoload=yes

; Resource modules currently not needed
noload => res_speech.so
noload => res_phoneprov.so
noload => res_ael_share.so
noload => res_clialiases.so
noload => res_adsi.so
```

```
; PBX modules currently not needed
noload => pbx_ael.so
noload => pbx_dundi.so

; Channel modules currently not needed
noload => chan_oss.so
noload => chan_mgcp.so
noload => chan_skinny.so
noload => chan_phone.so
noload => chan_agent.so
noload => chan_unistim.so
noload => chan_alsa.so

; Application modules currently not needed
noload => app_nbscat.so
noload => app_amd.so
noload => app_minivm.so
noload => app_zapateller.so
noload => app_ices.so
noload => app_sendtext.so
noload => app_speech_utils.so
noload => app_mp3.so
noload => app_flash.so
noload => app_getcpeid.so
noload => app_setcallerid.so
noload => app_adsiprog.so
noload => app_forkcdr.so
noload => app_sms.so
noload => app_morsecode.so
noload => app_followme.so
noload => app_url.so
noload => app_alarmreceiver.so
noload => app_disa.so
noload => app_dahdiras.so
noload => app_senddtmf.so
noload => app_sayunixtime.so
noload => app_test.so
noload => app_externalivr.so
noload => app_image.so
noload => app_dictate.so
noload => app_festival.so
```

Ctrl + D

14. Configure *musiconhold.conf*:

```
$ cat >> musiconhold.conf

; musiconhold.conf
[default]
mode=files
directory=moh
```

Ctrl + D

15. Save your changes and your module configuration is done. Your system is ready to configure your dialplan and channels.

Distribution Installation

Because Asterisk relies so heavily on having priority access to the CPU, it is essential that you install Asterisk onto a server without any graphical interface, such as the X Windowing system (Gnome, KDE, etc.). Both CentOS and Ubuntu ship a GUI-free distribution designed for server usage. We will cover instructions for both distributions.

CentOS Server

CentOS means "Community Enterprise Operating System," and it is based on Red Hat Enterprise Linux (RHEL). For more information about what CentOS is and its history, see *http://www.centos.org*.

You will need to download an ISO from the CentOS website, located at *http://mirror .centos.org/centos/5/isos/*. Select either the *i386* or *x86_64* directory for 32-bit or 64-bit hardware, respectively. You will then be presented with a list of mirrors that appear to be close to you physically. Choose one of the mirrors, and you will be presented with a list of files to download. Likely you will want the first available selection, which is the first ISO file of a set. You will only need the first ISO file of the set as we'll be installing additional software with *yum*.

Once you've downloaded the ISO file, burn it to a CD or DVD and start the installation process. If you're installing into a virtual machine (which we don't recommend for production use,‖ but can be a great way to test out Asterisk), you should be able to mount the ISO file directly and install from there.

Base system installation

Upon booting from the CD, type *linux text* and then press ⌷Enter⌷.#

At this point the text installation interface will start. You will be asked whether you want to test the media. These instructions assume you've already done so, and therefore can skip that step.

CentOS will then welcome you to the installation. Press ⌷Enter⌷ to continue.

‖ Actually, some people have great success running Asterisk inside virtual machines. It does depend what you're planning on using it for though, as you'll have limited access to hardware, for example.

#You should test the media the first time you are using that particular CD/DVD.

Choose your language and make a keyboard selection.[*] If you're in North America, you will probably just select the defaults.

If you've previously formatted your hard drive, you will be asked to initialize the drive, which will erase all data. Select $\boxed{\text{Yes}}$.

The installer will ask if you want to remove the existing partitioning scheme and create a new one. Select *Remove all partitions on selected drives and create default layout*. If a more appropriate option exists, select that instead. In the drive window, verify that the correct disk drive is selected. (Pressing $\boxed{\text{Tab}}$ will cycle through the selections on the screen.) Once the drive window is selected, you can scroll up and down (presuming you have multiple drives) and select which hard drive you wish to install to. Toggle the selections by pressing $\boxed{\text{space bar}}$. Verify that the correct drive is selected, press $\boxed{\text{Tab}}$ until the $\boxed{\text{OK}}$ button is highlighted, and press $\boxed{\text{Enter}}$.

A message confirming that you want to remove all Linux partitions and create the new partition scheme will be presented. Select $\boxed{\text{Yes}}$.

You will be asked to review the partitioning layout. Feel free to modify the partition scheme if you prefer something different (see the following sidebar for some advice on this); however, the default answer $\boxed{\text{No}}$ is fine for light production use where storage requirements will be low.[†]

Separating the /var Mount Point to Its Own Partition

On a system dedicated to Asterisk, the directory with the largest storage requirement is */var*. This is where Asterisk will store recordings, voicemails, log files, prompts, and a myriad of constantly growing information. In normal operation, it is unlikely that Asterisk will fill the hard disk. However, if you have extensive logging turned on or are recording all calls, this could, in theory, occur. (This is likely to happen several months after you've completed the install and to take your entire staff by surprise.)

If the drive the operating system is mounted on fills up, there is the potential for a kernel panic. By separating */var* from the rest of the hard drive, you significantly lower the risk of a system failure.

 Having a full volume is still a major problem; however, you will at least be able to log into the system to rectify the situation.

---

[*] Bear in mind that Asterisk is developed using the US keyboard and language, and we're not aware of any testing having been done on anything other than US English.

[†] Due to the ever-increasing size of hard drives, capacity is becoming less of a problem. A system with a 1 terabyte drive can store somewhere in the range of 2 million minutes of telephony-quality recordings.

At the *Review and modify the partitioning layout* screen, you can create a separate volume for */var*. Selecting Yes will bring up the *Partitioning* tool. To partition the drive accurately, you need to know what the hard drive size is; this may not jibe with what is stamped on the outside of the drive because you have to tell the tool how to chop up the drive. A limitation of the tool is that there is no option to say "use all available space"; that is, you can't simply could use 500 MB on the / partition and then say "use the rest for */var*". The workaround is to make a note of the size it has selected for / currently, as that is the full space, subtract 500 MB from that, and make that the size for the / partition. The subtracted amount will then be reserved for */var*.

A message will appear asking if you'd like to configure the *eth0* network interface on your system. Select Yes. Be sure the *Activate on boot* and *Enable IPv4 support* options are enabled, then select OK.

If your network provides automatic IP provisioning via DHCP, you can just select OK. Otherwise, select *Manual address configuration*, enter the appropriate information, and then select OK.

Next, you'll be asked to provide a hostname. You can either allow the DHCP server to provide one for you (if your network assigns hostnames automatically) or enter one manually, then select OK.

You will be presented with a list of time zones. Highlight your local time zone and select OK.

At this point, you will be asked for a root password. Enter a secure password and type it again to confirm. After entering your secure password, select OK.

Next up will be the package selection. Several packages that you don't need to install (and that require additional ISO files you probably haven't downloaded) are selected by default. Deselect all options in the list using the space bar, then select the *Customize software selection* option. Once you've done that, select OK.

You will then be presented with the *Package Group Selection* screen. Scroll through the whole list, deselecting each item. If any packages are selected, you'll be prompted for additional CDs that you have not downloaded. We'll be installing additional packages with the *yum* application after the operating system is installed. Once you've deselected all packages, select OK.

A dependency check will then be performed and a confirmation that installation is ready to begin will be presented. Select OK to start the installation. The filesystem will then be formatted, the installation image transferred to the hard drive, and installation of the system packages performed. Upon installation, you will be asked to reboot. Remove any media in the drives and select the Reboot button.

Base system update

Once you've rebooted your system, you need to run the *yum update* command to make sure you have the latest base packages. To do this, log in using the username *root* and the password you created during installation. Once logged in, run the following:

```
# yum update
Is this ok [y/N]: y
```

When prompted to install the latest packages, press y and wait for the packages to update. If you're asked to accept a GPG key, press y. When complete, reboot the system as it is likely the kernel will have been updated[‡]:

```
# reboot
```

 If you're running CentOS Server 64-bit, you'll need to remove all the 32-bit libraries manually. Once you've rebooted, or just prior to reboot, run the following command:

```
# yum remove *.i386 -y
```

This will remove all the 32-bit libraries on your 64-bit system, which can otherwise cause conflicts and issues when compiling Asterisk and other software.

Congratulations! You've successfully installed and updated the base CentOS system.

Enabling NTP for accurate system time

Keeping accurate time is essential on your Asterisk system, both for maintaining accurate call detail records and for synchronization with your other programs. You don't want the times of your voicemail notifications to be off by 10 or 20 minutes, as this can lead to confusion and panic from those who might think their voicemail notifications are taking took too long to be delivered. The *ntpd* command can be used to ensure that the time on your Asterisk server remains in sync with the rest of the world:

```
# yum install ntp
...
Is this ok [y/N]: y
...
# ntpdate pool.ntp.org
# chkconfig ntpd on
# service ntpd start
```

The defaults shipped with CentOS are sufficient to synchronize the time and keep the machine's time in sync with the rest of the world.

---

‡ This reboot step is essential prior to installing Asterisk.

Adding a system user

The Ubuntu server install process asks you to add a system user other than root, but CentOS does not. In order to be consistent in the book and to be more secure, we're going to add another system user and provide it *sudo* access.[§] To add the new user, execute the *adduser* command:

```
# adduser asteriskpbx
# passwd asteriskpbx
Changing password for user asteriskpbx.
New UNIX password:
Retype new UNIX password:
```

Now we need to provide the *asteriskpbx* user *sudo* access. We do this by modifying the *sudoers* file with the *visudo* command. You'll need to install *visudo* the first time you use it:

```
# yum install sudo
```

With the *sudo*-related applications and file installed, we can modify the *sudoers* file. Execute the *visudo* command and look for the lines shown below:

```
# visudo

## Allows people in group wheel to run all commands
%wheel  ALL=(ALL)       ALL
```

With the %wheel line uncommented as shown in our example, save the file by pressing Esc, then typing :wq and pressing Enter. Now open the */etc/group* file in your favorite editor (*nano* is easy to use) and find the line that starts with the word wheel. Modify it like so:

```
wheel:x:10:root,asteriskpbx
```

Save the file, log out from *root* by typing exit, and log in as the *asteriskpbx* user you created. Test your *sudo* access by running the following command:

```
$ sudo ls /root/
[sudo] password for asteriskpbx:
```

After typing your password, you should get the output of the */root/* directory. If you don't, go back and verify the steps to make sure you didn't skip or mistype anything. The rest of the instructions in this chapter will assume that you're the *asteriskpbx* user and that you have *sudo* access.

One last thing needs to done, which will allow you to enter commands without having to enter the full path. By default only *root* has */sbin/* and */usr/sbin/* in the default system PATH, but we'll add it to our *asteriskpbx* user as well since we'll be running many applications located in those directories.

[§] *sudo* is an application that allows a user to execute commands as another user, such as root, or the superuser.

Start by opening the hidden file *.bash_profile* located within the *asteriskpbx* home directory with an editor. We're then going to append `:/usr/sbin:/sbin` to the end of the line starting with `PATH`:

```
$ vim ~/.bash_profile
PATH=$PATH:$HOME/bin:/usr/sbin:/sbin
```

As previously, save the file by pressing Esc and then typing `:wq` and pressing Enter.

With the operating system installed, you're ready to install the dependencies required for Asterisk. The next section deals with Ubuntu, so you can skip ahead to the section "Software Dependencies" on page 44, which provides an in-depth review of the installation process. Alternatively, if you've already reviewed the information in that section, you may want to refer back to the "Installation Cheat Sheet" on page 31 for a high-level review of how to install Asterisk.

Ubuntu Server

Ubuntu Server is a popular Linux distribution loosely based on Debian. There is also a popular desktop version of the software. The Ubuntu Server package contains no GUI and is ideal for Asterisk installations.

To get the latest version of Ubuntu Server,‖ visit *http://www.ubuntu.com* and select the *Server* tab at the top of the page. You will be provided with a page that contains information about Ubuntu Server Edition. Clicking the orange *Download* button in the upper-right corner will take you to a page where you can select either the 32-bit or 64-bit version of Ubuntu Server. After selecting one of the options, you can press the *Start download* button.

Once you've downloaded the ISO file, burn it to a CD and start the installation process. If you're installing into a virtual machine (which we don't recommend for production use, but can be a great way to test out Asterisk), you should be able to mount the ISO file directly and install from there.

Base system installation

Upon booting from the CD, you will be presented with a screen where you can select your language of choice. By default English is the selected language, and after a timeout period, it will be automatically selected. After selecting your language, press Enter.

The next screen will give you several options, the first of which is *Install Ubuntu Server*. Select it by pressing Enter.

You will then be asked which language to use for the installation (yes, this is slightly redundant). Select your language of choice (the default is English), and press Enter.

‖ Of course, projects can change their websites whenever they want. Hopefully the instructions we've provided here are accurate enough to help guide you through the site even in the event of changes.

You will be presented with a list of countries. Once you've found your country and highlighted it, press Enter.

You will then be asked if you would like to use the keyboard layout detector. If you know which keyboard type you have, you can select No and then pick it from a list of formats.

If you are utilizing the keyboard layout detector, you will be prompted to press a series of keys. If you use the keyboard detector and it does not detect your keyboard correctly (typical when installing into a virtual machine via a remote console), you can go back and select from a list manually.

Once you've picked your keyboard, the installation will continue by attempting to set up your network automatically. If all goes well, you will be prompted to enter a host-name for your system. You can pick anything you want here, unless your network requires your system to a have a specific hostname. Input it now and then press Enter.

The installer will attempt to contact a Network Time Protocol (NTP) server to synchronize your clock. Ubuntu will then try to autodetect your time zone and present you with its choice. If correct, select Yes, otherwise, select No and you'll be presented with a list of time zones to select from. Select your time zone, or select from the world-wide list if your time zone is not shown. Once you've selected your time zone, press Enter to continue.

The installer will then ask you some questions about partitioning your system. Typically the default is fine, which is to use the guided system, utilizing the entire disk, and to set up the Logical Volume Manager (LVM). Press Enter once you've made your selection. Then you'll be asked which partition to install to, which likely is the only one on your system. Press Enter to continue, at which point you'll be asked to confirm the changes to the partition table. Select Yes and press Enter to continue.

You will now be asked how much space to use (the default value will be to use the entire disk). Press Enter once you've entered and confirmed the amount of space you want to use. The installer will then request one last confirmation before making the changes to the disk. Select Yes to write the changes to disk. The installer will now format the hard disk, write the partitioning scheme to disk, copy the files, and perform the file installation.

When the file installation is complete, you'll be asked to enter the *Full name of the new user*, from which a username will be generated. The system will suggest a username, but you are free to change the username to whatever you like.

After entering your username, you'll be asked to supply a password, and then asked to confirm the password you've entered. Ubuntu does a good job of providing a secure system by not providing direct access to root, but rather using the *sudo* application, which allows you to run commands as root without being the root user. Enter a username,# such as *asteriskpbx*, and a secure password to continue. You'll use these to log into the system once the installer ends. The installer will then ask you if you want to encrypt your home directory. This is not necessary and will add CPU overhead.

> The rest of the installation instructions will assume that *asteriskpbx* was chosen as the username.

If your system is behind a web proxy, enter the proxy information now. If you're not behind a proxy, or don't know if you are, simply press Enter.

You will then be asked if you want to install updates automatically. The default is to perform no automatic updates, which is what we recommend. Should a system reboot occur, an update to the kernel will render Asterisk nonstartable until you recompile it* (which won't make you popular). It is better practice to identify updates on a regular basis and perform them manually in a controlled manner. Normally, you would want to advise your users of the expected downtime and schedule the downtime to happen after business hours (or while a redundant system is running). Select *No automatic updates* and press Enter.

Since we'll be installing our dependencies with *apt-get*, we only need to select one package during the install: OpenSSH server. SSH is essential if you wish to perform remote work on the system. However, if your local policy states that your server needs to be managed directly, you may not want to install the OpenSSH server.

> Pressing the Enter key will accept the current selections and move on with the install. You need to use space bar to toggle your selections.

After you've selected OpenSSH server, press Enter.

If this is the only operating system on the machine (which it likely is), Ubuntu will give you the option to install the GRUB bootloader on your system. It provides this prompt in order to give you the option of skipping the GRUB installation, as it will modify the master boot record (MBR) on your system. If there is another operating system it has

---

#Ubuntu has reserved the username *asterisk* internally.

* While we say Asterisk here, specifically it is DAHDI that is the problem. DAHDI is a set of Linux kernel modules used with Asterisk.

failed to detect that has information loaded into the MBR, it's nice to be able to skip modifying it. If this is the only operating system installed on your server, select $\boxed{\text{Yes}}$.

When the system has finished the install, you'll be asked to remove any media in the drives and to reboot the system by selecting *Continue*, at which point the installation will be complete and the system will reboot.

Base system update

Now that we've completed installing Ubuntu Server, we need to perform a system update with *apt-get* to make sure we have the latest packages installed. You'll be presented with a login prompt where you'll log in with the username and password you created in the installer (e.g., *asteriskpbx*). Once logged in, run the following command:

```
$ sudo apt-get update
[sudo] password for asteriskpbx:
...
Reading package lists... Done

$ sudo apt-get upgrade
Reading state information... Done
...
Do you want to continue [Y/n]? y
```

 The password that *sudo* wants is the password you just logged in with.

Press $\boxed{\text{Enter}}$ when prompted to continue, at which point the latest package updates will be installed. When complete, reboot the system for the changes to take effect as the kernel has probably been updated.

```
$ sudo reboot
```

Congratulations! You've successfully installed and updated the base Ubuntu Server system.

Enable NTP for accurate system time

Keeping accurate time is essential on your Asterisk system, both for maintaining accurate call detail records as well as for synchronization with your other programs. You don't want the times of your voicemail notifications to be off by 10 or 20 minutes, as this can lead to confusion and panic from those who might think their voicemail notification took too long to be delivered:

```
$ sudo apt-get install ntp
```

The default on Ubuntu is to run a time sync server without ever changing the time on your own machine. This won't work for our needs, so we'll need to change the configuration file slightly. Because of this, we need to guide you through using a command line editor. The *nano* editor is already installed on your Ubuntu machine and is remarkably easy to use†:

```
$ sudo nano /etc/ntp.conf
```

Your terminal will switch to full-screen output.

Use your arrow keys to move down to the section that looks like

```
# By default, exchange time with everybody, but don't allow configuration.
restrict -4 default kod notrap nomodify nopeer noquery
restrict -6 default kod notrap nomodify nopeer noquery
```

Add two new lines after this section, to allow *ntpd* to synchronize your time with the outside world, such that the above section now looks like

```
# By default, exchange time with everybody, but don't allow configuration.
restrict -4 default kod notrap nomodify nopeer noquery
restrict -6 default kod notrap nomodify nopeer noquery

restrict -4 127.0.0.1
restrict -6 ::1
```

That's everything we need to change, so exit the editor by pressing Ctrl+X. When prompted whether to save the modifications, press Y; *nano* will additionally ask you for the filename. Just hit Enter to confirm the default */etc/ntp.conf*.

Now restart the NTP daemon:

```
$ sudo /etc/init.d/ntp restart
```

With the operating system installed, you're ready to install the dependencies required for Asterisk. The next section provides an in-depth review of the installation process. If you've already reviewed the information in "Software Dependencies" on page 44, you may want to refer back to "Installation Cheat Sheet" on page 31 for a high-level review of how to install Asterisk.

Software Dependencies

The first thing you need to do once you've completed the installation of your operating system is to install the software dependencies required by Asterisk. The commands listed in Table 3-1 have been split into two columns, for Ubuntu Server and CentOS Server. These packages will allow you to build a basic Asterisk system, along with DAHDI and LibPRI. Not every module will be available at compile time with these

---

† If you're already familiar with another editor, go ahead and use it. The *nano* editor has been selected for its ease of use and its handy on-screen instructions. We even know a developer at Digium who uses it while writing code for Asterisk, though most people tend to use more complex editors such as *emacs* or *vim*.

dependencies; only the most commonly used modules will be built. If additional dependencies are required for other modules used later in the book, instructions will be provided as necessary.

 Please be aware that the dependency information on CentOS 64-bit does not take into account that 32-bit libraries should not be installed. If such libraries are installed, you will end up with additional packages that use disk space and can cause conflicts if the system attempts to compile against a 32-bit library instead of its 64-bit counterpart. In order to resolve this problem, add *.x86_64* to the end of each package name when installing it. So, for example, instead of executing *yum install ncurses-devel*, you will execute *yum install ncurses-devel.x86_64*. This is not necessary on a 32-bit platform.

Table 3-1. Software dependencies for Asterisk on Ubuntu Server and CentOS Server

Ubuntu	CentOS
`sudo apt-get install build-essential \`	`sudo yum install gcc gcc-c++ make wget \`
`subversion libncurses5-dev libssl-dev \`	`subversion libxml2-devel ncurses-devel \`
`libxml2-dev vim-nox`	`openssl-devel vim-enhanced`

These packages will get you most of what you'll need to get started with installing Asterisk, DAHDI, and LibPRI. Note that you will also require the software dependencies for each package that we indicate needs to be installed. These will be resolved automatically for you when you use either *yum* or *apt-get*.

We have also included the OpenSSL development libraries, which are not strictly necessary to compile Asterisk, but are good to have: they enable key support and other encryption functionality.

We have installed *vim* as our editor, but you can choose anything you want, such as *nano*, *joe*, or *emacs*.

Asterisk contains a script that will install the dependencies for all features in Asterisk. At this time it is complete for Ubuntu but does not list all required packages for CentOS. Once you have downloaded Asterisk using the instructions in "Downloading What You Need" on page 46, use the following commands if you would like to run it:

```
$ cd ~/src/asterisk-complete/asterisk/1.8
$ sudo ./contrib/scripts/install_prereq install
$ sudo ./contrib/scripts/install_prereq install-unpackaged
```

Third-Party Repositories

For certain software dependencies, a third-party repository may be necessary. This appears to be most often the case when using CentOS. A couple of repositories that seem to be able to provide all the extra dependencies required are RPMforge (*http://dag .wieers.com/rpm/*) and EPEL (Extra Packages for Enterprise Linux, *http://fedoraproject .org/wiki/EPEL*).

We may occasionally refer to these third-party repositories when they are required to obtain a dependency for a module we are trying to build and use.

Downloading What You Need

There are several methods of getting Asterisk: via the Subversion code repository, via *wget* from the downloads site, or via a package-management system such as *apt-get* or *yum*. We're only going to cover the first two methods, since we're interested in building the latest version of Asterisk from source. Typically, package-management systems will have versions that are older than those available from Subversion or the downloads site, and we want to make sure we have the most recent fixes available to us, so we tend to avoid them.

 The official packages from Digium do tend to stay up to date. There are currently packages for CentOS/RHEL available at *http://www.asterisk .org/downloads/yum*.

Before we start getting the source files, let's create a directory structure to house the downloaded code. We're going to create the directory structure within the home directory for the *asteriskpbx* user on the system. Once everything is built, it will be installed with the *sudo* command. We'll then go back and change the permissions and ownership of the installed files in order to build a secure system. To begin, issue the following command:

```
$ mkdir -p ~/src/asterisk-complete/asterisk
```

Now that we've created a directory structure to hold everything, let's get the source code. Choose one of the following two methods to get your files:

1. Subversion
2. *wget*

Getting the Latest Version

Asterisk is a constantly evolving project, and there are many different versions of the software that you can implement.

In Chapter 2, we talked about Asterisk versioning. The concept of how Asterisk is versioned is important to understand because the versioning system for Asterisk has undergone a few changes of methodology over the years. So, if you're not up to speed on Asterisk versioning, we strongly recommend that you go back and read "Asterisk Versioning" on page 26.

Having said all that, in most cases all you need to do is grab the latest version from the *http://www.asterisk.org/downloads* website. We will be installing and using Asterisk 1.8 throughout this book.

Getting the Source via Subversion

Subversion is a version control system that is used by developers to track changes to code over a period of time. Each time the code is modified, it must first be checked out of the repository; then it must be checked back in, at which point the changes are logged. Thus, if a change creates a regression, the developers can go back to that change and remove it if necessary. This is a powerful and robust system for development work. It also happens to be useful for Asterisk administrators seeking to retrieve the software. To download the source code to the latest version of Asterisk 1.8, use these commands:

```
$ cd ~/src/asterisk-complete/asterisk
$ svn co http://svn.asterisk.org/svn/asterisk/branches/1.8
```

You can now skip directly to "How to Install It" on page 48.

> The preceding commands will retrieve the latest changes to the source in that particular branch, which are changes that have been made after the latest release. If you would prefer to use a released version, please refer to the next section.

Getting the Source via wget

To obtain the latest released versions of DAHDI, LibPRI, and Asterisk using the *wget* application, issue the following commands:

```
$ cd ~/src/asterisk-complete/asterisk
$ wget \
http://downloads.asterisk.org/pub/telephony/asterisk/asterisk-1.8-current.tar.gz
$ tar zxvf asterisk-1.8-current.tar.gz
```

The next step is to compile and install the software, so onward to the next section.

How to Install It

With the source files downloaded you can compile the software and install it. The order for installing is:

1. LibPRI‡
2. DAHDI§
3. Asterisk‖

Installing in this order ensures that any dependencies for DAHDI and Asterisk are installed prior to running the configuration scripts, which will subsequently ensure that any modules dependent on LibPRI or DAHDI will be built.

So, let's get started.

LibPRI

LibPRI is a library that adds support for ISDN (PRI and BRI). The use of LibPRI is optional, but since it takes very little time to install, doesn't interfere with anything, and will come in handy if you ever want to add cards to a system at a later point, we recommend that you install it now.

Check out the latest version of LibPRI and compile it like so:

```
$ cd ~/src/asterisk-complete/
$ mkdir libpri
$ cd libpri/
$ svn co http://svn.asterisk.org/svn/libpri/tags/1.4.<your version number>
$ cd 1.4.<your version number>
$ make
$ sudo make install
```

 You can also download the source via *wget* from *http://downloads.aster isk.org/pub/telephony/libpri/*.

With LibPRI installed, we can now install DAHDI.

---

‡ Strictly speaking, if you are not going to be using any ISDN connections (BRI and PRI), you can install Asterisk without LibPRI. However, we are going to install it for the sake of completeness.

§ This package contains the kernel drivers to allow Asterisk to connect to traditional PSTN circuits. It is also required for the `MeetMe()` conferencing application. Again, we will install this for completeness.

‖ If you don't install this, none of the examples in this book will work, but it could still make a great bathroom reader. Just sayin'.

DAHDI

The *Digium Asterisk Hardware Device Interface*, or DAHDI (formerly known as Zaptel), is the software Asterisk uses to interface with telephony hardware. We recommend that you install it even if you have no hardware installed, because DAHDI is a dependency required for building the timing module `res_timing_dahdi` and is used for Asterisk dialplan applications such as `MeetMe()`.

DAHDI-tools and DAHDI-linux

DAHDI is actually a combination of two separate code bases: *DAHDI-tools*, which provides various administrator tools such as *dahdi_cfg*, *dahdi_scan*, etc.; and *DAHDI-linux*, which provides the kernel drivers. Unless you're only updating one or the other, you'll be installing both at the same time, which is referred to as *DAHDI-linux-complete*. The version numbering for *DAHDI-linux-complete* will look something like `2.4.0+2.4.0`, where the number to the left of the plus sign is the version of *DAHDI-linux* included, and the version number to the right of the plus sign is the *DAHDI-tools* version included.

There are also FreeBSD drivers for DAHDI, which are maintained by the community. These drivers are available at *http://downloads.asterisk.org/pub/telephony/dahdi-freebsd -complete/*.

Another dependency is required for installing DAHDI, and that is the kernel source. It is important that the kernel version being used match exactly that of the kernel source being installed. You can use *uname -a* to verify the currently running kernel version:

- CentOS: `sudo yum install kernel-devel-`uname -r``
- Ubuntu: `sudo apt-get install linux-headers-`uname -r``

The use of *uname -r* surrounded by backticks (`` ` ``) is for filling in the currently running kernel version so the appropriate package is installed.

The following commands show how to install *DAHDI-linux-complete 2.4.0+2.4.0*. There may be a newer version available by the time you are reading this, so check *downloads.asterisk.org* first. If there is a newer version available, just replace the version number in the commands:

```
$ cd ~/src/asterisk-complete/
$ mkdir dahdi
$ cd dahdi/
$ svn co http://svn.asterisk.org/svn/dahdi/linux-complete/tags/2.4.0+2.4.0
$ cd 2.4.0+2.4.0
$ make
$ sudo make install
$ sudo make config
```

You will need to have Internet access when running the *make all* command, as it will attempt to download the latest hardware firmware from the Digium servers.

After installing DAHDI, we can move on to installing Asterisk.

You can also download the source via *wget* from *http://downloads.aster isk.org/pub/telephony/dahdi-linux-complete/*.

Asterisk

With both DAHDI and LibPRI installed, we can now install Asterisk:

```
$ cd ~/src/asterisk-complete/asterisk/1.8
$ ./configure
$ make
$ sudo make install
$ sudo make config
```

With the files now installed in their default locations, we need to modify the permissions of the directories and their contents.

There is an additional step that is not strictly required, but is quite common (and arguably important): the *make menuselect* command, which provides a graphical interface that allows detailed selection of which modules and features will be compiled. We will discuss this in "make menuselect" on page 59.

Setting File Permissions

In order to run our system more securely, we'll be installing Asterisk and then running it as the *asteriskpbx* user. After installing the files into their default locations, we need to change the file permissions to match those of the user we're going to be running as. Execute the following commands after running *make install* (which we did previously):

```
$ sudo chown -R asteriskpbx:asteriskpbx /usr/lib/asterisk/
$ sudo chown -R asteriskpbx:asteriskpbx /var/lib/asterisk/
$ sudo chown -R asteriskpbx:asteriskpbx /var/spool/asterisk/
$ sudo chown -R asteriskpbx:asteriskpbx /var/log/asterisk/
$ sudo chown -R asteriskpbx:asteriskpbx /var/run/asterisk
$ sudo chown asteriskpbx:asteriskpbx /usr/sbin/asterisk
```

In order to use MeetMe() and DAHDI with Asterisk as non-root, you must change the */etc/udev/rules.d/dahdi.rules* so that the OWNER and GROUP fields match the non-root user Asterisk will be running as. In this case, we're using the *asteriskpbx* user.

Change the last line of the *dahdi.rules* file to the following:

```
SUBSYSTEM=="dahdi", OWNER="asteriskpbx", GROUP="asteriskpbx", MODE="0660"
```

With that out of the way, we can move on to performing the base configuration that should be done after all installations.

Base Configuration

Now that we've got Asterisk installed, we can get our system up and running. The purpose here is to get Asterisk loaded up and ready to go, as it isn't doing anything useful yet. These are the steps that all system administrators will need to start out with when installing a new system. If the commands that need to be run differ on CentOS and Ubuntu, you will see a table with rows labeled for each distribution; otherwise, you will see a single command that should be run regardless of which Linux distribution you have chosen.

Disable SELinux

This section applies only to CentOS users, so if you're using Ubuntu, you can skip to the next section.

In CentOS, the Security-Enhanced Linux (SELinux) system is enabled by default, and it often gets in the way of Asterisk. Sometimes the issues are quite subtle, and at least one of the authors has spent a good number of hours debugging issues in Asterisk that turned out to be resolved by disabling SELinux. There are many articles on the Internet that describe the correct configuration of SELinux, but we're going to disable it for the sake of simplicity.

While disabling SELinux is not the ideal situation, the configuration of SELinux is beyond the scope of this book, and frankly, we just don't have enough experience with it to configure it correctly.

To temporarily switch off SELinux, perhaps in order to verify whether an issue you're having is being caused by SELinux, run the following command as root:

```
$ sudo echo 0 > /selinux/enforce
```

You can reenable SELinux by doing the same thing, but replacing the *0* with a *1*:

```
$ sudo echo 1 > /selinux/enforce
```

To disable SELinux permanently, modify the */etc/selinux/config* file:

```
$ cd /etc/selinux/
$ sudo vim config
```

Change the SELINUX option from enforcing to disabled.

 Alternatively, you can change the value of enforcing to permissive, which simply logs the errors instead of enforcing the policy.

When you're done modifying the configuration file, you'll have the following:

```
# This file controls the state of SELinux on the system.
# SELINUX= can take one of these three values:
#       enforcing - SELinux security policy is enforced.
#       permissive - SELinux prints warnings instead of enforcing.
#       disabled - SELinux is fully disabled.
SELINUX=disabled
# SELINUXTYPE= type of policy in use. Possible values are:
#       targeted - Only targeted network daemons are protected.
#       strict - Full SELinux protection.
SELINUXTYPE=targeted

# SETLOCALDEFS= Check local definition changes
SETLOCALDEFS=0
```

Since you can't disable SELinux without rebooting, you'll need to do that now:

```
$ sudo reboot
```

Initial Configuration

In order to get Asterisk up and running cleanly, we need to create some configuration files. We could potentially install the sample files that come with Asterisk (by executing the *make samples* command in our Asterisk source) and then modify those files to suit our needs, but the *make samples* command installs many sample files, most of them for modules that you will never use. We want to limit which modules we are loading, and we also believe that it's easier to understand Asterisk configuration if you build your config files from scratch, so we're going to create our own minimal set of configuration files.#

The first thing we need to do (assuming it does not already exist) is create the */etc/asterisk/* directory where our configuration files will live:

```
$ sudo mkdir /etc/asterisk/
$ sudo chown asteriskpbx:asteriskpbx /etc/asterisk/
```

#If your */etc/asterisk/* folder has files in it already, move those files to another directory, or delete them if you are sure you don't need what is there.

Running *make samples* on a system that already has configuration files will overwrite the existing files.

Using make samples to Create Sample Configuration Files for Future Reference

Even though we are not going to use the sample configuration files that come with Asterisk, the fact is that they are an excellent reference. If there is a module that you are not currently using but wish to put into production, the sample file will show you exactly what syntax to use, and what options are available for that module.

Running the *sudo make samples* command in your Asterisk source directory* is harmless on a new system that has just been built, but it is very dangerous to run on a system that already has configuration files, as this command will overwrite any existing files (which would be a disaster for you if you do not have a current backup).

If you've run the *sudo make samples* command, you will want to move the files that it has created in */etc/asterisk/* to another folder. We like to create a folder called */etc/asterisk/unused/* and put any sample/unused configuration files in there, but feel free to store them wherever you like.

We're now going to step through all the files that are required to get a simple Asterisk system up and running.

indications.conf and asterisk.conf

The first file needed is *indications.conf*, a file that contains information about how to detect different telephony tones for different countries. There is a perfectly good sample file that we can use in the Asterisk source, so let's copy it into our */etc/asterisk/* directory:

```
$ cp ~/src/asterisk-complete/asterisk/1.8/configs/indications.conf.sample \
/etc/asterisk/indications.conf
```

Because we're running Asterisk as non-root, we need to tell Asterisk which user to run as. This is done with the *asterisk.conf* file. We can copy a sample version of it from the Asterisk source to */etc/asterisk*:

```
$ cp ~/src/asterisk-complete/asterisk/1.8/configs/asterisk.conf.sample \
/etc/asterisk/asterisk.conf
```

The *asterisk.conf* file contains many options that we won't go over here (they are covered in "asterisk.conf" on page 71), but we do need to make an adjustment. Near the end of the [options] section, there are two options we need to enable: runuser and rungroup.

---

* */usr/src/asterisk-complete/asterisk/asterisk-1.8.<your version>/*

Open the *asterisk.conf* file with an editor such as *nano* or *vim*: Uncomment the `run user` and `rungroup` lines, and modify them so that they each contain *asteriskpbx* as the assigned value. Open the */etc/asterisk/asterisk.conf* file with vim:

```
$ vim /etc/asterisk/asterisk.conf
```

Then modify the file by uncommenting the two lines starting with `runuser` and `run group` and modifying the value to *asteriskpbx*.

```
runuser=asteriskpbx
rungroup=asteriskpbx
```

We now have all the configuration files required to start a very minimal version of Asterisk.[†] Give it a shot by starting Asterisk up in the foreground:

```
$ /usr/sbin/asterisk -cvvv
```

 We are specifying the full path to the *asterisk* binary, but if you modify your `PATH` system variable to include the */usr/sbin/* directory you don't need to specify the full path. See "Adding a system user" on page 39 for information about modifying the `$PATH` environment variable.

Asterisk will start successfully without any errors or warnings (although it does warn you that some files are missing), and present to you the Asterisk command-line interface (CLI). At this point there are no modules, minimal core functionality, and no channel modules with which to communicate, but Asterisk is up and running.

Executing the *module show* command at the Asterisk CLI shows that there are no external modules loaded:

```
*CLI> module show

Module                    Description                      Use Count
0 modules loaded
```

We've done this simply to demonstrate that Asterisk can be run in a very minimal state, and doesn't require the dozens of modules that a default install will enable. Let's stop Asterisk with the *core stop now* CLI command:

```
*CLI> core stop now
```

---

† So minimal, in fact, that it's completely useless at this point. But we digress.

The Asterisk Shell Command

Asterisk can be run either as a daemon or as an application. In general, you will want to run it as an application when you are building, testing, and troubleshooting, and as a daemon when you put it into production.

The command to start Asterisk is the same regardless of whether you're running it as a daemon or an application:

asterisk

However, without any arguments, this command will assume certain defaults and start Asterisk as a background application. In other words, you never want to run the command *asterisk* on its own, but rather will want to pass some options to it to better define the behavior you are looking for. The following list provides some examples of common usages.

`-h`

> This command displays a helpful list of the options you can use. For a complete list of all the options and their descriptions, run the command *man asterisk*.

`-c`

> This option starts Asterisk as an application (in the foreground). This means that Asterisk is tied to your user session. In other words, if you close your user session by logging out or losing the connection, Asterisk dies. This is the option you will typically use when building, testing, and debugging, but you would not want to use this option in production. If you started Asterisk in this manner, type *core stop now* at the CLI prompt to stop Asterisk and exit.

`-v, -vv, -vvv, -vvvv,` *etc.*

> This option can be used with other options (e.g., *-cvvv*) in order to increase the verbosity of the console output. It does exactly the same thing as the CLI command *core set verbose n* where *n* is any integer between 0 and 5 (any integer greater than 5 will work, but will not provide any more verbosity). Sometimes it's useful to not set the verbosity at all. For example, if you are looking to see only startup errors, notices, and warnings, leaving verbosity off will prevent all the other startup messages from being displayed.

`-d, -dd, -ddd, -dddd,` *etc.*

> This option can be used in the same way as *-v*, but instead of normal output, this will specify the level of debug output (which is primarily useful for developers who wish to troubleshoot problems with the code). You will also need to enable output of debugging information in the *logger.conf* file (which we will cover in more detail in Chapter 24).

`-r`

> This command is essential if you want to connect to the CLI of an Asterisk process running as a daemon. You will probably use this option more than any other for Asterisk systems that are in production. This option will only work if you have a daemonized instance of Asterisk already running. To exit the CLI when this option has been used, type *exit*.

-T

> This option will add a timestamp to CLI output.

-x

> This command allows you to pass a string to Asterisk that will be executed as if it had been typed at the CLI. As an example, to get a quick listing of all the channels in use without having to start the Asterisk console, simply type *asterisk -rx 'core show channels'* from the shell, and you'll get the output you are looking for.

-g

> This option instructs Asterisk to dump a core file if it crashes.

We recommend you try out a few combinations of these commands to see what they do.

---

safe_asterisk

When you install Asterisk using the *make config* directive it will create a script called *safe_asterisk*, which is run during the *init* process of Linux each time you boot.

The *safe_asterisk* script provides the following benefits:

- Restarts Asterisk automatically after a crash
- Can be configured to email the administrator if a crash has occurred
- Defines where crash files are stored (*/tmp* by default)
- Executes a script if a crash has occurred

You don't need to know too much about this script, other than to understand that it should normally be running. In most environments this script works fine in its default format.

---

modules.conf

So, we've managed to get Asterisk running, but it's not able to do anything useful for us yet. To tell Asterisk what modules we expect it to load, we'll need a *modules.conf* file.

Create the file *modules.conf* in your */etc/asterisk/* directory with the following command (replace the >> with > if you instead want to overwrite an existing file):

```
$ cat >> /etc/asterisk/modules.conf
```

Type (or paste) the following lines, and press Ctrl+D on a new line when you're finished:

```
; The modules.conf file, used to define which modules Asterisk should load (or
; not load).
;
[modules]
autoload=yes
```

The autoload=yes line will tell Asterisk to automatically load all modules located in the */usr/lib/asterisk/modules/* directory. If you wanted to, you could leave the file like this, and Asterisk would simply load any modules it found in the *modules* folder.

With your new *modules.conf* file in place, starting Asterisk will cause a whole slew of modules to be loaded. You can verify this by starting Asterisk and running the *module show* command:

```
$ asterisk -c
*CLI> module show

Module                      Description                        Use Count
res_speech.so               Generic Speech Recognition API     0
res_monitor.so              Call Monitoring Resource           0
...
func_math.so                Mathematical dialplan function     0
171 modules loaded
```

We now have many modules loaded, and many additional dialplan applications and functions at our disposal. We don't need all these resources loaded, though, so let's filter out some of the more obscure modules that we don't need at the moment. Modify your *modules.conf* file to contain the following noload lines, which will tell Asterisk to skip loading the identified modules:

```
; Resource modules
noload => res_speech.so
noload => res_phoneprov.so
noload => res_ael_share.so
noload => res_clialiases.so
noload => res_adsi.so

; PBX modules
noload => pbx_ael.so
noload => pbx_dundi.so
```

```
; Channel modules
noload => chan_oss.so
noload => chan_mgcp.so
noload => chan_skinny.so
noload => chan_phone.so
noload => chan_agent.so
noload => chan_unistim.so
noload => chan_alsa.so

; Application modules
noload => app_nbscat.so
noload => app_amd.so
noload => app_minivm.so
noload => app_zapateller.so
noload => app_ices.so
noload => app_sendtext.so
noload => app_speech_utils.so
noload => app_mp3.so
noload => app_flash.so
noload => app_getcpeid.so
noload => app_setcallerid.so
noload => app_adsiprog.so
noload => app_forkcdr.so
noload => app_sms.so
noload => app_morsecode.so
noload => app_followme.so
noload => app_url.so
noload => app_alarmreceiver.so
noload => app_disa.so
noload => app_dahdiras.so
noload => app_senddtmf.so
noload => app_sayunixtime.so
noload => app_test.so
noload => app_externalivr.so
noload => app_image.so
noload => app_dictate.so
noload => app_festival.so
```

There are, of course, other modules that you could remove, and others that you may find extremely useful, so feel free to tweak this file as you wish. Ideally, you should be loading only the modules that you need for the system you are running. The examples in this book assume that your *modules.conf* file looks like our example here.

Additional information about the *modules.conf* file can be found in the section "modules.conf" on page 75.

musiconhold.conf

The *musiconhold.conf* file defines the classes for music on hold in your Asterisk system. By defining different classes, you can specify different hold music to be used in various situations, such as different announcements to be played while holding in a queue, or different hold music if you have multiple PBXs hosted on the same system. For now, we'll just create a default music on hold class so that we have at a minimum some hold music when placing callers on hold:

```
$ cd /etc/asterisk/
$ cat >> musiconhold.conf

; musiconhold.conf
[default]
mode=files
directory=moh
```

Ctrl+D

We've created a *musiconhold.conf* file and defined our [default] hold music class. We're also assuming you installed the hold music from the *menuselect* system; by default there is at least one music on hold package installed, so unless you disabled it, you should have music in at least one format.

Additional information about *musiconhold.conf* can be found in the section "musiconhold.conf" on page 79.

make menuselect

menuselect is a text-based menu system in Asterisk used to configure which modules to compile and install. The modules are what give Asterisk its power and functionality. New modules are constantly being created.

In the installation sections, we conveniently skipped over using the *menuselect* system in order to keep the instructions simple and straightforward. However, it is important enough that we have given *menuselect* its own section.

In addition to specifying which modules to install, *menuselect* also allows you to set flags that can aid in debugging issues (see Chapter 2), set optimization flags, choose different sound prompt files and formats, and do various other nifty things.

Uses for menuselect

We would need a whole chapter in order to fully explore *menuselect*, and for the most part you won't need to make many changes to it. However, the following example will give you an idea of how *menuselect* works, and is recommend for any installation.

By default Asterisk only installs the core sound prompt files, and only in GSM format. Also, the three OpSound (*http://www.opsound.org/*) music on hold files available for download are only selected in *.wav* format.‡

---

‡ A good way to put the final touches on your new system is to install some appropriate sound files to be used as music on hold. There are only three songs installed by default, and callers will quickly tire of listening to the same three songs over and over again. We'll discuss this more in "musiconhold.conf" on page 79.

We're going to want extra sound prompts installed instead of just the default core sound prompts, and in a better-sounding format than GSM. We can do this with the *menuselect* system by running *make menuselect* in the Asterisk source directory. Before exploring that, though, let's talk about the different *menuselect* interfaces.

menuselect interfaces

There are two interfaces available for *menuselect*: curses and newt. If the libnewt libraries are installed, you will get the blue and red interface shown in Figure 3-1. Otherwise, by default *menuselect* will use the curses (black and white) interface shown in Figure 3-2.

 The minimum screen size for the curses interface is 80x27, which means it may not load if you're using the default terminal size for a simple distribution installation. This is not a problem when you're using SSH to reach the server remotely, as typically your terminal can be resized, but if you're working at the terminal directly you may need to have screen buffers installed to enable a higher resolution, which is not recommended for a system running Asterisk. The solution is to use the newt-based *menuselect* system.

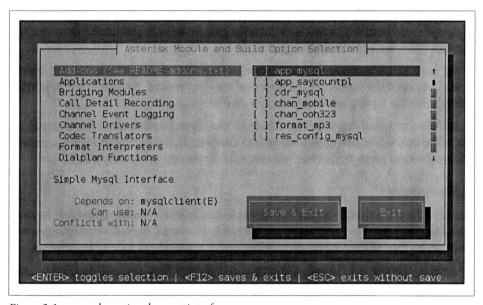

Figure 3-1. menuselect using the newt interface

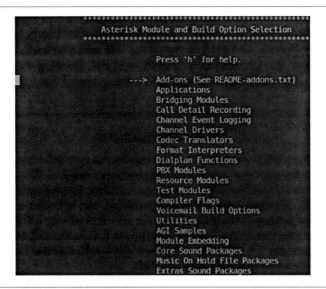

Figure 3-2. menuselect using the curses interface

Installing Dependencies for newt-Based menuselect

To get the newt-based *menuselect* working, you need to have the `libnewt` development libraries installed:

- CentOS: `sudo yum install libnewt-devel`
- Ubuntu: `sudo apt-get install libnewt-dev`

If you've previously used *menuselect* with the curses interface, you need to rebuild. You can do this with the following commands:

```
$ cd ~/src/asterisk-complete/asterisk/1.8.<your version>/
$ cd menuselect
$ make clean
$ ./configure
$ cd ..
$ make menuselect
```

After that you should have the newt-based interface available to you.

Using menuselect

Run the following commands to start *menuselect*:

```
$ cd ~/src/asterisk-complete/asterisk/1.8.<your version>/
$ make menuselect
```

You will be presented with a screen such as that in Figure 3-1 or Figure 3-2. You can use the arrow keys on your keyboard to move up and down. The right arrow key will take you into a submenu, and the left arrow key will take you back. You can use the space bar or Enter key to select and deselect modules. Pressing the q key will quit without saving, while the x key will save your selections and then quit.

Module Dependencies

Modules that have XXX in front of them are modules that cannot be compiled because the *configure* script was not able to find the dependencies required (for example, if you don't have the *unixODBC* development package installed, you will not be able to compile func_odbc[§]). Whenever you install a dependency, you will always need to rerun *configure* before you run *menuselect*, so that the new dependency will be properly located. The dependant module will at that point be available in *menuselect*. If the module selection still contains XXX, either the *configure* script is still unable to find the dependency or not all dependencies have been satisfied.

Once you've started *menuselect*, scroll down to Core Sound Packages and press the right arrow key (or Enter) to open the menu. You will be presented with a list of available options. These options represent the core sound files in various languages and formats. By default, the only set of files selected is CORE-SOUNDS-EN-GSM, which is the English-language Core Sounds package in GSM format.

Select CORE-SOUNDS-EN-WAV and CORE-SOUNDS-EN-ULAW (or ALAW if you're outside of North America or Japan[‖]), and any other sound files that may be applicable in your network.

The reason we have multiple formats for the same files is that Asterisk can play back the appropriate format depending on which codec is negotiated by an endpoint. This can lower the CPU load on a system significantly.

After selecting the appropriate sound files, press the left arrow key to go back to the main menu. Then scroll down two lines to the Extra Sound Packages menu and press the right arrow key (or Enter). You will notice that by default there are no packages selected. As with the core sound files, select the appropriate language and format to be installed. A good option is probably to install the English sound files in the WAV, ULAW, and ALAW formats.

---

§ Which we will cover in Chapter 16, along with many other cool things.

‖ If you want to understand all about mu-law and A-law, you can read the section "Logarithmic companding" on page 607. All you need to know here is that outside of North America and Japan, A-law is used.

Once you've completed selecting the sound files, press the $\boxed{\text{x}}$ key to save and exit *menuselect*. You then need to install your new prompts by downloading them from the Asterisk downloads site. This is done simply by running *make install* again:

```
$ sudo make install
$ sudo chown -R asteriskpbx:asteriskpbx /var/lib/asterisk/sounds/
```

The files will be downloaded, extracted, and installed into the appropriate location (*/var/lib/asterisk/sounds/<language>/* by default). Your Asterisk server will need to have a working Internet connection in order to retrieve the files.

Scripting menuselect

Administrators often build tools when performing installations on several machines, and Asterisk is no exception. If you need to install Asterisk onto several machines, you may wish to build a set of scripts to help automate this process. The *menuselect* system contains command-line options that you can use to enable or disable the modules that are built and installed by Asterisk.

If you are starting with a fresh checkout of Asterisk, you must first execute the *configure* script in order to determine what dependencies are installed on the system. Then you need to build the *menuselect* application and run the *make menuselect-tree* command to build the initial tree structure:

```
$ cd ~/src/asterisk-complete/asterisk/1.8.<your version>/
$ ./configure
$ cd menuselect
$ make menuselect
$ cd ..
$ make menuselect-tree
Generating input for menuselect ...
```

For details about the options available, run *menuselect/menuselect --help* from the top level of your Asterisk source directory. You will be returned output like the following:

```
Usage: menuselect/menuselect [--enable <option>] [--disable <option>]
    [--enable-category <category>] [--enable-all]
    [--disable-category <category>] [--disable-all] [...]
    [<config-file> [...]]
Usage: menuselect/menuselect { --check-deps | --list-options
    | --list-category <category> | --category-list | --help }
    [<config-file> [...]]
```

The options displayed can then be used to control which modules are installed via the *menuselect* application. For example, if you wanted to disable all modules and install a base system (which wouldn't be of much use) you could use the command:

```
$ menuselect/menuselect --disable-all menuselect.makeopts
```

If you then look at the *menuselect.makeopts* file, you will see a large amount of text that displays all the modules and categories that have been disabled. Let's say you now want to enable the SIP channel and the `Dial()` application. Enabling those modules can be done with the following command, but before doing that look at the current *menuselect.makeopts* (after disabling all the modules) and locate `app_dial` in the `MENUSE LECT_APPS` category and `chan_sip` in the `MENUSELECT_CHANNELS` category. After executing the following command, look at the *menuselect.makeopts* file again, and you will see that those modules are no longer listed:

```
$ menuselect/menuselect --disable-all --enable chan_sip \
--enable app_dial menuselect.makeopts
```

 The modules listed in the *menuselect.makeopts* file are those that will not be built—modules that are not listed will be built when the *make* application is executed.

You can then build the *menuselect.makeopts* file in any way you want by utilizing the other commands, which will allow you to build custom installation scripts for your system using any scripting language you prefer.

Updating Asterisk

If this is your first installation, you can skip ahead to the section "Base Configuration" on page 51. If you're in the process of updating your system, however, there are a couple of things you should be aware of.

 When we say *updating* your system, that is quite different from *upgrading* your system. Updating your system is the process of installing new minor versions of the same branch. For example, if your system is running Asterisk 1.8.2 and you need to upgrade to the latest bug fix version for the 1.8 branch, which was version 1.8.3, you'd be *updating* your system to 1.8.3. In contrast, we use the term *upgrade* to refer to changes between Asterisk branches (major version number increases). So, for example, an upgrade would be going from Asterisk 1.4.34 to Asterisk 1.8.0.

When performing an update, you follow the same instructions outlined in the section "How to Install It" on page 48.

 Additionally, if you've checked out a new directory for this version of Asterisk (versus running *svn up* on a checked-out branch), and previously used *menuselect* to tweak the modules to be compiled, you can copy the *menuselect.makeopts* file from one directory to another prior to running *./configure*. By copying *menuselect.makeopts* from the old version to the new version, you save the step of having to (de)select all your modules again.

The basic steps are:

```
$ cd ~/src/asterisk-complete/asterisk/1.8.<your version number>/
$ ./configure
$ make
$ make install
```

 You don't need to run *sudo make install* because we've already set the directory ownership to the *asteriskpbx* user. You should be able to install new files directly into the appropriate directories.

Upon installation, however, you may get a message like the following:

```
WARNING WARNING WARNING

Your Asterisk modules directory, located at
/usr/lib/asterisk/modules
contains modules that were not installed by this
version of Asterisk. Please ensure that these
modules are compatible with this version before
attempting to run Asterisk.

    chan_mgcp.so
    chan_oss.so
    chan_phone.so
    chan_skinny.so
    chan_skype.so
    codec_g729a.so
    res_skypeforasterisk.so

WARNING WARNING WARNING
```

This warning message is indicating that modules installed in the */usr/lib/asterisk/modules/* directory are not compatible with the version you've just installed. This most often occurs when you have installed modules in one version of Asterisk, and then installed a new version of Asterisk without compiling those modules (as the installation process will overwrite any modules that existed previously, replacing them with their upgraded versions).

To get around the warning message, you can clear out the */usr/lib/asterisk/modules/* directory prior to running *make install*. There is a caveat here, though: if you've installed third-party modules, such as commercial modules from Digium (including chan_skype, codec_g729a, etc.), you will need to reinstall those if you've cleared out your modules directory.

It is recommended that you keep a directory with your third-party modules in it that you can reinstall from upon update of your Asterisk system. So, for example, you might create the */usr/src/asterisk-complete/thirdparty/1.8* directory as follows:

```
$ cd ~/src/asterisk-complete/
$ mkdir thirdparty/
$ mkdir thirdparty/1.8/
```

Downloading third-party modules into this directory allows you to easily reinstall those modules when you upgrade. Just follow the installation instructions for your module, many of which will be as simple as rerunning *make install* from the *modules* source directory or copying the precompiled binary to the */usr/lib/asterisk/modules/* directory.

Be sure to change the file permissions to match those of the user running Asterisk!

Common Issues

In this section we're going to cover some common issues you may run into while compiling Asterisk, DAHDI, or LibPRI. Most of the issues you'll run into have to do with missing dependencies. If that is the case, please review "Software Dependencies" on page 44 to make sure you've installed everything you need.

Any time you install additional packages, you will need to run the *./configure* script in your Asterisk source in order for the new package to be detected.

-bash: wget: command not found

This message means you have not installed the *wget* application, which is required for you to download packages from the Asterisk downloads site, for Asterisk to download sound files, or for DAHDI to download firmware for hardware.

Ubuntu	CentOS
$ sudo apt-get install wget	$ sudo yum -y install wget

configure: error: no acceptable C compiler found in $PATH

This means that the Asterisk *configure* script is unable to find your C compiler, which typically means you have not yet installed one. Be sure to install the *gcc* package for your system.

Ubuntu	CentOS
$ sudo apt-get install gcc	$ sudo yum install gcc

make: gcc: command not found

This means that the Asterisk *configure* script is unable to find your C compiler, which typically means you have not yet installed one. Be sure to install the *gcc* package for your system.

Ubuntu	CentOS
$ `sudo apt-get install gcc`	$ `sudo yum install gcc`

configure: error: C++ preprocessor "/lib/cpp" fails sanity check

This error is presented by the Asterisk *configure* script when you have not installed the GCC C++ preprocessor.

Ubuntu	CentOS
$ `sudo apt-get install g++`	$ `sudo yum install gcc-c++`

configure: error: *** Please install GNU make. It is required to build Asterisk!

This error is encountered when you have not installed the *make* application, which is required to build Asterisk.

Ubuntu	CentOS
$ `sudo apt-get install make`	$ `sudo yum install make`

configure: *** XML documentation will not be available because the 'libxml2' development package is missing.

You will encounter this error when the XML parser libraries are not installed. These are required by Asterisk 1.8 and later, since console documentation (e.g., when you run *core show application dial* on the Asterisk CLI) is generated from XML.

Ubuntu	CentOS
$ `sudo apt-get install libxml2-dev`	$ `sudo yum install libxml2-devel`

configure: error: *** termcap support not found

This error happens when you don't have the `ncurses` development library installed, which is required by *menuselect* and for other console output in Asterisk.

Ubuntu	CentOS
$ `sudo apt-get install ncurses-dev`	$ `sudo yum install ncurses-devel`

You do not appear to have the sources for the 2.6.18-164.6.1.el5 kernel installed.

You will get this error when attempting to build DAHDI without having installed the Linux headers, which are required for building Linux drivers.

Ubuntu	CentOS
`$ sudo apt-get install linux-headers-`uname -r``	`$ sudo yum install kernel-devel`

E: Unable to lock the administration directory (/var/lib/dpkg/), are you root?

If you encounter this error it's likely that you forgot to prepend *sudo* to the start of the command you were running, which requires root permissions.

Upgrading Asterisk

Upgrading Asterisk between major versions, such as from 1.2 to 1.4 or from 1.6.2 to 1.8 is akin to upgrading an operating system. Once a phone switch is in production, it is terribly disruptive for that system to be unavailable for nearly any length of time, and the upgrade of that phone system needs to be well thought-out, planned, and tested as much as possible prior to deployment. And because every deployment is different, it is difficult, if not impossible, for us to walk you through a real system upgrade. However, we can certainly point you in the right direction for the information you require in order to perform such an upgrade, thereby giving you the tools you need to be successful.

A production Asterisk system should never be upgraded between major versions without first deploying it into a development environment where the existing configuration files can be tested and reviewed against new features and syntax changes between versions. For example, it may be that your dialplan relies on a deprecated command and should be updated to use a new command that contains more functionality, has a better code base, and will be updated on a more regular basis. Commands that are deprecated are typically left in the code for backward-compatibility, but issues reported about these deprecated commands will be given lower priority than issues to do with the newer preferred methods.

There exist two files that should be read prior to any system upgrade: *CHANGES* and *UPGRADE.txt*, which are shipped with the Asterisk source code. These files contain details on changes to syntax and other things to be aware of when upgrading between major versions. The files are broken into different sections that reference things such as dialplan syntax changes, channel driver syntax changes, functionality changes, and deprecation of functionality, with suggestions that you update your configuration files to use the new methods.

Another thing to consider when performing an upgrade is whether you really need to perform the upgrade in the first place. If you're using a long-term support (LTS)# version of Asterisk and that version is happily working along for you, perhaps there is no reason to upgrade your existing production system. An alternative to upgrading the entire system is simply to add functionality to your system by running two versions simultaneously on separate systems. By running separate boxes, you can access the functionality added to a later version of Asterisk without having to disrupt your existing production system. You can then perform the migration more gradually, rather than doing a complete system upgrade instantly.

Two parts of Asterisk should be thoroughly tested when performing an upgrade between major versions: the Asterisk Manager Interface (AMI) and the Asterisk Gateway Interface (AGI).

These two parts of Asterisk rely on testing your code to make sure any cleanup of syntax changes in either the AMI or the AGI, or added functionality, does not interfere with your existing code. By performing a code audit on what your program is expecting to send or receive against what actually happens, you can save yourself a headache down the road.

The testing of call detail records (CDRs) is also quite important, especially if they are relied upon for billing. The entire CDR structure is really designed for simple call flows, but it is often employed in complex call flows, and when someone reports an issue to the tracker and it is fixed, it can sometimes have an effect on others who are relying on the same functionality for different purposes. Asterisk 1.8 now includes channel event logging (CEL), which is a system designed to get around some of the limitations of CDR in more complex call flows (such as those that involve transfers, etc.). More information about CEL is available in "CEL (Channel Event Logging)" on page 537.

Upgrading Asterisk can be a successful endeavor as long as sufficient planning and testing are carried out prior to the full rollout. In some cases migrating to a separate physical machine on which you've performed testing is preferred, as it can give you a system to roll back to in case of some failure that can't be resolved immediately. It's the planning, and particularly having a backup plan, that is the most important aspect of an Asterisk upgrade.

Conclusion

In this chapter we looked at how to install an operating system (one of Ubuntu or CentOS) and Asterisk itself. We did this securely by installing via *sudo* and running Asterisk as the non-root user *asteriskpbx*. We are well on our way to building a functional Asterisk system that will serve us well. In the following chapters we will explore

---

#More information about Asterisk releases and their support schedule is available at *https://wiki.asterisk.org/wiki/display/AST/Asterisk+Versions*.

how to connect devices to our Asterisk system in order to start placing calls internally and how to connect Asterisk to outside services in order to place phone calls to endpoints connected to the PSTN and accept calls from those endpoints.

## CHAPTER 4

Initial Configuration Tasks

Careful. We don't want to learn from this.

—Calvin & Hobbes

In the last chapter, we covered how to install Asterisk. But where should you get started with configuration? That is the question this chapter answers. There are a few common configuration files that are relevant regardless of what you are using Asterisk to accomplish. In some cases they may not require any modification, but you need to be aware of them.

asterisk.conf

The *asterisk.conf* configuration file allows you to tweak various settings that can affect how Asterisk runs as a whole.

There is a sample *asterisk.conf* file included with the Asterisk source. It is not necessary to have this file in your */etc/asterisk* folder in order to have a working system, but you may find that some of the possible options will be of use to you.

 Asterisk will look for *asterisk.conf* in the default configuration location, which is usually */etc/asterisk*. To specify a different location for *asterisk.conf*, use the -C command-line option:

```
$ sudo asterisk -C /custom/path/to/asterisk.conf
```

The [directories] Section

For most installations of Asterisk, changing the directories is not necessary. However, this can be useful for running more than one instance of Asterisk at the same time, or if you would like files stored in nonstandard locations.

The default directory locations and the options you can use to modify them are listed in Table 4-1. For additional information about the usage of these directories, see the File Structure section of Chapter 2.

Table 4-1. asterisk.conf [directories] section

Option	Value/Example	Notes
astetcdir	/etc/asterisk	The location where the Asterisk configuration files are stored.
astmoddir	/usr/lib/asterisk/modules	The location where loadable modules are stored.
astvarlibdir	/var/lib/asterisk	The base location for variable state information used by various parts of Asterisk. This includes items that are written out by Asterisk at runtime.
astdbdir	/var/lib/asterisk	Asterisk will store its internal database in this directory as a file called *astdb*.
astkeydir	/var/lib/asterisk	Asterisk will use a subdirectory called *keys* in this directory as the default location for loading keys for encryption.
astdatadir	/var/lib/asterisk	This is the base directory for system-provided data, such as the sound files that come with Asterisk.
astagidir	/var/lib/asterisk/agi-bin	Asterisk will use a subdirectory called *agi-bin* in this directory as the default location for loading AGI scripts.
astspooldir	/var/spool/asterisk	The Asterisk spool directory, where voicemail, call recordings, and the call origination spool are stored.
astrundir	/var/run/asterisk	The location where Asterisk will write out its UNIX control socket as well as its process ID (PID) file.
astlogdir	/var/log/asterisk	The directory where Asterisk will store its log files.

The [options] Section

This section of the *asterisk.conf* file configures defaults for global runtime options. The available options are listed in Table 4-2. Most of these are also controllable via command-line parameters to the *asterisk* application. For a complete list of the command-line options that relate to these options, see the Asterisk manpage:

```
$ man asterisk
```

Table 4-2. asterisk.conf [options] section

Option	Value/Example	Notes
verbose	3	Sets the default verbose setting for the Asterisk logger. This value is also set by the *-v* command-line option. The verbose level is 0 by default.
debug	3	Sets the default debug setting for the Asterisk logger. This value is also set by the *-d* command-line option. The debug level is 0 by default.
alwaysfork	yes	Forking forces Asterisk to always run in the background. This option is set to no by default.
nofork	yes	Forces Asterisk to always run in the foreground. This option is set to no by default.

Option	Value/Example	Notes
quiet	yes	Quiet mode reduces the amount of output seen at the console when Asterisk is run in the foreground. This option is set to no by default.
timestamp	yes	Adds timestamps to all output except output from a CLI command. This option is set to no by default.
execincludes	yes	Enables the use of #exec in Asterisk configuration files. This option is set to no by default.
console	yes	Runs Asterisk in console mode. Asterisk will run in the foreground and will present a prompt for CLI commands. This option is set to no by default.
highpriority	yes	Runs the Asterisk application with real-time priority. This option is set to no by default.
initcrypto	yes	Loads keys from the astkeydir at startup. This option is set to no by default.[a]
nocolor	yes	Suppresses color output from the Asterisk console. This is useful when saving console output to a file. This option is set to no by default.
dontwarn	yes	Disables a few warning messages. This option was put in place to silence warnings that are generally correct, but may be considered to be so obvious that they become an annoyance. This option is set to no by default.
dumpcore	yes	Tells Asterisk to generate a core dump in the case of a crash. This option is set to no by default.[b]
languageprefix	yes	Configures how the prompt language is used in building the path for a sound file. By default, this is yes, which places the language before any subdirectories, such as *en/digits/1.gsm*. Setting this option to no causes Asterisk to behave as it did in previous versions, placing the language as the last directory in the path, (e.g. *digits/en/1.gsm*).
internal_timing	yes	Uses a timing source to synchronize audio that will be sent out to a channel in cases such as file playback or music on hold. This option is set to yes by default and should be left that way; its usefulness has greatly diminished over the last few major versions of Asterisk.
systemname	my_system_name	Gives this instance of Asterisk a unique name. When this has been set, the system name will be used as part of the uniqueid field for channels. This is incredibly useful if more than one system will be logging CDRs to the same database table. By default, this option is not set.
autosystemname	yes	Automatically sets the system name by using the hostname of the system. This option is set to no by default.
maxcalls	100	Sets a maximum number of simultaneous inbound channels. No limit is set by default.
maxload	0.9	Sets a maximum load average. If the load average is at or above this threshold, Asterisk will not accept new calls. No threshold is set by default.
maxfiles	1000	Set the maximum number of file descriptors that Asterisk is allowed to have open. The default limit imposed by the system is commonly 1024, which is not enough for heavily loaded systems. It is common to set this limit to a very high number. The default system-imposed limit is used by default.

Option	Value/Example	Notes
minmemfree	1	Sets the minimum number of megabytes of free memory required for Asterisk to continue accepting calls. If Asterisk detects that there is less free memory available than this threshold, new calls will not be accepted. This option is not set by default.
cache_record_files	yes	When doing recording, stores the file in the `record_cache_dir` until recording is complete. Once complete, it will be moved into the originally specified destination. The default for this option is no.
record_cache_dir	/tmp	Sets the directory to be used when `cache_record_files` is set to yes. The default location is a directory called *tmp* within the `astspooldir`.
transmit_silence	yes	Transmits silence to the caller in cases where there is no other audio source. This includes call recording and the `Wait()` family of dialplan applications, among other things. The default for this option is no.[c]
transcode_via_sln	yes	When building a codec translation path, forces signed linear to be one of the steps in the path. The default for this option is yes.
runuser	asterisk	Sets the system user that the Asterisk application should run as. This option is not set by default, meaning that the application will continue to run as the user that executed the application.
rungroup	asterisk	Sets the system group that the Asterisk application should run as. This option is not set by default.
lightbackground	yes	When using colors in the Asterisk console, it will output colors that are compatible with a light-colored background. This option is set to no by default, in which case Asterisk uses colors that look best on a black background.
documentation_language	en_US	The built-in documentation for Asterisk applications, functions, and other things is included in an external XML document. This option specifies the preferred language for documentation. If it is not available, the default of en_US will be used.
hideconnect	yes	Setting this option to yes causes Asterisk to not display notifications of remote console connections and disconnections at the Asterisk CLI. This is useful on systems where there are scripts that use remote consoles heavily. The default setting is no.
lockconfdir	no	When this option is enabled, the Asterisk configuration directory will be protected with a lock. This helps protect against having more than one application attempting to write to the same file at the same time. The default value is no.

[a] If any of the keys require a passphrase, this will block the startup process of Asterisk. An alternative is to run *keys init* at the Asterisk command line.

[b] This is critical for debugging crashes. However, Asterisk must be compiled with the DONT_OPTIMIZE option enabled in *menuselect* for the core dump to be useful.

[c] There is an important caveat to note when this option is enabled. The silence is generated in uncompressed signed linear format, which means that it will have to be transcoded into the format that the caller's channel expects. The result may be that transcoding is required for a call that would not normally require it.

The [files] Section

This section of *asterisk.conf* includes options related to the Asterisk control socket. It is primarily used by remote consoles (*asterisk -r*). The available options are listed in Table 4-3.

Table 4-3. asterisk.conf [files] section

Option	Value/Example	Notes
astctlpermissions	0660	Sets the permissions for the Asterisk control socket.
astctlowner	root	Sets the owner for the Asterisk control socket.
astctlgroup	apache	Sets the group for the Asterisk control socket.
astctl	asterisk.ctl	Sets the filename for the Asterisk control socket. The default is *asterisk.ctl*.

The [compat] Section

Occasionally the Asterisk development team decides that the best way forward involves making a change that is not backward-compatible. This section contains some options (listed in Table 4-4) that allow reverting behavior of certain modules back to previous behavior.

Table 4-4. asterisk.conf [compat] section

Option	Value/Example	Notes
pbx_realtime	1.6	In versions earlier than Asterisk 1.6.x, the pbx_realtime module would automatically convert pipe characters into commas for arguments to Asterisk applications. This is no longer done by default. To enable this previous behavior, set this option to 1.4.
res_agi	1.6	In versions earlier than Asterisk 1.6.x, the EXEC AGI command would automatically convert pipe characters into commas for arguments to Asterisk applications. This is no longer done by default. To enable this previous behavior, set this option to 1.4.
app_set	1.6	Starting with the Asterisk 1.6.x releases, the Set() application only allows setting the value of a single variable. Previously, Set() would allow setting more than one variable by separating them with a &. This was done to allow any characters in the value of a variable, including the & character, which was previously used as a separator. MSet() is a new application that behaves like Set() used to. However, setting this option to 1.4 makes Set() behave like MSet().

modules.conf

This file is not strictly required in an Asterisk installation; however, without any modules Asterisk won't really be able to do anything, so for all practical purposes, you need a *modules.conf* file in your */etc/asterisk* folder. If you simply define autoload=yes in your *modules.conf* file, Asterisk will search for all modules in the */usr/lib/asterisk/modules* folder and load them at startup.

Although most modules do not use much in the way of resources, and they all load very quickly, it just seems cleaner to our minds to load only those modules that you are planning on using in your system. Additionally, there are security benefits to not loading modules that accept connections over a network.

In the past we felt that explicitly loading each desired module was the best way to handle this, but we have since found that this practice creates extra work. After every upgrade we found ourselves having to edit the *modules.conf* file to correct all the module differences between releases, and the whole process ended up being needlessly complicated. What we prefer to do these days is to allow Asterisk to automatically load the modules that it finds, but to explicitly tell Asterisk not to load any modules we do not want loaded by use of the `noload` directive. A sample *modules.conf* file can be found in "modules.conf" on page 56.

Using menuselect to Control Which Modules Are Compiled and Installed

One other way that you can control which modules Asterisk loads is to simply not compile and install them in the first place. During the Asterisk installation process, the *make menuselect* command provides you with a menu interface that allows you to specify many different directives to the compiler, including which modules to compile and install. If you never compile and install a module, the effect of this at load time is that it won't exist, and therefore won't be loaded. If you are new to Linux and Asterisk, this may create confusion for you if you later want to use a module and discover that it doesn't exist on your system.

More information about *menuselect* is available in "make menuselect" on page 59.

The [modules] Section

The *modules.conf* file contains a single section. The options available in this section are listed in Table 4-5. With the exception of `autoload`, all of the options may be specified more than once.

 A list of all loadable modules is available in Chapter 2, with notes on our opinion regarding the popularity/status of each of them.

Table 4-5. modules.conf [modules] section

Option	Value/Example	Notes
autoload	yes	Instead of explicitly listing which modules to load, you can use this directive to tell Asterisk to load all modules that it finds in the modules directory, with the exception of modules listed as not to be loaded using the noload directive. The default, and our recommendation, is to set this option to yes.
preload	res_odbc.so	Indicates that a module should be loaded at the beginning of the module load order. This directive is much less relevant than it used to be; modules now have a load priority built into them that solves the problems that this directive was previously used to solve.
load	chan_sip.so	Defines a module that should be loaded. This directive is only relevant if autoload is set to no.
noload	chan_alsa.so	Defines a module that should not be loaded. This directive is only relevant if autoload is set to yes.
require	chan_sip.so	Does the same thing as load; additionally, Asterisk will exit if this module fails to load for some reason.
preload-require	res_odbc.so	Does the same thing as preload; additionally, Asterisk will exit if this module fails to load for some reason.

indications.conf

The sounds that people expect from the telephone network vary in different parts of the world. Different countries or regions present different sounds for events such as dialtone, busy signal, ringback, congestion, and so forth.

The *indications.conf* file defines the parameters for the various sounds that a telephone system might be expected to produce, and allows you to customize them. In the early days of Asterisk this file only contained sounds for a limited number of countries, but it is now quite comprehensive.

To assign the tones common for your region to channels, you can simply assign the tonezone using the CHANNEL() function, and that tonezone will apply for the duration of the call (unless changed later):

```
Set(CHANNEL(tonezone)=[yourcountry]) ; i.e., uk, de, etc.
```

However, since signaling from a call could come from various places (from the carrier, from Asterisk, or even from the set itself), you should note that simply setting the tonezone in your dialplan does not guarantee that those tones will be presented in all situations.

Hacking indications.conf for Fun and Profit

If you have too much time on your hands, you can do all sorts of pointless but entertaining things with your indications. For example, fans of *Star Wars* can make the following change to the end of their *indications.conf* files:

```
[starwars](us)
description = Star Wars Theme Song
ring = 262/400,392/500,0/100,349/400,330/400,294/400,524/400,392/500,0/100,349/400, \
       330/400,294/400,524/400,392/500,0/100,349/400,330/400,349/400,294/500,0/2000
```

If you then use the country named **'starwars'** in your configuration files or dialplan, any ringing you pass back will sound quite different from the standard ring you are used to. Try the following dialplan code to test out your new ringing sound:

```
exten => 500,1,Answer()
    same => n,Set(CHANNEL(tonezone)=starwars)
    same => n,Dial(SIP/0000FFFF0002) ; or whatever your channel is named in sip.conf
```

 Depending on the type of device used to call into this example, you may wonder if it will actually work. SIP phones, for example, typically generate their own tones instead of having Asterisk generate them. This example was carefully crafted to ensure that Asterisk will generate a ringback tone to the caller. The key is the `Answer()` that is executed first. Later, when an outbound call is made to another device, the only method Asterisk has available to pass back a ringing indication to the caller is by generating inband audio, since as far as the caller's phone is concerned, this call has already been answered.

While Asterisk will run without an *indications.conf* file, it is strongly recommended that you include one: copy the sample over from */usr/src/asterisk-complete/1.8/configs/indications.conf.sample*, modify the `country` parameter in the [general] section to match your region, and restart Asterisk.

chan_dahdi Ignores indications.conf

DAHDI does not use the *indications.conf* file from Asterisk, but rather has the tones compiled in. For more information, see Chapter 7.

If your system supports multiple countries (for example, if you have a centralized Asterisk system that has users from different regions), you may not be able to simply define the default country. In this case, you have a couple of options:

1. Define the country in the channel definition file for the user.
2. Define the country in the dialplan using the `CHANNEL(tonezone)` function.

For more information about using Asterisk in different countries, see Chapter 9.

musiconhold.conf

If you plan on selling Asterisk-based telephone systems and you do not change the default music on hold that ships with Asterisk, you are sending the message, loud and clear, that you don't really know what you are doing.*

Part of the problem with music on hold is that while in the past it was common to just plug a radio or CD player into the phone system, the legal reality is that most music licenses do not actually allow you to do this. That's right: if you want to play music on hold, somebody, somewhere, typically wants you to pay them for the privilege.

So how to deal with this? There are two legal ways: 1) pay for a music on hold license from the copyright holder, or 2) find a source of music that is released under a license suitable for Asterisk.

We're not here to give you legal advice; you are responsible for understanding what is required of you in order to use a particular piece of music as your music on hold source. What we will do, however, is show you how to take the music you have and make it work with Asterisk.

Getting Free Music

There are several websites that offer music that has been released under Creative Commons or other licenses. Lately, we've been enjoying music from Jamendo (*http://www .jamendo.com*). Each song may have its own licensing requirements, and just because you can download a song for free does not mean you have permission to use it as music on hold. Be aware of the licensing terms for the music you are planning to use for your music on hold.

Converting Music to a Format That Works Best with Asterisk

It's quite common to have music in MP3 format these days. While Asterisk can use MP3s as a music source, this method is not at all ideal. MP3s are heavily compressed, and in order to play them the CPU has to do some serious work to decompress them in real time. This is fine when you are only playing one song and want to save space on your iPod, but for music on hold, the proper thing to do is convert the MP3 to a format that is easier on the CPU.

---

* Note that Leif uses the default music, but his excuse is that he's lazy and wants to go and play Forza on his Xbox. The cobbler's kids have no shoes.

> # CentOS Prerequisite
>
> Since CentOS does not have MP3 capability installed with *sox*, you will have to install *mpg123* before you can convert MP3 files for use with Asterisk.
>
> First you will need to install the *rpmforge* repository. To find out which version you need, open your web browser and go to *http://dag.wieers.com/rpm/FAQ.php#B*. Select the text for the version/architecture you want to install and paste it into your shell:
>
> ```
> $ rpm -Uhv http://apt.sw.be/redhat ...
> ```
>
> You need to make sure this new repository is used correctly, so run the following:
>
> ```
> $ yum install yum-priorities
> ```
>
> (If you want to know more about *yum* priorities, see this site: *http://wiki.centos.org/PackageManagement/Yum/Priorities*.)
>
> Once the repository has been added, you can proceed to get *mpg123*:
>
> ```
> $ yum install mpg123
> ```
>
> Once that's done, your CentOS system is ready to convert MP3 files for use with Asterisk.

If you are familiar with the file formats and have some experience working with audio engineering software such as Audacity, you can convert the files on your PC and upload them to Asterisk. We find it is simpler to upload the source MP3 files to the Asterisk server (say, to the */tmp* folder), and then convert them from the command line.

To convert your MP3 files to a format that Asterisk understands, you need to run the commands outlined here (in this example we are using a file named *SilentCity.mp3*).

CentOS

First, convert the MP3 file to a WAV file:

```
$ mpg123 -w SilentCity.wav SilentCity.mp3
```

Then, downsample the resulting WAV file to a sampling rate that Asterisk understands:

```
$ sox SilentCity.wav -t raw -r 8000 -s -w -c 1 SilentCity.sln
```

Ubuntu

If you have not done so already, install *sox*, and the `libsox-fmt-all` package:

```
# sudo apt-get install sox libsox-fmt-all
```

Then, convert your MP3 file directly to the uncompressed SLN format:

```
$ sox SilentCity.mp3 -t raw -r 8000 -s -w -c 1 SilentCity.sln
```

In newer versions of *sox* (e.g., version 14.3.0, which shipped with Ubuntu 10.10), the *-w* option has changed to *-2*.

Completing file conversion

The resulting file will exist in the */tmp* folder (or wherever you uploaded to) and needs to be copied to the */var/lib/asterisk/moh* folder:

```
$ cp *.sln /var/lib/asterisk/moh
```

You now need to reload *musiconhold* in Asterisk in order to have it recognize your new files:

```
$ asterisk -rx "module unload res_musiconhold.so"
$ asterisk -rx "module load res_musiconhold.so"
```

To test that your music is working correctly, add the following to the [UserServices] context in your dialplan:

```
exten => 664,1,NoOp()
    same => n,Progress()
    same => n,MusicOnHold()
```

Dialing 664 from one of your sets should play a random file from your *moh* directory.

Conclusion

This chapter helped you complete some initial configuration of Asterisk. From here you can move on to setting up some phones and taking advantage of the many features Asterisk has to offer.

## CHAPTER 5

User Device Configuration

I don't always know what I'm talking about,
but I know I'm right.

—Muhammad Ali

In this chapter we'll delve into the user devices that you might want to connect to Asterisk, typically VoIP telephones of some sort. Configuring a channel in Asterisk for the device to connect through is relatively straightforward, but you also need to configure the device itself so it knows where to send its calls.* In other words, there are two parts to configuring a device on Asterisk: 1) telling Asterisk about the device, and 2) telling the device about Asterisk.

How Asterisk Relates to the SIP Protocol

SIP is a peer-to-peer protocol, and while it is common to have a setup where endpoints act as clients and some sort of gateway acts as a server, the protocol still thinks in terms of peer-to-peer relationships. What this means is that a SIP telephone expects to make a direct connection to another SIP telephone, without a PBX in between.

The reality is that many SIP transactions happen through a server, and in the case of Asterisk, it is common to have the PBX in the middle of all connections. When a SIP call is made from a telephone to another telephone through Asterisk, there are actually two calls happening: one from the originating set to Asterisk, and another separate call from Asterisk to the destination set. Asterisk bridges the two channels together.

From the perspective of the SIP telephone, therefore, you need to configure it to send all its calls to Asterisk, even though the device is quite capable of directly connecting to another SIP endpoint without the Asterisk server. The SIP protocol is complex and very flexible, and configuring endpoints can seem difficult because they have much more flexibility than we require of them for an Asterisk implementation.

---

* This has nothing to do with Asterisk configuration, and each hardware manufacturer will have its own tools to allow you to configure its devices.

While most devices will have a web-based interface for defining parameters, if you're putting more than one or two phones into production we recommend using a server-based configuration process, wherein the set is only told the location of a file server. The set will identify itself and download customized files that define the required parameters for that telephone. As an example, these could be XML files on an FTP server. The exact download process and syntax of these files will differ from manufacturer to manufacturer. In this chapter we will only talk about the configuration of sets from the perspective of Asterisk.

Telephone Naming Concepts

Before we get started with configuring Asterisk for our telephones, we are going to recommend some best practices regarding telephone naming, abstracting the concepts of users, extension numbers, and telephones from each other.

In Asterisk, all the system cares about is the channel name. There is really no concept of a user at all,[†] and extensions are simply ways of directing call flow through the system. For example, your dialplan might inform Asterisk that when extension number 100 is requested it should call the phone on my desk, but extension 100 could just as easily call a company voicemail box, play back a prompt, or join a conference room. We can even specify that extension 100 should ring the device on my desk from Monday to Friday between 9 A.M. and 5 P.M., but ring a device on someone else's desk the rest of the time. Inversely, when a call is made from a device during business hours, the callerID could show a daytime number, and the rest of the time could show an after-hours number (many reception desks become security desks at night).

Asterisk Extensions

The concept of an extension in Asterisk is crucial. In most PBXs, an extension is a number that you dial to cause a phone or service to ring. In Asterisk, an extension is the name of a grouping of instructions in the dialplan. Think of an Asterisk extension as a script name, and you're on the right track. Yes, an Asterisk extension could be a number (such as 100) that rings a phone, but it could just as easily be a name (such as voicemail) that runs a sequence of dialplan applications.

We'll be going into Asterisk extensions in far more detail throughout this book, but before we do that we want to get some phones set up.

The abstraction between the name of an extension and what that extension does is a powerful concept in Asterisk, as extension 100 could do a number of things depending

---

† Actually, Asterisk does try to implement and abstract the concepts of users and devices internally by using the *users.conf* file; however, it is typically only used by the Asterisk GUI. Abstracting the concepts logically using the dialplan is easier to understand and far more flexible.

on any number of variables that are programmed into the system. This is especially relevant in the context of features such as hot-desking.

Hot-desking is a feature that allows someone to log into a device and receive his calls at that device. Let's say we have three sales agents who typically work outside of the office, but spend a couple of days each month in the office to do paperwork. Since they are unlikely to be on-site at the same time, instead of having a separate telephone for each of those three sales agents, they could share a single office phone (or on a larger scale, a dozen folks could share a pool of, say, three phones). This scenario illustrates the convenience (and necessity) of allowing the system to separate the concept of a user and extension from the physical phone.

So what are some examples of bad names for telephone devices? Something like a person's name, such as [SimonLeBon], would be a poor name for a telephone as the phone may also be used by Joan Jett and Rick Astley. The same reasoning can be applied to why you would not want to name a phone based on an extension number: a phone name of [100] would be a poor choice since you might want to reprovision the device for extension 160 in the future, or it might be used by several people with different extensions in a hot-desking solution. Using numeric account names is also very bad from a security perspective and is discussed in more detail in Chapter 26.

A popular way to name a phone is using the MAC address of the device. This is a unique identifier specific to the phone that follows it where it goes and doesn't directly relate to the user operating the phone or the extension number currently associated with it. Some corporations have stickers they place on their equipment with a bar code and other information that allows them to keep stock of provisioned equipment; these unique codes would also be an acceptable choice to use for phone names as they don't provide any logical relation to a particular person, but do provide specific information about the devices themselves.

The choice is yours as to how you want to name your phones, but we primarily want to abstract any concept of the telephone being owned by a person, or even its location in the network, since these concepts are outside the realm of Asterisk and can change at any time.

Throughout this book, you'll see us using phone names that look like MAC addresses (such as 0000FFFF0001 and 0000FFFF0002) to differentiate between devices. You will want to use phone names that match the hardware you are using (or some other string that is unique to the device you are registering).

As a final consideration, we should make it clear that what we are suggesting regarding device names is not a technical requirement. You are free to name your devices anything you want, as long as they meet the requirements of Asterisk's naming conventions for devices (stay with alphanumeric characters with no spaces and you'll be fine).

Hardphones, Softphones, and ATAs

There are three types of endpoints you would typically provide your users with that could serve as a telephone set. They are popularly referred to as hardphones, softphones, and Analog Terminal Adaptors (ATAs).

A *hardphone* is a physical device. It looks just like an office telephone: it has a handset, numbered buttons, etc. It connects directly to the network, and it's what people are referring to when they talk about a VoIP telephone (or a SIP telephone).

A *softphone* is a software application that runs on a laptop or desktop. The audio must pass through the PC's sound system, so you normally need a headset that will work well with telephony applications. More recently, softphone applications have been written for smart phones that allow you to connect to other networks other than just the cellular network. The interface of the softphone is often styled to look like a physical telephone, but this is not necessary.

An *ATA* is designed to allow traditional analog telephones (and other analog devices, such as fax machines, cordless phones, paging amplifiers, and such) to connect to a SIP network,[‡] and will typically be a sandwich-sized box that contains an RJ-11 connector for the phone (commonly referred to as an FXS port), an RJ-45 connector for the network, and a power connector. Some ATAs may support more than one phone.

Hardphones have the advantage that the handsets have good acoustic properties for voice communications. Any decent-quality telephone is engineered to pick up the frequencies of the human voice, filter out unwanted background noise, and normalize the resulting waveform. People have been using telephones for as long as the telephone network has existed, and we tend to like what is familiar, so having a device that communicates with Asterisk using a familiar interface will be attractive to many users. Also, a hardphone does not require your computer to be running all the time.

Disadvantages to hardphones include that they are nonportable and expensive, relative to the many quality softphones on the market today that are available for free. Also, the extra clutter on your desk may not be desirable if you have limited work space, and if you move around a lot and are not generally at the same location, a hardphone is not likely to suit your needs (although, one at each location you frequent might be a valid solution).

Softphones solve the portability issue by being installed on a device that is likely already moving with you, such as your laptop or smart phone. Also, their minimal cost (typically free, or around the $30 price range for a fully featured one) is attractive. Because many softphones are free, it is likely that the first telephone set you connect to Asterisk will be a softphone. Also, because softphones are just software, they are easy to install and upgrade, and they commonly have other features that utilize other peripherals, like

---

‡ Or any other network, for that matter. ATAs could more formally be said to be analog-to-digital gateways, where the nature of the digital protocol may vary (e.g., proprietary ATAs on traditional PBXs).

a webcam for video calling, or perhaps an ability to load files from your desktop for faxing.

Some of the disadvantages of softphones are the not-always-on nature of the devices, the necessity to put on a headset each time you take a call, and the fact that many PCs will at random times during the day choose to do something other than what the user wants them to do, which might cause the softphone to stop working while some background task hogs the CPU.

ATAs have the advantage of allowing you to connect to your SIP network analog devices,[§] such as cordless phones (which are still superior in many cases to more advanced types of wireless phones[||]), paging amplifiers, and ringers. ATAs can also sometimes be used to connect to old wiring, where a network connection might not function correctly.

The main disadvantage of an ATA is that you will not get the same features through an analog line as you would from a SIP telephone. This is technology that is over a century old.

With Asterisk, we don't necessarily need to make the choice between having a softphone, a hardphone, or an ATA; it's entirely possible and quite common to have a single extension number that rings multiple devices at the same time, such as a desk phone, the softphone on a laptop, a cell phone, and perhaps a strobe light in the back of the factory (where there is too much noise for a ringer to be heard).

Asterisk will happily allow you to interact with the outside world in ways that were scarcely dreamed of only a few years ago. As we see more unification of communications applications with the popularity of social networks, communities such as Skype, and more focus on network-based services such as those provided by Google, the flexibility and popularity of software-based endpoints will continue to grow. The blurring of the lines between voice and applications is constantly evolving, and softphones are well positioned to rapidly respond to these changes.

We still like a desk phone, though.

Configuring Asterisk

In this section we'll cover how to create the *sip.conf* and *iax.conf* configuration files in the */etc/asterisk/* directory, which are used for defining the parameters by which SIP and IAX2 devices can communicate with your system.

---

§ An ATA is not the only way to connect analog phones. Hardware vendors such as Digium sell cards that go in the Asterisk server and provide analog telephony ports.

‖ For a really awesome cordless analog phone, you want to check out the EnGenius DuraFon devices, which are expensive, but impressive.

Asterisk allows devices using many different protocols to speak to it (and therefore to each other). However, the SIP and IAX2 protocols are the most popular and mature VoIP modules, so we will focus our attention on them. For your first Asterisk build, you might be best off not bothering with the other protocols (such as Skinny/SCCP, Unistim, H.323, and MGCP), and getting comfortable working with SIP and IAX2 first. The configuration for the other protocols is similar, and the sample configuration files are full of information and examples, so once you have the basics down, other protocols should be fairly easy to work with.

The channel configuration files, such as *sip.conf* and *iax.conf*, contain the configuration for the channel driver, such as *chan_iax2.so* or *chan_sip.so*, along with the information and credentials required for a telephony device to contact and interact with Asterisk.

Common information about the channel driver is contained at the top of the configuration file, in the [general] section. All section names are encased in square brackets, including device names. Anything that follows a section name (or device definition, which for our purposes is essentially the same thing) is applied to that section. The [general] section can also contain information to define defaults for device configurations, which are overridden in the section for each device, or in a template. Asterisk also comes with defaults that are hardcoded, so while some settings are mandatory, many other settings can be ignored as long as you are happy with the defaults.

Asterisk will check for parameters in the following order:

1. Check the specific section for the relevant channel.
2. Check the template for the section.
3. Check the [general] section.
4. Use the hardcoded defaults.

This means that just because you didn't specify a setting for a particular parameter doesn't mean your device isn't going to have a setting for that parameter. If you are not sure, set the parameter explicitly in the section of the configuration file that deals with that specific channel, or in the relevant template.

This concept should make more sense as you read on.

How Channel Configuration Files Work with the Dialplan

While we haven't discussed Asterisk dialplans yet, it is useful to be able to visualize the relationship between the channel configuration files (*sip.conf*, *iax.conf*) and the dialplan (*extensions.conf*). The dialplan is the heart of an Asterisk system: it controls how call logic is applied to any connection from any channel, such as what happens when a device dials extension 101 or an incoming call from an external provider is routed. Both the relevant channel configuration file and the *extensions.conf* file play a role in most calls routed through the system. Figure 5-1 provides a graphical representation of the relationship between the *sip.conf* and *extensions.conf* files.

When a call comes into Asterisk, the identity of the incoming call is matched in the channel configuration file for the protocol in use (e.g., *sip.conf*). The channel configuration file also handles authentication and defines where that channel will enter the dialplan.

Once Asterisk has determined how to handle the channel, it will pass call control to the correct context in the dialplan. The context parameter in the channel configuration file tells the channel where it will enter the dialplan (which contains all the information about how to handle and route the call).

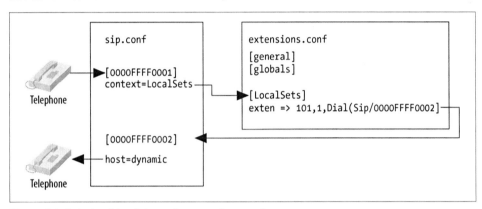

Figure 5-1. Relationship of sip.conf to extensions.conf

Conversely, if the dialplan has been programmed to dial another device when the request for extension number 101 is being processed, a request to dial telephony device 0000FFFF0002 will use the channel configuration file to determine how to pass the call back out of the dialplan to the telephone on the network (including such details as authentication, codec, and so forth).

A key point to remember is that the channel configuration files control not only how calls enter the system, but also how they leave the system. So, for example, if one set calls another set, the channel configuration file is used not only to pass the call through to the dialplan, but also to direct the call from the dialplan to the destination.

sip.conf

The SIP# channel module is arguably the most mature and feature-rich of all the channel modules in Asterisk. This is due to the enormous popularity of the SIP protocol, which

---

#The SIP RFC is a long read, but about the first 25 pages are a good introduction. Check it out at *http://www .ietf.org/rfc/rfc3261.txt.*

has taken over the VoIP/telecom industry and been implemented in thousands of devices and PBXs. If you look through the *sip.conf.sample* file in the *./configs* subdirectory of your Asterisk source you will notice a wealth of options available. Fortunately, the default options are normally all you need, and therefore you can create a very simple configuration file that will allow most standard SIP telephones to connect with Asterisk.

The first thing you need to do is create a configuration file in your */etc/asterisk* directory called *sip.conf*.

Paste or type the following information into the file:

```
[general]
context=unauthenticated         ; default context for incoming calls
allowguest=no                   ; disable unauthenticated calls
srvlookup=yes                   ; enabled DNS SRV record lookup on outbound calls
udpbindaddr=0.0.0.0             ; listen for UDP requests on all interfaces
tcpenable=no                    ; disable TCP support

[office-phone](!)   ; create a template for our devices
type=friend         ; the channel driver will match on username first, IP second
context=LocalSets   ; this is where calls from the device will enter the dialplan
host=dynamic        ; the device will register with asterisk
nat=yes             ; assume device is behind NAT
                    ; *** NAT stands for Network Address Translation, which allows
                    ; multiple internal devices to share an external IP address.
secret=s3CuR#p@s5   ; a secure password for this device -- DON'T USE THIS PASSWORD!
dtmfmode=auto       ; accept touch-tones from the devices, negotiated automatically
disallow=all        ; reset which voice codecs this device will accept or offer
allow=ulaw          ; which audio codecs to accept from, and request to, the device
allow=alaw          ; in the order we prefer

; define a device name and use the office-phone template
[0000FFFF0001](office-phone)

; define another device name using the same template
[0000FFFF0002](office-phone)
```

Open the *sip.conf* file you've just created, and we'll go over each item.

We've created four sections, the first one being the [general] section. This is a standard section that appears at the top of the configuration file for all channel modules, and must always be named in this way. The [general] section contains general configuration options for how that protocol relates to your system, and can be used to define default parameters as well.

For example, we've defined the default context as unauthenticated, to ensure that we have explicitly declared where unauthenticated guest calls will enter the dialplan (rather than leaving that to chance). We've named it unauthenticated to make it obvious that calls processed in this context are not trusted, and thus should not be able to do things such as make outbound calls to the PSTN (which could potentially cost money, or represent identity theft). You should be aware that we could have used any name we wanted, and also that there needs to be an identically named context in *extensions.conf* to define the call flow for unauthenticated calls.

The next option is `allowguest`, which we've disabled as we don't want to accept any unauthenticated calls at this time. Keep in mind that for some channels you may actually want to accept unauthenticated calls. A common use for allowing unauthenticated calls is for companies that allow dialing by uniform resource identifiers (URIs), like email addresses. If we wanted to allow customers to call us from their phones without having to authenticate, we could enable guest calls and handle them in the `unauthenticated` context defined by the previous option.

> You may be wondering why you might ever want to allow unauthenticated calls. The reason is that if you publish your SIP URI on your business cards (e.g., *sip:leif.madsen@shifteight.org*), calls to that URI will fail if your unauthenticated context simply hangs up. What you want instead is for your unauthenticated context to put incoming calls into a controlled environment. You may wish to allow the calls, but you won't necessarily trust them.[*]

The `srvlookup` option is used to enable Asterisk to perform a lookup via a DNS SRV record, which is typically used for outbound connections to service providers. We'll talk more about Asterisk and DNS in Chapter 12.

The `udpbindaddr`[†]option takes the value of an IP address or `0.0.0.0` to tell Asterisk which network interface it should listen to for requests carried by the UDP network transport protocol (which is the protocol that actually carries the voice channels). By defining `0.0.0.0`, we're instructing the channel driver to listen on all available interfaces. Alternatively, we could limit VoIP connections for this protocol to a single interface by defining the IP address of a specific network interface on our system.

> Currently in Asterisk the `udpbindaddr` and `tcpbindaddr` options are an all-or-one proposition. In other words, if you have three NICs in your system, you can't restrict VoIP traffic to two of them: it's either one only, or all of them.

---

[*] The whole concept of security and trust on a VoIP network is something that can become quite complex. Spammers are already hard at work figuring out this technology, and you need to be aware of the concepts. We'll cover this in more depth later in the book, such as in Chapter 7 and Chapter 26.

[†] The complement to this option is `tcpbindaddr`, used for listening for requests carried via the TCP network transport protocol.

The `tcpenable` option allows us to accept requests via the TCP network transport protocol. For now we've disabled it, as the UDP method is currently more mature (and more popular) and we're attempting to eliminate as many barriers as possible. Having said that, feel free to test TCP support once you're comfortable configuring your devices.

> There are also `tlsenable` and `tlsbindaddr` options for enabling SIP over TLS (encrypted SIP). We'll cover the configuration of SIP with TLS in Chapter 7.

The next section we've defined is a template we have chosen to name `[office-phone]` (!). We've created it as a template so that we can use the values within it for all of our devices.

> Following the section name with (!) tells Asterisk to treat this section as a template. By doing this we eliminate the need to repetitively add and change configuration options for every device we choose to define. Templates are extremely useful and are available in all of Asterisk's configuration files. If you want to change something for an individual device that was previously defined in the template for that device, you can do that under the section header, and it will override what was defined by the template. It is not necessary to use templates, but they are extremely handy, and we use them extensively.

In the [office-phone] template we've defined several options required for authentication and control of calls to and from devices that use that template. The first option we've configured is the type, which we've set to friend. This tells the channel driver to attempt to match on name first, and then IP address.

SIP Configuration Matching and the type Option

In the example we have provided, the configuration for SIP phones is set with type=friend. There are two other type definitions you can use: user and peer. The difference between them has to do with how Asterisk interprets incoming SIP requests. The rules are covered in this table:

type =	Description
peer	Match incoming requests to a configuration entry using the source IP address and port number.
user	Match incoming requests to a configuration entry using the username in the From header of the SIP request. This name is matched to a section in *sip.conf* with the same name in square brackets.
friend	This enables matching rules for both peer and user. This is the setting most commonly used for SIP phones.

When a request from a telephone is received and authenticated by Asterisk, the requested extension number is handled by the dialplan in the context defined in the device configuration; in our case, the context named LocalSets.

The host option is used when we need to send a request to the telephone (such as when we want to call someone). Asterisk needs to know where the device is on the network. By defining the value as dynamic, we let Asterisk know that the telephone will tell us where it is on the network instead of having its location defined statically. If we wanted to define the address statically, we could replace dynamic with an IP address such as 192.168.128.30.

The nat option is used to tell Asterisk to enable some tricks to make phone calls work when a SIP phone may be located behind a NAT. This is important because the SIP protocol includes IP addresses in messages. If a phone is on a private network, it may end up placing private addresses in SIP messages, which are often not useful.

The password for the device is defined by the secret parameter. While this is not strictly required, you should note that it is quite common for unsavory folks to run phishing scripts that look for exposed VoIP accounts with insecure passwords and simple device names (such as a device name of 100 with a password of 1234). By utilizing an uncommon device name such as a MAC address, and a password that is a little harder to guess, we can significantly lower the risk to our system should we need to expose it to the outside world.

You can generate a secure password using one of several password generators available on the Internet and on your operating system. Here is a simple script that you can run at your console to generate one:

```
$ dd if=/dev/random count=1 bs=8 2>/dev/null | base64 | sed -e 's/=*$//'
```

Do not use the password we have defined in the example. This book will be available online and can be downloaded by anyone, and that particular password is almost certain to become one of the first passwords added to the list employed by VoIP phishing scripts in brute-force password attacks. If your SIP ports are exposed to the Internet and you use simple passwords, rest assured that you will eventually be defrauded.

The dtmfmode option is used to define which DTMF (touch-tone) format Asterisk should expect to be sent by the telephone. The four options are: info, inband, rfc2833, and auto. The info value means to use the SIP INFO method, inband is for inband audio tones, and rfc2833 is for the out-of-band method defined by that RFC. Using auto allows Asterisk to automatically determine which DTMF mode to use (it prefers rfc2833 if available).

The last two options, disallow and allow (sip.conf), are used to control which audio codecs are accepted from and offered to the telephone. By defining disallow=all first, we're telling Asterisk to reset any previously defined codec definitions in the [general] section (or the internal defaults); then we explicitly declare which codecs we'll accept (and the order we prefer). In our example we've enabled both ulaw and alaw, with ulaw most preferred (if you are outside of Canada or the US, you'll likely want to declare alaw first).

Now that we're finished with our template, we can define our device names and, utilizing the office-phone template, greatly simplify the amount of details required under each device section. The device name is defined in the square brackets, and the template to be applied is defined in the parentheses following the device name. We can add additional options to the device name by specifying them below the device name:

```
[0000FFFF0003](office-phone) ; template must be on same line and no space between
secret=@NOth3rP4S5
allow=gsm
```

As far as the SIP module in Asterisk can tell, it will end up reading this in as if it was a section defined like this, which is a configuration section with all of the options specified from the template first:

```
[0000FFFF0003]

;
; These options came from the template section.
;
type=friend
```

```
context=LocalSets
host=dynamic
nat=yes
secret=s3CuR#p@s5
dtmfmode=auto
disallow=all
allow=ulaw
allow=alaw

;
; These options came from the specific section.
;
secret=@N0th3rP4S5 ; Overrides the secret from the template
allow=gsm          ; Adds gsm to the list of allowed codecs
```

Note that if you specify an option that was also specified in the template, the module will see that option twice. For most options, it will override the value in the template, but for some, such as type, allow, and disallow, it will not.

iax.conf

IAX2 stands for the Inter-Asterisk eXchange protocol, version 2. It was originally developed to connect Asterisk systems together across different physical networks, and especially those behind firewalls and NAT devices. IAX2 was designed to limit the number of ports required to carry VoIP calls across a firewall, and to easily traverse networks that employed NAT devices (which historically have been problematic for the SIP protocol). As Asterisk has developed over the years, the IAX2 protocol has matured. It was standardized in an IETF RFC in 2009,[‡] but IAX2 has not become popular with hardware vendors, possibly due to the relative recency of the RFC, and certainly due to the fact that SIP is far and away the most recognized VoIP protocol in terms of mind-share (i.e., nontechnical folks have probably heard of SIP). IAX2, however, has advantages that make it a protocol worth discussing.

One of the primary advantages of IAX is single-port firewall penetration. All traffic, including signaling and audio data, is transferred over a single UDP port (4569), which can greatly simplify securing the VoIP network.[§]

Another advantage is IAX2 trunking, which encapsulates packets for several voice frames into the same datagram using a single IAX2 header. The benefit of this is a reduction in the amount of overhead required to send many simultaneous calls between two endpoints. The amount of bandwidth saved with IAX2 trunking when sending just a couple of calls between locations is pretty insignificant, but when you start scaling to the size of dozens or hundreds of calls, the savings can be substantial.

[‡] *http://www.rfc-editor.org/rfc/rfc5456.txt.*

[§] SIP, which has separate signaling and voice data protocols and ports, requires port 5060 for signaling, and at least two RTP ports for every active call for voice. By default, Asterisk uses ports 5060 for SIP and 10,000 through 20,000 for RTP, although that can be tuned with the *rtp.conf* file.

For now, we're only interested in the minimal configuration required to get our IAX2 endpoints talking to each other, so let's explore what we need to configure in *iax.conf* to do so.

First, we need to create our *iax.conf* file in the */etc/asterisk* configuration directory and add the following configuration information to the file:

```
[general]
autokill=yes     ; don't stall for a long time if other endpoint doesn't respond
srvlookup=yes    ; enable DNS SRV lookups for outbound calls

[office-phone](!)     ; template for IAX-based office phones
type=friend           ; Asterisk will allow calls to and from this phone
host=dynamic          ; inform Asterisk that the phone will tell us where
                      ;    it is on the network
secret=my5UP3rp@s5!   ; a secure password -- which means DON'T USE THIS ONE!
context=LocalSets     ; the context where incoming requests will enter the dialplan
disallow=all          ; reset the available voice codecs
allow=ulaw            ; prefer the ulaw codec
allow=alaw            ; but also allow the alaw codec

[0000FFFF0004](office-phone) ; define our first phone with the office-phone template

[0000FFFF0005](office-phone) ; define another phone with our office-phone template
```

Let's go over the options we added to this file, starting first with the [general] section. This is where we define our default configuration, our global options, and our general channel driver setup. There are many options we can define here, and we encourage you to check out the *iax.conf.sample* file in the *configs* directory of your Asterisk source,‖ but since we're looking for a straightforward configuration, we're going to allow many of the default options to be applied.

One option we have added is autokill, which is used to instruct the *chan_iax2.so* channel driver not to wait around too long on stalled connections. Also, srvlookup has been enabled to allow DNS SRV lookups for outgoing calls.

The next section we've defined is named [office-phone](!), which is a template that contains the options we'll apply to our IAX phones and the authentication information required for communication between the phones and Asterisk.

 As mentioned in the preceding section, following the section name with (!) tells Asterisk to treat this section as a template. Templates are useful, so use them(!).

‖ We also encourage you to read "sip.conf" on page 89 in this chapter, as it contains many concepts that apply equally to other Asterisk configuration files.

The first option we've configured in our template is the **type**. We've defined the **type** to be **friend**, which informs Asterisk that we plan on placing calls to the phone and that we expect to receive requests (calls) from the phone.

The other two types are **user** and **peer**. In IAX2, a **friend** is a combination of both a **user** and a **peer**, which is common for telephones because we expect to send calls to them, and to receive requests for placing calls from them. We could alternatively define two separate sections with the same name with types of **user** and **peer**, and then define only the required information for each of those types; or if we only ever expected to send calls to or place calls using a section (such as in the case of an inbound- or outbound-only service provider), we would define it as only either a **user** or a **peer**. However, in the vast majority of cases, just utilizing the **type** of **friend** is the most logical choice.

Note the difference between the meaning of the **type** option in *iax.conf* versus *sip.conf* (see the sidebar "SIP Configuration Matching and the type Option" on page 93). In *iax.conf*, the meaning is much simpler: it only has to do with the direction of the phone calls.

Following the **type** option, we've set **host** to **dynamic**, which means the phone will register with us to identify where it exists on the network (so that we know where to send its calls). Alternatively, we could statically define an IP address such as **192.168.128.50** where all calls will be sent to and accepted from. This will only work if the device will always have the same IP address.

The next option is **secret**, which defines the password.

You should make sure you are implementing a secure password here, especially if you plan on opening your system up to the outside world at all. Do not use something stupid such as **1234**, or you'll regret it. No, seriously. We're not kidding. Not even in the lab.

The number of successful attacks on VoIP-enabled telephone systems (not just Asterisk) is on the rise, and will continue to get worse. Commonly, successful intrusions are due to weak passwords. If you get in the habit of using strong passwords now, you'll have that much more protection in the future.

You can generate a complex password using a password generator like those available on the Internet and on your operating system. Here is a simple script that you can run at the Linux shell prompt to generate a string that'll be suitable as a password:

```
$ dd if=/dev/random count=1 bs=8 2>/dev/null | base64 | sed -e 's/=*$//'
```

Do not use the password we have defined in the example. You can be sure that shortly after this book is published, that will be one of the first passwords a brute-force bot will try to use on your system.

The `context` option defines the context at which this channel will enter the dialplan. When a call is placed by a telephone and the request is received by Asterisk, it is handled by the logic defined in the context configured here within the dialplan (*extensions.conf*).

Following that are the `disallow` and `allow` options. These define the codecs that will be allowed for this channel (in order of preference). The directive `disallow=all` resets any default codecs that may have been permitted in the `[general]` section (or as part of the channel defaults). Then, we define the specific codecs we wish to permit with `allow`.

If we wanted to use specific codecs for one of our devices, we could define that codec within the device configuration section for that channel simply by adding it below the device section header:

```
[0004F2FF0006](office-phone)
disallow=all ; Clear previous codec settings.
allow=gsm    ; Allow only gsm.
```

Modifying Your Channel Configuration Files for Your Environment

Our examples so far have been based on hypothetical device names. To create actual channels based on whatever you have in your environment, you will want to change the device names in your *sip.conf* and *iax.conf* files to something that makes more sense.

For example, if you have a Polycom IP 430 set with a MAC address of `0004f2119698`, you'll want to define a device identifier in *sip.conf* for that device:

```
[0004f2119698](office-phone)
```

If you have an IAX softphone on your PC that you wish to use, your *iax.conf* file may want something like this:

```
[001b63a28ccc](office-phone) ; you could have a comment like Leif's Laptop
```

Remember that you can name your devices anything you want. Asterisk doesn't care, but for ease of management, make sure you choose a naming convention that is logical, scalable, and sustainable.

Loading Your New Channel Configurations

In order to inform Asterisk of the new configurations, you will need to pass it a command that instructs it to reload the relevant configuration file. The Asterisk CLI is where you can pass various commands to a running Asterisk system.

The Asterisk CLI

The best way to see what is happening with your Asterisk system is through the Asterisk CLI. This interface provides various levels of output to let you know what is happening on your system, and offers a wealth of useful utilities to allow you to affect your running system. Begin by calling up the Asterisk CLI and reloading the configuration files for your channel modules:

```
$ sudo asterisk -r
*CLI> module reload chan_sip.so
*CLI> module reload chan_iax2.so
```

Verify that your new channels have been loaded:

```
*CLI> sip show peers
*CLI> sip show users
*CLI> iax2 show peers
*CLI> iax2 show users
```

> At this point your Asterisk system should be configured to handle registrations from the defined devices. Calls to and from the sets will not work until the configuration on the devices has been completed. Since each device is different in this regard, detailed configuration instructions for each model are outside of the scope of this book.

Testing to Ensure Your Devices Have Registered

Once your device has registered to Asterisk, you will be able to query the location and state of the device from the Asterisk CLI.

> It is a common misconception that registration is how a device authenticates itself for the purpose of obtaining permission to make calls. This is incorrect. The only purpose of registration is to allow a device to identify its location on the network, so that Asterisk# knows where to send calls intended for that device.
>
> Authentication for outgoing calls is an entirely separate process and always happens on a per-call basis, regardless of whether a set has registered. This means that your set may be able to make calls, but not receive them. This will normally happen when the set has not registered successfully (so Asterisk does not know where it is), and yet has the correct credentials for making calls (so Asterisk is willing to accept calls from it).

To check the registration status of a device, simply call up the Asterisk CLI:

```
$ sudo asterisk -r
```

#Or any other SIP registrar server, for that matter.

Typing the following command returns a listing of all the peers that Asterisk knows about (regardless of their state):

```
*CLI> sip show peers
Name/username             Host             Dyn Nat ACL Port      Status
0000FFFF0001/0000FFFF0001  192.168.1.100    D   N       5060      Unmonitored
0000FFFF0002/0000FFFF0002  192.168.1.101    D   N       5060      Unmonitored
```

 You may notice that the `Name/username` field does not always show the full name of the device. This is because this field is limited to 25 characters.

Note that the `Status` in our example is set to `Unmonitored`. This is because we are not using the `qualify=yes` option in our *sip.conf* file.

Analog Phones

There are two popular methods for connecting analog phones to Asterisk. The first is by using an ATA that most commonly connects to Asterisk using the SIP protocol. The Asterisk configuration for an ATA is the same as it would be for any other SIP-based handset. The other method is to directly connect the phones to the Asterisk server using telephony hardware from a vendor such as Digium. Digium sells telephony cards that can be added to your server to provide FXS ports for connecting analog phones (or fax machines). For the purposes of demonstrating the configuration, we're going to show the configuration required if you had a Digium AEX440E card, which is an AEX410 half-length PCI-Express with four FXS modules and hardware-based echo cancellation.

 Regardless of which hardware you are using, consult your vendor's documentation for any hardware-specific configuration requirements.

First, ensure that both Asterisk and DAHDI are installed (refer back to "How to Install It" on page 48 for instructions). Note that DAHDI must be installed before you install Asterisk. When you install DAHDI, be sure to install the *init* script as well. This will ensure that your hardware is properly initialized when the system boots up. The *init* script is installed from the *DAHDI-tools* package.

The *init* script uses the */etc/dahdi/modules* file to determine which modules should be loaded to support the hardware in the system. The installation of the *init* script attempts to automatically set up this file for you, but you should check it to make sure it is correct:

```
# Autogenerated by tools/xpp/dahdi_genconf (Dahdi::Config::Gen::Modules) on
# Tue Jul 27 10:31:46 2010
# If you edit this file and execute tools/xpp/dahdi_genconf again,
# your manual changes will be LOST.
wctdm24xxp
```

There is one more configuration file required for DAHDI: */etc/dahdi/system.conf*. It looks like this:

```
# Specify that we would like DAHDI to generate tones that are
# used in the United States.
loadzone = us
defaultzone = us

# We have 4 FXS ports; configure them to use FXO signaling.
fxoks = 1-4
```

> This configuration assumes the card is being used in the United States. For some tips on internationalization, see Chapter 9.

If the card you are configuring does not have hardware-based echo cancellation, another line will need to be added to */etc/dahdi/system.conf* to enable software-based echo cancellation:

```
echocanceller = mg2,1-4
```

> MG2 is the recommended echo canceller that comes with the official DAHDI package. There is another open source echo canceller out there that is compatible with DAHDI, called OSLEC (Open Source Line Echo Canceller). Many people report very good results with the use of OSLEC. For more information about the installation of OSLEC on your system, see the website at *http://www.rowetel.com/blog/oslec.html*.

Now, use the *init* script to load the proper modules and initialize the hardware:

```
$ sudo /etc/init.d/dahdi start
Loading DAHDI hardware modules:
  wctdm24xxp:                                          [  OK  ]

  Running dahdi_cfg:                                   [  OK  ]
```

Now that DAHDI has been configured, it is time to move on to the relevant configuration of Asterisk. Once Asterisk is installed, ensure that the chan_dahdi module has been installed. If it is not loaded in Asterisk, check to see if it exists in */usr/lib/asterisk/*

modules/. If it is there, edit */etc/asterisk/modules.conf* to load *chan_dahdi.so*. If the module is not present on disk, DAHDI was not installed before installing Asterisk; go back and install it now (see "DAHDI" on page 49 for details). You can verify its presence using the following command:

```
*CLI> module show like chan_dahdi.so
Module                          Description                          Use Count
chan_dahdi.so                   DAHDI Telephony Driver               0
1 modules loaded
```

Next, you must configure */etc/asterisk/chan_dahdi.conf*. This is the configuration file for the chan_dahdi module, which is the interface between Asterisk and DAHDI. It should look like this:

```
[trunkgroups]

; No trunk groups are needed in this configuration.

[channels]

; The channels context is used when defining channels using the
; older deprecated method.  Don't use this as a section name.

[phone](!)
;
; A template to hold common options for all phones.
;
usecallerid = yes
hidecallerid = no
callwaiting = no
threewaycalling = yes
transfer = yes
echocancel = yes
echotraining = yes
immediate = no
context = LocalSets
signalling = fxo_ks ; Uses FXO signaling for an FXS channel

[phone1](phone)
callerid = "Mark Michelson" <(256)555-1212>
dahdichan = 1

[phone2](phone)
callerid = "David Vossel" <(256)555-2121>
dahdichan = 2

[phone3](phone)
callerid = "Jason Parker" <(256)555-3434>
dahdichan = 3

[phone4](phone)
callerid = "Matthew Nicholson" <(256)555-4343>
dahdichan = 4
```

You can verify that Asterisk has loaded your configuration by running the *dahdi show channels* CLI command:

```
*CLI> dahdi show channels
    Chan Extension  Context      Language  MOH Interpret  Blocked  State
  pseudo            default                default                 In Service
       1            LocalSets              default                 In Service
       2            LocalSets              default                 In Service
       3            LocalSets              default                 In Service
       4            LocalSets              default                 In Service
```

For detailed information on a specific channel, you can run *dahdi show channel 1*.

A Basic Dialplan to Test Your Devices

We're not going to dive too deeply into the dialplan just yet, but an initial dialplan that you can use to test your newly registered devices will be helpful. Place the following contents in */etc/asterisk/extensions.conf*:

```
[LocalSets]

exten => 100,1,Dial(SIP/0000FFFF0001) ; Replace 0000FFFF0001 with your device name

exten => 101,1,Dial(SIP/0000FFFF0002) ; Replace 0000FFFF0002 with your device name


;
; These will allow you to dial each of the 4 analog phones configured
; in the previous section.
;
exten => 102,1,Dial(DAHDI/1)
exten => 103,1,Dial(DAHDI/2)
exten => 104,1,Dial(DAHDI/3)
exten => 105,1,Dial(DAHDI/4)

exten => 200,1,Answer()
     same => n,Playback(hello-world)
     same => n,Hangup()
```

This basic dialplan will allow you to dial your SIP devices using extensions 100 and 101. The four lines of the analog card can be dialed with extensions 102 through 105, respectively. You can also listen to the hello-world prompt that was created for this book by dialing extension 200. All of these extensions are arbitrary numbers, and could be anything you want. Also, this is by no means a complete dialplan; we'll develop it further in later chapters.

You will need to reload your dialplan before changes will take effect in Asterisk. You can reload it from the Linux shell:

```
$ sudo asterisk -rx "dialplan reload"
```

or from the Asterisk CLI:

```
*CLI> dialplan reload
```

You should now be able to dial between your two new extensions. Open up the CLI in order to see the call progression. You should see something like this (and the set you are calling should ring):

```
-- Executing [100@LocalSets:1] Dial("SIP/0000FFFF0001-0000000c",
   "SIP/0000FFFF0001") in new stack
-- Called 0000FFFF0001
-- SIP/0000FFFF0001-0000000d is ringing
```

If this does not happen, you are going to need to review your configuration and ensure you have not made any typos.

Under the Hood: Your First Call

In order to get you thinking about what is happening under the hood, we're going to briefly cover some of what is actually happening with the SIP protocol when two sets on the same Asterisk system call each other.

Asterisk as a B2BUA

Bear in mind that there are actually two calls going on here: one from the originating set to Asterisk, and another from Asterisk to the destination set. SIP is a peer-to-peer protocol, and from the perspective of the protocol there are two calls happening. The SIP protocol is not aware that Asterisk is bridging the calls; each set understands its connection to Asterisk, with no real knowledge of the set on the other side. It is for this reason that Asterisk is often referred to as a B2BUA (Back to Back User Agent). This is also why it is so easy to bridge different protocols together using Asterisk.

For the call you just made, the dialogs shown in Figure 5-2 will have taken place.

For more details on how SIP messaging works, please refer to Appendix B and the SIP RFC at *http://www.ietf.org/rfc/rfc3261.txt*.

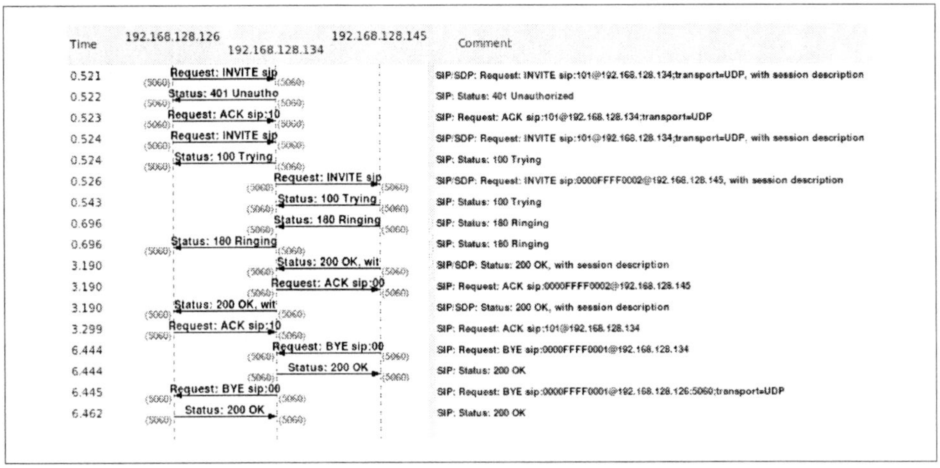

Figure 5-2. SIP dialogs

Conclusion

In this chapter we learned best practices for device naming by abstracting the concepts of users, extension numbers, and devices, and how to define the device configuration and authentication parameters in the channel configuration files. Next, we'll delve into the magic of Asterisk that is the dialplan, and see how simple things can create great results.

## CHAPTER 6

Dialplan Basics

> *Everything should be made as simple as possible,*
> *but not simpler.*
>
> —Albert Einstein

The dialplan is the heart of your Asterisk system. It defines how calls flow into and out of the system. A form of scripting language, the dialplan contains instructions that Asterisk follows in response to external triggers. In contrast to traditional phone systems, Asterisk's dialplan is fully customizable.

This chapter introduces the essential concepts of the dialplan. The information presented here is critical to your understanding of dialplan code and will form the basis of any dialplan you write. The examples have been designed to build upon one another, and we recommend that you do not skip too much of this chapter, since it is so fundamentally important to Asterisk. Please also note that this chapter is by no means an exhaustive survey of all the possible things dialplans can do; our aim is to cover just the essentials. We'll cover more advanced dialplan topics in later chapters. You are encouraged to experiment.

Dialplan Syntax

The Asterisk dialplan is specified in the configuration file named *extensions.conf*.

 The *extensions.conf* file usually resides in the */etc/asterisk/* directory, but its location may vary depending on how you installed Asterisk. Other common locations for this file include */usr/local/etc/asterisk/* and */opt/etc/asterisk/*.

The dialplan is made up of four main concepts: contexts, extensions, priorities, and applications. After explaining the role each of these elements plays in the dialplan, we'll have you build a basic but functioning dialplan.

Contexts

Dialplans are broken into sections called *contexts*. Contexts keep different parts of the dialplan from interacting with one another. An extension that is defined in one context is completely isolated from extensions in any other context, unless interaction is specifically allowed. (We'll cover how to allow interaction between contexts near the end of the chapter. See "Includes" on page 129 for more information.)

As a simple example, let's imagine we have two companies sharing an Asterisk server. If we place each company's automated attendant in its own context, they will be completely separated from each other. This allows us to independently define what happens when, say, extension 0 is dialed: Callers dialing 0 from Company A's voice menu will get Company A's receptionist, while callers dialing 0 at Company B's voice menu will get Company B's receptionist. (This assumes, of course, that we've told Asterisk to transfer the calls to the receptionists when callers press 0.*)

Contexts are defined by placing the name of the context inside square brackets ([ ]). The name can be made up of the letters A through Z (upper- and lowercase), the numbers 0 through 9, and the hyphen and underscore.† A context for incoming calls might look like this:

```
[incoming]
```

---

* This is a very important consideration. With traditional PBXs, there are generally a set of defaults for things like reception, which means that if you forget to define them, they will probably work anyway. In Asterisk, the opposite is true. If you do not tell Asterisk how to handle every situation, and it comes across something it cannot handle, the call will typically be disconnected. We'll cover some best practices later that will help ensure this does not happen. See "Handling Invalid Entries and Timeouts" on page 119 for more information.

† Please note that the space is conspicuously absent from the list of allowed characters. Don't use spaces in your context names—you won't like the result!

Context names have a maximum length of 79 characters (80 characters – 1 terminating null).

All of the instructions placed after a context definition are part of that context, until the next context is defined. At the beginning of the dialplan, there are two special contexts named [general] and [globals]. The [general] section contains a list of general dialplan settings (which you'll probably never have to worry about), and we will discuss the [globals] context in the section "Global variables" on page 123. For now, it's just important to know that these two labels are not really contexts. Avoid the use of [general], [default], and [globals] as context names, but otherwise name your contexts anything you wish.

When you define a channel (which is not done in the *extensions.conf* file, but rather in files such as *sip.conf*, *iax.conf*, *chan_dahdi.conf*, etc.), one of the required parameters in each channel definition is **context**. *The context is the point in the dialplan where connections from that channel will begin.* The context setting for the channel is how you plug the channel into the dialplan. Figure 6-1 illustrates the relationship between channel configuration files and contexts in the dialplan.

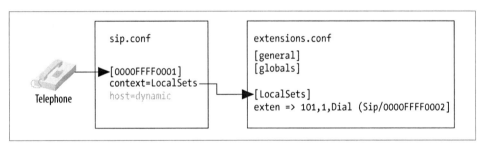

Figure 6-1. Relation between channel configuration files and contexts in the dialplan

 This is one of the most important concepts to understand when dealing with channels and dialplans. Once you understand the relationship of the context definition in a channel to the matching context in the dialplan, you will find it much easier to troubleshoot the call flow through an Asterisk system.

An important use of contexts (perhaps the most important use) is to provide security. By using contexts correctly, you can give certain callers access to features (such as long-distance calling) that aren't made available to others. If you do not design your dialplan carefully, you may inadvertently allow others to fraudulently use your system. Please keep this in mind as you build your Asterisk system; there are many bots on the Internet that were specifically written to identify and exploit poorly secured Asterisk systems.

The Asterisk wiki at *https://wiki.asterisk.org/wiki/display/AST/Important+Security+Considerations* outlines several steps you should take to keep your Asterisk system secure. (Chapter 26 in this book also deals with security.) It is vitally important that you read and understand this page. If you ignore the security precautions outlined there, you may end up allowing anyone and everyone to make long-distance or toll calls at your expense!

If you don't take the security of your Asterisk system seriously, you may end up paying—literally. *Please* take the time and effort to secure your system from toll fraud.

Extensions

In the world of telecommunications, the word *extension* usually refers to a numeric identifier that, when dialed, will ring a phone (or system resource such as voicemail or a queue). In Asterisk, an extension is far more powerful, as it defines the unique series of steps (each step containing an application) through which Asterisk will take that call.

Within each context, we can define as many (or few) extensions as required. When a particular extension is triggered (by an incoming call or by digits being dialed on a channel), Asterisk will follow the steps defined for that extension. It is the extensions, therefore, that specify what happens to calls as they make their way through the dialplan. Although extensions can, of course, be used to specify phone extensions in the traditional sense (i.e., extension 153 will cause the SIP telephone set on John's desk to ring), in an Asterisk dialplan, they can be used for much more.

The syntax for an extension is the word `exten`, followed by an arrow formed by the equals sign and the greater-than sign, like this:

```
exten =>
```

This is followed by the name (or number) of the extension. When dealing with traditional telephone systems, we tend to think of extensions as the numbers you would dial to make another phone ring. In Asterisk, you get a whole lot more; for example, extension names can be any combination of numbers and letters. Over the course of this chapter and the next, we'll use both numeric and alphanumeric extensions.

Assigning names to extensions may seem like a revolutionary concept, but when you realize that many VoIP transports support (or even actively encourage) dialing by name or email address rather than just by number, it makes perfect sense. This is one of the features that makes Asterisk so flexible and powerful.

Each step in an extension is composed of three components:

- The name (or number) of the extension
- The priority (each extension can include multiple steps; the step number is called the "priority")
- The application (or command) that will take place at that step

These three components are separated by commas, like this:

```
exten => name,priority,application()
```

Here's a simple example of what a real extension might look like:

```
exten => 123,1,Answer()
```

In this example, the extension name is `123`, the priority is `1`, and the application is `Answer()`.

Priorities

Each extension can have multiple steps, called *priorities*. The priorities are numbered sequentially, starting with 1, and each executes one specific application. As an example, the following extension would answer the phone (in priority number 1), and then hang it up (in priority number 2):

```
exten => 123,1,Answer()
exten => 123,2,Hangup()
```

It's pretty obvious that this code doesn't really do anything useful. We'll get there. The key point to note here is that for a particular extension, Asterisk follows the priorities in order. This style of dialplan syntax is still seen from time to time, although (as you'll see momentarily) it is not generally used anymore for new code:

```
exten => 123,1,Answer()
exten => 123,2,do something
exten => 123,3,do something else
exten => 123,4,do one last thing
exten => 123,5,Hangup()
```

Unnumbered priorities

In older releases of Asterisk, the numbering of priorities caused a lot of problems. Imagine having an extension that had 15 priorities, and then needing to add something at step 2: all of the subsequent priorities would have to be manually renumbered. Asterisk does not handle missing steps or misnumbered priorities, and debugging these types of errors was pointless and frustrating.

Beginning with version 1.2, Asterisk addressed this problem: it introduced the use of the n priority, which stands for "next." Each time Asterisk encounters a priority named n, it takes the number of the previous priority and adds 1. This makes it easier to make changes to your dialplan, as you don't have to keep renumbering all your steps. For example, your dialplan might look something like this:

```
exten => 123,1,Answer()
exten => 123,n,do something
exten => 123,n,do something else
exten => 123,n,do one last thing
exten => 123,n,Hangup()
```

Internally, Asterisk will calculate the next priority number every time it encounters an n.‡ Bear in mind that you must always specify priority number 1. If you accidentally put an n instead of 1 for the first priority (a common mistake even among experienced dialplan coders), you'll find after reloading the dialplan that the extension will not exist.

The 'same =>' operator

In the never-ending effort to simplify coding effort, a new construct was created to make extension building and management even easier. As long as the extension remains the same, rather than having to type the full extension on each line, you can simply type same => , followed by the priority and application:

```
exten => 123,1,Answer()
    same => n,do something
    same => n,do something else
    same => n,do one last thing
    same => n,Hangup()
```

The indentation is not required, but it may make for easier reading. This style of dialplan will also make it easier to copy code from one extension to another. We prefer this style ourselves, and highly recommend it.

Priority labels

Priority labels allow you to assign a name to a priority within an extension. This is to ensure that you can refer to a priority by something other than its number (which probably isn't known, given that dialplans now generally use unnumbered priorities). The reason it is important to be able to address a particular priority in an extension is that you will often want to send calls from other parts of the dialplan to a particular priority in a particular extension. We'll talk about that more later. To assign a text label to a priority, simply add the label inside parentheses after the priority, like this:

```
exten => 123,n(label),application()
```

Later, we'll cover how to jump between different priorities based on dialplan logic. You'll see a lot more of priority labels, and you'll use them often in your dialplans.

---

‡ Asterisk permits simple arithmetic within the priority, such as n+200, and the priority s (for same), but their usage is somewhat deprecated due to the existence of priority labels. Please note that *extension* s and *priority* s are two distinct concepts.

A very common mistake when writing labels is to insert a comma between the n and the (, like this:

```
exten => 123,n,(label),application() ;<-- THIS IS NOT GOING TO WORK
```

This mistake will break that part of your dialplan, and you will get an error stating that the application cannot be found.

Applications

Applications are the workhorses of the dialplan. Each application performs a specific action on the current channel, such as playing a sound, accepting touch-tone input, looking something up in a database, dialing a channel, hanging up the call, and so forth. In the previous example, you were introduced to two simple applications: `Answer()` and `Hangup()`. You'll learn more about how these work momentarily.

Some applications, including `Answer()` and `Hangup()`, need no other instructions to do their jobs. Most applications, however, require additional information. These additional elements, or *arguments*, are passed on to the applications to affect how they perform their actions. To pass arguments to an application, place them between the parentheses that follow the application name, separated by commas.

Occasionally, you may also see the pipe character (|) being used as a separator between arguments, instead of a comma. Starting in Asterisk 1.6.0, support for the pipe as a separator character has been removed.§

The Answer(), Playback(), and Hangup() Applications

The `Answer()` application is used to answer a channel that is ringing. This does the initial setup for the channel that receives the incoming call. As we mentioned earlier, `Answer()` takes no arguments. `Answer()` is not always required (in fact, in some cases it may not be desirable at all), but it is an effective way to ensure a channel is connected before performing further actions.

The Progress() Application

Sometimes it is useful to be able to pass information back to the network before answering a call. The `Progress()` application attempts to provide call progress information to the originating channel. Some carriers expect this, and thus you may be able to resolve strange signaling problems by inserting `Progress()` into the dialplan where your incoming calls arrive.

---

§ Except in some parts of *voicemail.conf*.

The Playback() application is used for playing a previously recorded sound file over a channel. Input from the user is ignored, which means that you would not use Playback() in an auto attendant, for example, unless you did not want to accept input at that point.‖

 Asterisk comes with many professionally recorded sound files, which should be found in the default sounds directory (usually */var/lib/aster-isk/sounds/*). When you compile Asterisk, you can choose to install various sets of sample sounds that have been recorded in a variety of languages and file formats. We'll be using these files in many of our examples. Several of the files in our examples come from the Extra Sound Package, so please take the time to install it (see Chapter 3). You can also have your own sound prompts recorded in the same voices as the stock prompts by visiting *http://www.theivrvoice.com/*. Later in the book we'll talk more about how you can use a telephone and the dialplan to create and manage your own system recordings.

To use Playback(), specify a filename (without a file extension) as the argument. For example, Playback(filename) would play the sound file called *filename.wav*, assuming it was located in the default sounds directory. Note that you can include the full path to the file if you want, like this:

    Playback(/home/john/sounds/filename)

The previous example would play *filename.wav* from the */home/john/sounds/* directory. You can also use relative paths from the Asterisk sounds directory, as follows:

    Playback(custom/filename)

This example would play *filename.wav* from the *custom/* subdirectory of the default sounds directory (probably */var/lib/asterisk/sounds/custom/filename.wav*). Note that if the specified directory contains more than one file with that filename but with different file extensions, Asterisk automatically plays the best file.#

The Hangup() application does exactly as its name implies: it hangs up the active channel. You should use this application at the end of a context when you want to end the current call, to ensure that callers don't continue on in the dialplan in a way you might not have anticipated. The Hangup() application does not require any arguments, but you can pass an ISDN cause code if you want (e.g., Hangup(16)).

‖ There is another application called Background() that is very similar to Playback(), except that it does allow input from the caller. You can read more about this application in Chapter 15 and Chapter 17.

Asterisk selects the best file based on translation cost—that is, it selects the file that is the least CPU-intensive to convert to its native audio format. When you start Asterisk, it calculates the translation costs between the different audio formats (they often vary from system to system). You can see these translation costs by typing *show translation* at the Asterisk command-line interface. The numbers shown represent how many milliseconds it takes Asterisk to transcode one second of audio. We'll talk more about the different audio formats (known as *codecs*) in "Codecs" on page 625.

As we work through the book, we will be introducing you to many more Asterisk applications.

A Simple Dialplan

OK, enough theory. Open up the file */etc/asterisk/extensions.conf*, and let's take a look at your first dialplan (which was created in Chapter 5). We're going to add to that.

Hello World

As is typical in many technology books (especially computer programming books), our first example is called "Hello World!"

In the first priority of our extension, we answer the call. In the second, we play a sound file named *hello-world*, and in the third we hang up the call. The code we are interested in for this example looks like this:

```
exten => 200,1,Answer()
    same => n,Playback(hello-world)
    same => n,Hangup()
```

If you followed along in Chapter 5, you'll already have a channel or two configured, as well as the sample dialplan that contains this code. If not, what you need is an *extensions.conf* file in your */etc/asterisk/* directory that contains the following code:

```
[LocalSets] ; this is the context name
exten => 100,1,Dial(SIP/0000FFFF0001) ; Replace 0000FFFF0001 with your device name

exten => 101,1,Dial(SIP/0000FFFF0002) ; Replace 0000FFFF0002 with your device name

exten => 200,1,Answer()
    same => n,Playback(hello-world)
    same => n,Hangup()
```

 If you don't have any channels configured, now is the time to do so. There is real satisfaction that comes from passing your first call into an Asterisk dialplan on a system that you've built from scratch. People get this funny grin on their faces as they realize that they have just created a telephone system. This pleasure can be yours as well, so please, don't go any further until you have made this little bit of dialplan work. If you have any problems, get back to Chapter 5 and work through the examples there.

If you don't have this dialplan code built yet, you'll need to add it and reload the dialplan with this CLI command:

```
*CLI> dialplan reload
```

or from the shell with:

```
$ sudo /usr/sbin/asterisk -rx "dialplan reload"
```

Calling extension 200 from either of your configured phones should reward you with the voice of Allison Smith saying "Hello World."

If it doesn't work, check the Asterisk console for error messages, and make sure your channels are assigned to the `LocalSets` context.

 We do not recommend that you move forward in this book until you have verified the following:

1. Calls between extension 100 and 101 are working
2. Calling extension 200 plays "Hello World"

Even though this example is very short and simple, it emphasizes the core concepts of contexts, extensions, priorities, and applications. You now have the fundamental knowledge on which all dialplans are built.

Building an Interactive Dialplan

The dialplan we just built was static; it will always perform the same actions on every call. Many dialplans will also need logic to perform different actions based on input from the user, so let's take a look at that now.

The Goto(), Background(), and WaitExten() Applications

As its name implies, the `Goto()` application is used to send a call to another part of the dialplan. The syntax for the `Goto()` application requires us to pass the destination context, extension, and priority on as arguments to the application, like this:

```
same => n,Goto(context,extension,priority)
```

We're going to create a new context called `TestMenu`, and create an extension in our `LocalSets` context that will pass calls to that context using `Goto()`:

```
exten => 201,1,Goto(TestMenu,start,1) ; add this to the end of the
                                       ; [LocalSets] context

[TestMenu]
exten => start,1,Answer()
```

Now, whenever a device enters the `LocalSets` context and dials `201`, the call will be passed to the `start` extension in the `TestMenu` context (which currently won't do anything interesting because we still have more code to write).

We used the extension start in this example, but we could have used anything we wanted as an extension name, either numeric or alpha. We prefer to use alpha characters for extensions that are not directly dialable, as this makes the dialplan easier to read. Point being, we could have used 123 or xyz123, or 99luftballons, or whatever we wanted instead of start. The word "start" doesn't actually mean anything to the dialplan; it's just another extension.

One of the most useful applications in an interactive Asterisk dialplan is the Background()* application. Like Playback(), it plays a recorded sound file. Unlike Playback(), however, when the caller presses a key (or series of keys) on her telephone keypad, it interrupts the playback and passes the call to the extension that corresponds with the pressed digit(s). If a caller presses 5, for example, Asterisk will stop playing the sound prompt and send control of the call to the first priority of extension 5 (assuming there is an extension 5 to send the call to).

The most common use of the Background() application is to create voice menus (often called *auto attendants*† or *phone trees*). Many companies use voice menus to direct callers to the proper extensions, thus relieving their receptionists from having to answer every single call.

Background() has the same syntax as Playback():

```
[TestMenu]
exten => start,1,Answer()
    same => n,Background(main-menu)
```

If you want Asterisk to wait for input from the caller after the sound prompt has finished playing, you can use WaitExten(). The WaitExten() application waits for the caller to enter DTMF digits and is used directly following the Background() application, like this:

```
[TestMenu]
exten => start,1,Answer()
    same => n,Background(main-menu)
    same => n,WaitExten()
```

If you'd like the WaitExten() application to wait a specific number of seconds for a response (instead of using the default timeout‡), simply pass the number of seconds as the first argument to WaitExten(), like this:

```
    same => n,WaitExten(5) ; We recommend always passing a time argument to WaitExten()
```

---

* It should be noted that some people expect that Background(), due to its name, will continue onward through the next steps in the dialplan while the sound is being played. In reality, its name refers to the fact that it is playing a sound in the background, while waiting for DTMF in the foreground.

† More information about auto attendants can be found in Chapter 15.

‡ See the dialplan function TIMEOUT() for information on how to change the default timeouts. See Chapter 10 for information on what dialplan functions are.

Both `Background()` and `WaitExten()` allow the caller to enter DTMF digits. Asterisk then attempts to find an extension in the current context that matches the digits that the caller entered. If Asterisk finds a match, it will send the call to that extension. Let's demonstrate by adding a few lines to our dialplan example:

```
[TestMenu]
exten => start,1,Answer()
    same => n,Background(main-menu)
    same => n,WaitExten(5)

exten => 1,1,Playback(digits/1)

exten => 2,1,Playback(digits/2)
```

After making these changes, save and reload your dialplan:

```
*CLI> dialplan reload
```

If you call into extension 201, you should hear a sound prompt that says "main menu." The system will then wait 5 seconds for you to enter a digit. If the digit you press is either 1 or 2, Asterisk will match the relevant extension, and read that digit back to you. Since we didn't provide any further instructions, your call will then end. You'll also find that if you enter a different digit (such as 3), the dialplan will be unable to proceed.

Let's embellish things a little. We're going to use the `Goto()` application to have the dialplan repeat the greeting after playing back the number:

```
[TestMenu]
exten => start,1,Answer()
    same => n,Background(main-menu)
    same => n,WaitExten(5)

exten => 1,1,Playback(digits/1)
    same => n,Goto(TestMenu,start,1)

exten => 2,1,Playback(digits/2)
    same => n,Goto(TestMenu,start,1)
```

These new lines will send control of the call back to the **start** extension after playing back the selected number. This is generally considered friendlier than just hanging up.

> If you look up the details of the `Goto()` application, you'll find that you can actually pass either one, two, or three arguments to the application. If you pass a single argument, Asterisk will assume it's the destination priority in the current extension. If you pass two arguments, Asterisk will treat them as the extension and the priority to go to in the current context.
>
> In this example, we've passed all three arguments for the sake of clarity, but passing just the extension and priority would have had the same effect, since the destination context is the same as the source context.

Handling Invalid Entries and Timeouts

Now that our first voice menu is starting to come together, let's add some additional special extensions. First, we need an extension for invalid entries. In Asterisk, when a context receives a request for an extension that is not valid within that context (e.g., pressing 9 in the preceding example), the call is sent to the i extension. We also need an extension to handle situations when the caller doesn't give input in time (the default timeout is 10 seconds). Calls will be sent to the t extension if the caller takes too long to press a digit after WaitExten() has been called. Here is what our dialplan will look like after we've added these two extensions:

```
[TestMenu]
exten => start,1,Answer()
    same => n,Background(main-menu)
    same => n,WaitExten(5)

exten => 1,1,Playback(digits/1)
    same => n,Goto(TestMenu,start,1)

exten => 2,1,Playback(digits/2)
    same => n,Goto(TestMenu,start,1)

exten => i,1,Playback(pbx-invalid)
    same => n,Goto(TestMenu,start,1)

exten => t,1,Playback(vm-goodbye)
    same => n,Hangup()
```

Using the i and t extensions makes our menu a little more robust and user-friendly. That being said, it is still quite limited, because outside callers still have no way of connecting to a live person. To do that, we'll need to learn about another application, called Dial().

Using the Dial() Application

One of Asterisk's most valuable features is its ability to connect different callers to each other. This is especially useful when callers are using different methods of communication. For example, caller A might be communicating over the traditional analog telephone network, while user B might be sitting in a café halfway around the world and speaking on an IP telephone. Luckily, Asterisk takes much of the hard work out of connecting and translating between disparate networks. All you have to do is learn how to use the Dial() application.

The syntax of the Dial() application is more complex than that of the other applications we've used so far, but don't let that scare you off. Dial() takes up to four arguments, which we'll look at next.

Argument 1: Destination

The first argument is the destination you're attempting to call, which (in its simplest form) is made up of a technology (or transport) across which to make the call, a forward slash, and the address of the remote endpoint or resource. Common technology types include DAHDI (for analog and T1/E1/J1 channels), SIP, and IAX2.

For example, let's assume that we want to call a DAHDI endpoint identified by DAHDI/ 1, which is an FXS channel with an analog phone plugged into it. The technology is DAHDI, and the resource (or channel identifier) is 1. Similarly, a call to a SIP device (as defined in *sip.conf*) might have a destination of SIP/0004F2001122, and a call to an IAX device (defined in *iax.conf*) might have a destination of IAX2/Softphone.[§] If we wanted Asterisk to ring the DAHDI/1 channel when extension 105 is reached in the dialplan, we'd add the following extension:

```
exten => 105,1,Dial(DAHDI/1)
```

We can also dial multiple channels at the same time, by concatenating the destinations with an ampersand (&), like this:

```
exten => 105,1,Dial(DAHDI/1&SIP/0004F2001122&IAX2/Softphone)
```

The Dial() application will ring all of the specified destinations simultaneously, and bridge the inbound call with whichever destination channel answers first (the other channels will immediately stop ringing). If the Dial() application can't contact any of the destinations, Asterisk will set a variable called DIALSTATUS with the reason that it couldn't dial the destinations, and continue on with the next priority in the extension.[‖]

The Dial() application also allows you to connect to a remote VoIP endpoint not previously defined in one of the channel configuration files. The full syntax is:

```
Dial(technology/user[:password]@remote_host[:port][/remote_extension])
```

As an example, you can dial into a demonstration server at Digium using the IAX2 protocol by using the following extension:

```
exten => 500,1,Dial(IAX2/guest@misery.digium.com/s)
```

The full syntax for the Dial() application is slightly different for DAHDI channels:

```
Dial(DAHDI/[gGrR]channel_or_group[/remote_extension])
```

For example, here is how you would dial 1-800-555-1212 on DAHDI channel number 4[#]:

```
exten => 501,1,Dial(DAHDI/4/18005551212)
```

[§] If this were a production environment, this would not actually be a good name for this device. If you have more than one softphone on your system (or add another in the future), how will you tell them apart?

[‖] We'll cover variables in the upcoming section "Using Variables" on page 122. In future chapters we'll discuss how to have your dialplan make decisions based on the value of DIALSTATUS.

[#] Bear in mind that this assumes that this channel connects to something that knows how to reach external numbers.

Argument 2: Timeout

The second argument to the `Dial()` application is a timeout, specified in seconds. If a timeout is given, `Dial()` will attempt to call the specified destination(s) for that number of seconds before giving up and moving on to the next priority in the extension. If no timeout is specified, `Dial()` will continue to dial the called channel(s) until someone answers or the caller hangs up. Let's add a timeout of 10 seconds to our extension:

```
exten => 201,1,Dial(DAHDI/1,10)
```

If the call is answered before the timeout, the channels are bridged and the dialplan is done. If the destination simply does not answer, is busy, or is otherwise unavailable, Asterisk will set a variable called `DIALSTATUS` and then continue on with the next priority in the extension.

Let's put what we've learned so far into another example:

```
exten => 201,1,Dial(DAHDI/1,10)
    same => n,Playback(vm-nobodyavail)
    same => n,Hangup()
```

As you can see, this example will play the *vm-nobodyavail.gsm* sound file if the call goes unanswered.

Argument 3: Option

The third argument to `Dial()` is an option string. It may contain one or more characters that modify the behavior of the `Dial()` application. While the list of possible options is too long to cover here, one of the most popular is the `m` option. If you place the letter `m` as the third argument, the calling party will hear hold music instead of ringing while the destination channel is being called (assuming, of course, that music on hold has been configured correctly). To add the `m` option to our last example, we simply change the first line:

```
exten => 201,1,Dial(DAHDI/1,10,m)
    same => n,Playback(vm-nobodyavail)
    same => n,Hangup()
```

Argument 4: URI

The fourth and final argument to the `Dial()` application is a URI. If the destination channel supports receiving a URI at the time of the call, the specified URI will be sent (for example, if you have an IP telephone that supports receiving a URI, it will appear on the phone's display; likewise, if you're using a softphone, the URI might pop up on your computer screen). This argument is very rarely used.

 Few (if any) phones support URI information being passed to them. If you're looking for something like a screen pop, you might want to check out Chapter 18, and more specifically the section on Jabber in "Using XMPP (Jabber) with Asterisk" on page 418.

Updating the dialplan

Let's modify extensions 1 and 2 in our menu to use the Dial() application:

```
[TestMenu]
exten => start,1,Answer()
    same => n,Background(main-menu)
    same => n,WaitExten(5)

exten => 1,1,Dial(SIP/0000FFFF0001,10) ; Replace 0000FFFF0001 with your device name
    same => n,Playback(vm-nobodyavail)
    same => n,Hangup()

exten => 2,1,Dial(SIP/0000FFFF0002,10) ; Replace 0000FFFF0002 with your device name
    same => n,Playback(vm-nobodyavail)
    same => n,Hangup()

exten => i,1,Playback(pbx-invalid)
    same => n,Goto(TestMenu,start,1)

exten => t,1,Playback(vm-goodbye)
    same => n,Hangup()
```

Blank arguments

Note that the second, third, and fourth arguments may be left blank; only the first argument is required. For example, if you want to specify an option but not a timeout, simply leave the timeout argument blank, like this:

```
exten => 1,1,Dial(DAHDI/1,,m)
```

Using Variables

Variables can be used in an Asterisk dialplan to help reduce typing, improve clarity, or add logic. If you have some computer programming experience, you already understand what a variable is. If not, we'll briefly explain what variables are and how they are used. They are a vitally important Asterisk dialplan concept (and something you will not find in the dialplan of any proprietary PBX).

A *variable* is a named container that can hold a value. The advantage of a variable is that its contents may change, but its name does not, which means you can write code that references the variable name and not worry about what the value will be. So, for example, we might create a variable called JOHN and assign it the value of DAHDI/1. This way, when we're writing our dialplan we can refer to John's channel by name, instead of remembering that John is using the channel named DAHDI/1. If at some point we change John's channel to something else, we don't have to change any of our code that references the JOHN variable; we only have to change the value assigned to the variable.

There are two ways to reference a variable. To reference the variable's name, simply type the name of the variable, such as LEIF. If, on the other hand, you want to reference the contents of the value, you must type a dollar sign, an opening curly brace, the name of the variable, and a closing curly brace (in the case of LEIF, we would reference the value of the variable with ${LEIF}). Here's how we might use a variable inside the Dial() application:

```
exten => 301,1,Set(LEIF=SIP/0000FFFF0001)
    same => n,Dial(${LEIF})
```

In our dialplan, whenever we refer to ${LEIF}, Asterisk will automatically replace it with whatever value has been assigned to the variable named LEIF.

 Note that variable names are case-sensitive. A variable named LEIF is different than a variable named Leif. For readability's sake, all our variable names in the examples will be written in uppercase. You should also be aware that any variables set by Asterisk will be uppercase. Some variables, such as CHANNEL and EXTEN, are reserved by Asterisk. You should not attempt to set these variables. It is popular to write global variables in uppercase and channel variables in Pascal/Camel case.

There are three types of variables we can use in our dialplan: global variables, channel variables, and environment variables. Let's take a moment to look at each type.

Global variables

As their name implies, *global* variables are visible to all channels at all times. Global variables are useful in that they can be used anywhere within a dialplan to increase readability and manageability. Suppose for a moment that you had a large dialplan and several hundred references to the SIP/0000FFFF0001 channel. Now imagine you had to go through your dialplan and change all of those references to SIP/0000FFFF0002. It would be a long and error-prone process, to say the least.

On the other hand, if you had defined a global variable that contained the value SIP/0000FFFF0001 at the beginning of your dialplan and then referenced that instead, you would have to change only one line of code to affect all places in the dialplan where that channel was used.

Global variables should be declared in the [globals] context at the beginning of the *extensions.conf* file. As an example, we will create a global variable named LEIF with a value of SIP/0000FFFF0001. This variable is set at the time Asterisk parses the dialplan:

```
[globals]
LEIF=SIP/0000FFFF0001
```

Channel variables

A *channel* variable is a variable that is associated only with a particular call. Unlike global variables, channel variables are defined only for the duration of the current call and are available only to the channels participating in that call.

There are many predefined channel variables available for use within the dialplan, which are explained in the Asterisk wiki at *https://wiki.asterisk.org/wiki/display/AST/Channel+Variables*. Channel variables are set via the Set() application:

```
exten => 202,1,Set(MagicNumber=42)
    same => n,SayNumber(${MagicNumber})
```

You're going to be seeing a lot more channel variables. Read on.

Environment variables

Environment variables are a way of accessing Unix environment variables from within Asterisk. These are referenced using the ENV() dialplan function.* The syntax looks like ${ENV(*var*)}, where *var* is the Unix environment variable you wish to reference. Environment variables aren't commonly used in Asterisk dialplans, but they are available should you need them.

Adding variables to our dialplan

Now that we've learned about variables, let's put them to work in our dialplan. We're going to add three global variables that will associate a variable name to a channel name:

```
[globals]
LEIF=SIP/0000FFFF0001
JIM=SIP/0000FFFF0002
RUSSELL=SIP/0000FFFF0003

[LocalSets]
exten => 100,1,Dial(${LEIF})
exten => leif,1,Dial(${LEIF})

exten => 101,1,Dial(${JIM})
exten => jim,1,Dial(${JIM})

exten => 102,1,Dial(${RUSSELL})
exten => russell,1,Dial(${RUSSELL})

[TestMenu]
exten => 201,1,Answer()
    same => n,Background(enter-ext-of-person)
    same => n,WaitExten()

exten => 1,1,Dial(DAHDI/1,10)
    same => n,Playback(vm-nobodyavail)
    same => n,Hangup()

exten => 2,1,Dial(SIP/Jane,10)
    same => n,Playback(vm-nobodyavail)
    same => n,Hangup()
```

---

* We'll get into dialplan functions later. Don't worry too much about environment variables right now. They are not important to understanding the dialplan.

```
exten => i,1,Playback(pbx-invalid)
    same => n,Goto(incoming,123,1)

exten => t,1,Playback(vm-goodbye)
    same => n,Hangup()
```

You'll notice we've added pseudonym extension names for our extension numbers. In "Extensions" on page 110, we explained that Asterisk does not care which naming scheme you use to identify an extension. We've simply added both numeric and named extension identifiers for reaching the same endpoint; extensions 100 and leif both reach the device located at *SIP/0000FFFF0001*, extensions 101 and jim both reach the device located at *SIP/0000FFFF0002*, and both 102 and russell reach the device located at *SIP/0000FFFF0003*. The devices are identified with the global variables ${LEIF}, $ {JIM}, and ${RUSSELL}, respectively, and we're dialing those locations using the Dial() application.

In our test menu we've simply picked a couple of random endpoints to dial, such as DAHDI/1 and SIP/Jane. These could be replaced with any available endpoints that you wish. Our TestMenu context has been built to start giving you an idea as to what an Asterisk dialplan looks like.

Pattern Matching

If we want to be able to allow people to dial *through* Asterisk and have Asterisk connect them to outside resources, we need a way to match on any possible phone number that the caller might dial. For situations like this, Asterisk offers *pattern matching*. Pattern matching allows you to create one extension in your dialplan that matches many different numbers. This is enormously useful.

Pattern-matching syntax

When using pattern matching, certain letters and symbols represent what we are trying to match. *Patterns always start with an underscore (_).* This tells Asterisk that we're matching on a pattern, and not on an explicit extension name.

 If you forget the underscore at the beginning of your pattern, Asterisk will think it's just a named extension and won't do any pattern matching. This is one of the most common mistakes people make when starting to learn Asterisk.

After the underscore, you can use one or more of the following characters:

X

Matches any single digit from 0 to 9.

Z

Matches any single digit from 1 to 9.

N
    Matches any single digit from 2 to 9.

[15-7]
    Matches a single character from the range of digits specified. In this case, the pattern matches a single 1, as well as any number in the range 5, 6, 7.

. *(period)*
    Wildcard match; matches *one or more* characters, no matter what they are.

 If you're not careful, wildcard matches can make your dialplans do things you're not expecting (like matching built-in extensions such as i or h). You should use the wildcard match in a pattern only after you've matched as many other digits as possible. For example, the following pattern match should probably never be used:

```
_.
```

In fact, Asterisk will warn you if you try to use it. Instead, if you really need a catch-all pattern match, use this one to match all strings that start with a digit:

```
_X.
```

Or this one, to match any alphanumeric string:

```
_[0-9a-zA-Z].
```

! *(bang)*
    Wildcard match; matches *zero or more* characters, no matter what they are.

To use pattern matching in your dialplan, simply put the pattern in the place of the extension name (or number):

```
exten => _NXX,1,Playback(silence/1&auth-thankyou)
```

In this example, the pattern matches any three-digit extension from 200 through 999 (the N matches any digit between 2 and 9, and each X matches a digit between 0 and 9). That is to say, if a caller dialed any three-digit extension between 200 and 999 in this context, he would hear the sound file *auth-thankyou.gsm*.

One other important thing to know about pattern matching is that if Asterisk finds more than one pattern that matches the dialed extension, it will use the *most specific* one (going from left to right). Say you had defined the following two patterns, and a caller dialed 555-1212:

```
exten => _555XXXX,1,Playback(silence/1&digits/1)
exten => _55512XX,1,Playback(silence/1&digits/2)
```

In this case the second extension would be selected, because it is more specific.

Pattern-matching examples

This pattern matches any seven-digit number, as long as the first digit is 2 or higher:

```
_NXXXXXX
```

The preceding pattern would be compatible with any North American Numbering Plan local seven-digit number.

In areas with 10-digit dialing, that pattern would look like this:

```
_NXXNXXXXXX
```

Note that neither of these two patterns would handle long-distance calls. We'll cover those shortly.

The NANP and Toll Fraud

The North American Numbering Plan (NANP) is a shared telephone numbering scheme used by 19 countries in North America and the Caribbean. All of these countries share country code 1.

In the United States and Canada, telecom regulations are similar (and sensible) enough that you can place a long-distance call to most numbers in country code 1 and expect to pay a reasonable toll. However, many people don't realize that 17 other countries, many of which have very different telecom regulations, share the NANP. (More information can be found at *http://www.nanpa.com*.)

One popular scam using the NANP tries to trick naïve North Americans into calling expensive per-minute toll numbers in a Caribbean country; the callers believe that since they dialed 1-NPA-NXX-XXXX to reach the number, they'll be paying their standard national long-distance rate for the call. Since the country in question may have regulations that allow for this form of extortion, the caller is ultimately held responsible for the call charges.

The only way to prevent this sort of activity is to block calls to certain area codes (809, for example) and remove the restrictions only on an as-needed basis.

Let's try another:

```
_1NXXNXXXXXX
```

This one will match the number 1, followed by an area code between 200 and 999, then any seven-digit number. In the NANP calling area, you would use this pattern to match any long-distance number.[†]

[†] If you grew up in North America, you may believe that the 1 you dial before a long-distance call is "the long-distance code." This is incorrect. The number 1 is the international country code for NANP. Keep this in mind if you send your phone number to someone in another country. The recipient may not know your country code, and thus be unable to call you with just your area code and phone number. Your full phone number with country code is +1 NPA NXX XXXX (where NPA is your area code)–e.g., +1 416 555 1212.

And finally this one:

```
_011.
```

Note the period on the end. This pattern matches any number that starts with 011 and has at least one more digit. In the NANP, this indicates an international phone number. (We'll be using these patterns in the next section to add outbound dialing capabilities to our dialplan.)

Pattern Matches in Other Countries

The examples in this section were NANPA-centric, but the basic logic applies in any country. Here are some examples for other countries (note that we were not able to test these, and they are almost certainly incomplete):

```
; UK, Germany, Italy, China, etc.
_00. ; international dialing code
_0. ; national dialing prefix

; Australia
_0011. ; international dialing code
_0. ; national dialing prefix
```

This is by no means comprehensive, but it should give you a general idea of the patterns you'll want to consider for your own country.

Using the ${EXTEN} channel variable

So what happens if you want to use pattern matching but need to know which digits were actually dialed? Enter the ${EXTEN} channel variable. Whenever you dial an extension, Asterisk sets the ${EXTEN} channel variable to the digits that were dialed. We can use an application called SayDigits() to test this out:

```
exten => _XXX,1,Answer()
    same => n,SayDigits(${EXTEN})
```

In this example, the SayDigits() application will read back to you the three-digit extension you dialed.

Often, it's useful to manipulate the ${EXTEN} by stripping a certain number of digits off the front of the extension. This is accomplished by using the syntax ${EXTEN:*x*}, where *x* is where you want the returned string to start, from left to right. For example, if the value of ${EXTEN} is 95551212, ${EXTEN:1} equals 5551212. Let's try another example:

```
exten => _XXX,1,Answer()
    same => n,SayDigits(${EXTEN:1})
```

In this example, the SayDigits() application would start at the second digit, and thus read back only the last two digits of the dialed extension.

More Advanced Digit Manipulation

The ${EXTEN} variable properly has the syntax ${EXTEN:x:y}, where *x* is the starting position and *y* is the number of digits to return. Given the following dial string:

94169671111

we can extract the following digit strings using the ${EXTEN:x:y} construct:

- ${EXTEN:1:3} would contain 416
- ${EXTEN:4:7} would contain 9671111
- ${EXTEN:-4:4} would start four digits from the end and return four digits, giving
  us 1111
- ${EXTEN:2:-4} would start two digits in and exclude the last four digits, giving us
  16967
- ${EXTEN:-6:-4} would start six digits from the end and exclude the last four digits,
  giving us 67
- ${EXTEN:1} would give us everything after the first digit, or 4169671111 (if the num-
  ber of digits to return is left blank, it will return the entire remaining string)

This is a very powerful construct, but most of these variations are not very common in
normal use. For the most part, you will be using ${EXTEN} (or perhaps ${EXTEN:1} if you
need to strip off an external access code).

Includes

Asterisk has an important feature that allows extensions from one context to be available from within another context. This is accomplished through use of the include
directive. The include directive allows us to control access to different sections of the
dialplan.

The include statement takes the following form, where *context* is the name of the
remote context we want to include in the current context:

    include => context

Including one context within another context allows extensions within the included
context to be dialable.

When we include other contexts within our current context, we have to be mindful of
the order in which we are including them. Asterisk will first try to match the dialed
extension in the current context. If unsuccessful, it will then try the first included context (including any contexts included in that context), and then continue to the other
included contexts in the order in which they were included.

We will discuss the include directive more in Chapter 7.

Conclusion

And there you have it—a basic but functional dialplan. There is still much we have not covered, but you've got all of the fundamentals. In the following chapters, we'll continue to build on this foundation.

If parts of this dialplan don't make sense, you may want to go back and reread a section or two before continuing on to the next chapter. It's imperative that you understand these principles and how to apply them, as the next chapters build on this information.

## CHAPTER 7

Outside Connectivity

*You cannot always control what goes on outside. But
you can always control what goes on inside.*

—Wayne Dyer

In the previous chapters, we have covered a lot of important information that is essential to a working Asterisk system. However, we have yet to accomplish the one thing that is vital to any useful PBX: namely, connecting it to the outside world. In this chapter we will rectify that situation.

The architecture of Asterisk is significant, due in large part to the fact that it treats all channel types as equal. This is in contrast to a traditional PBX, where trunks (which connect to the outside world) and extensions (which connect to users and resources) are very different. The fact that the Asterisk dialplan treats all channels in a similar manner means that in an Asterisk system you can accomplish very easily things that are much more difficult (or impossible) to achieve on a traditional PBX.

This flexibility does come with a price, however. Since the system does not inherently know the difference between an internal resource (such as a telephone set) and an external resource (for example, a telco circuit), it is up to you to ensure that your dialplan handles each type of resource appropriately.

The Basics of Trunking

The purpose of *trunking* is to provide a shared connection between two entities. For example, a trunk road would be a highway that connects two towns together. Railroads used the term "trunk" extensively, to refer to a major line that connected feeder lines together.

Similarly, in telecom, trunking is used to connect two systems together. Carriers use telecom trunks to connect their networks together, and in a PBX, the circuits that connect the PBX to the outside world are commonly referred to as trunks (although the carriers themselves do not generally consider these to be trunks). From a technical perspective, the definition of a trunk is not as clear as it used to be (PBX trunks used totally different technology from station circuits), but as a concept, trunks are still very important. For example, with VoIP, everything is actually peer-to-peer (so from a technology perspective there isn't really such a thing as a trunk anymore), but it is still useful to be able to differentiate between VoIP resources that connect to the outside world and VoIP resources that connect to user endpoints (such as SIP telephones).

It's probably easiest to think of a trunk as a collection of circuits that service a route. So, in an Asterisk PBX, you might have trunks that go to your VoIP provider for long-distance calls, trunks for your PSTN circuits, and trunks that connect your various offices together. These trunks might actually run across the same network connection, but in your dialplan you could treat them quite differently.

While we believe that VoIP will eventually completely replace the PSTN, many of the concepts that are in use on VoIP circuits (such as a "phone number") owe their existence more to history than any technical requirement, and thus we feel it will be helpful to discuss using traditional PSTN circuits with Asterisk before we get into VoIP.

If the system you are installing will use VoIP circuits only, that is not a problem. Go straight to the VoIP section of this chapter,[*] and we'll take you through what you need to do. We do recommend reading the PSTN sections at your convenience, since there may be general knowledge in them that could be of use to you, but it is not strictly required in order to understand and use Asterisk.

Fundamental Dialplan for Outside Connectivity

In a traditional PBX, external lines are generally accessed by way of an access code that must be dialed before the number.[†] It is common to use the digit 9 for this purpose.

In Asterisk, it is similarly possible to assign 9 for routing of external calls, but since the Asterisk dialplan is so much more intelligent, it is not really necessary to force your users to dial 9 before placing a call. Typically, you will have an extension range for your system (say, 100–199), and a feature code range (*00 to *99). Anything outside those ranges that matches the dialing pattern for your country or region can be treated as an external call.

If you have one carrier providing all of your external routing, you can handle your external dialing through a few simple pattern matches. The example in this section is valid for the North American Numbering Plan (NANP). If your country is not within the NANP (which serves Canada, the US, and several Caribbean countries), you will need a different pattern match.

---

[*] But do not collect $200.

[†] In a key system, each line has a corresponding button on each telephone, and lines are accessed by pressing the desired line key.

The [globals] section contains two variables, named LOCAL and TOLL.[‡] The purpose of these variables is to simplify management of your dialplan should you ever need to change carriers. They allow you to make one change to the dialplan that will affect all places where the specified channel is referred to:

```
[globals]
LOCAL=DAHDI/G0          ; assuming you have a PSTN card in your system
TOLL=SIP/YourVoipCarrier ; as defined in sip.conf
```

The [external] section contains the actual dialplan code that will recognize the numbers dialed and pass them to the Dial() application[§]:

```
[external]
exten => _NXXNXXXXXX,1,Dial(${LOCAL}/$EXTEN})  ; 10-digit pattern match for NANP
exten => _NXXXXXX,1,Dial(${LOCAL}/${EXTEN})    ; 7-digit pattern match for NANP
exten => _1NXXNXXXXXX,1,Dial(${TOLL}/${EXTEN}) ; Long-distance pattern match for NANP
exten => _011.,1,Dial(${TOLL}/${EXTEN})        ; International pattern match for
                                               ; calls made from NANP

; This section is functionally the same as the above section.
; It is for people who like to dial '9' before their calls
exten => _9NXXNXXXXXX,1,Dial(${LOCAL}/${EXTEN:1})
exten => _9NXXXXXX,1,Dial(${LOCAL}/${EXTEN:1})
exten => _91NXXNXXXXXX,1,Dial(${TOLL}/${EXTEN:1})
exten => _9011.,1,Dial(${TOLL}/${EXTEN:1})
```

In any context that would be used by sets or user devices, you would use an include=> directive to allow access to the external context:

```
[LocalSets]
include => external
```

It is critically important that you do not include access to the external lines in any context that might process an incoming call. The risk here is that a phishing bot could eventually gain access to your outgoing trunks (you'd be surprised at how common these phishing bots are).

We cannot stress enough how important it is that you ensure that no external resource can access your toll lines.

PSTN Circuits

The Public Switched Telephone Network (PSTN) has existed for over a century. It is the precursor to many of the technologies that shape our world today, from the Internet to MP3 players.

[‡] You can name these anything you wish. The words "local" and "toll" do not have any built-in meaning to the Asterisk dialplan.

[§] For more information on pattern matches, see Chapter 6.

Traditional PSTN Trunks

There are two types of fundamental technology that phone carriers use to deliver telephone circuits: analog and digital.

Analog telephony

The first telephone networks were all analog. The audio signal that you generated with your voice was used to generate an electrical signal that was carried to the other end. The electrical signal had the same characteristics as the sound being produced.

Analog circuits have several characteristics that differentiate them from other circuits you might wish to connect to Asterisk:

- No signaling channel exists—most signaling is electromechanical.
- Disconnect supervision is usually delayed by several seconds, and is not completely reliable.
- Far-end supervision is minimal (for example, answer supervision is lacking).
- Differences in circuits means that audio characteristics will vary from circuit to circuit, and will require tuning.

Analog circuits that you wish to connect to your Asterisk system will need to connect to a Foreign eXchange Office (FXO) port. Since there is no such thing as an FXO port in any standard computer, an FXO card must be purchased and installed in the system in order to connect traditional analog lines.‖

FXO and FXS

For any analog circuit, there are two ends: the office (typically the central office of the PSTN), and the station (typically a phone, but could also be a card such as a modem or line card in a PBX).

The central office is responsible for:

- Power on the line (nominally 48 VDC)
- Ringing voltage (nominally 90 VAC)
- Providing dial tone
- Detecting hook state (off-hook and on-hook)
- Sending supplementary signaling such as caller ID

‖ You would use the exact same card if you wanted to connect a traditional home telephone line to your Asterisk system.

The station is responsible for:

- Providing a ringer (or at least being able to handle ringing voltage in some manner)
- Providing a dialpad (or some way of sending DTMF)
- Providing a hook switch to indicate the status of the line

A Foreign eXchange (FX) port *is named by what it connects to*, not by what it does. So, for example, a Foreign eXchange Office (FXO) port is actually a station: it connects to the central office. A Foreign eXchange Station (FXS) port is actually a port that provides the services of a central office (in other words, you would plug an analog set into an FXS port).

It is for this reason that the signaling settings in the Asterisk config files seem backwards: FXO ports use FXS signaling; FXS ports use FXO signaling. When you understand that the name of the physical port type is based on what it connects to, the signaling names in Asterisk make a bit more sense: if an FXO port connects to the central office, it needs to be able to behave as a station, and therefore needs FXS signaling.

Note that changing from FXO to FXS is not something you can simply do with a settings change. FXO and FXS ports require completely different electronics. Most analog cards available for Asterisk use some form of daughter card that connects to the main card and provides the correct channel type, meaning that you have some flexibility in defining what types of ports you have on your card.

Analog ports are not generally used in medium to large systems. They are most commonly used in smaller offices (less than 10 lines; less than 30 phones). Your decision to use analog might be based on some of the following factors:

- Availability of digital trunks in your area
- Cost (analog is less expensive at smaller densities, but more expensive at higher densities)
- Logistics (if you already have analog lines installed, you may wish to keep them)

From a technical perspective, you would normally want to have digital rather than analog circuits. Reality does not always accommodate, though, so analog will likely be around for a few more years yet.

Digital telephony

Digital telephony was developed in order to overcome many of the limitations of analog. Some of the benefits of digital circuits include:

- No loss of amplitude over long distances
- Reduced noise on circuits (especially long-distance circuits)
- Ability to carry more than one call per circuit
- Faster call setup and teardown

- Richer signaling information (especially if using ISDN)
- Lower cost for carriers
- Lower cost for customers (at higher densities)

In an Asterisk system (or any PBX, for that matter), there are several types of digital circuits you might want to connect:

T1 *(24 channels)*
    Used in Canada and the United States (mostly for ISDN-PRI)

E1 *(32 channels)*
    Used in the rest of the world (ISDN-PRI or MFC/R2)

BRI *(2 channels)*
    Used for ISDN-BRI circuits (Euro-ISDN)

Note that the physical circuit can be further defined by the protocol running on the circuit. For example, a T1 could be used for either ISDN-PRI, or CAS, and an E1 could be used for ISDN-PRI, CAS, or MFC/R2. We'll discuss the different protocols in the next section.

Installing PSTN Trunks

Depending on the hardware you have installed, the process for installing your PSTN cards will vary. We will discuss installation in general terms, which will apply to all Digium PSTN cards. Other manufacturers tend to provide installation scripts with their hardware, which will automate much of this for you.

Downloading and installing DAHDI

The Digium Asterisk Hardware Device Interface, a.k.a. DAHDI (*DAW-dee*)# is the software framework required to enable communication between PSTN cards and Asterisk. Even if you do not have any PSTN hardware, we recommend installing DAHDI since it is a simple, reliable way to get a valid timing source.* Complete DAHDI installation instructions can be found in Chapter 3.

Disable Loading Extra DAHDI Modules

By default DAHDI will load all compiled modules into memory. As this is unnecessary, let's disable loading any of the hardware modules for now. If no modules are loaded in the configuration files, DAHDI will load the `dahdi_dummy` driver, which provides an interface for Asterisk to get timing from the kernel so that timing-dependent modules such as MeetMe and IAX2 trunking work correctly.

#Don't ask.

* There are other ways of getting a timing source, and if you want a really tight system it is possible to run Asterisk without DAHDI, but it's not something we're going to cover here.

As of DAHDI 2.3.0, the requirement to load dahdi_dummy for a timing interface no longer exists. The same functionality has now been integrated into the main dahdi kernel module.

The configuration file defining which modules DAHDI will load is in */etc/dahdi/modules*. To disable loading of extra modules, all we need to do is edit the *modules* file and comment out all the modules by placing a hash (#) at the start of each line. When you're done, your *modules* configuration file should look similar to the following:

```
# Contains the list of modules to be loaded / unloaded by /etc/init.d/dahdi.
#
# NOTE: Please add/edit /etc/modprobe.d/dahdi or /etc/modprobe.conf if you
# would like to add any module parameters.
#
# Format of this file: list of modules, each in its own line.
# Anything after a '#' is ignored, likewise trailing and leading
# whitespace and empty lines.
# Digium TE205P/TE207P/TE210P/TE212P: PCI dual-port T1/E1/J1
# Digium TE405P/TE407P/TE410P/TE412P: PCI quad-port T1/E1/J1
# Digium TE220: PCI-Express dual-port T1/E1/J1
# Digium TE420: PCI-Express quad-port T1/E1/J1
#wct4xxp
# Digium TE120P: PCI single-port T1/E1/J1
# Digium TE121: PCI-Express single-port T1/E1/J1
# Digium TE122: PCI single-port T1/E1/J1
#wcte12xp
# Digium T100P: PCI single-port T1
# Digium E100P: PCI single-port E1
#wct1xxp
# Digium TE110P: PCI single-port T1/E1/J1
#wcte11xp
# Digium TDM2400P/AEX2400: up to 24 analog ports
# Digium TDM800P/AEX800: up to 8 analog ports
# Digium TDM410P/AEX410: up to 4 analog ports
#wctdm24xxp
# X100P - Single port FXO interface
# X101P - Single port FXO interface
#wcfxo
# Digium TDM400P: up to 4 analog ports
#wctdm
# Digium B410P: 4 NT/TE BRI ports
#wcb4xxp
# Digium TC400B: G729 / G723 Transcoding Engine
#wctc4xxp
# Xorcom Astribank Devices
#xpp_usb
```

You can also use *dahdi_genconf modules* to generate a proper empty configuration file. The *dahdi_genconf* application will search your system for hardware and, if none is found, create a *modules* file that does not load any hardware modules.

You can then restart your DAHDI process to unload any existing drivers that were loaded, and load just the dahdi_dummy module with the init script:

```
$ sudo /etc/init.d/dahdi restart
Unloading DAHDI hardware modules: done
Loading DAHDI hardware modules:
No hardware timing source found in /proc/dahdi, loading dahdi_dummy
Running dahdi_cfg: [ OK ]
```

Before you can start using your hardware, though, you'll need to configure the */etc/dahdi/system.conf* file; this process is described in "Configuring digital circuits" on page 138 and "Configuring analog circuits" on page 142.

Configuring digital circuits

Digital telephony was developed by carriers as a way to reduce the cost of long-distance circuits, as well as improve transmission quality. The entire PSTN backbone has been fully digital for many years now. The essence of a digital circuit is the digitization of the audio, but digital trunks also allow for more complex and reliable signaling. Several standards have been developed and deployed, and for each standard there may be regional differences as well.

> You can use *dahdi_hardware* and *lsdahdi* to help you determine what telephony hardware your system contains. You can also use *dahdi_genconf modules* to build an */etc/asterisk/modules* file for you based on the found hardware.

PRI ISDN. Primary Rate Interface ISDN (commonly known as PRI) is a protocol designed to run primarily on a DS1 circuit (a T1 or E1, depending on where you are in the world) between a carrier and a customer. PRI uses one of the DS0 channels as a signaling channel (referred to as the *D-channel*). A typical PRI circuit is therefore broken down into a group of *B-channels* (the bearer channels that actually carry the calls), and a D-channel for signaling. Although it is most common to find a PRI circuit being carried across a single physical circuit (such as a T1 or E1), it is possible to have a PRI circuit span multiple DS1s, and even to have multiple D-channels.[†]

While there are many different ways to configure PRI circuits, we are hoping to avoid confusing you with all of the options (many of which are obsolete or at least no longer in common use), and instead provide examples of the more common configurations.

---

† Sometimes circuits are referenced by the number of B- and D-channels they contain, so a single T1 running the PRI protocol in North America might be referred to as 23B+D, and a dual T1 circuit with a backup D-channel would be a 46B+2D. We've even seen PRI referenced as nB+nD, although this can get a little bit pedantic.

When installing telephony hardware, be sure you update the */etc/dahdi/ modules* file to enable the appropriate modules for your hardware and then reload DAHDI with the init script (*/etc/init.d/dahdi*). You can use the *dahdi_genconf modules* command to generate the *modules* file for your system as well.

Most PRI circuits in North America will use a T1 with the following characteristics:

- Line code: B8ZS (bipolar with 8-zeros substitution)
- Framing: ESF (extended superframe)

You will need to configure two files. The */etc/dahdi/system.conf* file should look something like this:

```
loadzone = us
defaultzone = us

span = 1,1,0,esf,b8zs
bchan = 1-23
echocanceller = mg2,1-23
hardhdlc = 24
```

And the */etc/asterisk/chan_dahdi.conf* file should look like this:

```
[trunkgroups]

[channels]

usecallerid = yes
hidecallerid = no
callwaiting = yes
usecallingpres = yes
callwaitingcallerid = yes
threewaycalling = yes
transfer = yes
canpark = yes
cancallforward = yes
callreturn = yes
echocancel = yes
echocancelwhenbridged = yes
relaxdtmf = yes
rxgain = 0.0
txgain = 0.0
group = 1
callgroup = 1
pickupgroup = 1
immediate = no

switchtype = national ; commonly referred to as NI2
context = from-pstn
group = 0
echocancel = yes
```

```
signalling = pri_cpe
channel => 1-23
```

Some carriers will use Nortel's DMS switch, which commonly uses the DMS100 protocol instead of National ISDN 2. In this case you would set the switchtype to DMS100:

```
switchtype = dms100
```

Outside of Canada and the US, PRI circuits will be carried on an E1 circuit.

In Europe, an E1 circuit used for PRI will normally have the following characteristics:

- Line code: CCS
- Framing: HDB3 (high-density bipolar)

The */etc/dahdi/system.conf* file might look something like this:

```
span = 1,0,0,ccs,hdb3,crc4
bchan = 1-15,17-31
hardhdlc = 16
```

And the */etc/asterisk/chan_dahdi.conf* file should look something like this:

```
[trunkgroups]

[channels]

usecallerid = yes
hidecallerid = no
callwaiting = yes
usecallingpres = yes
callwaitingcallerid = yes
threewaycalling = yes
transfer = yes
canpark = yes
cancallforward = yes
callreturn = yes
echocancel = yes
echocancelwhenbridged = yes
relaxdtmf = yes
rxgain = 0.0
txgain = 0.0
group = 1
callgroup = 1
pickupgroup = 1
immediate = no

switchtype = qsig
context = pri_incoming
group = 0
signalling = pri_cpe
channel => 1-15,17-31
```

BRI ISDN. Basic Rate Interface ISDN (commonly known as BRI, or sometimes even just ISDN) was intended to be the smaller sibling to PRI. BRI only provides two 64K B-channels and a 16K D-channel. The use of BRI has been somewhat limited in North

America (we don't recommend using it for any reason), but in some countries in Europe it is widely used and has almost completely replaced analog.

BRI support under Asterisk will be different depending on the BRI card you are installing. The manufacturer of your BRI card will provide specific installation instructions for its hardware.

> When installing telephony hardware, be sure you update the */etc/dahdi/ modules* file to enable the appropriate modules for your hardware and then reload DAHDI with the init script (*/etc/init.d/dahdi*). You can use the *dahdi_genconf modules* command to generate the *modules* file for your system as well.

MFC/R2. The MFC/R2 protocol could be thought of as a precursor to ISDN. It was at first used on analog circuits, but it is now mostly deployed on the same E1 circuits that also carry ISDN-PRI. This protocol is not typically found in Canada, the US, or Western Europe, but it is very popular in some parts of the world (especially Latin America and Asia), mostly because it tends to be a less expensive service offering from the carriers.

There are many different flavors of this protocol, each country having a different regional variant.

The OpenR2 project provides the `libopenr2` library, which needs to be installed on your system in order for Asterisk to support your R2 circuits. Before installing `libopenr2`, however, you need to have DAHDI installed.

The compilation and installation order, therefore, is:

1. DAHDI
2. `libopenr2`
3. Asterisk

Once OpenR2 has been installed, you can use the *r2test* application to see a list of variants that are supported:

```
$ r2test -l
Variant Code        Country
AR                  Argentina
BR                  Brazil
CN                  China
CZ                  Czech Republic
CO                  Colombia
EC                  Ecuador
ITU                 International Telecommunication Union
MX                  Mexico
PH                  Philippines
VE                  Venezuela
```

For additional information on configuring R2 support in Asterisk, see the *configs/ chan_dahdi.conf.sample* file included in the Asterisk source tree (search for "mfcr2").

Additionally, OpenR2 contains some sample configuration files for connecting Asterisk to networks in various countries. To read information about some of the country variants, search the */doc/asterisk* folder and refer to the documents inside the appropriate subdirectory:

```
$ ls doc/asterisk/
ar  br  ec  mx  ve
```

As an example, OpenR2 provides a sample configuration for connecting to Telmex or Axtel in Mexico. We'll step you through this to give you an idea of the process. First, you must configure DAHDI by modifying */etc/dahdi/system.conf* as shown here:

```
loadzone = us
defaultzone = us

span = 1,1,0,cas,hdb3
cas = 1-15:1101
cas = 17-31:1101

span = 2,1,0,cas,hdb3
cas = 32-46:1101
cas = 48-62:1101
```

Next, you must configure Asterisk by modifying */etc/asterisk/chan_dahdi.conf* as follows:

```
signalling = mfcr2
mfcr2_variant = mx
mfcr2_get_ani_first = no
mfcr2_max_ani = 10
mfcr2_max_dnis = 4
mfcr2_category = national_subscriber
mfcr2_mfback_timeout = -1
mfcr2_metering_pulse_timeout = -1
; this is for debugging purposes
mfcr2_logdir = log
mfcr2_logging = all
; end debugging configuration
channel => 1-15
channel => 17-31
```

Configuring analog circuits

There are many companies producing PSTN cards for Asterisk. The card will need to have its drivers installed so that Linux can recognize it (DAHDI ships with these drivers for Digium cards). From that point, configuration is handled by the Asterisk module chan_dahdi.

> You can use *dahdi_hardware* and *lsdahdi* to determine what telephony hardware your system contains.

When installing telephony hardware, be sure you update the */etc/dahdi/
modules file to enable the appropriate modules for your hardware and
then reload DAHDI with the init script (*/etc/init.d/dahdi*). You can use
the *dahdi_genconf modules* command to generate the *modules* file for
your system as well.

In order to configure an FXO card to work with Asterisk, two files are required.

The first is not an Asterisk configuration file, and is thus located in the */etc/dahdi* folder
on your system.[‡] This file, *system.conf* allows you to define some basic parameters, as
well as specify the channels that will be available to your system. Our example assumes
a four-port FXO card, but many different combinations are possible, depending on
your hardware.

```
loadzone = us        ; tonezone defines sounds the interface must produce
                     ; (dialtone, busy signal, ringback, etc.)
defaultzone = us     ; define a default tonezone
fxsks = 1-4          ; which channels on the card will have these parameters
```

Once your card and channels are known to the operating system, you must configure
them for Asterisk by means of the file */etc/asterisk/chan_dahdi.conf*:

```
[channels]

;
; To apply other options to these channels, put them before "channel".
;
signalling = fxs_ks  ; in Asterisk, FXO channels use FXS signaling
                     ; (and yes, FXS channels use FXO signaling)
channel => 1-4       ; apply all the previously defined settings to this channel
```

In this example, we have told Asterisk that the first four DAHDI channels in the system
are FXO ports.

The s extension. If you are connecting to the PSTN using analog channels, we need to
explain extension **s**. When calls enter a context without a specific destination extension
(for example, a ringing FXO line from the PSTN), they are passed to the **s** extension.
(The **s** stands for "start," as this is where a call will start if no extension information
was passed with the call). This extension can also be useful for accepting calls that have
been redirected from other parts of the dialplan. For example, if we had a list of DID
numbers that were all going to the same place, we might want to point each DID to the
s extension, rather than having to code duplicate dialplan logic for each DID.

Since this is exactly what we need for our dialplan, let's begin to fill in the pieces. We
will be performing three actions on the call (answer it, play a sound file, and hang it
up), so our **s** extension will need three priorities. We'll place the three priorities below

---

‡ In theory, these cards could be used for any software that supports DAHDI; therefore, the basic card
  configuration is not a part of Asterisk.

[incoming], because we have decided that all incoming calls should start in this context[§]:

```
[incoming]
exten => s,1,Answer()
    same => n,Playback(tt-weasels)
    same => n,Hangup()
```

Obviously, you would not normally want to answer a call and then hang up. Typically, an incoming call will either be answered by an automated attendant, or ring directly to a phone (or group of phones).

VoIP

In the world of telecom, VoIP is still a relatively new concept. For the century or so prior to VoIP, the only way to connect your site to the PSTN was through the use of circuits provided for that purpose by your local telephone company. VoIP now allows for connections between endpoints without the PSTN having to be involved at all (although in most VoIP scenarios, there will still be a PSTN component at some point, especially if there is a traditional E.164 phone number involved).

PSTN Termination

Until VoIP totally replaces the PSTN, there will be a need to connect calls from VoIP networks to the public telephone network. This process is referred to as *termination*. What it means is that at some point a gateway connected to the PSTN needs to accept calls from the VoIP network and connect them to the PSTN network. From the perspective of the PSTN, the call will appear to have originated at the termination point.

Asterisk can be used as a PSTN termination point. In fact, given that Asterisk handles protocol conversion with ease, this can be an excellent use for an Asterisk system.

In order to provide termination, an Asterisk box will need to be able to handle all of the protocols you wish to connect to the PSTN. In general, this means that your Asterisk box will need a PRI circuit to handle the PSTN connection, and SIP channels to handle the calls coming from the VoIP network. The underlying principle is the same regardless of whether you're running a small system providing PSTN trunks to an office full of VoIP telephones, or a complex network of gateway machines deployed in strategic locations, offering termination to thousands of subscribers.

[§] There is nothing special about any context name. We could have named this context [stuff_that_comes_in], and as long as that was the context assigned in the channel definition in *sip.conf*, *iax.conf*, *chan_dahdi.conf*, et al., the channel would enter the dialplan in that context. Having said that, it is strongly recommended that you give your contexts names that help you to understand their purpose. Some good context names might include [incoming], [local_calls], [long_distance], [sip_telephones], [user_services], [experimental], [remote_locations], and so forth. Always remember that a context determines how a channel enters the dialplan, so name accordingly.

Calls from the VoIP network will arrive in the dialplan in whatever context you assigned to the incoming SIP channels, and the dialplan will relay the calls out through the PSTN interface. At its very simplest, a portion of a dialplan that supports termination could look like this:

```
[from-voip-network]
exten => _X.,1,Verbose(2, Call from VoIP network to ${EXTEN})
    same => n,Dial(DAHDI/g0/${EXTEN})
```

In reality, though, you will often have to handle a more complex routing plan that takes into consideration things like geography, corporate policy, cost, available resources, and so forth.

Given that most PSTN circuits will allow you to dial any number, anywhere in the world, and given that you will be expected to pay for all incurred charges, we cannot stress enough the importance of security on any gateway machine that is providing PSTN termination. Criminals put a lot of effort into cracking phone systems (especially poorly secured Asterisk systems), and if you do not pay careful attention to all aspects of security, you will be the victim of toll fraud. It's only a matter of time.

Do not allow any unsecured VoIP connections into any context that contains PSTN termination.

PSTN Origination

Obviously, if you want to pass calls from your VoIP network to the PSTN, you might also want to be able to accept calls from the PSTN into your VoIP network. The process of doing this is commonly referred to as *origination*. This simply means that the call originated in the PSTN.

In order to provide origination, a phone number is required. You will therefore need to obtain a circuit from your local phone company, which you will connect to your Asterisk system. Depending on where you are in the world, there are several different types of circuits that could provide this functionality, from a basic analog POTS line to a carrier-grade SS7 circuit.

Phone numbers as used for the purpose of origination are commonly called direct inward dialing numbers (DIDs). This is not strictly the case in all situations (for example, the phone number on a traditional analog line would not be considered a DID), but the term is useful enough that it has caught on. Historically, a DID referred to a phone number associated with a trunk connected to customer premise equipment (CPE).

Since phone numbers are controlled by the traditional telecom industry, you will need to obtain the number either from a carrier directly, or from one of the many companies that purchase numbers in bulk and resell them in smaller blocks. If you obtain a circuit such as a PRI circuit, you will normally be able to order DID numbers to be delivered with that circuit.

In order to accept a call from a circuit you are using for origination, you will normally need to handle the passing of the phone number that was called. This is because PSTN trunks can typically handle more than one phone number, and thus the carrier needs to identify which number was called so that your Asterisk system will know how to route the call. The number that was dialed is commonly referred to as the Dialed Number Identification Service (DNIS) number. The DNIS number and the DID do not have to match,‖ but typically they will. If you are ordering a circuit from the carrier, you will want to ask that they send the DNIS (if they don't understand that, you may want to consider another carrier).

In the dialplan, you associate the incoming circuit with a context that will know how to handle the incoming digits. As an example, it could look something like this:

```
[from-pstn]
; This is the context that would be listed in the config file
; for the circuit (i.e. chan_dahdi.conf)

exten => _X.,1,Verbose(2,Incoming call to ${EXTEN})
    same => n,Goto(number-mapping,${EXTEN},1)

[number-mapping]
; This context is not strictly required, but will make it easier
; to keep track of your DIDs in a single location in your dialplan.
; From here you can pass the call to another part of the dialplan
; where the actual dialplan work will take place.

exten => 4165551234,1,Dial(SIP/0000FFFF0001)
exten => 4165554321,1,Goto(autoattendant-context,start,1)
exten => 4165559876,1,VoiceMailMain() ; a handy back door for listening
                                      ; to voice messages

exten => i,1,Verbose(2,Incoming call to invalid number)
```

In the number-mapping context you explicitly list all of the DIDs that you expect to handle, plus an invalid handler for any DIDs that are not listed (you could send invalid numbers to reception, or to an automated attendant, or to some context that plays an invalid prompt).

‖ In traditional PBXs, the purpose of DIDs was to allow connection directly to an extension in the office. Many PBXs could not support concepts such as number translation or flexible digit lengths, and thus the carrier had to pass the extension number as the DID digits, rather than the number that was dialed (the DNIS number). For example, the phone number 416-555-1234 might have been mapped to extension 100, and thus the carrier would have sent the digits 100 to the PBX instead of the DNIS of 4165551234. If you ever replace an old PBX with an Asterisk system, you may find this translation in place, and you'll need to obtain a list of mappings between the numbers that the caller dials and the numbers that are sent to the PBX. It is also common to see the carrier only pass the last four digits of the DNIS number, which the PBX then translates into an internal number.

VoIP to VoIP

Eventually, the need for the PSTN will likely vanish, and most voice communications will take place over network connections.

The original thinking behind the SIP protocol was that it was to be a peer-to-peer protocol. Technically, this is still the case. What has happened, however, is that things have gotten a bit more complicated. Issues such as security, privacy, corporate policies, integration, centralization, and so forth have made things a bit more involved than simply putting a URI into a SIP phone and having a SIP phone somewhere else ring in response.

The SIP protocol has become bloated and complex. Implementing SIP-based systems and networks has arguably become even more complicated than implementing traditional phone PBXs and networks.#

We are not going to get into the complexities of designing and implementing VoIP networks in this book, but we will discuss some of the ways you can configure Asterisk to support VoIP connectivity to other VoIP systems.

Configuring VoIP Trunks

In Asterisk, there is no need to explicitly install your VoIP modules (unless for some reason you did not compile Asterisk with the required modules). There are several VoIP protocols that you can choose to use with Asterisk, but we will focus on the two most popular: SIP and IAX.

Configuring SIP trunks between Asterisk systems

SIP is far and away the most popular of the VoIP protocols—so much so that many people would consider other VoIP protocols to be obsolete (they are not, but it cannot be denied that SIP has dominated VoIP for several years now).

The SIP protocol is peer-to-peer and does not really have a formal trunk specification. This means that whether you are connecting a single phone to your server or connecting two servers together, the SIP connections will be similar.

Connecting two Asterisk systems together with SIP. The need to be able to connect two Asterisk systems together to allow calls to be sent between them is a fairly common requirement. Perhaps you have a company with two physical locations and want to have a PBX at each location, or maybe you're the administrator of the company PBX and you like Asterisk so much that you would also like to install it at home. This section provides a quick guide on configuring two Asterisk servers to be able to pass calls to each other

---

#There are many proprietary PBX systems in the market that have a basic configuration that will work right out of the box. Asterisk deployments are far more flexible, but seldom as simple.

over SIP. In our example, we will creatively refer to the two servers as serverA and serverB.

The first file that must be modified is */etc/asterisk/sip.conf*. This is the main configuration file for setting up SIP accounts. First, this entry must be added to *sip.conf* on serverA. It defines a SIP peer for the other server:

```
[serverB]

;
; Specify the SIP account type as 'peer'.  This means that incoming
; calls will be matched on IP address and port number.  So, when Asterisk
; receives a call from 192.168.1.102 and the standard SIP port of 5060,
; it will match this entry in sip.conf.  It will then request authentication
; and expect the password to match the 'secret' specified here.
;
type = peer
;
; This is the IP address for the remote box (serverB). This option can also
; be provided a hostname.
;
host = 192.168.1.102
;
; When we send calls to this SIP peer and must provide authentication,
; we use 'serverA' as our username.
;
username = serverA
;
; This is the shared secret with serverB.  It will be used as the password
; when either receiving a call from serverB, or sending a call to serverB.
;
secret = apples
;
; When receiving a call from serverB, match it against extensions
; in the 'incoming' context of extensions.conf.
;
context = incoming
;
; Start by clearing out the list of allowed codecs.
;
disallow = all
;
; Only allow the ulaw codec.
;
allow = ulaw
```

 Be sure to change the host option to match the appropriate IP address for your own setup.

Now put the following entry in */etc/asterisk/sip.conf* on **serverB**. It is nearly identical to the contents of the entry we put on **serverA**, but the name of the peer and the IP address were changed:

```
[serverA]

type = peer
host = 192.168.1.101
username = serverB
secret = apples
context = incoming
disallow = all
allow = ulaw
```

At this point you should be able to verify that the configuration has been successfully loaded into Asterisk using some CLI commands. The first command to try is *sip show peers*. As the name implies, it will show all SIP peers that have been configured:

```
*CLI> sip show peers
Name/username      Host           Dyn Forcerport ACL Port    Status
serverB/serverA    192.168.1.101                       5060    Unmonitored
1 sip peers [Monitored: 0 online, 0 offline Unmonitored: 1 online, 0 offline]
```

 You can also try *sip show peer serverB*. That command will show much more detail.

The last step in setting up SIP calls between two Asterisk servers is to modify the dialplan in */etc/asterisk/extensions.conf*. For example, if you wanted any calls made on **serverA** to extensions 6000 through 6999 to be sent over to **serverB**, you would use this line in the dialplan:

```
exten => _6XXX,1,Dial(SIP/${EXTEN}@serverB)
```

Connecting an Asterisk system to a SIP provider. When you sign up for a SIP provider, you may have service for sending and/or receiving phone calls. The configuration will differ slightly depending on your usage of the SIP provider. Further, the configuration will differ between each provider. Ideally, the SIP provider that you sign up with will provide Asterisk configuration examples to help get you connected as quickly as possible. In case they do not, though, we will attempt to give you a common setup that will help you get started.

If you will be receiving calls from your service provider, the service provider will most likely require your server to register with one of its servers. To do so, you must add a registration line to the [general] section of */etc/asterisk/sip.conf*:

```
[general]
...
register => username:password@your.provider.tld
...
```

Next, you will need to create a peer entry in *sip.conf* for your service provider. Here is a sample peer entry:

```
[myprovider]

type = peer
host = your.provider.tld
username = username
secret = password
; Most providers won't authenticate when they send calls to you,
; so you need this line to just accept their calls.
insecure = invite
dtmfmode = rfc2833
disallow = all
allow = ulaw
```

Now that the account has been defined, you must add some extensions in the dialplan to allow you to send calls to your service provider:

```
exten => _1NXXNXXXXXX,1,Dial(SIP/${EXTEN}@myprovider)
```

Encrypting SIP calls. Asterisk supports TLS for encryption of the SIP signaling and SRTP for encryption of the media streams of a phone call. In this section we will set up calls using SIP TLS and SRTP between two Asterisk severs. The first step is to ensure the proper dependencies have been installed. Ensure that you have both OpenSSL and LibSRTP installed. If either one of these was not installed, reinstall Asterisk after installing these dependencies to ensure that support for TLS and SRTP are included. Once complete, make sure that the `res_srtp` module was compiled and installed. To install OpenSSL, the package is `openssl-devel` on CentOS and `libssl-dev` on Ubuntu. To install LibSRTP, the package is `libsrtp-devel` on CentOS and `libsrtp0-dev` on Ubuntu.

Next we will configure SIP TLS. You must enable TLS using the global `tlsenable` option in the `[general]` section of */etc/asterisk/sip.conf* on both servers. You can optionally specify an address to bind to if you would like to limit listening for TLS connections to a single IP address on the system. In this example, we have the IPv6 wildcard address specified to allow TLS connections on all IPv4 and IPv6 addresses on the system:

```
[general]

tlsenable = yes
tlsbindaddr = ::
```

The next step is to get certificates in place. For the purposes of demonstrating the configuration and functionality, we are going to generate self-signed certificates using a helper script distributed with Asterisk. If you were setting this up in a production environment, you might not want to use self-signed certificates. However, if you do, there are a number of applications out there that help make it easier to manage your own certificate authority (CA), such as TinyCA.

The script that we are going to use is *ast_tls_cert*, which is in the *contrib/scripts/* directory of the Asterisk source tree. We need to generate a CA certificate and two server certificates. The first invocation of *ast_tls_cert* will generate the CA cert and the server cert for serverA. The second invocation of *ast_tls_cert* will generate the server cert for serverB:

```
$ cd contrib/scripts
$ mkdir certs
$ ./ast_tls_cert -d certs -C serverA -o serverA
$ ./ast_tls_cert -d certs -C serverB -o serverB -c certs/ca.crt -k certs/ca.key
$ ls certs
ca.cfg  ca.crt  ca.key  serverA.crt  serverA.csr  serverA.key  serverA.pem
serverB.crt  serverB.csr  serverB.key  serverB.pem  tmp.cfg
```

Now that the certificates have been created, they need to be moved to the appropriate locations on serverA and serverB. We will use the */var/lib/asterisk/keys/* directory to hold the certificates. Move the following files to serverA:

- *ca.crt*
- *serverA.pem*

And move these files to serverB:

- *ca.crt*
- *serverB.pem*

With the certificates in place, we can complete the Asterisk configuration. We need to point Asterisk to the server certificate that we just created. Since we're using self-signed certificates, we also need to point to the CA certificate. In the [general] section of */etc/asterisk/sip.conf* on serverA, add these options:

```
[general]

tlscertfile = /var/lib/asterisk/keys/serverA.pem
tlscafile = /var/lib/asterisk/keys/ca.crt
```

Make the same changes to *sip.conf* on serverB:

```
[general]

tlscertfile = /var/lib/asterisk/keys/serverB.pem
tlscafile = /var/lib/asterisk/keys/ca.crt
```

 When you create the server certificates, the Common Name field must match the hostname of the server. If you use the *ast_tls_cert* script, this is the value given to the -C option. If there is a problem verifying the server certificate when you make a call, you may need to fix the Common Name field. Alternatively, for the sake of testing you can set the tlsdont verifyserver option to yes in the [general] section of */etc/asterisk/sip.conf*, and Asterisk will allow the call to proceed even if it fails verification of the server certificate.

In "Connecting two Asterisk systems together with SIP" on page 147, we created the configuration necessary to pass calls between serverA and serverB. We are now going to modify that configuration so that Asterisk knows that the calls between the two servers should be encrypted. The only change required is to add the `transport = tls` option to the peer entry for the other server.

On serverA:

```
[serverB]

type = peer
host = 192.168.1.102
username = serverA
secret = apples
context = incoming
disallow = all
allow = ulaw
transport = tls
```

On serverB:

```
[serverA]

type = peer
host = 192.168.1.101
username = serverB
secret = apples
context = incoming
disallow = all
allow = ulaw
transport = tls
```

Now when you make a call using `Dial(SIP/serverA)` or `Dial(SIP/serverB)`, the SIP signaling will be encrypted. You can modify the dialplan to force outgoing calls to have encrypted signaling by setting the `CHANNEL(secure_bridge_signaling)` function to 1:

```
[default]

exten => 1234,1,Set(CHANNEL(secure_bridge_signaling)=1)
    same => n,Dial(SIP/1234@serverB)
```

On the side receiving the call, you can check whether the signaling on an incoming call is encrypted using the `CHANNEL(secure_signaling)` dialplan function. Consider the following example dialplan:

```
[incoming]

exten => _X.,1,Answer()
    same => n,GotoIf($["${CHANNEL(secure_signaling)}" = "1"]?secure:insecure)
    same => n(secure),NoOp(Signaling is encrypted.)
    same => n,Hangup()
    same => n(insecure),NoOp(Signaling is not encrypted.)
    same => n,Hangup()
```

When a call is sent from serverA to serverB using this configuration, you can see from the output on the Asterisk console that the dialplan determines that the signaling of the incoming call is encrypted:

```
-- Executing [1234@incoming:1] Answer("SIP/serverA-00000000", "") in new stack
-- Executing [1234@incoming:2] GotoIf("SIP/serverA-00000000",
   "1?secure:insecure") in new stack
-- Goto (incoming,1234,3)
-- Executing [1234@incoming:3] NoOp("SIP/serverA-00000000",
   "Signaling is encrypted.") in new stack
-- Executing [1234@incoming:4] Hangup("SIP/serverA-00000000", "") in new stack
```

Now that SIP TLS has been set up for calls between serverA and serverB, we will set up SRTP so that the media streams associated with the call are encrypted as well. Luckily, it is quite easy to configure, compared to what was required to get SIP TLS working. First, make sure that you have the res_srtp module loaded in Asterisk:

```
*CLI> module show like res_srtp.so
Module                        Description                    Use Count
res_srtp.so                   Secure RTP (SRTP)              0
1 modules loaded
```

To enable SRTP, set the CHANNEL(secure_bridge_media) function to 1:

```
[default]

exten => 1234,1,Set(CHANNEL(secure_bridge_signaling)=1)
    same => n,Set(CHANNEL(secure_bridge_media)=1)
    same => n,Dial(SIP/1234@serverB)
```

This indicates that encrypted media is required for an outbound call. When the call is sent out via SIP, Asterisk will require that SRTP be used, or the call will fail.

With all of these tools in place, you can ensure that calls between two Asterisk servers are fully encrypted. The same techniques should be applied for encrypting calls between Asterisk and a SIP phone.

The dialplan functions provide a mechanism for verifying the encryption status of an incoming call and forcing encryption on an outgoing call. However, keep in mind that these tools only provide the means for controlling encryption for one hop of the call path. If the call goes through multiple servers, these tools do not guarantee that the call is encrypted through the entire call path. It is important to carefully consider what your requirements are for secure calls and take all of the necessary steps to ensure that those requirements are respected throughout the entire call path. Security is complicated, hard work.

Configuring IAX trunks between Asterisk systems

The Inter-Asterisk eXchange protocol, version 2 (most commonly known as IAX*) is Asterisk's own VoIP protocol. It is different from SIP in that the signaling and media are carried in the same connection. This difference is one of the advantages of the IAX protocol, as it makes getting IAX to work across NAT connections much simpler.

---

* Pronounced "eeks."

IAX trunking. One of the more unique features of the IAX protocol is IAX trunking. Trunking an IAX connection could be useful on any network link that will often be carrying multiple simultaneous VoIP calls between two systems. By encapsulating multiple audio streams in one packet, IAX trunking cuts down on the overhead on the data connection, which can save bandwidth on a heavily used network link.

IAX encryption. The principal advantage of IAX encryption is that it requires one simple change to the */etc/asterisk/iax.conf* file:

```
[general]
encryption = yes
```

For extra protection, you can set the following option to ensure that no IAX connection can happen without encryption:

```
forceencryption = yes
```

Both of these options can be specified in the [general] section, as well as in peer/user/ friend sections in *iax.conf*.

Emergency Dialing

In North America, people are used to being able to dial 911 in order to reach emergency services. Outside of North America, well-known emergency numbers are 112 and 999. If you make your Asterisk system available to people, you are obligated (in many cases regulated) to ensure that calls can be made to emergency services from any telephone connected to the system (even those phones that otherwise are restricted from making calls).

One of the essential pieces of information the emergency response organization needs to know is where the emergency is (i.e., where to send the fire trucks). In a traditional PSTN trunk this information is already known by the carrier and is subsequently passed along to the Public Safety Answering Point (PSAP). With VoIP circuits things can get a bit more complicated, by virtue of the fact that VoIP circuits are not physically tied to any geographical location.

You need to ensure that your system will properly handle 911 calls from any phone connected to it, and you need to communicate what is available to your users. As an example, if you allow users to register to the system from softphones on their laptops, what happens if they are in a hotel room in another country, and they dial 911?[†]

The dialplan for handling emergency calls does not need to be complicated. In fact, it's far better to keep it simple. People are often tempted to implement all sorts of fancy functionality in the emergency services portions of their dialplans, but if a bug in one of your fancy features causes an emergency call to fail, lives could be at risk. This is no

---

† Don't assume this can't happen. When somebody calls 911 it's because they have an emergency, and it's not safe to assume that they're going to be in a rational state of mind.

place for playing around. The [emergency-services] section of your dialplan might look something like this:

```
[emergency-services]
exten => 911,Goto(dialpsap,1)
exten => 9911,Goto(dialpsap,1) ; some people will dial '9' because
                               ; they're used to doing that from the PBX
exten => 999,Goto(dialpsap,1)
exten => 112,Goto(dialpsap,1)

exten => dialpsap,1,Verbose(1,Call initiated to PSAP!)
    same => n,Dial(${LOCAL}/911) ; REPLACE 911 HERE WITH WHATEVER
                                 ; IS APPROPRIATE TO YOUR AREA

[internal]
include => emergency-services   ; you have to have this in any context
                                ; that has users in it
```

In contexts where you know the users are not on-site (for example, remote users with their laptops), something like this might be best instead:

```
[no-emergency-services]
exten => 911,Goto(nopsap,1)
exten => 9911,Goto(nopsap,1) ; for people who dial '9' before external calls
exten => 999,Goto(nopsap,1)
exten => 112,Goto(nopsap,1)

exten => nopsap,1,Verbose(1,Call initiated to PSAP!)
    same => n,Playback(no-emerg-service) ; you'll need to record this prompt

[remote-users]
include => no-emergency-services
```

In North America, regulations have obligated many VoIP carriers to offer what is popularly known as *E911.*[‡] When you sign up for their services, they will require address information for each DID that you wish to associate with outgoing calls. This address information will then be sent to the PSAP appropriate to that address, and your emergency calls should be handled the same as they would be if they were dialed on a traditional PSTN circuit.

Handling emergency calls does not have to be complicated (in fact, it is best to keep this as simple as possible). The bottom line is that you need to make sure that the phone system you create allows emergency calls.

[‡] It's not actually the carrier that's offering this; rather it's a capability of the PSAP. E911 is also used on PSTN trunks, but since that happens without any involvement on your part (the PSTN carriers handle the paperwork for you), you are generally not aware that you have E911 on your local lines.

Conclusion

Eventually, we believe that the PSTN will disappear entirely. Before that happens, however, a distributed mechanism that is widely used and trusted will be needed to allow organizations and individuals to publish addressing information so that they can be found. We'll explore some of the ways this is already possible in Chapter 12.

Voicemail

Just leave a message, maybe I'll call.

—Joe Walsh

Before email and instant messaging became ubiquitous, voicemail was a popular method of electronic messaging. Even though most people prefer text-based messaging systems, voicemail remains an essential component of any PBX.

Comedian Mail

One of the most popular (or, arguably, unpopular) features of any modern telephone system is voicemail. Asterisk has a reasonably flexible voicemail system named Comedian Mail.* Some of the features of Asterisk's voicemail system include:

- Unlimited password-protected voicemail boxes, each containing mailbox folders for organizing voicemail
- Different greetings for busy and unavailable states
- Default and custom greetings
- The ability to associate phones with more than one mailbox and mailboxes with more than one phone
- Email notification of voicemail, with the voicemail optionally attached as a sound file†
- Voicemail forwarding and broadcasts
- Message-waiting indicator (flashing light or stuttered dialtone) on many types of phones
- Company directory of employees, based on voicemail boxes

---

* This name was a play on words, inspired in part by Nortel's voicemail system Meridian Mail.

† No, you really don't have to pay for this—and yes, it really does work.

And that's just the tip of the iceberg!

The default version of the */etc/asterisk/voicemail.conf* file requires a few tweaks in order to provide a configuration that will be suitable to most situations.

We'll begin by going through the various options you can define in *voicemail.conf*, and then we'll provide a sample configuration file with the settings we recommend for most deployments.

The *voicemail.conf* file contains several sections where parameters can be defined. The following sections detail all the options that are available.

The [general] Section

The first section, [general], allows you to define global settings for your voicemail system. The available options are listed in Table 8-1.

Table 8-1. [general] section options for voicemail.conf

Option	Value/Example	Notes
format	wav49|gsm|wav	For each format listed, Asterisk will create a separate recording in that format whenever a message is left. The benefit of this is that some transcoding steps may be saved if the stored format is the same as the codec used on the channel. We like wav because it is the highest quality, and wav49 because it is nicely compressed and easy to email. We don't like gsm due to it's scratchy sound, but it enjoys some popularity.[a]
serveremail	user@domain	When an email is sent from Asterisk, this is the email address that it will appear to come from.[b]
attach	yes,no	If an email address is specified for a mailbox, this determines whether the messages is attached to the email (if not, a simple message notification is sent).
maxmsg	9999	By default Asterisk will only allow a maximum of 100 messages to be stored per user. For users who delete messages, this is no problem. For people who like to save their messages, this space can get eaten up quickly. With the size of hard drives these days, you could easily store thousands of messages for each user, so our current thinking is to set this to the maximum and let the users manage things from there.
maxsecs	0	This type of setting was useful back in the days when a large voicemail system might have only 40 MB[c] of storage: it was necessary to limit the system because it was easy to fill up the hard drive. This setting can be annoying to callers (although it does force them to get to the point, so some people like it). Nowadays, with terabyte drives becoming common, there is no reason not to set this to a high value. Two considerations are: 1) if a channel gets hung in a mailbox, it's good to set some sort of value so it doesn't stay there for days, but 2) if a user wants to use her mailbox to record notes to herself, she

Option	Value/Example	Notes
		won't appreciate it if you cut her off after three minutes. A setting somewhere between 600 seconds (10 minutes) and 3600 seconds (1 hour) will probably be about right.
minsecs	4	Many folks will hang up instead of leaving a message when they call somebody and get voicemail. Sometimes this hangup happens after recording has started, so the mailbox owner gets an annoying two-second message of somebody hanging up. This setting ensures that Asterisk will ignore messages that are shorter than the configured minimum length. You should take care not to set this to a value that is too high, though, because then a message like "Hey it's me give me a call" (which can be said in less than one second) will get lost, and you'll get complaints of messages disappearing. Three seconds seems to be about right. To discourage people from leaving ultra-short messages that might be discarded, you can request callers to identify themselves and leave some information about why they called.
maxgreet	1800	You can define the maximum greeting length if you want. Again, since storage is not a problem and setting this too low will annoy your more verbose users, we suggest setting this to a high value and letting your users figure it out an appropriate length for themselves.
skipms	3000	When listening to messages, users can skip ahead or backwards by pressing (by default) * and #. This setting indicates the length of the jump (in milliseconds).
maxsilence	5	This setting defines the maximum time for which the caller can remain silent before the recording is stopped. We like to set this setting to one second longer than minsecs (if you set it equal to or less than minsecs, you will get a warning).
silencethreshold	128	You can fine-tune the silence sensitivity of Asterisk to better define what qualifies as silence. In practice, this is seldom a good idea, since you cannot control the volumes of all the calls you'll be getting from different places. It's best to leave this at the default.
maxlogins	3	This little security feature is intended to make brute-force attacks on your mailbox passwords more time-consuming. If a bad password is received this many times, voicemail will hang up and you'll have to call back in to try again. Note that this will not lock up the mailbox. Patient snoopers can continue to try to log into your mailbox as many times as they like, they'll just have to call back every third attempt. If you have a lot of ham-fingered users, you can set this to something like 5.
moveheard	yes	This setting will move listened-to messages to the *Old* folder. We recommend leaving this at the default.
forward_urgent_auto	no	Setting this to yes will preserve the original urgency setting of any messages the user receives and then forwards on. If you leave it at no, users can set the urgency level themselves on messages that they forward.

Option	Value/Example	Notes
userscontext	default	If you use the *users.conf* file (we don't), you can define here the context where entries are registered.
externnotify	*/path/to/script*	If you wish to run an external app whenever a message is left, you can define it here.
smdienable	no	If you are using Asterisk as a voicemail server on a PBX that supports SMDI, you can enable it here.
smdiport	*/dev/ttyS0*	Here is where you would define the SMDI port that messages between Asterisk and the external PBX would pass across.
externpass	*/path/to/script*	Any time the password on a mailbox is changed, the script you define here will be notified of the context, mailbox, and new password. The script will then be responsible for updating *voicemail.conf* (the Asterisk voicemail app will not update the password if this parameter is defined).
externpassnotify	*/path/to/script*	Any time the password on a mailbox is changed, the script you define here will be notified of the context, mailbox, and new password. Asterisk will handle updating the password in *voicemail.conf*. If you have defined externpass, this option will be ignored.
externpasscheck	*/usr/local/bin/voice-mailpwcheck.py*	See the sidebar following this table for a description of this option.
directoryintro	dir-intro	The Directory() dialplan application uses the *voicemail.conf* file to search by name from an auto attendant. There is a default prompt that plays, called dir-intro. If you want, you can specify a different file to play instead.
charset	ISO-8859-1	If you need a character set other than ISO-8859-1 (a.k.a Latin 1) to be supported, you can specify it here.
adsifdn	0000000F	Use this option to configure the Feature Descriptor Number.[d]
adsisec	9BDBF7AC	Use this option to configure the security lock code.
adsiver	1	This specifies the ADSI voicemail application version number.
pbxskip	yes	If you do not want emails from your voicemail to have the string [PBX] added to the subject, you can set this to yes.
fromstring	The Asterisk PBX	You can use this setting to configure the From: name that will appear in emails from your PBX.
usedirectory	yes	This option allows users composing messages from their mailboxes to take advantage of the Directory.
odbcstorage	*<item from res_odbc.conf>*	If you want to store voice messages in a database, you can do that using the Asterisk res_odbc connector. Here, you would set the name of the item in the *res_odbc* file. For details, see Chapter 22.
odbctable	*<table name>*	This setting specifies the table name in the database that the odbc storage setting refers to. For details, see Chapter 22.

Option	Value/Example	Notes
emailsubject	[PBX]: New message $ {VM_MSGNUM} in mailbox $ {VM_MAILBOX}	When Asterisk sends an email, you can use this setting to define what the Subject: line of the email will look like. See the *voicemail.conf.sample* file for more details.
emailbody	Dear $ {VM_NAME}:\n\n \tjust wanted to let you know you were just left a ${VM_DUR} long message (number ${VM_MSGNUM}) \nin mailbox $ {VM_MAILBOX u might\nwant to check it when you get a chance. Thanks! \n\n\t\t\t\t-- Asterisk\n	When Asterisk sends an email, you can use this setting to define what the body of the email will look like. See the *voicemail.conf.sample* file for more details.
pagerfromstring	The Asterisk PBX	We don't actually know anybody who uses pagers anymore (nor can we recall having seen one in many years), but if you have one of these historical oddities and you want to customize what Asterisk sends with its pager notification, presumably you can do that with this. A very practical usage of this feature for short message voicemail notifications is to send a message to an email to SMS gateway.
pagersubject	New VM	As above.
pagerbody	New ${VM_DUR} long msg in box ${VM_MAILBOX} \nfrom $ {VM_CALLERID}, on ${VM_DATE}	The formatting for this uses the same rules as emailbody.
emaildateformat	%A, %d %B %Y at %H:%M:%S	This option allows you to specify the date format in emails. Uses the same rules as the C function STRFTIME.
pagerdateformat	%A, %d %B %Y at %H:%M:%S	This option allows you to specify the date format in pager. Uses the same rules as the C function STRFTIME.
mailcmd	/usr/sbin/send mail -t	If you want to override the default operating system application for sending mail, you can specify it here.
pollmailboxes	no, yes	If the contents of mailboxes are changed by anything other than app_voicemail (such as external applications or another Asterisk system), setting this to yes will cause app_voicemail to poll all

Option	Value/Example	Notes
		the mailboxes for changes, which will trigger proper message waiting indication (MWI) updates.
pollfreq	30	Used in concert with pollmailboxes, this option specifies the number of seconds to wait between mailbox polls.
imapgreetings	no, yes	This enables/disables remote storage of greetings in the IMAP folder. For more details, see Chapter 18.
greetingsfolder	INBOX	If you've enabled imapgreetings, this parameter allows you to define the folder your greetings will be stored in (defaults to *INBOX*).
imapparentfolder	INBOX	IMAP servers can handle parent folders in different ways. This field allows you to specify the parent folder for your mailboxes. For more details, see Chapter 7.

[a] The separator that is used for each format option must be the pipe (|) character.

[b] Sending email from Asterisk can require some careful configuration, because many spam filters will find Asterisk messages suspicious and will simply ignore them. We talk more about how to set email for Asterisk in Chapter 18.

[c] Yes, you read that correctly: megabytes.

[d] The Analog Display Services Interface is a standard that allows for more complex feature interactions through the use of the phone display and menus. With the advent of VoIP telephones, ADSI's popularity has decreased in recent years.

External Validation of Voicemail Passwords

By default, Asterisk does not validate user passwords to ensure they are at least somewhat secure. Anyone who maintains voicemail systems will tell you that a large percentage of mailbox users set their passwords to something like 1234 or 1111, or some other string that's easy to guess. This represents a huge security hole in the voicemail system.

Since the *app_voicemail.so* module does not have the built-in ability to validate passwords, the settings externpass, externpassnotify, and externpasscheck allow you to validate them using an external program. Asterisk will call the program based on the path you specify, and pass it the following arguments:

```
mailbox context oldpass newpass
```

The script will then evaluate the arguments based on rules that you defined in the external script and, based on your rules, it should return to Asterisk a value of VALID for success or INVALID for failure (actually, the return value for a failed password can be anything except the words VALID or FAILURE). This value is typically printed to *stdout*. If the script returns INVALID, Asterisk will play an invalid-password prompt and the user will need to attempt something different.

Ideally, you would want to implement rules such as the following:

- Passwords must be a minimum of six digits in length
- Passwords must not be strings of repeated digits (e.g., 111111)
- Passwords must not be strings of contiguous digits (e.g., 123456 or 987654)

Asterisk comes with a simple script that will greatly improve the security of your voicemail system. It is located in the source code under the folder: */contrib/scripts/voicemailpwcheck.py*.

We strongly recommend that you copy it to your */usr/local/bin* folder (or wherever you prefer to put such things), and then uncomment the `externpasscheck=` option in your *voicemail.conf* file. Your voicemail system will then enforce the password security rules you have established.

Part of the `[general]` section is an area that is referred to as *advanced options*. These options (listed in Table 8-2) are defined in the same way as the other options in the `[general]` section, but they can also be defined on a per-mailbox basis, overriding whatever is defined under `[general]` for that particular setting.

Table 8-2. Advanced options for voicemail.conf

Option	Value/Example	Notes
tz	eastern, europ ean, etc.	Specifies the zonemessages name, as defined in the [zonemessages] section, discussed in the next section.
locale	de_DE.utf8, es_US.utf8, etc.	Used to define how Asterisk generates date/time strings in different locales. To determine the locales that are valid on your Linux system, type *locale -a* at the shell.
attach	yes, no	If an email address is specified for a mailbox, this determines whether the messages are attached to the email notifications (otherwise, a simple message notification is sent).
attachfmt	wav49, wav, etc.	If attach is enabled and messages are stored in different formats, this defines which format is sent with the email notifications. Often wav49 is a good choice, as it uses a better compression algorithm and thus will use less bandwidth.
saycid	yes, no	This command will state the caller ID of the person who left the message.
cidinternalcontexts	<context>, <another context>	Any dialplan contexts listed here will be searched in an attempt to locate the mailbox context, so that the name associated with the mailbox number can be spoken. The voicemail box number needs to match the extension number that the call came from, and the voicemail context needs to match the dialplan context.[a]
sayduration	yes, no	This command will state the length of the message.
saydurationm	2	Use this to specify the minimum duration of a message to qualify for its length being played back. For example, if you set this to 2, any message less than 2 minutes in length will not have its length stated.
dialout	<context>	If allowed, users can dial out from their mailboxes. This is considered a very dangerous feature in a phone system (mainly because many voicemail users like to use 1234 as their password), and is therefore not recommended. If you insist on allowing this, make sure you have

Option	Value/Example	Notes
		a second level of password in the dialplan where another password is specified. Even so, this is not a safe practice.
sendvoicemail	yes, no	This allows users to compose messages to other users from within their mailboxes.
searchcontexts	yes, no	This allows voicemail applications in the dialplan to not have to specify the voicemail context, since all contexts will be searched. This is not recommended.
callback	<context>	This specifies which dialplan context to use to call back to the sender of a message. The specified context will need to be able to handle dialing of numbers in the format in which they are received (for example, the country code may not be received with the caller ID, but might be required for the outgoing call).
exitcontext	<context>	There are options that allow the callers to exit the voicemail system when they are in the process of leaving a message (for example, pressing 0 to get an operator). By default, the context the caller came from will be used as the exit context. If desired, this setting will define a different context for callers exiting the voicemail system.
review	yes, no	This should almost always be set to yes (even though it defaults to no). People get upset if your voicemail system does not allow them to review their messages prior to delivering them.
operator	yes, no	Best practice dictates that you should allow your callers to "zero out" from a mailbox, should they not wish to leave a message. Note that an o extension (not "zero," "oh") is required in the exitcontext in order to handle these calls.
envelope	no, yes	You can have voicemail play back the details of the message before it plays the actual message. Since this information can also be accessed by pressing 5, we generally set this to no.
delete	no, yes	After an email message notification is sent (which could include the message itself), the message will be deleted. This option is risky, because even though a message was emailed, it is no guarantee that it was received (spam filters seem to love to delete Asterisk voicemail messages). Point being: on a new system, leave this at no until you are certain that no messages are being lost due to spam filters.
volgain	0.0	This setting allows you to increase the volume of received messages. Volume used to be a problem in older releases of Asterisk, but has not been an issue for many years. We recommend leaving this at the default. The *sox* utility is required for this to work.
nextaftercmd	yes, no	This handy little setting will save you some time, as it takes you directly to the next message once you've finished dealing with the current message.
forcename	yes, no	This strange little setting will check if the mailbox password is the same as the mailbox number. If it is, it will force the user to change his voicemail password and record his name.

Option	Value/Example	Notes
forcegreetings	yes, no	As above, but for greetings.
hidefromdir	no, yes	If you wish, you can hide specific mailboxes from the Directory() application using this setting.
tempgreetwarn	yes, no	Setting this to yes will warn the mailbox owner that she has a temporary greeting set. This can be a useful reminder when people return from trips or vacations.
passwordlocation	spooldir	If you want, you can have mailbox passwords stored in the spool folder for each mailbox.[b]
messagewrap	no, yes	If this is set to yes, when the user has listened to the last message, pressing next (6) will take him to the first message. Also, pressing previous (4) when at the first message will take the user to the last message.
minpassword	6	This option enforces a minimum password length. Note that this does not prevent the users from setting their passwords to something that's easy to guess (such as 123456).
vm-password	custom_sound	If you want, you can specify a custom sound here to use for the password prompt in voicemail.
vm-newpassword	custom_sound	If you want, you can specify a custom sound here to use for the "Please enter your new password followed by the pound key" prompt in voicemail.
vm-passchanged	custom_sound	If you want, you can specify a custom sound here to use for the "Your password has been changed" prompt in voicemail.
vm-reenterpassword	custom_sound	If you want, you can specify a custom sound here to use for the "Please reenter your password followed by the pound key" prompt in voicemail.
vm-mismatch	custom_sound	If you want, you can specify a custom sound here to use for the "The passwords you entered and reentered did not match" prompt in voicemail.
vm-invalid-password	custom_sound	If you want, you can specify a custom sound here to use for the "That is not a valid password. Please try again" prompt in voicemail.
vm-pls-try-again	custom_sound	If you want, you can specify a custom sound here to use for the "Please try again" prompt in voicemail.
listen-control-forward-key	#	You can use this setting to customize the fast forward key.
listen-control-reverse-key	*	You can use this setting to customize the rewind key.
listen-control-pause-key	0	You can use this setting to customize the pause/unpause key.
listen-control-restart-key	2	You can use this setting to customize the replay key.

Option	Value/Example	Notes
listen-control-stop-key	13456789	You can use this setting to customize the interrupt playback key.
backupdeleted	0	This setting will allow you to specify how many deleted messages are automatically stored by the system. This is similar to a recycle bin. Setting this to 0 disables this feature. Up to 9999 messages can be stored, after which the oldest message will be erased each time another message is deleted.

[a] Yes, we found this a bit confusing too.

[b] Typically the spool folder is */var/spool/asterisk/*, and it can be defined in */etc/asterisk/asterisk.conf*.

The [zonemessages] Section

The next section of the *voicemail.conf* file is the [zonemessages] section. The purpose of this section is to allow time zone–specific handling of messages, so you can play back to the user messages with the correct timestamps. You can set the name of the zone to whatever you need. Following the zone name, you can define which time zone you want the name to refer to, as well as some options that define how timestamps are played back. You can look at the */usr/src/asterisk-complete/asterisk/1.8/configs/voicemail.conf.sample* file for syntax details. Asterisk includes the examples shown in Table 8-3.

Table 8-3. [zonemessages] section options for voicemail.conf

Zone name	Value/Example	Notes
eastern	America/New_York|'vm-received' Q 'digits/at' IMp	This value would be suitable for the eastern time zone (EST/EDT).
central	America/Chicago|'vm-received' Q 'digits/at' IMp	This value would be suitable for the central time zone (CST/CDT).
central24	America/Chicago|'vm-received' q 'digits/at' H N 'hours'	This value would also be suitable for CST/CDT, but would play back the time in 24-hour format.
military	Zulu|'vm-received' q 'digits/at' H N 'hours' 'phonetic/z_p'	This value would be suitable for Universal Time Coordinated (Zulu time, formerly GMT).
european	Europe/Copenhagen|'vm-received' a d b 'digits/at' HM	This value would be suitable for Central European time (CEST).

The Contexts Section

All the remaining sections in the *voicemail.conf* file will be the voicemail contexts, which allow you to segregate groups of mailboxes.

In many cases, you will only need one voicemail context, commonly named [default]. This is worth noting, as it will make things simpler in the dialplan: all the voicemail-related applications assume the context default if no context is specified. In other words, if you don't require separation of your voicemail users, use default as your one and only voicemail context.

The format for the mailboxes is as follows:

```
mailbox => password[,FirstName LastName[,email addr[,pager addr[,options[|options]]]]]
```

 The pipe character (|) used to be more popular in Asterisk. For the first few years, it was used as the standard delimiter. More recently, it has almost completely been replaced by the comma; however, there are still a few places where the pipe is used. One of them is in *voicemail.conf*: for example, as a separator for any mailbox-specific options, and also as the separator character in the format= declarative. You'll see this in our upcoming example, as well as in the *voicemail.conf.sample* file.

The parts of the mailbox definition are:

mailbox
> This is the mailbox number. It usually corresponds with the extension number of the associated set.

password
> This is the numeric password that the mailbox owner will use to access her voicemail. If the user changes her password, the system will update this field in the *voicemail.conf* file.

FirstName LastName
> This is the name of the mailbox owner. The company directory uses the text in this field to allow callers to spell usernames.

email address
> This is the email address of the mailbox owner. Asterisk can send voicemail notifications (including the voicemail message itself, as an attachment) to the specified email box.

pager address
> This is the email address of the mailbox owner's pager or cell phone. Asterisk can send a short voicemail notification message to the specified email address.

options
> This field is a list of options for setting the mailbox owner's time zone and overriding the global voicemail settings. There are nine valid options: **attach,**

serveremail, tz, saycid, review, operator, callback, dialout, and exitcontext. These options should be in *option* = *value* pairs, separated by the pipe character (|). The tz option sets the user's time zone to a time zone previously defined in the [zonemessages] section of *voicemail.conf*, and the other eight options override the global voicemail settings with the same names.

The mailboxes you define in your *voicemail.conf* file might look like the following examples:

```
[default]
100 => 5542,Mike Loukides,mike@shifteight.org
101 => 67674,Tim OReilly,tim@shifteight.org
102 => 36217,Mary JonesSmith,mary.jones-smith@shifteight.org

; *** This needs to all be on the same line
103 => 5426,Some Guy,,,dialout=fromvm|callback=fromvm
|review=yes|operator=yes|envelope=yes

[shifteight]
100 => 0107,Leif Madsen,leif@shifteight.org
101 => 0523,Jim VanMeggelen,jim@shifteight.org,,attach=no|maxmsg=100
102 => 11042,Tilghman Lesher,,,attach=no|tz=central
```

 The Asterisk directory cannot handle the concept of a family name that is anything other than a simple word. This means that family names such as *O'Reilly*, *Jones-Smith*, and yes, even *Van Meggelen* must have any punctuation characters and spaces removed before being added to *voicemail.conf*.

The contexts in *voicemail.conf* are an excellent and powerful concept, but you will likely find that the default context will be all that you need in normal use.

An Initial voicemail.conf File

We recommend the following sample as a starting point. You can refer to *~/asterisk-complete/asterisk/1.8/configs/voicemail.conf.sample* for details on the various settings:

```
; Voicemail Configuration

[general]
format=wav49|wav
serveremail=voicemail@shifteight.org
attach=yes
skipms=3000
maxsilence=10
silencethreshold=128
maxlogins=3
emaildateformat=%A, %B %d, %Y at %r
pagerdateformat=%A, %B %d, %Y at %r
sendvoicemail=yes ; Allow the user to compose and send a voicemail while inside
```

```
[zonemessages]
eastern=America/New_York|'vm-received' Q 'digits/at' IMp
central=America/Chicago|'vm-received' Q 'digits/at' IMp
central24=America/Chicago|'vm-received' q 'digits/at' H N 'hours'
military=Zulu|'vm-received' q 'digits/at' H N 'hours' 'phonetic/z_p'
european=Europe/Copenhagen|'vm-received' a d b 'digits/at' HM

[shifteight.org]
100 => 1234,Leif Madsen,leif@shifteight.org
101 => 1234,Jim Van Meggelen,jim@shifteight.org
102 => 1234,Russell Bryant,russell@shifteight.org
103 => 1234,Jared Smith,jared@shifteight.org
```

 Setting up a Linux server to handle the sending of email is a Linux administration task that is beyond the scope of this book. You will need to test your voicemail to email service to ensure that the email is being handled appropriately by the Mail Transfer Agent (MTA),‡ and that downstream spam filters are not rejecting the messages (one reason this might happen is if your Asterisk server is using a hostname in the email body that does not in fact resolve to it).

Dialplan Integration

There are two primary dialplan applications that are provided by the *app_voicemail.so* module in Asterisk. The first, simply named VoiceMail(), does exactly what you would expect it to, which is to record a message in a mailbox. The second one, VoiceMailMain(), allows a caller to log into a mailbox to retrieve messages.

The VoiceMail() Dialplan Application

When you want to pass a call to voicemail, you need to provide two arguments: the mailbox (or mailboxes) in which the message should be left, and any options relating to this, such as which greeting to play or whether to mark the message as urgent. The structure of the VoiceMail() command is this:

```
VoiceMail(mailbox[@context][&mailbox[@context][&...]][,options])
```

The options you can pass to VoiceMail() to provide a higher level of control are detailed in Table 8-4.

---

‡ Also sometimes called a Message Transfer Agent.

Table 8-4. VoiceMail() optional arguments

Argument	Purpose
b	Instructs Asterisk to play the busy greeting for the mailbox (if no busy greeting is found, the unavailable greeting will be played).
d([c])	Accepts digits to be processed by context c. If the context is not specified, it will default to the current context.
g(#)	Applies the specified amount of gain (in decibels) to the recording. Only works on DAHDI channels.
s	Suppresses playback of instructions to the callers after playing the greeting.
u	Instructs Asterisk to play the unavailable greeting for the mailbox (this is the default behavior).
U	Indicates that this message is to be marked as urgent. The most notable effect this has is when voicemail is stored on an IMAP server. In that case, the email will be marked as urgent. When the mailbox owner calls in to the Asterisk voicemail system, he should also be informed that the message is urgent.
P	Indicates that this message is to be marked as priority.

The `VoiceMail()` application sends the caller to the specified mailbox, so that he can leave a message. The mailbox should be specified as *mailbox@context*, where *context* is the name of the voicemail context. The option letters b or u can be added to request the type of greeting. If the letter b is used, the caller will hear the mailbox owner's *busy* message. If the letter u is used, the caller will hear the mailbox owner's *unavailable* message (if one exists).

Consider this simple example extension 101, which allows people to call John:

```
exten => 101,1,Dial(${JOHN})
```

Let's add an unavailable message that the caller will be played if John doesn't answer the phone. Remember, the second argument to the `Dial()` application is a timeout. If the call is not answered before the timeout expires, the call is sent to the next priority. Let's add a 10-second timeout, and a priority to send the caller to voicemail if John doesn't answer in time:

```
exten => 101,1,Dial(${JOHN},10)
exten => 101,n,VoiceMail(101@default,u)
```

Now, let's change it so that if John is busy (on another call), the caller will be sent to his voicemail, where he will hear John's busy message. To do this, we will make use of the `${DIALSTATUS}` variable, which contains one of several status values (type *core show application dial* at the Asterisk console for a listing of all the possible values):

```
exten => 101,1,Dial(${JOHN},10)
    same => n,GotoIf($["${DIALSTATUS}" = "BUSY"]?busy:unavail)
    same => n(unavail),VoiceMail(101@default,u)
    same => n,Hangup()
    same => n(busy),VoiceMail(101@default,b)
    same => n,Hangup()
```

Now callers will get John's voicemail (with the appropriate greeting) if John is either busy or unavailable. A slight problem remains, however, in that John has no way of retrieving his messages. We will remedy that in the next section.

The VoiceMailMain() Dialplan Application

Users can retrieve their voicemail messages, change their voicemail options, and record their voicemail greetings using the `VoiceMailMain()` application. `VoiceMailMain()` accepts two arguments: the mailbox number (and optionally the context) to be accessed, and some options. Both arguments are optional.

The structure of the `VoiceMailMain()` application looks like this:

```
VoiceMailMain([mailbox][@context][,options])
```

If you do not pass any arguments to `VoiceMailMain()`, it will play a prompt asking the caller to provide her mailbox number. The options that can be supplied are listed in Table 8-5.

Table 8-5. VoiceMailMain() optional arguments

Argument	Purpose
p	Allows you to treat the <*mailbox*> parameter as a prefix to the mailbox number.
g(#)	Increases the gain by # decibels when playing back messages.
s	Skips the password check.
a(*folder*)	Starts the session in one of the following voicemail folders (defaults to 0): 0 - INBOX, 1 - Old, 2 - Work, 3 - Family, 4 - Friends, 5 - Cust1, 6 - Cust2, 7 - Cust3, 8 - Cust4, 9 - Cust5

To allow users to dial 8500 to check their voicemail or modify their voicemail options, you would add an extension to the dialplan like this:

```
[Services]

exten => *98,1,VoiceMailMain()
```

Creating a Dial-by-Name Directory

One last feature of the Asterisk voicemail system that we should cover is the dial-by-name directory. This is created with the `Directory()` application. This application uses the names defined in the mailboxes in *voicemail.conf* to present the caller with a dial-by-name directory of the users.

`Directory()` takes up to three arguments: the voicemail context from which to read the names, the optional dialplan context in which to dial the user, and an option string (which is also optional). By default, `Directory()` searches for the user by last name, but passing the f option forces it to search by first name instead. Let's add two dial-by-name directories to the `incoming` context of our sample dialplan, so that callers can search by either first or last name:

```
exten => 8,1,Directory(default,incoming,f)
exten => 9,1,Directory(default,incoming)
```

If callers press 8, they'll get a directory by first name. If they dial 9, they'll get the directory by last name.

Using a Jitterbuffer

When using Asterisk as a voicemail server,[§] you may want to add a *jitterbuffer* in between voicemail and the caller. The purpose of a jitterbuffer is to help deal with the fact that when a call traverses an IP network, the traffic may not arrive with perfect timing and in perfect order. If packets occasionally arrive with a bit of delay (jitter) or if they arrive out of order, a jitterbuffer can fix it so that the voicemail system receives the voice stream on time and in order. If the jitterbuffer detects that a packet was lost (or may arrive so late that it will no longer matter), it can perform packet loss concealment. That is, it will attempt to make up a frame of audio to put in place of the lost audio to make it harder to hear that audio was lost.

In Asterisk, jitterbuffer support can only be enabled on a bridge between two channels. In the case of voicemail, there is generally only a single channel connected to one of the voicemail applications. To enable the use of a jitterbuffer in front of voicemail, we create a bridge between two channels by using a `Local` channel and specifying the j option. Specifying the n option for the `Local` channel additionally ensures that the `Local` channel is not optimized out of the call path in Asterisk:

```
[Services]

exten => *98,1,Dial(Local/vmm@Services/nj)

exten => vmm,1,VoiceMailMain()
```

Storage Backends

The storage of messages on a traditional voicemail system has always tended to be overly complicated.[‖] Asterisk, on the other hand, not only provides you with a simple, logical, filesystem-based storage mechanism, but also offers a few extra message storage options.

Linux Filesystem

By default, Asterisk will store voice messages in the spool folder, at */var/spool/asterisk/voicemail/<context>/<mailbox>*. The messages can be stored in multiple formats (such

---

[§] This advice applies to any situation where Asterisk is the endpoint of a call. Another example would be when using the `MeetMe()` or `ConfBridge()` applications for conferencing.

[‖] Nortel used to store its messages in a sort of special partition, in a proprietary format, which made it pretty much impossible to extract messages from the system, or email them, or archive them, or really do anything with them.

---

as WAV and GSM), depending on what you specified as the `format` in the `[general]` section of your *voicemail.conf* file. Your greetings will also be stored in this folder.

 Asterisk will not create a folder for any mailboxes that do not have any recordings yet (as would be the case with a new mailbox), so this folder cannot be used as a reliable method of determining which mailboxes exist on the system.

Here's an example of what might be in a mailbox folder. This mailbox has no new messages in the *INBOX*, has two saved messages in the *Old* folder, and has *busy* and *unavailable* greetings recorded:

```
/var/spool/asterisk/voicemail/default
./INBOX
./Old
./Old/msg0000.WAV
./Old/msg0000.txt
./Old/msg0001.WAV
./Old/msg0001.txt
./Urgent
./busy.WAV
./unavail
./unavail.WAV
```

 For each message, there is a matching *msg####.txt* file, which contains the envelope information for the message. The *msg####.txt* file is also critically important for message waiting indication (MWI), as this is the file that Asterisk looks for in the *INBOX* to determine whether the message light for a user should be on or off.

ODBC

In a centralized or distributed system, you may find it desirable to store messages as binary objects in a database, instead of as files on the filesystem. We'll discuss this in detail in "ODBC Voicemail" on page 378.

IMAP

Many people would prefer to manage their voicemail as part of their email. This has been called *unified messaging* by the telecom industry, and its implementation has traditionally been expensive and complex. Asterisk allows for a fairly simple integration between voicemail and email, either through its built-in voicemail to email handler, or through a relationship with an IMAP server. We'll discuss IMAP integration in detail in "VoiceMail IMAP Integration" on page 411.

Using Asterisk As a Standalone Voicemail Server

In a traditional telecom environment, the voicemail server was typically a standalone unit (provided either as a separate server altogether, or as an add-in card to the system). Very few PBXs had fully integrated voicemail (in the sense that voicemail was an integral part of the PBX rather than a peripheral device).

Asterisk is quite capable of serving as a standalone voicemail system. The two most common reasons one might want to do this are:

1. If you are building a large, centralized system and have several servers each providing a specific function (proxy server, media gateway, voicemail, conferencing, etc.)

2. If you wish to replace the voicemail system on a traditional PBX with an Asterisk voicemail·

Asterisk can serve in either of these roles.

Integrating Asterisk into a SIP Environment As a Standalone Voicemail Server

If you want to have Asterisk act as a dedicated voicemail server (i.e., with no sets registered to it and no other types of calls passing through it), the process from the dialplan perspective is quite simple. Getting message waiting to work can be a bit more difficult, though.

Let's start with a quick diagram. Figure 8-1 shows an overly simplified example of a typical SIP enterprise environment. We don't even have an Asterisk server in there (other than for the voicemail), in order to give you a generic representation of how Asterisk could serve as a standalone voicemail server in an otherwise non-Asterisk environment.

Unfortunately, Asterisk cannot send message notification to an endpoint if it doesn't know where that endpoint is. In a typical Asterisk system, where set registration and voicemail are handled on the same machine, this is never a problem, since Asterisk knows where the sets are. But in an environment where the sets are not registered to Asterisk, this can become a complex problem.

There are several solutions on the Internet that recommend using the `externnotify` option in *voicemail.conf*, triggering an external script whenever a message is left in a mailbox (or deleted). While we can't say that's a bad approach, we find it a bit kludgy, and it requires the administrator to understand how to write an external script or program to handle the actual passing of the message.

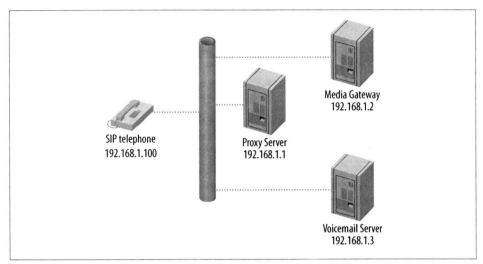

Figure 8-1. Simplified SIP enterprise environment

Instead you can statically define an entry for each mailbox in the voicemail server's *sip.conf* file, indicating where the message notifications are to be sent. Rather than defining the address of each endpoint, however, you can have the voicemail server send all messages to the proxy, which will handle the relay of the message notifications to the appropriate endpoints.

The voicemail server still needs to know about the SIP endpoints, even though the devices are not registered directly to it. This can be done either through a *sip.conf* file that identifies each SIP endpoint, or through a static real-time database that does the same thing. Whether you use *sip.conf* or the Asterisk Realtime Architecture (ARA), each endpoint will require an entry similar to this:

```
[messagewaiting](!)              ; a template to handle the settings common
                                 ; to all mailboxes
type=peer
subscribecontext=voicemailbox    ; the dialplan context on the voicemail server
context=voicemailbox             ; the dialplan context on the voicemail server
host=192.168.1.1                 ; ip address of presence server

[0000FFFF0001](messagewaiting)   ; this will need to match the subscriber
                                 ; name on the proxy
mailbox=0000FFFF0001@DIR1        ; this has to be in the form mailbox@mailboxcontext
defaultuser=0000FFFF0001         ; this will need to match the subscriber
                                 ; name on the proxy
```

Note that Asterisk's dynamic realtime will not work with this configuration, as a peer's information is only loaded into memory when there is an actual call involving that peer. Since message notification is not a call as far as Asterisk is concerned, using dynamic realtime will not allow message waiting to happen for any peers not registered to Asterisk.

You will *not* want to implement this unless you have prototyped the basic operation of the solution. Although we all agree that SIP is a protocol, not everyone agrees as to the correct way to implement the protocol. As a result, there are many interoperability challenges that need to be addressed in a solution like this. We have provided a basic introduction to this concept in this book, but the implementation details will depend on other factors external to Asterisk, such as the capabilities of the proxy.

The fact that no device has to register with Asterisk will significantly reduce the load on the Asterisk server, and as a result this design should allow for a voicemail server that can support several thousand subscribers.

Dialplan requirements

The dialplan of the voicemail server can be fairly simple. Two needs must be satisfied:

1. Receive incoming calls and direct them to the appropriate mailbox
2. Handle incoming calls from users wishing to check their messages

The system that is passing calls to the voicemail server should set some SIP headers in order to pass additional information to the voicemail server. Typically, this information would include the mailbox/username that is relevant to the call. In our example, we are going to set the headers X-Voicemail-Mailbox and X-Voicemail-Context, which will contain information we wish to pass to the voicemail server.#

If the source system is also an Asterisk system, you might set the headers using the SIPAddHeader() voicemail application, in a manner similar to this:

```
exten => sendtovoicemail,1,Verbose(2,Set SIP headers for voicemail)
    same => n,SipAddHeader(X-Voicemail-Mailbox: <mailbox number>)
    same => n,SipAddHeader(X-Voicemail-Context: voicemailbox)
```

Note that this dialplan does not go on the voicemail server. It would only be useful if one of the other servers in your environment was also an Asterisk server. If you were using a different kind of server, you would need to find out how to set custom headers in that platform, or find out if it already uses specific headers for this sort of thing, and possibly modify the dialplan on the voicemail server to handle those headers.

#As far as we know, there aren't any specific SIP headers that are standardized for this sort of thing, so you should be able to name the headers whatever you want. We chose these header names simply because they make some sort of sense. You may find that other headers would suit your needs better.

The voicemail server will need an *extensions.conf* file containing the following:

```
[voicemailbox]
; direct incoming calls to a mailbox
exten => Deliver,1,NoOp()
    same => n,Set(Mailbox=${SIP_HEADER(X-Voicemail-Mailbox)})
    same => n,Set(MailboxContext=${SIP_HEADER(X-Voicemail-Context)})
    same => n,VoiceMail(${Mailbox}@${MailboxContext})
    same => n,Hangup()

; connect users to their mailbox so that they can retrieve messages exten =>
Retrieve,1,NoOp()
    same => n,Set(Mailbox=${SIP_HEADER(X-Voicemail-Mailbox)})
    same => n,Set(MailboxContext=${SIP_HEADER(X-Voicemail-Context)})
    same => n,VoiceMailMain(${Mailbox}@${MailboxContext})
    same => n,Hangup()
```

sip.conf requirements

In the *sip.conf* file on the voicemail server, not only are entries required for all the mailboxes for message waiting notification, but some sort of entry is required to define the connection between the voicemail server and the rest of the SIP environment:

```
[VOICEMAILTRUNK]
type=peer
defaultuser=voicemail
fromuser=voicemail
secret=sOm3th1ngs3cur3
canreinvite=no
host=<address of proxy/registrar server>
disallow=all
allow=ulaw
dtmfmode=rfc2833
context=voicemailbox
```

The other end of the connection (probably your proxy server) must be configured to pass voicemail connections to the voicemail server.

Running Asterisk as a standalone voicemail server requires some knowledge of clustering and integration, but you can't beat the price.

SMDI (Simplified Message Desk Interface)

The Simplified Message Desk Interface (SMDI) protocol is intended to allow communication of basic message information between telephone systems and voicemail systems.

Asterisk supports SMDI, but given that this is an old protocol that runs across a serial connection, there are likely to be integration challenges. Support in various PBXs and other devices may be spotty. Still, it's a fairly simple protocol, so for sure it's worth testing out if you are considering using Asterisk as a voicemail replacement.

The following is not a detailed explanation of how to configure SMDI for Asterisk, but rather an introduction to the concepts, with some basic examples. If you are planning on implementing SMDI, you will need to write some complex dialplan logic and have a good understanding of how to interconnect systems via serial connections.

SMDI is enabled in Asterisk by the use of two options in the [general] section of the *voicemail.conf* file:

```
smdienable=yes
smdiport=/dev/ttyS0; or whatever serial port you are connecting your SMDI service to
```

Additionally, you will need an *smdi.conf* file in your */etc/asterisk* folder to define the details of your SMDI configuration. It should look something like this (see the *smdi.conf.sample* file for more information on the available options):

```
[interfaces]
charsize=7
paritybit=even
baudrate=1200           ; hopefully a higher bitrate is supported
smdiport=/dev/ttyS0     ; or whatever serial port you'll be using to handle
                        ; SMDI messages on asterisk

[mailboxes]             ; map incoming digit strings (typically DID numbers)
                        ; to a valid mailbox@context in voicemail.conf
smdiport=/dev/ttyS0     ; first declare which SMDI port the following mailboxes
                        ; will use
4169671111=1234@default
4165551212=9999@default
```

In the dialplan there are two functions that will be wanted in an SMDI configuration. The SMDI_MSG_RETRIEVE() function pulls the relevant message from the SMDI message queue. You need to pass the function a search key (typically the DID that is referred to in the message), and it will pass back an ID number that can be referenced by the SMDI_MSG() function:

```
SMDI_MSG_RETRIEVE(<smdi port>,<search key>[,timeout[,options]])
```

Once you have the SMDI message ID, you can use the SMDI_MSG() function to access various details about the message, such as the station, callerID, and type (the SMDI message type):

```
SMDI_MSG(<message_id>,<component>)
```

In your dialplan, you will need to handle the lookup of the SMDI messages that come in, in order to ensure that calls are handled correctly. For example, if an incoming call is intended for delivery to a mailbox, the message type might be one of B (for busy) or N (for unanswered calls). If, on the other hand, the call is intended to go to VoiceMail Main() because the caller wants to retrieve his messages, the SMDI message type would be D, and that would have to be handled.

Conclusion

While the Asterisk voicemail system is quite old in terms of Asterisk code, it is never-theless a powerful application that can (and does) compete quite successfully with expensive, proprietary voicemail systems.

Internationalization

David Duffett

> *I traveled a good deal all over the world, and I got along*
> *pretty good in all these foreign countries, for I have a*
> *theory that it's their country and they got a right*
> *to run it like they want to.*
>
> —Will Rogers

Telephony is one of those areas of life where, whether at home or at work, people do not like surprises. When people use phones, anything outside of the norm is an expectation not met, and as someone who is probably in the business of supplying telephone systems, you will know that expectations going unmet can lead to untold misery in terms of the extra work, lost money, and so forth that are associated with customer dissatisfaction.

In addition to ensuring that the user experience is in keeping with what users expect, there is also the need to make your Asterisk feel "at home." For example, if an outbound call is placed over an analog line (FXO), Asterisk will need to interpret the tones that it "hears" on the line (busy, ringing, etc.).

By default (and maybe as one might expect since it was "born in the USA"), Asterisk is configured to work within North America. However, since Asterisk gets deployed in many places and (thankfully) people from all over the world make contributions to it, it is quite possible to tune Asterisk for correct operation just about anywhere you choose to deploy it.

If you have been reading this book from the beginning, chapter by chapter, you will have already made some choices during the process of installation and initial configuration that will have set up your Asterisk to work in your local area (and live up to your customers' expectations).

Quite a few of the chapters in this book contain information that will help you internationalize* or (perhaps more properly) localize your Asterisk implementation. The purpose of this chapter is to provide a single place where all aspects of the changes that need to be made to your Asterisk-based telephone system in this context can be referenced, discussed, and explained. The reason for using the phrase "Asterisk-based telephone system" rather than just "Asterisk" is that some of the changes will need to be made in other parts of the system (IP phones, ATAs, etc.), while other changes will be implemented within Asterisk and DAHDI configuration files.

Let's start by getting a list together (in no particular order) of the things that may need to be changed in order to optimise your Asterisk-based telephone system for a given location outside of North America. You can shout some out if you like...

- Language/accent of the prompts
- Physical connectorization for PSTN interfaces (FXO, BRI, PRI)
- Tones heard by users of IP phones and/or ATAs
- Caller ID format sent and/or received by analog interfaces
- Tones for analog interfaces to be supplied or detected by Asterisk
- Format of time/date stamps for voicemail
- The way the above time/date stamps are announced by Asterisk
- Patterns within the dialplan (of IP phones, ATAs, and Asterisk itself if you are using the sample dialplan)
- The way to indicate to an analog device that voicemail is waiting (MWI)
- Tones supplied to callers by Asterisk (these come into play once a user is "inside" the system; e.g., the tones heard during a call transfer)

We'll cover everything in this list, adopting a strategy of working from the outer edge of the system toward the very core (Asterisk itself). We will conclude with a handy checklist of what you may need to change and where to change it.

Although the principles discussed in this chapter will allow you to adapt your Asterisk specifically for your region (or that of your customer), for the sake of consistency all of our examples will focus on how to adapt Asterisk for one region: the United Kingdom.

Devices External to the Asterisk Server

There are massive differences between a good old fashioned analog telephone and any one of the large number of IP phones out there, and we need to pick up on one of the

---

* *i18n* is a term used to abbreviate the word *internationalization*, due to its length. The format is *<first_letter><number><last_letter>*, where *<number>* is the number of letters between the first and last letters. Other words, such as localization (L10n), modularization (m12n), etc. have also found a home with this scheme, which Leif finds a little bit ridiculous. More information can be found here: *http://www.w3.org/2001/12/Glossary#I18N*.

really fundamental differences in order to throw light on the next explanation, which covers the settings we might need to change on devices external to Asterisk, such as IP phones.

Have you ever considered the fact that an analog phone is a totally dumb device (we know that a basic model is very, very cheap) that needs to connect to an intelligent network (the PSTN), whereas an IP phone (e.g., SIP or IAX2) is a very intelligent device that connects to a dumb network (the Internet, or any regular IP network)? Figures 9-1 and 9-2 illustrate the difference.

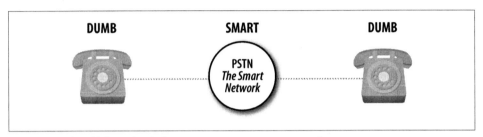

Figure 9-1. The old days: dumb devices connect to a smart network

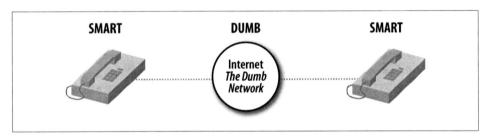

Figure 9-2. The situation today: smart devices connect through a dumb network

Could we take two analog phones, connect them directly to each other and have the functionality we would normally associate with a regular phone? No, of course not, because the network supplies everything: the actual power to the phone, the dialtone (from the local exchange or CO), the caller ID information, the ringing tone (from the remote [closest to the destination phone] exchange or CO), all the signaling required, and so on.

Conversely, could we take two IP phones, connect them directly to each other, and get some sensible functionality? Sure we could, because all the intelligence is inside the IP phones themselves—they provide the tones we hear (dialtone, ringing, busy) and run the protocol that does all the required signaling (usually SIP). In fact, you can try this for yourself—most mid-price IP phones have a built-in Ethernet switch, so you can actually connect the two IP phones directly to each other with a regular (straight-through) Ethernet patch cable, or just connect them through a regular switch. They will need to have fixed IP addresses in the absence of a DHCP server, and you can usually dial the IP address of the other phone just by using the * key for the dots in the address.

Figure 9-2 points to the fact that on an IP phone, we are responsible for setting all of the tones that the network would have provided in the old days. This can be done in one of (at least) two ways. The first is to configure the tones provided by the IP phone on the device's own web GUI. This is done by browsing to the IP address of the phone (the IP address can usually be obtained by a menu option on the phone) and then selecting the appropriate options. For example, on a Yealink IP phone, the tones are set on the *Phone* page of the web GUI, under the *Tones* tab (where you'll find a list of the different types of tone that can be changed—in the case of the Yealink, these are Dial, Ring Back, Busy, Congestion, Call Waiting, Dial Recall, Record, Info, Stutter, Message, and Auto Answer).

The other way that this configuration can be applied is to auto-provision the phone with these settings. A full explanation of the mechanism for auto-provisioning is beyond the scope of this book, but you can usually set up the tones in the appropriate attributes of the relevant elements in the XML file.

While we are changing settings on the IP phones, there are two other things that may need to be changed in order for the phones to look right and to function correctly as part of the system.

Most phones display the time when idle and, since many people find it particularly annoying when their phones show the wrong time, we need to ensure that the correct local time is displayed. It should be fairly easy to find the appropriate page of the web GUI (or XML attributes) to specify the time server. You will also find that there are settings for daylight saving time and other relevant stuff nearby.

The last thing to change is a potential show-stopper as far as the making of a phone call is concerned—the dialplan. We're not talking about the dialplan we find in */etc/asterisk/extensions.conf*, but the dialplan of the phone. Not everyone realizes that IP phones have dialplans too—although these dialplans are more concerned with which dial strings are permitted than with what to do on a given dial.

The general rule seems to be that if you dial on-hook the built-in dialplan is bypassed, but if you pick up the handset the dialplan comes into play, and it just might happen that the dialplan will not allow the dial string you need to be dialed. Although this problem can manifest itself with a refusal by the phone to pass certain types of numbers through to Asterisk, it can also affect any feature codes you plan to use. This can easily be remedied by Googling the model number of the phone along with "UK dialplan" (or the particular region you need), or you can go to the appropriate page on the web GUI and either manually adjust the dialplan or pick the country you need from a drop-down box (depending on the type of phone you are working with).

The prior discussion of IP phone configuration also applies to any analog telephone adaptors (ATAs) you plan to use—specifically, to those supporting an FXS interface. In addition, you may need to specify some of the electrical characteristics of the telephony interface, like line voltage and impedance, together with the caller ID format that will work with local phones. All that differs is the way you obtain the IP address for the web GUI—this is usually done by dialing a specific code on the attached analog phone, which results in the IP address being read back to the caller.

Of course, an ATA may also feature an FXO interface, which will also need to be configured to properly interact with the analog line provided in your region. The types of things that need to be changed are similar to the FXS interface.

What if you are connecting your analog phone or line to a Digium card? We'll cover this next.

PSTN Connectivity, DAHDI, Digium Cards, and Analog Phones

Before we get to DAHDI and Asterisk configuration, we need to physically connect to the PSTN. Unfortunately, there are no worldwide standards for these connections; in fact, there are often variations from one part of a given country to another.

PRI connections are generally terminated in an RJ45 connection these days, although the impedance of the connections can vary. In some countries (notably in South America), it is still possible to find PRIs terminated in two BNC connectors, one for transmit and one for receive.

Generally speaking, a PRI terminated in an RJ45 will be an ISDN connection, and if you find the connection is made by a pair of BNC connectors (push-and-twist coaxial connectors), the likelihood is that you are dealing with a CAS-based protocol (like R2).

Figure 9-3 shows the adaptor required if your telco has supplied BNC connectors (the Digium cards require an RJ45 connection). It is called a *balun*, as it converts from a balanced connection (RJ45) to an unbalanced connection (the BNCs), in addition to changing the connection impedance.

 Basic Rate Interfaces (BRIs) are common in continental Europe and are almost always supplied via an RJ45 connection.

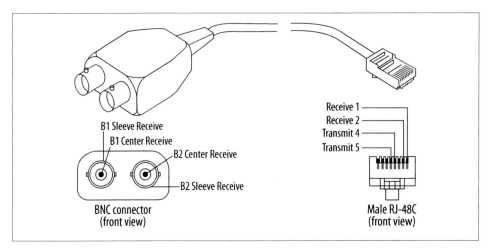

Figure 9-3. A balun

Analog connections vary massively from place to place—you will know what kind of connector is used in your locality. The important thing to remember is that the analog line is only two wires, and these need to connect to the middle two pins of the RJ11 plug that goes into the Digium card—the other end is the local one. Figure 9-4 shows the plug used in the UK, where the two wires are connected to pins 2 and 5.

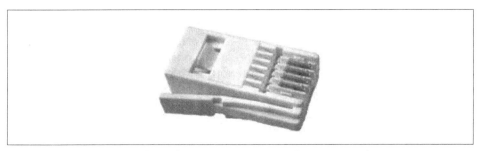

Figure 9-4. The BT plug used for analog PSTN connections in the UK (note only pins 2–5 are present)

The Digium Asterisk Hardware Device Interface, or DAHDI, actually covers a number of things. It contains the kernel drivers for telephony adaptor cards that work within the DAHDI framework, as well as automatic configuration utilities and test tools. These parts are contained in two separate packages (*dahdi-linux* and *dahdi-tools*), but we can also use one complete package, called *dahdi-linux-complete*. All three packages are available at *http://downloads.digium.com/pub/telephony/*. The installation of DAHDI was covered in Chapter 3.

Chapter 7 covered the use of analog and digital PSTN connections, and we will not reiterate those details here. If you are using digital PSTN connections, your job is to find out what sort of connection the telco is giving you. Generally, if you have requested a primary rate interface (PRI), this will be a T1 in North America, a J1 in Japan, or an E1 in pretty much the rest of the world.

Once you have established the type of PRI connection the telco has given you, there are some further details that you will require in order to properly configure DAHDI and Asterisk (e.g., whether the connection is ISDN or a CAS-based protocol). Again, you will find these in Chapter 7.

DAHDI Drivers

The connections where some real localization will need to take place are those of analog interfaces. For the purposes of configuring your Asterisk-based telephone system to work best in a given locality, you will first need to specifically configure some low-level aspects of the way the Digium card interacts with the connected device or line. This is done through the DAHDI kernel driver(s), in a file called */etc/dahdi/system.conf*.

In the following lines (taken from the sample configuration that you get with a fresh install of DAHDI), you will find both the `loadzone` and `defaultzone` settings. The `load zone` setting allows you to choose which tone set(s) the card will both generate (to feed to analog telephones) and recognize (on the connected analog telephone lines):

```
# Tone Zone Data
# ^^^^^^^^^^^^^^
# Finally, you can preload some tone zones, to prevent them from getting
# overwritten by other users (if you allow non-root users to open /dev/dahdi/*
# interfaces anyway). Also this means they won't have to be loaded at runtime.
# The format is "loadzone=<zone>" where the zone is a two letter country code.
#
# You may also specify a default zone with "defaultzone=<zone>" where zone
# is a two letter country code.
#
# An up-to-date list of the zones can be found in the file zonedata.c
#
loadzone = us
#loadzone = us-old
#loadzone=gr
#loadzone=it
#loadzone=fr
#loadzone=de
#loadzone=uk
#loadzone=fi
#loadzone=jp
#loadzone=sp
#loadzone=no
#loadzone=hu
#loadzone=lt
#loadzone=pl
defaultzone=us
#
```

The */etc/dahdi/system.conf* file uses the hash symbol (#) to indicate a comment instead of a semicolon (;) like the files in */etc/asterisk/*.

Although it is possible to load a number of different tone sets (you can see all the sets of tones in detail in *zonedata.c*) and to switch between them, in most practical situations you will only need:

```
loadzone=uk     # to load the tone set
defaultzone=uk  # to default DAHDI to using that set
```

...or whichever tones you need for your region.

If you perform a *dahdi_genconf* to automatically (or should that be auto-magically?) configure your DAHDI adaptors, you will notice that the newly generated */etc/dahdi/system.conf* will have defaulted both loadzone and defaultzone to being us. Despite the warnings not to hand-edit the file, it is fine to change these settings to what you need.

In case you were wondering how we tell whether there are any voicemails in the mailbox associated with the channel an analog phone is plugged into, it is done with a stuttered dialtone. The format of this stuttered dialtone is decided by the loadzone/default zone combination you have used.

As a quick aside, analog phones that have a message waiting indicator (e.g., an LED or lamp that flashes to indicate there is new voicemail) achieve this by automatically going off-hook periodically and *listening* for the stuttered dialtone. You can witness this by watching the Asterisk command line to see the DAHDI channel go active (if you have nothing better to do!).

That's it at the DAHDI level. We chose the protocol(s) for PRI or BRI connections, the type of signaling for the analog channels (all covered in Chapter 7), and the tones for the analog connections that have just been discussed.

Once you have completed your configuration at the DAHDI level (in */etc/dahdi/system.conf*), you need to perform a *dahdi_cfg -vvv* to have DAHDI reread the configuration. This is also a good time to use *dahdi_tool* to check that everything appears to be in order at the Linux level.

This way, if things do not work properly after you have configured Asterisk to work with the DAHDI adaptors, you can be sure that the problem is confined to *chan_dahdi.conf* (or an *#included dahdi-channels.conf* if you are using this part of the *dahdi_genconf* output).

The relationship between Linux, DAHDI, and Asterisk (and therefore */etc/dahdi/system.conf* and */etc/asterisk/chan_dahdi.conf*) is shown in Figure 9-5.

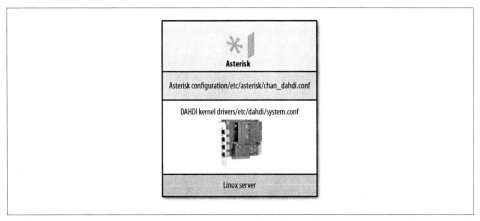

Figure 9-5. The relationship between Linux, DAHDI, and Asterisk

Asterisk

With everything set at the Linux level, we now only need to configure Asterisk to make use of the channels we just enabled at the Linux level and to customize the way that Asterisk interprets and generates information that comes in from, or goes out over, these channels. This work is done in */etc/asterisk/chan_dahdi.conf*.

In this file we will not only tell Asterisk what sort of channels we have (these settings will fit with what we already did in DAHDI), but also configure a number of things that will ensure Asterisk is well suited to its new home.

Caller ID

A key component of this change is caller ID. While caller ID delivery methods are pretty much standard within the BRI and PRI world, they vary widely in the analog world; thus, if you plugged an American analog phone into the UK telephone network, it would actually work as a phone, but caller ID information would not be displayed. This is because that information is transmitted in different ways in different places around the world, and an American phone would be looking for caller ID signaling in the US format, while the UK telephone network would be supplying it (if it is enabled— it is not standard in the UK; you have to pay for caller ID!) in the UK format.

Not only is the format different, but the method of telling a telephone (or Asterisk) to look out for the caller ID may vary from place to place too. This is important, as we do not want Asterisk to waste time looking for caller ID information if it is not being presented on the line.

Again, Asterisk defaults to the North American caller ID format (no entries in */etc/asterisk/chan_dahdi.conf* describe this, it's just the default), and in order to change it we will need to make some entries that describe the technical details of the caller ID system. In the case of the UK, the delivery of caller ID information is signaled by a polarity reversal on the telephone line (in other words, the A and B legs of the pair of telephone wires are temporarily switched over), and the actual caller ID information is delivered in a format known as V.23 (frequency shift keying, or FSK). So, the entries in *chan_dahdi.conf* to receive UK-style caller ID on any FXO interfaces will look like this:

```
cidstart=polarity      ; the delivery of caller ID will be
                       ; signaled by a polarity reversal
cidsignalling=v23      ; the delivery of the called ID information
                       ; will be in V23 format
```

Of course, you may also need to send caller ID using the same local signaling information to any analog phones that are connected to FXS interfaces, and one more entry may be needed as in some locations the caller ID information is sent after a specified number of rings. If this is the case, you can use this entry:

```
sendcalleridafter=2
```

Before you can make these entries, you will need to establish the details of your local caller ID system (someone from your local telco or Google could be your friend here, but there is also some good information in the sample */etc/asterisk/chan_dahdi.conf* file).

Language and/or Accent of Prompts

As you may know, the prompts (or recordings) that Asterisk will use are stored in */var/lib/asterisk/sounds/*. In older versions of Asterisk all the sounds were in this actual directory, but these days you will find a number of subdirectories that allow the use of different languages or accents. The names of these subdirectories are arbitrary; you can call them whatever you want.

Note that the filenames in these directories must be what Asterisk is expecting—for example, in */var/lib/asterisk/sound/en/* the file *hello.gsm* would contain the word "Hello" (spoken by the lovely Allison), whereas *hello.gsm* in */var/lib/asterisk/sounds/es/* (for Spanish in this case) would contain the word "Hola" (spoken by the Spanish equivalent of the lovely Allison†).

---

† Who is, in fact, the same Allison who does the English prompts; June Wallack does the French prompts. The male Australian-accented prompts are done by Cameron Twomey. All voiceover talent are available to record additional prompts as well. See *http://www.digium.com/en/products/ivr/* for more information.

The default directory used is */var/lib/asterisk/sounds/en*, so how do you change that?

There are two ways. One is to set the language in the channel configuration file that calls are arriving through using the language directive. For example, the line:

```
language=en_UK
```

placed in *chan_dahdi.conf*, *sip.conf*, and so on (to apply generally, or for just a given channel or profile) will tell Asterisk to use sound files found in */var/lib/asterisk/sounds/en_UK* (which could contain British-accented prompts) for all calls that come in through those channels.

The other way is to change the language during a phone call through the dialplan. This (along with many attributes of an individual call) can be set using the CHANNEL() dialplan function. See Chapter 10 for a full treatment of dialplan functions.

The following example would allow the caller to choose one of three languages in which to continue the call:

```
; gives the choice of (1) French, (2) Spanish, or (3) German
exten => s,1,Background(choose-language)
    same => n,WaitExten(5)

exten => 1,1,Set(CHANNEL(language)=fr)

exten => 2,1,Set(CHANNEL(language)=es)

exten => 3,1,Set(CHANNEL(language)=de)

; the next priority for extensions 1, 2, or 3 would be handled here
exten => _[123],n,Goto(menu,s,1)
```

If the caller pressed 1 sounds would be played from */var/lib/asterisk/sounds/fr*, if he pressed 2 the sounds would come from */var/lib/asterisk/sounds/es*, and so on.

As already mentioned, the names of these directories are arbitrary and do not need to be only two characters long—the main thing is that you match the name of the subdirectory you have created in the language directive in the channel configuration, or when you set the CHANNEL(language) argument in the dialplan.

Time/Date Stamps and Pronunciation

Asterisk uses the Linux system time from the host server, as you would expect, but we may have users of the system who are in different time zones, or even in different countries. Voicemail is where the rubber hits the road, as this is where users come into contact with time/date stamp information.

Consider a scenario where some users of the system are based in the US, while others are in the UK.

As well as the time difference, another thing to consider is the way people in the two locations are used to hearing date and time information—in the US, dates are usually ordered month, day, year and times are specified in 12-hour clock format (e.g., 2:54 P.M.).

In contrast, in the UK, dates are ordered day, month, year and times are often specified in 24-hour clock format (14:54 hrs)—although some people in the UK prefer 12-hour clock format, so we will cover that too.

Since all these things are connected to voicemail, you would be right to guess that we configure it in */etc/asterisk/voicemail.conf*—specifically, in the [zonemessages] section of the file.

Here is the [zonemessages] part of the sample *voicemail.conf* file that comes with Asterisk, with UK24 (for UK people that like 24-hour clock format times) and UK12 (for UK people that prefer 12-hour clock format) zones added:

```
[zonemessages]
; Users may be located in different timezones, or may have different
; message announcements for their introductory message when they enter
; the voicemail system. Set the message and the timezone each user
; hears here. Set the user into one of these zones with the tz=attribute
; in the options field of the mailbox. Of course, language substitution
; still applies here so you may have several directory trees that have
; alternate language choices.
;
; Look in /usr/share/zoneinfo/ for names of timezones.
; Look at the manual page for strftime for a quick tutorial on how the
; variable substitution is done on the values below.
;
; Supported values:
; 'filename' filename of a soundfile (single ticks around the filename
; required)
; ${VAR} variable substitution
; A or a Day of week (Saturday, Sunday, ...)
; B or b or h Month name (January, February, ...)
; d or e numeric day of month (first, second, ... thirty-first)
; Y Year
; I or l Hour, 12 hour clock
; H Hour, 24 hour clock (single digit hours preceded by "oh")
; k Hour, 24 hour clock (single digit hours NOT preceded by "oh")
; M Minute, with 00 pronounced as "o'clock"
; N Minute, with 00 pronounced as "hundred" (US military time)
; P or p AM or PM
; Q "today", "yesterday" or ABdY
; (*note: not standard strftime value)
; q " (for today), "yesterday", weekday, or ABdY
; (*note: not standard strftime value)
; R 24 hour time, including minute
;
eastern=America/New_York|'vm-received' Q 'digits/at' IMp
central=America/Chicago|'vm-received' Q 'digits/at' IMp
central24=America/Chicago|'vm-received' q 'digits/at' H N 'hours'
military=Zulu|'vm-received' q 'digits/at' H N 'hours' 'phonetic/z_p'
european=Europe/Copenhagen|'vm-received' a d b 'digits/at' HM
```

```
UK24=Europe/London|'vm-received' q 'digits/at' H N 'hours'
UK12=Europe/London|'vm-received' Q 'digits/at' IMp
```

These zones not only specify a time, but also dictate the way times and dates are ordered and read out.

Having created these zones, we can go to the voicemail context part of *voicemail.conf* to associate the appropriate mailboxes with the correct zones:

```
[default]
4001 => 1234,Russell Bryant,rb@shifteight.org,,|tz=central
4002 => 4444,David Duffett,dd@shifteight.org,,|tz=UK24
4003 => 4450,Mary Poppins,mp@shifteight.org,,|tz=UK12|attach=yes
```

As you can see, when we declare a mailbox, we also (optionally) associate it with a particular zone. Full details on voicemail can be found in Chapter 8.

The last thing to localize in our Asterisk configuration is the tones played to callers by Asterisk once they are inside the system (e.g., the tones a caller hears during a transfer).

As identified earlier in this chapter, the initial tones that people hear when they are calling into the system will come from the IP phone, or from DAHDI for analog channels.

These tones are set in */etc/asterisk/indications.conf*. Here is a part of the sample file, where you can see a given region specified by the country directive. We just need to change the country code as appropriate:

```
;
; indications.conf
;
; Configuration file for location specific tone indications
;
; NOTE:
; When adding countries to this file, please keep them in alphabetical
; order according to the 2-character country codes!
;
; The [general] category is for certain global variables.
; All other categories are interpreted as location specific indications
;
[general]
country=uk    ; default is US, so we have changed it to UK
```

Your dialplan will need to reflect the numbering scheme for your region. If you do not already know the scheme for your area, your local telecoms regulator will usually be able to supply details of the plan. Also, the example dialplan in */etc/asterisk/extensions.conf* is, of course, packed with North American numbers and patterns.

Conclusion—Easy Reference Cheat Sheet

As you can now see, there are quite a few things to change in order to fully localize your Asterisk-based telephone system, and not all of them are in the Asterisk, or even DAHDI, configuration—some things need to be changed on the connected IP phones or ATAs themselves.

Before we leave the chapter, have a look at Table 9-1: a cheat sheet for what to change and where to change it, for your future reference.

Table 9-1. Internationalization cheat sheet

What to change	Where to change it
Call progress tones	• IP phones—on the phone itself • ATAs—on the ATA itself • Analog phones—DAHDI (*/etc/dahdi/system.conf*)
Type of PRI/BRI and protocol	DAHDI—*/etc/dahdi/system.conf* and */etc/asterisk/chan_dahdi.conf*
Physical PSTN connections	• Balun if required for PRI • Get the analog pair to middle 2 pins of the RJ11 connecting to the Digium card
Caller ID on analog circuits	Asterisk—*/etc/asterisk/chan_dahdi.conf*
Prompt language and/or accent	• Channel—*/etc/asterisk/sip.conf*, */etc/asterisk/iax.conf*, */etc/asterisk/chan_dahdi.conf*, etc. • Dialplan—CHANNEL(language) function
Voicemail time/date stamps and pronunciation	Asterisk—*/etc/asterisk/voicemail.conf*
Tones delivered by Asterisk	Asterisk—*/etc/asterisk/indications.conf*

May all your Asterisk deployments feel at home...

Deeper into the Dialplan

*For a list of all the ways technology has failed to improve
the quality of life, please press three.*

—Alice Kahn

Alrighty. You've got the basics of dialplans down, but you know there's more to come. If you don't have Chapter 6 sorted out yet, please go back and give it another read. We're about to get into more advanced topics.

Expressions and Variable Manipulation

As we begin our dive into the deeper aspects of dialplans, it is time to introduce you to a few tools that will greatly add to the power you can exercise in your dialplan. These constructs add incredible intelligence to your dialplan by enabling it to make decisions based on different criteria you define. Put on your thinking cap, and let's get started.

Basic Expressions

Expressions are combinations of variables, operators, and values that you string together to produce a result. An expression can test values, alter strings, or perform mathematical calculations. Let's say we have a variable called COUNT. In plain English, two expressions using that variable might be "COUNT plus 1" and "COUNT divided by 2." Each of these expressions has a particular result or value, depending on the value of the given variable.

In Asterisk, expressions always begin with a dollar sign and an opening square bracket and end with a closing square bracket, as shown here:

```
$[expression]
```

Thus, we would write our two examples like this:

```
$[${COUNT} + 1]
$[${COUNT} / 2]
```

When Asterisk encounters an expression in a dialplan, it replaces the entire expression with the resulting value. It is important to note that this takes place *after* variable substitution. To demonstrate, let's look at the following code*:

```
exten => 321,1,Set(COUNT=3)
    same => n,Set(NEWCOUNT=$[${COUNT} + 1])
    same => n,SayNumber(${NEWCOUNT})
```

In the first priority, we assign the value of 3 to the variable named COUNT.

In the second priority, only one application—Set()—is involved, but three things actually happen:

1. Asterisk substitutes ${COUNT} with the number 3 in the expression. The expression effectively becomes this:

   ```
   exten => 321,n,Set(NEWCOUNT=$[3 + 1])
   ```

2. Asterisk evaluates the expression, adding 1 to 3, and replaces it with its computed value of 4:

   ```
   exten => 321,n,Set(NEWCOUNT=4)
   ```

3. The Set() application assigns the value 4 to the NEWCOUNT variable

The third priority simply invokes the SayNumber() application, which speaks the current value of the variable ${NEWCOUNT} (set to the value 4 in priority two).

Try it out in your own dialplan.

Operators

When you create an Asterisk dialplan, you're really writing code in a specialized scripting language. This means that the Asterisk dialplan—like any programming language—recognizes symbols called *operators* that allow you to manipulate variables. Let's look at the types of operators that are available in Asterisk:

Boolean operators
> These operators evaluate the "truth" of a statement. In computing terms, that essentially refers to whether the statement is something or nothing (nonzero or zero, true or false, on or off, and so on). The Boolean operators are:

expr1 | expr2
> This operator (called the "or" operator, or "pipe") returns the evaluation of *expr1* if it is true (neither an empty string nor zero). Otherwise, it returns the evaluation of *expr2*.

---

* Remember that when you *reference* a variable you can call it by its name, but when you refer to a variable's *value*, you have to use the dollar sign and brackets around the variable name.

expr1 & *expr2*

This operator (called "and") returns the evaluation of *expr1* if both expressions are true (i.e., neither expression evaluates to an empty string or zero). Otherwise, it returns zero.

expr1 {=, >, >=, <, <=, !=} *expr2*

These operators return the results of an integer comparison if both arguments are integers; otherwise, they return the results of a string comparison. The result of each comparison is 1 if the specified relation is true, or 0 if the relation is false. (If you are doing string comparisons, they will be done in a manner that's consistent with the current local settings of your operating system.)

Mathematical operators

Want to perform a calculation? You'll want one of these:

expr1 {+, -} *expr2*

These operators return the results of the addition or subtraction of integer-valued arguments.

expr1 {*, /, %} *expr2*

These return, respectively, the results of the multiplication, integer division, or remainder of integer-valued arguments.

Regular expression operator

You can also use the regular expression operator in Asterisk:

expr1 : *expr2*

This operator matches *expr1* against *expr2*, where *expr2* must be a regular expression.[†] The regular expression is anchored to the beginning of the string with an implicit ^.[‡]

If the match succeeds and the pattern contains at least one regular expression subexpression—\(... \)—the string corresponding to \1 is returned; otherwise, the matching operator returns the number of characters matched. If the match fails and the pattern contains a regular expression subexpression, the null string is returned; otherwise, 0 is returned.

In Asterisk version 1.0 the parser was quite simple, so it required that you put at least one space between the operator and any other values. Consequently, the following might not have worked as expected:

```
exten => 123,1,Set(TEST=$[2+1])
```

[†] For more on regular expressions, grab a copy of the ultimate reference, Jeffrey E. F. Friedl's *Mastering Regular Expressions* (O'Reilly), or visit *http://www.regular-expressions.info*.

[‡] If you don't know what a ^ has to do with regular expressions, you simply must read *Mastering Regular Expressions*. It will change your life!

This would have assigned the variable TEST to the string "2+1", instead of the value 3. In order to remedy that, we would put spaces around the operator, like so:

```
exten => 234,1,Set(TEST=$[2 + 1])
```

This is no longer necessary in current versions of Asterisk, as the expression parser has been made more forgiving in these types of scenarios. However, for readability's sake, we still recommend including the spaces around your operators.

To concatenate text onto the beginning or end of a variable, simply place them together, like this:

```
exten => 234,1,Set(NEWTEST=blah${TEST})
```

Dialplan Functions

Dialplan functions allow you to add more power to your expressions; you can think of them as intelligent variables. Dialplan functions allow you to calculate string lengths, dates and times, MD5 checksums, and so on, all from within a dialplan expression.

Syntax

Dialplan functions have the following basic syntax:

```
FUNCTION_NAME(argument)
```

You reference a function's name the same way as a variable's name, but you reference a function's *value* with the addition of a dollar sign, an opening curly brace, and a closing curly brace:

```
${FUNCTION_NAME(argument)}
```

Functions can also encapsulate other functions, like so:

```
${FUNCTION_NAME(${FUNCTION_NAME(argument)})}
  ^              ^ ^             ^      ^^^^
  1              2 3             4      4321
```

As you've probably already figured out, you must be very careful about making sure you have matching parentheses and braces. In the preceding example, we have labeled the opening parentheses and curly braces with numbers and their corresponding closing counterparts with the same numbers.

Examples of Dialplan Functions

Functions are often used in conjunction with the Set() application to either get or set the value of a variable. As a simple example, let's look at the LEN() function. This function calculates the string length of its argument. Let's calculate the string length of a variable and read back the length to the caller:

```
exten => 123,1,Set(TEST=example)
    same => n,SayNumber(${LEN(${TEST})})
```

This example will first evaluate $TEST as example. The string "example" is then given to the LEN() function, which will evaluate as the length of the string, 7. Finally, 7 is passed as an argument to the SayNumber() application.

Let's look at another simple example. If we wanted to set one of the various channel timeouts, we could use the TIMEOUT() function. The TIMEOUT() function accepts one of three arguments: absolute, digit, and response. To set the digit timeout with the TIMEOUT() function, we could use the Set() application, like so:

```
exten => s,1,Set(TIMEOUT(digit)=30)
```

Notice the lack of ${ } surrounding the function. Just as if we were assigning a value to a variable, we assign a value to a function without the use of the ${ } encapsulation.

A complete list of available functions can be found by typing *core show functions* at the Asterisk command-line interface.

Conditional Branching

Now that you've learned a bit about expressions and functions, it's time to put them to use. By using expressions and functions, you can add even more advanced logic to your dialplan. To allow your dialplan to make decisions, you'll use *conditional branching*. Let's take a closer look.

The GotoIf() Application

The key to conditional branching is the GotoIf() application. GotoIf() evaluates an expression and sends the caller to a specific destination based on whether the expression evaluates to true or false.

GotoIf() uses a special syntax, often called the *conditional syntax*:

```
GotoIf(expression?destination1:destination2)
```

If the expression evaluates to true, the caller is sent to *destination1*. If the expression evaluates to false, the caller is sent to the second destination. So, what is true and what is false? An empty string and the number 0 evaluate as false. Anything else evaluates as true.

The destinations can each be one of the following:

- A priority label within the same extension, such as weasels
- An extension and a priority label within the same context, such as 123,weasels
- A context, extension, and priority label, such as incoming,123,weasels

Either of the destinations may be omitted, but not both. If the omitted destination is to be followed, Asterisk simply goes on to the next priority in the current extension.

Let's use GotoIf() in an example:

```
exten => 345,1,Set(TEST=1)
    same => n,GotoIf($[${TEST} = 1]?weasels:iguanas)
    same => n(weasels),Playback(weasels-eaten-phonesys)
    same => n,Hangup()
    same => n(iguanas),Playback(office-iguanas)
    same => n,Hangup()
```

 You will notice that we have used the Hangup() application following each use of the Playback() application. This is done so that when we jump to the weasels label, the call stops before execution gets to the *office-iguanas* sound file. It is becoming increasingly common to see extensions broken up into multiple components (protected from each other by the Hangup() command), each one a distinct sequence of steps executed following a GotoIf().

Providing Only a False Conditional Path

If we wanted to, we could have crafted the preceding example like this:

```
exten => 345,1,Set(TEST=1)
    same => n,GotoIf($[${TEST} = 1]?:iguanas) ; we don't have the weasels label anymore,
                                              ; but this will still work
    same => n,Playback(weasels-eaten-phonesys)
    same => n,Hangup()
    same => n(iguanas),Playback(office-iguanas)
    same => n,Hangup()
```

There's nothing between the ? and the :, so if the statement evaluates to true, execution will continue at the next step. Since that's what we want, a label isn't needed.

We don't really recommend doing this, because it's hard to read, but you will see dialplans like this, so it's good to be aware that this syntax is totally correct.

Typically, when you have this type of layout where you end up wanting to prevent Asterisk from falling through to the next priority after you've performed that jump, it's probably better to jump to separate extensions instead of priority labels. If anything, it makes it a bit more clear when reading the dialplan. We could rewrite the previous bit of dialplan like this:

```
exten => 345,1,Set(TEST=1)
    same => n,GotoIf($[${TEST} = 1]?weasels,1:iguanas,1) ; now we're going to
                                                         ; extension,priority

exten => weasels,1,Playback(weasels-eaten-phonesys)      ; this is NOT a label.
                                                         ; It is a different extension
    same => n,Hangup()
```

```
exten => iguanas,1,Playback(office-iguanas)
    same => n,Hangup()
```

By changing the value assigned to TEST in the first line, you should be able to have your Asterisk server play a different greeting.

Let's look at another example of conditional branching. This time, we'll use both Goto() and GotoIf() to count down from 10 and then hang up:

```
exten => 123,1,Set(COUNT=10)
    same => n(start),GotoIf($[${COUNT} > 0]?:goodbye)
    same => n,SayNumber(${COUNT})
    same => n,Set(COUNT=$[${COUNT} - 1])
    same => n,Goto(start)
    same => n(goodbye),Hangup()
```

Let's analyze this example. In the first priority, we set the variable COUNT to 10. Next, we check to see if COUNT is greater than 0. If it is, we move on to the next priority. (Don't forget that if we omit a destination in the GotoIf() application, control goes to the next priority.) From there, we speak the number, subtract 1 from COUNT, and go back to priority label start. If COUNT is less than or equal to 0, control goes to priority label goodbye, and the call is hung up.

The classic example of conditional branching is affectionately known as the anti-girlfriend logic. If the caller ID number of the incoming call matches the phone number of the recipient's ex-girlfriend, Asterisk gives a different message than it ordinarily would to any other caller. While somewhat simple and primitive, it's a good example for learning about conditional branching within the Asterisk dialplan.

This example uses the CALLERID function, which allows us to retrieve the caller ID information on the inbound call. Let's assume for the sake of this example that the victim's phone number is 888-555-1212:

```
exten => 123,1,GotoIf($[${CALLERID(num)} = 8885551212]?reject:allow)
    same => n(allow),Dial(DAHDI/4)
    same => n,Hangup()
    same => n(reject),Playback(abandon-all-hope)
    same => n,Hangup()
```

In priority 1, we call the GotoIf() application. It tells Asterisk to go to priority label reject if the caller ID number matches 8885551212, and otherwise to go to priority label allow (we could have simply omitted the label name, causing the GotoIf() to fall through). If the caller ID number matches, control of the call goes to priority label reject, which plays back an uninspiring message to the undesired caller. Otherwise, the call attempts to dial the recipient on channel DAHDI/4.

Time-Based Conditional Branching with GotoIfTime()

Another way to use conditional branching in your dialplan is with the GotoIfTime() application. Whereas GotoIf() evaluates an expression to decide what to do, GotoIf Time() looks at the current system time and uses that to decide whether or not to follow a different branch in the dialplan.

The most obvious use of this application is to give your callers a different greeting before and after normal business hours.

The syntax for the GotoIfTime() application looks like this:

```
GotoIfTime(times,days_of_week,days_of_month,months?label)
```

In short, GotoIfTime() sends the call to the specified *label* if the current date and time match the criteria specified by *times*, *days_of_week*, *days_of_month*, and *months*. Let's look at each argument in more detail:

times
> This is a list of one or more time ranges, in a 24-hour format. As an example, 9:00 A.M. through 5:00 P.M. would be specified as 09:00-17:00. The day starts at 0:00 and ends at 23:59.

> It is worth noting that times will properly wrap around. So, if you wish to specify the times your office is closed, you might write 18:00-9:00 in the *times* parameter, and it will perform as expected. Note that this technique works as well for the other components of GotoIfTime(). For example, you can write sat-sun to specify the weekend days.

days_of_week
> This is a list of one or more days of the week. The days should be specified as mon, tue, wed, thu, fri, sat, and/or sun. Monday through Friday would be expressed as mon-fri. Tuesday and Thursday would be expressed as tue&thu.

> Note that you can specify a combination of ranges and single days, as in: sun-mon&wed&fri-sat, or, more simply: wed&fri-mon.

days_of_month
> This is a list of the numerical days of the month. Days are specified by the numbers 1 through 31. The 7th through the 12th would be expressed as 7-12, and the 15th and 30th of the month would be written as 15&30.

months

This is a list of one or more months of the year. The months should be written as jan-apr for a range, and separated with ampersands when wanting to include non-sequential months, such as jan&mar&jun. You can also combine them like so: jan-apr&jun&oct-dec.

If you wish to match on all possible values for any of these arguments, simply put an * in for that argument.

The *label* argument can be any of the following:

- A priority label within the same extension, such as time_has_passed
- An extension and a priority within the same context, such as 123,time_has_passed
- A context, extension, and priority, such as incoming,123,time_has_passed

Now that we've covered the syntax, let's look at a couple of examples. The following example would match from *9:00 A.M. to 5:59 P.M.*, on *Monday through Friday*, on *any day of the month*, in *any month of the year*:

```
exten => s,1,GotoIfTime(09:00-17:59,mon-fri,*,*?open,s,1)
```

If the caller calls during these hours, the call will be sent to the first priority of the s extension in the context named open. If the call is made outside of the specified times, it will be sent to the next priority of the current extension. This allows you to easily branch on multiple times, as shown in the next example (note that you should always put your most specific time matches before the least specific ones):

```
; If it's any hour of the day, on any day of the week,
; during the fourth day of the month, in the month of July,
; we're closed
exten => s,1,GotoIfTime(*,*,4,jul?closed,s,1)

; During business hours, send calls to the open context
    same => n,GotoIfTime(09:00-17:59,mon-fri,*,*?open,s,1)
    same => n,GotoIfTime(09:00-11:59,sat,*,*?open,s,1)

; Otherwise, we're closed
    same => n,Goto(closed,s,1)
```

 If you run into the situation where you ask the question, "But I specified 17:58 and it's now 17:59. Why is it still doing the same thing?" it should be noted that the granularity of the GotoIfTime() application is only to a two-minute period. So, if you specify 18:00 as the ending time of a period, the system will continue to perform the same way until 18:01:59.

Macros

Macros[§] are a very useful construct designed to avoid repetition in the dialplan. They also help in making changes to the dialplan. To illustrate this point, let's look at our sample dialplan again. If you remember the changes we made for voicemail, we ended up with the following for John's extension:

```
exten => 101,1,Dial(${JOHN},10)
    same => n,GotoIf($["${DIALSTATUS}" = "BUSY"]?busy:unavail)
    same => n(unavail),VoiceMail(101@default,u)
    same => n,Hangup()
    same => n(busy),VoiceMail(101@default,b)
    same => n,Hangup()
```

Now imagine you have a hundred users on your Asterisk system—setting up the extensions would involve a lot of copying and pasting. Then imagine that you need to make a change to the way your extensions work. That would involve a lot of editing, and you'd be almost certain to have errors.

Instead, you can define a macro that contains a list of steps to take, and then have all of the phone extensions refer to that macro. All you need to change is the macro, and everything in the dialplan that references that macro will change as well.

 If you're familiar with computer programming, you'll recognize that macros are similar to subroutines in many modern programming languages. If you're not familiar with computer programming, don't worry—we'll walk you through creating a macro.

The best way to appreciate macros is to see one in action, so let's move right along.

Defining Macros

Let's take the dialplan logic we used to set up voicemail for John and turn it into a macro. Then we'll use the macro to give John and Jane (and the rest of their coworkers) the same functionality.

[§] Although Macro() seems like a general-purpose dialplan subroutine, it has a stack overflow problem that means you should not try to nest Macro() calls more than five levels deep. If you plan to use a lot of macros within macros (and call complex functions within them), you may run into stability problems. You will know you have a problem with just one test call, so if your dialplan tests out, you're good to go. We also recommend that you take a look at the GoSub() and Return() applications (see "GoSub()" on page 207), as a lot of macro functionality can be implemented without actually using Macro(). Also, please note that we are not suggesting that you don't use Macro(). It is fantastic and works very well; it just doesn't nest efficiently.

Macro definitions look a lot like contexts. (In fact, you could argue that they really are small, limited contexts.) You define a macro by placing macro- and the name of your macro in square brackets, like this:

```
[macro-voicemail]
```

Macro names must start with macro-. This distinguishes them from regular contexts. The commands within the macro are built almost identically to anything else in the dialplan; the only limiting factor is that macros use only the s extension. Let's add our voicemail logic to the macro, changing the extension to s as we go:

```
[macro-voicemail]
exten => s,1,Dial(${JOHN},10)
    same => n,GotoIf($["${DIALSTATUS}" = "BUSY"]?busy:unavail)
    same => n(unavail),VoiceMail(101@default,u)
    same => n,Hangup()
    same => n(busy),VoiceMail(101@default,b)
    same => n,Hangup()
```

That's a start, but it's not perfect, as it's still specific to John and his mailbox number. To make the macro generic so that it will work not only for John but also for all of his coworkers, we'll take advantage of another property of macros: arguments. But first, let's see how we call macros in our dialplan.

Calling Macros from the Dialplan

To use a macro in our dialplan, we use the Macro() application. This application calls the specified macro and passes it any arguments. For example, to call our voicemail macro from our dialplan, we can do the following:

```
exten => 101,1,Macro(voicemail)
```

The Macro() application also defines several special variables for our use. They include:

${MACRO_CONTEXT}
: The original context in which the macro was called.

${MACRO_EXTEN}
: The original extension in which the macro was called.

${MACRO_PRIORITY}
: The original priority in which the macro was called.

${ARG n }
: The nth argument passed to the macro. For example, the first argument would be ${ARG1}, the second ${ARG2}, and so on.

As we explained earlier, the way we initially defined our macro was hardcoded for John, instead of being generic. Let's change our macro to use ${MACRO_EXTEN} instead of 101 for the mailbox number. That way, if we call the macro from extension 101 the voicemail messages will go to mailbox 101, if we call the macro from extension 102 messages will go to mailbox 102, and so on:

```
[macro-voicemail]
exten => s,1,Dial(${JOHN},10)
    same => n,GotoIf($["${DIALSTATUS}" = "BUSY"]?busy:unavail)
    same => n(unavail),VoiceMail(${MACRO_EXTEN}@default,u)
    same => n,Hangup()
    same => n(busy),VoiceMail(${MACRO_EXTEN}@default,b)
    same => n,Hangup()
```

Using Arguments in Macros

Okay, now we're getting closer to having the macro the way we want it, but we still have one thing left to change: we need to pass in the channel to dial, as it's currently still hardcoded for ${JOHN} (remember that we defined the variable JOHN as the channel to call when we want to reach John). Let's pass in the channel as an argument, and then our first macro will be complete:

```
[macro-voicemail]
exten => s,1,Dial(${ARG1},10)
    same => n,GotoIf($["${DIALSTATUS}" = "BUSY"]?busy:unavail)
    same => n(unavail),VoiceMail(${MACRO_EXTEN}@default,u)
    same => n,Hangup()
    same => n(busy),VoiceMail(${MACRO_EXTEN}@default,b)
    same => n,Hangup()
```

Now that our macro is done, we can use it in our dialplan. Here's how we can call our macro to provide voicemail to John, Jane, and Jack:

```
exten => 101,1,Macro(voicemail,${JOHN})
exten => 102,1,Macro(voicemail,${JANE})
exten => 103,1,Macro(voicemail,${JACK})
```

With 50 or more users, this dialplan will still look neat and organized; we'll simply have one line per user, referencing a macro that can be as complicated as required. We could even have a few different macros for various user types, such as executives, courtesy_phones, call_center_agents, analog_sets, sales_department, and so on.

A more advanced version of the macro might look something like this:

```
[macro-voicemail]
exten => s,1,Dial(${ARG1},20)
    same => n,Goto(s-${DIALSTATUS},1)

exten => s-NOANSWER,1,VoiceMail(${MACRO_EXTEN},u)
    same => n,Goto(incoming,s,1)

exten => s-BUSY,1,VoiceMail(${MACRO_EXTEN},b)
    same => n,Goto(incoming,s,1)

exten => _s-.,1,Goto(s-NOANSWER,1)
```

Since we know how to use dialplan functions now as well, here is another way of controlling which voicemail prompt (unavailable vs. busy) is played to the caller. In the following example, we'll be using the IF() dialplan function:

```
[macro-voicemail]
exten => s,1,Dial(${ARG1},20)
    same => n,VoiceMail(${MACRO_EXTEN},${IF($[${DIALSTATUS} = BUSY]?b:u)})
```

This macro depends on a nice side effect of the Dial() application: when you use the Dial() application, it sets the DIALSTATUS variable to indicate whether the call was successful or not. In this case, we're handling the NOANSWER and BUSY cases, and treating all other result codes as a NOANSWER.

GoSub()

The GoSub() dialplan application is similar to the Macro() application, in that the purpose is to allow you to call a block of dialplan functionality, pass information to that block, and return from it (optionally with a return value). GoSub() works in a different manner from Macro(), though, in that it doesn't have the stack space requirements, so it nests effectively. Essentially, GoSub() acts like Goto() with a memory of where it came from.

In this section we're going to reimplement what we learned in "Macros" on page 204. If necessary, you might want to review that section: it explains why we might use a subroutine, and the goal we're trying to accomplish.

Defining Subroutines

Unlike with Macro(), there are no special naming requirements when using GoSub() in the dialplan. In fact, you can use GoSub() within the same context and extension if you want to. In most cases, however, GoSub() is used in a similar fashion to Macro(), so defining a new context is common. When creating the context, we like to prepend the name with sub so we know the context is typically called from the GoSub() application (of course, there is no requirement that you do so, but it seems a sensible convention).

Here is a simple example of how we might define a subroutine in Asterisk:

```
[subVoicemail]
```

Let's take our example from "Macros" on page 204 and convert it to a subroutine. Here is how it is defined for use with Macro():

```
[macro-voicemail]
exten => s,1,Dial(${JOHN},10)
    same => n,GotoIf($["${DIALSTATUS}" = "BUSY"]?busy:unavail)
    same => n(unavail),VoiceMail(101@default,u)
    same => n,Hangup()
```

```
    same => n(busy),VoiceMail(101@default,b)
    same => n,Hangup()
```

If we were going to convert this to be used for a subroutine, it might look like this:

```
[subVoicemail]
exten => start,1,Dial(${JOHN},10)
    same => n,GotoIf($["${DIALSTATUS}" = "BUSY"]?busy:unavail)
    same => n(unavail),VoiceMail(101@default,u)
    same => n,Hangup()
    same => n(busy),VoiceMail(101@default,b)
    same => n,Hangup()
```

Not much of a change, right? All we've altered in this example is the context name, from [macro-voicemail] to [subVoicemail], and the extension, from s to start (since there is no requirement that the extension be called anything in particular, unlike with Macro(), which expects the extension to be s).

Of course, as in the example in the section "Macros" on page 204, we haven't passed any arguments to the subroutine, so whenever we call [subVoicemail], ${JOHN} will always be called, and the voicemail box 101 will get used. In the following sections, we'll dig a little deeper. First we'll look at how we would call a subroutine, and then we'll learn how to pass arguments.

Calling Subroutines from the Dialplan

Subroutines are called from the dialplan using the GoSub() application. The arguments to GoSub() differ slightly than those for Macro(), because GoSub() has no naming requirements for the context or extension (or priority) that gets used. Additionally, no special channel variables are set when calling a subroutine, other than the passed arguments, which are saved to ${ARG*n*} (where the first argument is ${ARG1}, the second argument is ${ARG2}, and so forth).

Now that we've updated our voicemail macro to be called as a subroutine, lets take a look at how we call it using GoSub():

```
exten => 101,1,GoSub(subVoicemail,start,1())
```

 You'll notice that we've placed a set of opening and closing parentheses within our GoSub() application. These are the placeholders for any arguments we might pass to the subroutine, and while it is optional for them to exist, it's a programming style we prefer to use.

Next, let's look at how we can pass arguments to our subroutine in order to make it more general.

Using Arguments in Subroutines

The ability to use arguments is one of the major features of using `Macro()` or `GoSub()`, because it allows you to abstract out code that would otherwise be duplicated across your dialplan. Without the need to duplicate the code, we can better manage it, and we can easily add functionality to large numbers of users by modifying a single location. You are encouraged to move code into this form whenever you find yourself creating duplicate code.

Before we start using our subroutine, we need to update it to accept arguments so that it is generic enough to be used by multiple users:

```
[subVoicemail]
exten => start,1,Dial(${ARG1},10)
    same => n,GotoIf($["${DIALSTATUS}" = "BUSY"]?busy:unavail)
    same => n(unavail),VoiceMail(${ARG2}@default,u)
    same => n,Hangup()
    same => n(busy),VoiceMail(${ARG2}@default,b)
    same => n,Hangup()
```

Recall that previously we had hardcoded the channel variable `${JOHN}` as the location to dial, and mailbox 101 as the voicemail box to be used if `${JOHN}` wasn't available. In this code, we've replaced `${JOHN}` and 101 with `${ARG1}` and `${ARG2}`, respectively. In more complex subroutines we might even assign the variables `${ARG1}` and `${ARG2}` to something like `${DESTINATION}` and `${VMBOX}`, to make it clear what the `${ARG1}` and `${ARG2}` represent.

Now that we've updated our subroutine, we can use it for several extensions:

```
[LocalSets]
exten => 101,1,GoSub(subVoicemail,start,1(${JOHN},${EXTEN}))
exten => 102,1,GoSub(subVoicemail,start,1(${JANE},${EXTEN}))
exten => 103,1,GoSub(subVoicemail,start,1(${JACK},${EXTEN}))
```

Again, our dialplan is nice and neat. We could even modify our subroutine down to just three lines:

```
[subVoicemail]
exten => start,1,Dial(${ARG1},10)
    same => n,VoiceMail(${ARG2}@default,${IF($[${DIALSTATUS} = BUSY]?b:u)})
    same => n,Hangup()
```

One difference to note between `GoSub()` and `Macro()`, however, is that if we left our subroutine like this, we'd never return. In this particular example that's not a problem, since after the voicemail is left, we would expect the caller to hang up anyway. In situations where we want to do more after the subroutine has executed, though, we need to implement the `Return()` application.

Returning from a Subroutine

Unlike Macro(), the GoSub() dialplan application does not return automatically once it is done executing. In order to return from whence we came, we need to use the Return() application. Now that we know how to call a subroutine and pass arguments, we can look at an example where we might need to return from the subroutine.

Using our previous example, we could break out the dialing portion and the voicemail portion into separate subroutines:

```
[subDialer]
exten => start,1,Dial(${ARG1},${ARG2})
    same => n,Return()

[subVoicemail]
exten => start,1,VoiceMail(${ARG1}@${ARG2},${ARG3})
    same => n,Hangup()
```

The [subDialer] context created here takes two arguments: ${ARG1}, which contains the destination to dial; and ${ARG2}, which contains the ring cycle, defined in seconds. We conclude the [subDialer] context with the dialplan application Return(), which will return to the priority following the one that called GoSub() (the next line of the dialplan).

The [subVoicemail] context contains the VoiceMail() application, which is using three arguments passed to it: ${ARG1} contains the mailbox number, ${ARG2} contains the voicemail context, and ${ARG3} contains a value to indicate which voicemail message (unavailable or busy) to play to the caller.

Calling these subroutines might look like this:

```
exten => 101,1,GoSub(subDialer,start,1(${JOHN},30))
    same => n,GoSub(subVoicemail,start,1(${EXTEN},default,u))
```

Here we've used the subDialer subroutine, which attempts to call ${JOHN}, ringing him for 30 seconds. If the Dial() application returns (e.g., if the line was busy, or there was no answer for 30 seconds), we Return() from the subroutine and execute the next line of our dialplan, which calls the subVoicemail subroutine. From there, we pass the extension that was dialed (e.g., 101) as the mailbox number, and pass the values default for the voicemail context and the letter u to play the unavailable message.

Our example has been hardcoded to play the unavailable voicemail message, but we can modify the Return() application to return the ${DIALSTATUS} so that we can play the busy message if its value is BUSY. To do this, we'll use the ${GOSUB_RETVAL} channel variable, which is set whenever we pass a value to the Return() application:

```
[subDialer]
exten => start,1,Dial(${ARG1},${ARG2})
    same => n,Return(${DIALSTATUS})

[subVoicemail]
```

```
exten => start,1,VoiceMail(${ARG1}@${ARG2},${ARG3})
    same => n,Hangup()
```

In this version we've made just the one change: `Return()` to `Return(${DIALSTATUS})`.

Now we can modify extension 101 to use the `${GOSUB_RETVAL}` channel variable, which will be set by `Return()`:

```
exten => 101,1,GoSub(subDialer,start,1(${JOHN},30))
    same => n,Set(VoicemailMessage=${IF($[${GOSUB_RETVAL} = BUSY]?b:u)})
    same => n,GoSub(subVoicemail,start,1(${EXTEN},default,${VoicemailMessage}))
```

Our dialplan now has a new line that sets the `${VoicemailMessage}` channel variable to a value of u or b, using the `IF()` dialplan function and the value of `${GOSUB_RETVAL}`. We then pass the value of `${VoicemailMessage}` as the third argument to our subVoice mail subroutine.

Before moving on, you might want to go back and review "Macros" on page 204 and "GoSub()" on page 207. We've given you a lot to digest here, but these concepts will save you a lot of work as you start building your dialplans.

Local Channels

Local channels are a method of executing dialplans from the `Dial()` application. They may seem like a bit of a strange concept when you first start using them, but believe us when we tell you they are a glorious and extremely useful feature that you will almost certainly want to make use of when you start writing advanced dialplans. The best way to illustrate the use of Local channels is through an example. Let's suppose we have a situation where we need to ring multiple people, but we need to provide delays of different lengths before dialing each of the members. The use of Local channels is the only solution to the problem.

With the `Dial()` application, you can certainly ring multiple endpoints, but all three channels will ring at the same time, and for the same length of time. Dialing multiple channels at the same time is done like so:

```
[LocalSets]
exten => 107,1,Verbose(2,Dialing multiple locations simultaneously)
    same => n,Dial(SIP/0000FFFF0001&DAHDI/g0/14165551212&SIP/MyITSP/12565551212,30)
    same => n,Hangup()
```

This example dials three destinations for a period of 30 seconds. If none of those locations answers the call within 30 seconds, the dialplan continues to the next line and the call is hung up.

However, let's say we want to introduce some delays, and stop ringing locations at different times. Using Local channels gives us independent control over each of the channels we want to dial, so we can introduce delays and control the period of time for which each channel rings independently. We're going to show you how this is done in the dialplan, both within a table that shows the delays visually, and all together in a box, like we've done for other portions of the dialplan. We'll be building the dialplan to match the time starts and stops described in Figure 10-1.

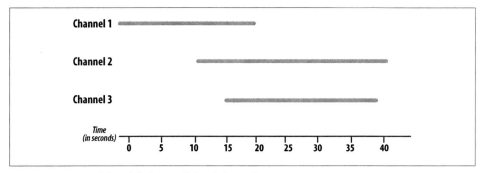

Figure 10-1. Time delayed dialing with local channels

First we need to call three Local channels, which will all execute different parts of the dialplan. We do this with the Dial() application, like so:

```
[LocalSets]
exten => 107,1,Verbose(2,Dialing multiple locations with time delay)

; *** This all needs to be on a single line
    same => n,Dial(Local/channel_1@TimeDelay&Local/channel_2@TimeDelay
&Local/channel_3@TimeDelay,40)
    same => n,Hangup()
```

Now our Dial() application will dial three Local channels. The destinations will be the channel_1, channel_2, and channel_3 extensions located within the TimeDelay dialplan context. Remember that Local channels are a way of executing the dialplan from within the Dial() application. Our master timeout for all the channels is 40 seconds, which means any Local channel that does not have a shorter timeout configured will be hung up if it does not answer the call within that period of time.

As promised, Table 10-1 illustrates the delay configurations.

Table 10-1. Delayed dialing using Local channels

Time period (in seconds)	channel_1	channel_2	channel_3
0	Dial(SIP/ 0000FFFF0001,20)	Wait(10)	Wait(15)
5			
10		Dial(DAHDI/ g0/14165551212)	
15			Dial(SIP/MyITSP/ 12565551212,15)
20	Hangup()		

Time period (in seconds)	channel_1	channel_2	channel_3
25			
30			Hangup()
35			
40			

In this table, we can see that channel_1 started dialing location SIP/0000FFFF0001 immediately and waited for a period of 20 seconds. After 20 seconds, that Local channel hung up. Our channel_2 waited for 10 seconds prior to dialing the endpoint DAHDI/g0/14165551212. There was no maximum time associated with this Dial(), so its dialing period ended when the master time out of 40 seconds (which we set when we initially called the Local channels) expired. Finally, channel_3 waited 15 seconds prior to dialing, then dialed SIP/MyITSP/12565551212 and waited for a period of 15 seconds prior to hanging up.

If we put all this together, we end up with the following dialplan:

```
[LocalSets]
exten => 107,1,Verbose(2,Dialing multiple locations with time delay)

; *** This all needs to be on a single line
    same => n,Dial(Local/channel_1@TimeDelay&Local/channel_2@TimeDelay
&Local/channel_3@TimeDelay,40)
    same => n,Hangup()

[TimeDelay]
exten => channel_1,1,Verbose(2,Dialing the first channel)
    same => n,Dial(SIP/0000FFFF0001,20)
    same => n,Hangup()

exten => channel_2,1,Verbose(2,Dialing the second channel with a delay)
    same => n,Wait(10)
    same => n,Dial(DAHDI/g0/14165551212)

exten => channel_3,1,Verbose(2,Dialing the third channel with a delay)
    same => n,Wait(15)
    same => n,Dial(SIP/MyITSP/12565551212,15)
    same => n,Hangup()
```

You'll see Local channels used throughout this book, for various purposes. Remember that the intention is simply to perform some dialplan logic from a location where you can only dial a location, but require some dialplan logic to be executed prior to dialing the endpoint you eventually want to get to. A good example of this is with the use of the Queue() application, which we'll discuss in "Using Local Channels" on page 293.

Additional scenarios and information about Local channels and the modifier flags (/n, /j, /m, /b) are available at *https://wiki.asterisk.org/wiki/display/AST/Local+Chan nel*. If you will be making any sort of regular use of Local channels, that is a very important document to read.

Using the Asterisk Database (AstDB)

Having fun yet? It gets even better!

Asterisk provides a powerful mechanism for storing values called the *Asterisk database* (AstDB). The AstDB provides a simple way to store data for use within your dialplan.

 For those of you with experience using relational databases such as PostgreSQL or MySQL, the Asterisk database is not a traditional relational database; it is a Berkeley DB version 1 database. There are several ways to store data from Asterisk in a relational database. Check out Chapter 16 for more about relational databases.

The Asterisk database stores its data in groupings called *families*, with values identified by *keys*. Within a family, a key may be used only once. For example, if we had a family called test, we could store only one value with a key called count. Each stored value must be associated with a family.

Storing Data in the AstDB

To store a new value in the Asterisk database, we use the Set() application,[||] but instead of using it to set a channel variable, we use it to set an AstDB variable. For example, to assign the count key in the test family with the value of 1, we would write the following:

```
exten => 456,1,Set(DB(test/count)=1)
```

If a key named count already exists in the test family, its value will be overwritten with the new value. You can also store values from the Asterisk command line, by running the command *database put <family> <key> <value>*. For our example, you would type *database put test count 1*.

Retrieving Data from the AstDB

To retrieve a value from the Asterisk database and assign it to a variable, we use the Set() application again. Let's retrieve the value of count (again, from the test family), assign it to a variable called COUNT, and then speak the value to the caller:

[||] Previous versions of Asterisk had applications called DBput() and DBget() that were used to set values in and retrieve values from the AstDB. If you're using an old version of Asterisk, you'll want to use those applications instead.

```
exten => 456,1,Set(DB(test/count)=1)
    same => n,Set(COUNT=${DB(test/count)})
    same => n,SayNumber(${COUNT})
```

You may also check the value of a given key from the Asterisk command line by running the command *database get <family> <key>*. To view the entire contents of the AstDB, use the *database show* command.

Deleting Data from the AstDB

There are two ways to delete data from the Asterisk database. To delete a key, you can use the DB_DELETE() application. It takes the path to the key as its arguments, like this:

```
; deletes the key and returns its value in one step
exten => 457,1,Verbose(0, The value was ${DB_DELETE(test/count)})
```

You can also delete an entire key family by using the DBdeltree() application. The DBdeltree() application takes a single argument: the name of the key family to delete. To delete the entire test family, do the following:

```
exten => 457,1,DBdeltree(test)
```

To delete keys and key families from the AstDB via the command-line interface, use the *database del <key>* and *database deltree <family>* commands, respectively.

Using the AstDB in the Dialplan

There are an infinite number of ways to use the Asterisk database in a dialplan. To introduce the AstDB, we'll look at two simple examples. The first is a simple counting example to show that the Asterisk database is persistent (meaning that it survives system reboots). In the second example, we'll use the BLACKLIST() function to evaluate whether or not a number is on the blacklist and should be blocked.

To begin the counting example, let's first retrieve a number (the value of the count key) from the database and assign it to a variable named COUNT. If the key doesn't exist, DB() will return NULL (no value). Therefore, we can use the ISNULL() function to verify whether or not a value was returned. If not, we will initialize the AstDB with the Set() application, where we will set the value in the database to 1. The next priority will send us back to priority 1. This will happen the very first time we dial this extension:

```
exten => 678,1,Set(COUNT=${DB(test/count)})
    same => n,GotoIf($[${ISNULL(${COUNT})}]?:continue)
    same => n,Set(DB(test/count)=1)
    same => n,Goto(1)
    same => n(continue),NoOp()
```

Next, we'll say the current value of COUNT, and then increment COUNT:

```
exten => 678,1,Set(COUNT=${DB(test/count)})
    same => n,GotoIf($[${ISNULL(${COUNT})}]?:continue)
    same => n,Set(DB(test/count)=1)
    same => n,Goto(1)
```

```
    same => n(continue),NoOp()
    same => n,SayNumber(${COUNT})
    same => n,Set(COUNT=$[${COUNT} + 1])
```

Now that we've incremented COUNT, let's put the new value back into the database. Remember that storing a value for an existing key overwrites the previous value:

```
exten => 678,1,Set(COUNT=${DB(test/count)})
    same => n,GotoIf($[${ISNULL(${COUNT})}]?:continue)
    same => n,Set(DB(test/count)=1)
    same => n,Goto(1)
    same => n(continue),NoOp()
    same => n,SayNumber(${COUNT})
    same => n,Set(COUNT=$[${COUNT} + 1])
    same => n,Set(DB(test/count)=${COUNT})
```

Finally, we'll loop back to the first priority. This way, the application will continue counting:

```
exten => 678,1,Set(COUNT=${DB(test/count)})
    same => n,GotoIf($[${ISNULL(${COUNT})}]?:continue)
    same => n,Set(DB(test/count)=1)
    same => n,Goto(1)
    same => n(continue),NoOp()
    same => n,SayNumber(${COUNT})
    same => n,Set(COUNT=$[${COUNT} + 1]
    same => n,Set(DB(test/count)=${COUNT})
    same => n,Goto(1)
```

Go ahead and try this example. Listen to it count for a while, and then hang up. When you dial this extension again, it should continue counting from where it left off. The value stored in the database will be persistent, even across a restart of Asterisk.

In the next example, we'll create dialplan logic around the BLACKLIST() function, which checks to see if the current caller ID number exists in the blacklist. (The blacklist is simply a family called blacklist in the AstDB.) If BLACKLIST() finds the number in the blacklist, it returns the value 1; otherwise, it will return 0. We can use these values in combination with a GotoIf() to control whether the call will execute the Dial() application:

```
exten => 124,1,GotoIf($[${BLACKLIST()}]?blocked,1)
    same => n,Dial(${JOHN})

exten => blocked,1,Playback(privacy-you-are-blacklisted)
    same => n,Playback(vm-goodbye)
    same => n,Hangup()
```

To add a number to the blacklist, run the *database put blacklist <number> 1* command from the Asterisk command-line interface.

Handy Asterisk Features

Now that we've gone over some more of the basics, let's look at a few popular functions that have been incorporated into Asterisk.

Zapateller()

Zapateller() is a simple Asterisk application that plays a special information tone at the beginning of a call, which causes auto-dialers (usually used by telemarketers) to think that the line has been disconnected. Not only will they hang up, but their systems will flag your number as out of service, which could help you avoid all kinds of telemarketing calls. To use this functionality within your dialplan, simply call the Zapateller() application.

We'll also use the optional nocallerid option so that the tone will be played only when there is no caller ID information on the incoming call. For example, you might use Zapateller() in the s extension of your [incoming] context, like this:

```
[incomimg]
exten => s,1,Zapateller(nocallerid)
    same => n,Playback(enter-ext-of-person)
```

Call Parking

Another handy feature is called *call parking*. Call parking allows you to place a call on hold in a "parking lot," so that it can be taken off hold from another extension. Parameters for call parking (such as the extensions to use, the number of spaces, and so on) are all controlled within the *features.conf* configuration file. The [general] section of the *features.conf* file contains four settings related to call parking:

parkext
> This is the parking lot extension. Transfer a call to this extension, and the system will tell you which parking position the call is in. By default, the parking extension is 700.

parkpos
> This option defines the number of parking slots. For example, setting it to 701-720 creates 20 parking positions, numbered 701 through 720.

context
> This is the name of the parking context. To be able to park calls, you must include this context.

parkingtime
> If set, this option controls how long (in seconds) a call can stay in the parking lot. If the call isn't picked up within the specified time, the extension that parked the call will be called back.

Also note that because the user needs to be able to transfer the calls to the parking lot extension, you should make sure you're using the t and/or T options to the Dial() application.

So, let's create a simple dialplan to show off call parking:

```
[incoming]
include => parkedcalls

exten => 103,1,Dial(SIP/Bob,,tT)
exten => 104,1,Dial(SIP/Charlie,,tT)
```

To illustrate how call parking works, say that Alice calls into the system and dials extension 103 to reach Bob. After a while, Bob transfers the call to extension 700, which tells him that the call from Alice has been parked in position 701. Bob then dials Charlie at extension 104, and tells him that Alice is at extension 701. Charlie then dials extension 701 and begins to talk to Alice. This is a simple and effective way of allowing callers to be transferred between users.

Conferencing with MeetMe()

Last but not least, let's cover setting up an audio conference bridge with the MeetMe() application.# This application allows multiple callers to converse together, as if they were all in the same physical location. Some of the main features include:

- The ability to create password-protected conferences
- Conference administration (mute conference, lock conference, kick participants)
- The option of muting all but one participant (useful for company announcements, broadcasts, etc.)
- Static or dynamic conference creation

Let's walk through setting up a basic conference room. The configuration options for the MeetMe conferencing system are found in *meetme.conf*. Inside the configuration file, you define conference rooms and optional numeric passwords. (If a password is defined here, it will be required to enter all conferences using that room.) For our example, let's set up a conference room at extension 600. First, we'll set up the conference room in *meetme.conf*. We'll call it 600, and we won't assign a password at this time:

```
[rooms]
conf => 600
```

Now that the configuration file is complete, we'll need to restart Asterisk so that it can reread the *meetme.conf* file. Next, we'll add support for the conference room to our dialplan with the MeetMe() application. MeetMe() takes three arguments: the name of

---

#In the world of legacy PBXs, this type of functionality is very expensive. Either you have to pay big bucks for a dial-in service, or you have to add an expensive conferencing bridge to your proprietary PBX.

the conference room (as defined in *meetme.conf*), a set of options, and the password the user must enter to join this conference. Let's set up a simple conference using room 600, the i option (which announces when people enter and exit the conference), and a password of 54321:

```
exten => 600,1,MeetMe(600,i,54321)
```

That's all there is to it! When callers enter extension 600, they will be prompted for the password. If they correctly enter 54321, they will be added to the conference. You can run *core show application MeetMe* from the Asterisk CLI for a list of all the options supported by the MeetMe() application.

Another useful application is MeetMeCount(). As its name suggests, this application counts the number of users in a particular conference room. It takes up to two arguments: the conference room in which to count the number of participants, and optionally a variable name to assign the count to. If the variable name is not passed as the second argument, the count is read to the caller:

```
exten => 601,1,Playback(conf-thereare)
    same => n,MeetMeCount(600)
    same => n,Playback(conf-peopleinconf)
```

If you pass a variable as the second argument to MeetMeCount(), the count is assigned to the variable, and playback of the count is skipped. You might use this to limit the number of participants, like this:

```
; limit the conference room to 10 participants
exten => 600,1,MeetMeCount(600,CONFCOUNT)
    same => n,GotoIf($[${CONFCOUNT} <= 10]?meetme:conf_full,1)
    same => n(meetme),MeetMe(600,i,54321)

exten => conf_full,1,Playback(conf-full)
```

Isn't Asterisk fun?

Conclusion

In this chapter, we've covered a few more of the many applications in the Asterisk dialplan, and hopefully we've given you some more tools that you can use to further explore the creation of your own dialplans. As with other chapters, we invite you to go back and reread any sections that require clarification.

Parking and Paging

I don't believe in angels, no. But I do have a wee parking
angel. It's on my dashboard and you wind it up. The
wings flap and it's supposed to give you a parking space.
It's worked so far.

—Billy Connolly

This chapter will focus on two important aspects of a PBX system: parking calls to allow them to be answered from a location different from where they were originally answered, and paging, which allows the announcement of who the call is for and how it can be retrieved.

In Asterisk, these two functionalities are exclusive to one another, and can be used independently of one another. Some businesses that contain large warehouses, or have employees who move around the office a lot and don't necessarily sit at a desk all day, utilize the paging and parking functionality of their systems to direct calls around the office. In this chapter we'll show you how to use both parking and paging in the traditional setting, along with a couple of more modern takes on this commonly used functionality.

features.conf

There are several features common to most modern PBXs that Asterisk also provides. Many of these features have optional parameters. The *features.conf* file is where you can adjust or define the various feature parameters in Asterisk.

DTMF-Based Features

Many of the parameters in *features.conf* only apply when invoked on calls that have been bridged by the dialplan applications `Dial()` or `Queue()`, with one or more of the options K, k, H, h, T, t, W, w, X, or x specified. Features accessed in this way are DTMF-based (meaning they can't be accessed via SIP messaging, but only through touch-tone

---

signals in the audio channel triggered by the users dialing the required digits on their dialpads).[*]

Transfers on SIP channels (for example from a SIP telephone) can be handled using the capabilities of the phone itself, and won't be affected by anything in the *features.conf* file.

The [general] section

In the [general] section of *features.conf*, you can define options that fine-tune the behavior of the park and transfer features in Asterisk. These options are listed in Table 11-1.

Table 11-1. features.conf [general] section

Option	Value/Example	Notes
parkext	700	Sets the default extension used to park calls.
parkpos	701-720	Sets the range of extensions used as the parking lot. Parked calls may be retrieved by dialing the numbers in this range.
context	parkedcalls	Sets the default dialplan context where the parking extension and the parking lot extensions are created.
parkinghints	no	Enables/disables automatic creation of dialplan hints for the parking lot extensions so that phones can subscribe to the state of extensions in the parking lot. The default is no.
parkingtime	45	Specifies the number of seconds a call will wait in the parking lot before timing out.
comebacktoorigin	yes	Configures the handling of timed-out parked calls. For more information on the behavior of this option, see the sidebar titled "Handling Timed-Out Parked Calls with the comebacktoorigin Option" on page 224.
courtesytone	beep	Specifies the sound file to be played to the parked caller when the parked call is retrieved from the parking lot.
parkedplay	caller	Indicates which side of the call to play the courtesytone to when a parked call is picked up. Valid options include callee, caller, both, or no. The default is no.
parkedcalltransfers	caller	Controls which side of a call has the ability to execute a DTMF-based transfer in the call that results from picking up a parked call. Valid options include callee, caller, both, or no. The default is no.
parkedcallreparking	caller	Controls which side of a call has the ability to execute a DTMF-based park in the call that results from picking up a parked call.[a] Valid options include callee, caller, both, or no. The default is no.

[*] Yes, we realize that a SIP INFO message is in fact a SIP message, and is not technically part of the audio channel, but the point is that you can't use the "transfer" or "park" button on your SIP phone to access these features while on a call. You'll have to send DTMF.

Option	Value/Example	Notes
parkedcallhangup	caller	Controls which side of a call has the ability to execute a DTMF-based hangup in the call that results from picking up a parked call. Valid options include callee, caller, both, or no. The default is no.
parkedcallrecording	caller	Controls which side of a call has the ability to initiate a DTMF-based one-touch recording in the call that results from picking up a parked call. Valid options include callee, caller, both, or no. The default is no.
parkeddynamic	yes	Enables the dynamic creation of parking lots in the dialplan. The channel variables PARKINGDYNAMIC, PARKINGDYNCONTEXT, and PARKINGDYNPOS need to be set.
adsipark	yes	Passes ADSI information regarding the parked call back to the originating set.
findslot	next	Configures the parking slot selection behavior. See ??? for more details.
parkedmusicclass	default	Specifies the class to be used for the music on hold played to a parked caller. A music class set in the dialplan using the CHANNEL(musicclass) dialplan function will override this setting.
transferdigittimeout	3	Sets the number of seconds to wait for each digit from the caller executing a transfer.
xfersound	beep	Specifies the sound to be played to indicate that an attended transfer is complete.
xferfailsound	beeperr	Specifies the sound to be played to indicate that an attended transfer has failed to complete.
pickupexten	*8	Configures the extension used for call pickup.
pickupsound	beep	Specifies the sound to be played to indicate a successful call pickup attempt. No sound is played by default.
pickupfailsound	beeperr	Specifies the sound to be played to indicate a failed call pickup attempt. No sound is played by default.
featuredigittimeout	1000	Sets the number of milliseconds to wait in between digits pressed during a bridged call when matching against DTMF activated call features.
atxfernoanswertimeout	15	Configures the number of seconds to wait for the target of an attended transfer to answer before considering the attempt timed out.
atxferdropcall	no	Configures behavior of attended transfer call handling when the transferer hangs up before the transfer is complete and the transfer fails. By default, this option is set to no and a call will be originated to attempt to connect the transferee back to the caller that initiated the transfer. If set to yes, the call will be dropped after the transfer fails.
atxferloopdelay	10	Sets the number of seconds to wait in between callback retries if atxferdropcall is set to no.
atxfercallbackretries	2	Sets the number of callback attempts to make if atxferdropcall is set to no. By default, this is set to 2 callback attempts.

[a] Read that again. It makes sense.

Handling Timed-Out Parked Calls with the comebacktoorigin Option

This option configures the behavior of call parking when the parked call times out (see the `parkingtime` option). `comebacktoorigin` can have one of two values:

yes *(default)*
>   When the parked call timeout is exceeded, Asterisk will attempt to send the call back to the peer that parked this call. If the channel is no longer available to Asterisk, the caller will be disconnected.

no
>   This option would be used when you want to perform custom dialplan functionality on parked calls that have exceeded their timeouts. The caller will be sent into a specific area of the dialplan where logic can be applied to gracefully handle the remainder of the call (this may involve simply returning the call to a different extension, or performing a lookup of some sort).

You also may need to take into account calls where the originating channel cannot handle a returned parked call. If, for example, the call was parked by a channel that is also a trunk to another system, there would not be enough information to send the call back to the correct person on that other system. The actions following a timeout would be more complex than `comebacktoorigin=yes` could handle gracefully.

Parked calls that time out with `comebacktoorigin=no` will always be sent into the `par kedcallstimeout` context.

 The dialplan (and contexts) were discussed in detail in Chapter 6.

The extension they will be sent to will be built from the name of the channel that parked the call. For example, if a SIP peer named `0004F2040808` parked this call, the extension will be `SIP_0004F2040808`.

If this extension does not exist, the call will be sent to the s extension in the `parked callstimeout` context instead. Finally, if the s extension of `parkedcallstimeout` does not exist, the call will be sent to the s extension of the `default` context.

Additionally, for any calls where `comebacktoorigin=no`, there will be an extension of `SIP_0004F2040808` created in the `park-dial` context. This extension will be set up to do a `Dial()` to `SIP/0004F2040808`.[†]

---

† We hope you realize that the actual extension will be related to the channel name that parked the call, and will not be `SIP_0004F2040808` (unless Leif sells you the Polycom phone from his lab).

The [featuremap] Section

This section allows you to define specific DTMF sequences, which will trigger various features on channels that have been bridged via options in the Dial() or Queue() application. The options are detailed in Table 11-2.

Table 11-2. features.conf [featuremap] section

Option	Value/Example	Notes	Dial()/Queue() Flags
blindxfer	#1	Invokes a blind (unsupervised) transfer	T, t
disconnect	*0	Hangs up the call	H, h
automon	*1	Starts recording of the current call using the Monitor() application (pressing this key sequence a second time stops the recording)	W, w
atxfer	*2	Performs an automated transfer	T, t
parkcall	#72	Parks a call	K, k
automixmon	*3	Starts recording of the current call using the MixMonitor() application (pressing this key sequence again stops the recording)	X, x

> The default blindxfer and disconnect codes are # and *, respectively. Normally you'll want to change them from the defaults, as they will interfere with other things that you might want to do (for example, if you use the Tt option in your Dial() command, every time you press the # key you'll initiate a transfer).

The [applicationmap] Section

This section of *features.conf* allows you to map DTMF codes to dialplan applications. The caller will be placed on hold until the application has completed execution.

The syntax for defining an application map is as follows (it must appear on a single line; line breaks are not allowed)[‡]:

```
<FeatureName> => <DTMF_sequence>,<ActivateOn>[/<ActivatedBy>]
,<Application>([<AppArguments>])[,MOH_Class]
```

What you are doing is the following:

1. Giving your map a name so that it can be enabled in the dialplan through the use of the DYNAMIC_FEATURES channel variable.
2. Defining the DTMF sequence that activates this feature (we recommend using at least two digits for this).

[‡] There is some flexibility in the syntax (you can look at the sample file for details), but our example uses the style we recommend, since it's the most consistent with typical dialplan syntax.

3. Defining which channel the feature will be activated on, and (optionally) which participant is allowed to activate the feature (the default is to allow both channels to use/activate this feature).

4. Giving the name of the application that this map will trigger, and its arguments.

5. Providing an optional music on hold (MOH) class to assign to this feature (which the opposite channel will hear when the application is executing). If you do not define any MOH class, the caller will hear only silence.

Here is an example of an application map that will trigger an AGI script:

```
agi_test => *6,self/callee,AGI(agi-test.agi),default
```

Since applications spawned from the application map are run outside the PBX core, you cannot execute any applications that trigger the dialplan (such as Goto(), Macro(), Background(), etc.). If you wish to use the application map to spawn external processes (including executing dialplan code), you will need to trigger an external application through an AGI() call or the System() application. Point being, if you want anything complex to happen through the use of an application map, you will need to test very carefully, as not all things will work as you might expect.

To use an application map, you must declare it in the dialplan by setting the DYNAMIC_FEATURES variable somewhere before the Dial() command that connects the channels. Use the double underscore modifier on the variable name in order to ensure that the application map is available to both channels throughout the life of the call. For example:

```
exten => 101,n,Set(__DYNAMIC_FEATURES=agi_test)
exten => 101,n,Dial(SIP/0000FFFF0002)
```

If you want to allow more than one application map to be available on a call, you will need to use the # symbol as a delimiter between multiple map names:

```
Set(__DYNAMIC_FEATURES=agi_test#my_other_map)
```

The reason why the # character was chosen instead of a simple comma is that older versions of the Set() application interpreted the comma differently than more recent versions, and the syntax for application maps has never been updated.

Don't forget to reload the features module after making changes to the *features.conf* file:

```
*CLI> features reload
```

You can verify that your changes have taken place through the CLI command *features show*. Make sure you test out your application map before you turn it over to your users!

Inheriting Channel Variables

Channel variables are always associated with the original channel that set them, and are no longer available once the channel is transferred.

In order to allow channel variables to follow the channel as it is transferred among the system, channel variable inheritance must be employed. There are two modifiers that can allow the channel variable to follow the channel: single underscore and double underscore.

The single underscore (_) causes the channel variable to be inherited by the channel for a single transfer, and is no longer available for additional transfers. If you use a double underscore (__), the channel variable will be inherited throughout the life of that channel.

Setting channel variables for inheritance simply requires you to prefix the channel name with a single or double underscore. The channel variables are then referenced exactly the same as they would be normally (e.g., do not attempt to read the values of channel variables with the underscores in the variable name).

Here's an example of setting a channel variable for single transfer inheritance:

```
exten => example,1,Set(_MyVariable=thisValue)
```

Here's an example of setting a channel variable for infinite transfer inheritance:

```
exten => example,1,Set(__MyVariable=thisValue)
```

To read the value of the channel variable, do not use underscore(s):

```
exten => example,1,Verbose(1,Value of MyVariable is: ${MyVariable})
```

Application Map Grouping

If you have a lot of features that you need to activate for a particular context or extension, you can group several features together in an application map grouping, so that one assignment of the DYNAMIC_FEATURES variable will assign all of the designated features of that map.

The application map groupings are added at the end of the *features.conf* file. Each grouping is given a name, and then the relevant features are listed.

```
[shifteight]
unpauseMonitor => *1     ; custom key mapping
pauseMonitor => *2       ; custom key mapping
agi_test =>              ; no custom key mapping
```

If you want to specify a custom key mapping to a feature in an application map grouping, simply follow the => with the key mapping you want. If you do not specify a key mapping, the default key map for that feature will be used (as found in the [featuremap] section). Regardless of whether you want to assign a custom key mapping or not, the => operator is required.

In the dialplan, you would assign this application map grouping with the Set() application:

```
Set(__DYNAMIC_FEATURES=shifteight) ; use the double underscore if you want to ensure
                                   ; both call legs have the variable assigned.
```

Parking Lots

A parking lot allows a call to be held in the system without being associated with a particular extension. The call can then be retrieved by anyone who knows the park code for that call. This feature is often used in conjunction with an overhead paging system (PA system, or Tannoy, for our UK readers). For this reason, it is often referred to as park-and-page; however, it should be noted that parking and paging are in fact separate.

To park a call in Asterisk, you need to transfer the caller to the feature code assigned to parking, which is assigned in the *features.conf* file with the parkext directive. By default, this is 700:

```
parkext => 700     ; What extension to dial to park (all parking lots)
```

You have to wait to complete the transfer until you get the number of the parking retrieval slot from the system, or you will have no way of retrieving the call. By default the retrieval slots, assigned with the parkpos directive in *features.conf*, are numbered from 701–720:

```
parkpos => 701-720  ; What extensions to park calls on (defafult parking lot)
```

Once the call is parked, anyone on the system can retrieve it by dialing the number of the retrieval slot (parkpos) assigned to that call. The call will then be bridged to the channel that dialed the retrieval code.

There are two common ways to define how retrieval slots are assigned. This is done with the findslot directive in the *features.conf* file. The default method (findslot => first) always uses the lowest-numbered slot if it is available, and only assigns higher-numbered codes if required. The second method (findslot => next) will rotate through the retrieval codes with each successive park, returning to the first retrieval code after the last one has been used. Which method you choose will depend on how busy your parking lots are. If you use parking rarely, the default findslot of first will be best (people will be used to their parked calls always being in the same slot). If you use parking a lot (for example, in an automobile dealership), on the other hand, it is far better for each successive page to assign the next slot, since you will often have more than one call parked at a time. Your users will get used to listening carefully to the actual parking lot number (instead of just always dialing 701), and this will minimize the chance of people accidentally retrieving the wrong call on a busy system.

If you are using parking, you are probably also going to need a way to announce the parked calls so that the intended parties know how to retrieve them. While you could just run down the hall yelling "Bob, there's a call for you on 701!," the more professional method is to use a paging system (more formally known as a public address system), which we will discuss in the next section.

Overhead and "Underchin" Paging (a.k.a. Public Address)

In many PBX systems, it is desirable to be able to allow a user to send his voice from a telephone into a public address system. This normally involves dialing a feature code or extension that makes a connection to a public address resource of some kind, and then making an announcement through the handset of the telephone that is broadcast to all devices associated with that paging resource. Often, this will be an external paging system consisting of an amplifier connected to overhead speakers; however, paging through the speakers of office telephones is also popular (mainly for cost reasons). If you have the budget (or an existing overhead paging system), overhead paging is generally better, but set paging (a.k.a. "underchin" paging) can work well in many environments. What is perhaps most common is to have a mix of set and overhead paging, where, for example, set-based paging might be in use for offices, but overhead paging would be used for warehouse, hallway, and public areas (cafeteria, reception, etc.).

In Asterisk, the Page() application is used for paging. This application simply takes a list of channels as its argument, calls all of the listed channels simultaneously, and, as they are answered, puts each one into a conference room. With this in mind, it becomes obvious that one requirement for paging to work is that *each destination channel must be able to automatically answer the incoming connection* and place the resultant audio onto a speaker of some sort (in other words, Page() won't work if all the phones just ring).

So, while the Page() application itself is painless and simple to use, getting all the destination channels to handle the incoming pages correctly is a bit trickier. We'll get to that shortly.

The Page() application takes three arguments, defining the group of channels the page is to be connected to, the options, and the timeout:

```
exten => *724,1,Page(${ChannelsToPage},i,120)
```

The options (outlined in Table 11-3) give you some flexibility with respect to how Page() works, but the majority of the configuration is going to have to do with how the target devices handle the incoming connection. We'll dive into the various ways you can configure devices to receive pages in the next section.

Table 11-3. Page() options

Option	Description	Discussion
d	Enables full-duplex audio	Sometimes referred to as "talkback paging," the use of this option implies that the equipment that receives the page has the ability to transmit audio back at the same time as it is receiving audio. Generally, you would not want to use this unless you had a specific need for it.
i	Ignores attempts to forward the call	You would normally want this option enabled.
q	Does not play beep to caller (quiet mode)	Normally you won't use this, but if you have an external amplifier that provides its own tone, you may want to set this option.
r	Records the page into a file	If you intended on using the same page multiple times in the future, you could record the page and then use it again later by triggering it using `Originate()` or using the `A(x)` option to `Page()`.
s	Dials a channel only if the device state is `NOT_INUSE`	This option is likely only useful (and reliable) on SIP-bound channels, and even so may not work if a single line is allowed multiple calls on it. Therefore, don't rely on this option in all cases.
A(x)	Plays announcement x to all participants	You could use a previously recorded file to be played over the paging system. If you combined this with `Originate()` and `Record()`, you could implement a delayed paging system.
n	Does not play announcement simultaneously to caller (implies `A(x)`)	By default the system will play the paged audio to both the caller and the callee. If this option is enabled, the paged audio will not be played to the caller (the person paging).

Because of how `Page()` works, it is very resource-intensive. We cannot stress this enough. Carefully read on, and we'll cover how to ensure that paging does not cause performance problems in a production environment (which it is almost certain to do if not designed correctly).

Places to Send Your Pages

As we stated before, `Page()` is in and of itself very simple. The trick is how to bring it all together. Pages can be sent to different kinds of channels, and they all require different configuration.

External paging

If a public address system is installed in the building, it is common to connect the telephone system to an external amplifier and send pages to it through a call to a channel. One way of doing this is to plug the sound card of your server into the amplifier and send calls to the channel named `Console/DSP`, but this assumes that the sound drivers on your server are working correctly and the audio levels are normalized correctly on that channel. Another, potentially simpler, and possibly more robust way to handle external paging is to use an FXS device of some kind (such as an ATA), which is connected to a paging interface such as a Bogen UTI1,§ which then connects to the paging amplifier.‖

In your dialplan, paging to an external amplifier would look like a simple `Dial()` to the device that is connected to the paging equipment. For example, if you had an ATA configured in *sip.conf* as `[PagingATA]`, and you plugged the ATA into a Bogen UTI1, you would perform paging by dialing:

```
exten => *724,1,Verbose(2,Paging to external amplifier) ; note the '*' in the
                                                         ; extension is part of
                                                         ; what you actually dial

    same => n,Set(PageDevice=SIP/PagingATA)
    same => n,Page(${PageDevice},i,120)
```

Note that for this to work you will have had to register your ATA as a SIP device under *sip.conf*, and in this case we named the device `[PagingATA]`. You can name this device anything you want (for example, we often use the MAC address as the name of a SIP device), but for anything that is not a user telephone, it can be helpful to use a name that makes it stand out from other devices.

If you had an FXS card in your system and you connected the UTI1 to that, you would `Dial()` to the channel for that FXS port instead:

```
    same => n,Dial(DAHDI/25)
```

The UTI1 answers the call and opens a channel to the paging system; you then make your announcement and hang up.

Set paging

Set-based paging first became popular in key telephone systems, where the speakers of the office telephones are used as a poor-man's public address system. Most SIP telephones have the ability to auto-answer a call on handsfree, which accomplishes what is required on a per-telephone basis. In addition to this, however, it is necessary to pass the audio to more than one set at the same time. Asterisk uses its built-in conferencing engine to handle the under-the-hood details. You use the `Page()` application to make it happen.

Like `Dial()`, the `Page()` application can handle several channels. Since you will generally want `Page()` to signal several sets at once (perhaps even all the sets on your system) you may end up with lengthy device strings that look something like this:

```
Page(SIP/SET1&SIP/SET2&SIP/SET3&SIP/SET4&SIP/SET5&SIP/SET6&SIP/SET7&...
```

---

§ The Bogen UTI1 is useful because it can handle all manner of different kinds of incoming and outgoing connections, which pretty nearly guarantees that you'll be able to painlessly connect your telephone system to any sort of external paging equipment, no matter how old or obscure.

‖ In this book we're assuming that the external paging equipment is already installed and was working with the old phone system.

Beyond a certain size, your Asterisk system will be unable to page multiple sets. For example, in an office with 200 telephones, using SIP to page every set would not be possible; the traffic and CPU load on your Asterisk server would simply be too much. In cases like this, you should be looking at either multicast paging or external paging.

Perhaps the trickiest part of SIP-based paging is the fact that you usually have to tell each set that it must auto-answer, but different manufacturers of SIP telephones use different SIP messages for this purpose. So, depending on the telephone model you are using, the commands needed to accomplish SIP-based set paging will be different. Here are some examples:

- For Aastra:

```
exten => *724,1,Verbose(2,Paging to Aastra sets)
    same => n,SIPAddHeader(Alert-Info: info=alert-autoanswer)
    same => n,Set(PageDevice=SIP/00085D000000)
    same => n,Page(${PageDevice},i)
```

- For Polycom:

```
exten => *724,1,Verbose(2,Paging to Polycom sets)
    same => n,SIPAddHeader(Alert-Info: Ring Answer)
    same => n,Set(PageDevice=SIP/0004F2000000)
    same => n,Page(${PageDevice},i)
```

- For Snom:

```
exten => *724,1,Verbose(2,Paging to Snom sets)
    same => n,Set(VXML_URL=intercom=true)

; replace 'domain.com' with the domain of your system
    same => n,SIPAddHeader(Call-Info: sip:domain.com\;answer-after=0)
    same => n,Set(PageDevice=SIP/000413000000)
    same => n,Page(${PageDevice},i)
```

- For Cisco SPA (the former Linksys phones, not the 79XX series):

```
exten => *724,1,Verbose(2,Paging to Cisco SPA sets -- but not Cisco 79XX sets)
    same => n,SIPAddHeader(Call-Info:\;answer-after=0)        ; Cisco SPA phones
    same => n,Set(PageDevice=SIP/0004F2000000)
    same => n,Page(${PageDevice},i)
```

Assuming you've figured that out, what happens if you have a mix of phones in your environment? How do you control which headers to send to which phones?[#]

Any way you slice it, it's not pretty.

[#]Hint: the local channel will be your friend here.

Fortunately, many of these sets support IP multicast, which is a far better way to send a page to multiple sets (read on for details). Still, if you only have a few phones on your system and they are all from the same manufacturer, SIP-based paging could be the simplest, so we don't want to scare you off it completely.

Multicast paging via the MulticastRTP channel

If you are serious about paging through the sets on your system, and you have more than a handful of phones, you will need to look at using IP multicast. The concept of IP multicast has been around for a long time,* but it has not been widely used. Nevertheless, it is ideal for paging within a single location.

Asterisk has a channel (`chan_multicast_rtp`) that is designed to create an RTP multicast. This stream is then subscribed to by the various phones, and the result is that whenever media appears on the multicast stream, the phones will pass that media to their speakers.

Since `MulticastRTP` is a channel driver, it does not have an application, but instead will work anywhere in the dialplan that you might otherwise use a channel. In our case, we'll be using the `Page()` application to initiate our multicast.

To use the multicast channel, you simply send a call to it the same as you would to any other channel. The syntax for the channel is as follows:

```
MulticastRTP/<type>/<ip address:port>[/<linksys address:port>]
```

The type can be either `basic` or `linksys`. The basic syntax of the `MulticastRTP` channel looks like this:

```
exten => *723,1,Page(MulticastRTP/basic/239.0.0.1:1234)
```

Not all sets support IP multicast, but we have tested it out on Snom,† Linksys/Cisco, and Aastra, and it works swell.‡

---

* It even has its own Class D reserved IP address space, from 224.0.0.0 to 239.255.255.255 (but read up on IP multicast before you just grab one of these and assign it). Parts of this address space are private, parts are public, and parts are designated for purposes other than what you might want to use them for. For information about multicast addressing, see *http://en.wikipedia.org/wiki/IP_multicast#IP_multicast_addressing_assignments*.

† Very loud, and no way to adjust gain.

‡ So far as we can tell, Polycom sets do not support multicast. We certainly were not able to find a way to use it.

Multicast Paging on Cisco SPA Telephones

The multicast paging feature on Cisco SPA phones is a bit strange, but once configured it works fine. The trick of it is that the address you put into the phone is not the multicast address that the page is sent across, but rather a sort of signaling channel.

What we have found is that you can make this address the same as the multicast address, but simply use a different port number.

The dialplan looks like this:

```
exten => *724,1,Page(MulticastRTP/linksys/239.0.0.1:1234/239.0.0.1:6061)
```

In the SPA phone, you need to log into the Administration interface and navigate to the SIP tab. At the very bottom of the page you will find the section called *Linksys Key System Parameters*. You need to set the following parameters:

- `Linksys Key System: Yes`
- `Multicast Address: 239.0.0.1:6061`

Note that the multicast address you assign to the phone is the one that comes second in the channel definition (in our example, the one using port 6061).

Note that you can write the `Page()` command in this format in an environment where there is a mix of SPA phones (FKA Linksys, now Cisco) phones and other types of phones. The other phones will use the first address and will work the same as if you had used `basic` instead of `linksys`.

VoIP paging adaptors

Recently, there have been some VoIP-based paging speakers introduced to the market. These devices are addressed in the dialplan in the exact same way as a SIP ATA connected to a UTI1, but they can be installed in the same manner as overhead speakers would be. Since they auto-answer, there is no need to pass them any extra information, the way you would need to with a SIP telephone set.

For smaller installations (where no more than perhaps half a dozen speakers are required), these devices may be cost-effective. However, for anything larger than that, (or installation in a complex environment such as a warehouse or parking lot), you will get better performance at far less cost with a traditional analog paging system connected to the phone system by an analog (FXS) interface.

We don't know if these devices support multicast. Keep this in mind if you are planning to use a large number of them.

Combination paging

In many organizations, there may be a need for both set-based and external paging. As an example, a manufacturing facility might want to use set-based paging for the office area but overhead paging for the plant and warehouse. From Asterisk's perspective, this is fairly simple to accomplish. When you call the Page() application, you simply specify the various resources you want to page, separated by the & character, and they will all be included in the conference that the Page() application creates.

Bringing it all together

At this point you should have a list of the various channel types that you want to page. Since Page() will nearly always want to signal more than one channel, we recommend setting a global variable that defines the list of channels to include, and then calling the Page() application with that string:

```
[global]

MULTICAST=MulticastRTP/linksys/239.0.0.1:1234
;MULTICAST=MulticastRTP/linksys/239.0.0.1:1234/239.0.0.1:6061 ; if you have SPA
                                                    ; (Linksys/Cisco)
                                                    ; phones

BOGEN=SIP/ATAforPaging   ; This assumes an ATA in your sip.conf file named
                         ; [ATAforPaging]
;BOGEN=DAHDI/25          ; We could do this too, assuming we have an analog
                         ; FXS card at DAHDI channel 25
PAGELIST=${MULTICAST}&${BOGEN} ; All of these variable names are arbitrary.
                               ; Asterisk doesn't care what you call these strings

[page_context] ; You don't need a page context, so long as the extension you
               ; assign to paging is dialable by your sets

exten => *724,1,Page(${PAGELIST},i,120)
```

This example offers several possible configurations, depending on the hardware. While it is not strictly required to have a PAGELIST variable defined, we have found that this will tend to simplify the management of multiple paging resources, especially during the configuration and testing process.

We created a context for paging for the purposes of this example. In order for this to work, you'll need to either include this context in the contexts where your sets enter the dialplan, or code a Goto() in those contexts to take the user to this context and extension (i.e., Goto(page_context,*724,1)) Alternatively, you could hardcode an extension for the Page() application in each context that services sets.

Zone Paging

Zone paging is popular in places such as automobile dealerships, where the parts department, the sales department, and perhaps the used car department all require paging, but have no need to hear each other's pages.

In zone paging, the person sending the page needs to select which zone she wishes to page into. A zone paging controller such as a Bogen PCM2000 is generally used to allow signaling of the different zones: the Page() application signals the zone controller, the zone controller answers, and then an additional digit is sent to select which zone the page is to be sent to. Most zone controllers will allow for a page to all zones, in addition to combining zones (for example, a page to both the new and used car sales departments).

You could also have separate extensions in the dialplan going to separate ATAs (or groups of telephones), but this may prove more complicated and expensive than simply purchasing a paging controller that is designed to handle this. Zone paging doesn't require any significantly different technology, but it does require a little more thought and planning with respect to both the dialplan and the hardware.

Conclusion

In this chapter we explored the *features.conf* file, which contains the functionality for enabling DTMF-based transfers, enabling the recording of calls during a call, and configuring parking lots for one or more companies. We also looked at various ways of announcing calls and information to people in the office using a multitude of paging methods, including traditional overhead paging systems and multicast paging to the phone sets on employees' desks. This exploration of the various methods of implementing the traditional parking and paging methods in a modern way will hopefully show you the flexibility Asterisk can offer.

Internet Call Routing

There ain't no such thing as a free lunch (TANSTAAFL).

—Robert Heinlein

One of the attractions of VoIP is the concept of avoiding the use of the PSTN altogether, and routing all calls directly between endpoints using the Internet at little or no cost. While the technology to do this has been around for some time, the reality is that most phone calls still cost money—even those that are routed across VoIP services.

From a technology standpoint, there are still many systems out there that cannot handle routing VoIP calls using anything other than a dialpad on a telephone.

From a cultural standpoint, we are still used to calling each other using a numerical string (a.k.a., a phone number). With VoIP, the concept of being able to phone somebody using *name@domain* (just as we do with email) makes sense, but there are a few things to consider before we can get there.

So what's holding everything up?

---

freenum.org (*http://www.freenum.org*)

The first few sections of this chapter may put you off the whole idea entirely, so we want to start off by saying that *freenum.org* proposes an interim solution to the whole mess that is so elegant, we can't see any reason why everyone in the VoIP community won't embrace it.[*]

DNS and SIP URIs

The Domain Name System (DNS) is designed to make it easier for humans to locate resources on the Internet. While ultimately all connections between endpoints are

---

[*] Seriously, get your butt over to *freenum.org* and get your ISN today. It's simple and free, and soon all the cool kids will have one.

handled through numerical IP addresses, it can be very helpful to associate a name (such as *www.google.com*) with what may in fact be multiple IP addresses.

In the case of VoIP, the use of a domain name can take something like `100@192.168.1.1` (*extension@server*) and make it available as `leif@shifteight.org` (which looks so much sexier on a business card).

The SIP URI

A SIP URI generally looks like `sip:endpoint@domain.tld`. Depending on your SIP client, you may be able to dial a SIP URI as *endpoint@domain.tld*, or even just as *endpoint* (if you have a proxy server and the endpoint you are calling is part of your domain).

For a SIP telephone, which often only has a numerical dialpad, it can be problematic to dial a SIP URI by name,[†] so it has become common to use numerical dialing to reach external resources. We are also used to making "phone calls" using "phone numbers." The SIP protocol itself, however, only understands *resource@address*, so whatever you dial must ultimately be converted to this format before SIP can do anything with it. Usually the only reason you can dial something by "phone number" from your SIP phone is because you are registered to a resource that understands how to convert the numerical strings you dial into SIP URIs.

In Asterisk, the *resource* part of the URI (the part before the @) must match an extension in the dialplan.[‡] The *address* portion will be the address (or hostname) of the Asterisk server itself. So, a URI of `sip:100@shifteight.org` will end up at an extension called `100`, somewhere in the dialplan of the server that provides SIP service for `shifteight.org`.

What is dialed (`100`) may not in any way relate to the actual identifier of the endpoint being connected to. For example, we might have a user named Leif whose phone may be a device that registers itself by its MAC address, and therefore could be something like `0000FFFF0001@192.168.1.99`.[§] Much of the purpose of the Asterisk dialplan is to simplify addressing for users and to handle the complexities of the various protocols that Asterisk supports.

SRV Records

A Service Record (SRV) is a somewhat new type of DNS record that provides information about available services. Defined in RFC 2782, it is often used by newer protocols

---

† Do you know where the @ symbol is on your dialpad?

‡ Bear in mind that an extension in Asterisk can be any alphanumeric string, such as `leif` or `100`.

§ You could actually dial this URI directly from your phone and bypass the Asterisk server, but you can see how dialing `100` is going to be a lot more popular than trying to figure out how to type `0004f2a1b2c3@192.168.1.99` into your phone using just the numeric dialpad (it can be done, by the way).

(SIP being one of them). If you want to support SIP lookups on your domain, you will require a relevant SRV record in order to properly respond.

When a SIP connection does a lookup on `leif@shifteight.org`, for the purposes of SIP, the SRV record can respond that the requested service (SIP) is actually found on the server `pbx.shifteight.org` (or possibly even on a completely different domain, such as `pbx.tothemoon.net`).

Internet hosting providers typically offer a web-based interface for setting up DNS records, but many of them do not provide a good interface for SRV records (assuming they offer anything at all). You can generally set up A records and MX records easily enough, but SRV records can be trickier. If your host does not support SRV records, you will need to move your DNS hosting to another provider if you want to be able to support SIP SRV lookups for your domain.

The majority of DNS servers run BIND (Berkeley Internet Name Daemon). The BIND record for an SRV entry for SIP will look something like this:

```
_sip._udp.shifteight.org. 86400 IN SRV 0 0 5060 pbx.shifteight.org.
```

The form of the record is detailed in Table 12-1.

Table 12-1. Components of a SIP SRV record

Name	Description	Example
Service	Symbolic name of service	`_sip.`
Proto	Transport protocol	`_udp.`
Name	Domain name for this record[a]	`shifteight.org.`
TTL	Time to live (in seconds)	`86400`
Class	DNS class field (always IN)	`IN`
Priority	Target host priority	`0`
Weight	Relative weight of this record	`0`
Port	TCP/UDP port number	`5060`
Target	Hostname of machine providing this service	`pbx.shifteight.org.`

[a] Note the trailing dot.

When you configure an SRV record, you can test it with the following Linux command:

```
# dig SRV _sip._udp.shifteight.org
```

The result will contain several components, but the section you are interested in is:

```
;; ANSWER SECTION:
_sip._udp.shifteight.org. 14210 IN SRV 0 0 5060 pbx.shifteight.org.
```

This means that your DNS server is responding correctly to an SRV lookup for SIP to your domain by responding with the hostname of your PBX (in this case, `pbx.shifteight.org`).

Any SIP requests to your domain will be referred to your Asterisk server, which will be responsible for handling incoming SIP connections.‖

If your dialplan does not understand the name/resource/endpoint portion of the SIP URI, calls will fail. This means that if you want to be able to offer resources in your Asterisk system by name, you will need relevant dialplan entries.

Accepting Calls to Your System

When a SIP URI comes into your Asterisk system, the resource portion of the URI will arrive in the dialplan as an `${EXTEN}`. So, for example, `leif@shifteight.org` would arrive in the dialplan as `leif` within the `${EXTEN}` channel variable in whatever context you use to handle unauthenticated SIP calls (if you are building your dialplan using the examples in this book, that will be the `unauthenticated` dialplan context).

Modifying sip.conf

Once you are familiar with the security implications of allowing unauthenticated SIP connections, you will need to ensure that your *sip.conf* file allows for them. While Asterisk allows them by default, in earlier chapters of this book we have instructed you to disable unauthenticated SIP calls. The logic for this is simple: if you don't need it, don't enable it.

Since we are now interested in allowing calls from the Internet, we will need to allow unauthenticated SIP calls. We do that by setting a general variable in the */etc/asterisk/sip.conf* file, as follows:

```
[general]
context=unauthenticated      ; default context for incoming calls
allowguest=yes               ; enable unauthenticated calls
```

After making this change, don't forget to reload SIP, using this command from the command line:

```
$ sudo asterisk -rx "sip reload"
```

or this one from the Asterisk CLI:

```
*CLI> sip reload
```

You can verify that the changes have succeeded using the Asterisk CLI command *sip show settings*. What you want to see is `Allow unknown access: Yes` under the `Global Settings` section, and `Context: unauthenticated` under the `Default Settings` header.

Standard dialplan

In order to handle an incoming name, your dialplan needs to contain an extension that matches that name.

‖ This could just as easily be a proxy server, or any other server capable of handling incoming SIP connections.

A dialplan entry on the `pbx.shifteight.org` system might look like this:

```
[unauthenticated]
exten => leif,1,Goto(PublicExtensions,100,1)

exten => jim,1,Goto(PublicExtensions,101,1)

exten => tilghman,1,Goto(PublicExtensions,102,1)

exten => russell,1,Goto(PublicExtensions,103,1)
```

This is by far the simplest way to implement name dialing, but it is also complex to maintain, especially in systems with hundreds of users.

In order to implement name handling in a more powerful way, you could add something like the following to your *extensions.conf* file. Note that some lines have been wrapped in this example due to space restrictions. These lines must appear on a single line in the dialplan. All lines should start with **exten** =>, **same** =>, or a comment indicator (;).

```
[unauthenticated]
exten => _[A-Za-z0-9].,1,Verbose(2,UNAUTHENTICATED REQUEST TO ${EXTEN} FROM
${CALLERID(all)})
    same => n,Set(FilteredExtension=${FILTER(A-Za-z0-9,${EXTEN})})
    same => n,Set(CheckPublicExtensionResult=${DIALPLAN_EXISTS(PublicExtensions,
${FilteredExtension},1)})
    same => n,GotoIf($["${CheckPublicExtensionResult}" = "0"]?CheckEmailLookup)
    same => n,Goto(PublicExtensions,${FilteredExtension},1)

; This is our handler for when someone dials a SIP URI with a name
    same => n(CheckEmailLookup),GoSub(subEmailToExtensionLookup,start,1
(${TOLOWER(${FilteredExtension})}))
    same => n,GotoIf($["${GOSUB_RETVAL}" = "NoResult"]?i,1:PublicExtensions,
${GOSUB_RETVAL},1)
    same => n,Goto(i,1)

; This handles invalid numbers/names
exten => i,1,Verbose(2,Incoming call from ${CALLERID(all)} to context ${CONTEXT}
found no result)
    same => n,Playback(silence/1&invalid)
    same => n,Hangup()

; These are explicit extension matches (useful on small systems)
exten => leif,1,Goto(PublicExtensions,100,1)

exten => jim,1,Goto(PublicExtensions,101,1)

exten => tilghman,1,Goto(PublicExtensions,102,1)

exten => russell,1,Goto(PublicExtensions,103,1)
```

When a call enters the dialplan, it can match in one of two places: it can match our pattern match at the top, or it can match the explicit named extensions closer to the bottom of our example (i.e., `leif`, `jim`, `tilghman`, or `russell`).

If the call does not explicitly match our named extensions, the pattern match will be utilized. Our pattern match of _[A-Za-z0-9]. matches any string starting with an alphanumeric character followed by one or more other characters.

The incoming string needs to be made safe, so we utilize the FILTER() function to remove nonalphanumeric characters, and assign the result to the FilteredExtension channel variable.

The DIALPLAN_EXISTS() function will be used to see if the request matches anything in the PublicExtensions context. This function will return either a 0 (if no match is found) or a 1 (when a match is found) and assign the result to the CheckPublicExtensionResult channel variable.

The next line is a GotoIf() that checks the status of the CheckPublicExtensionResult variable. If the result returned was 0, the dialplan will continue at the CheckEmail Lookup priority label. If the result was anything other than 0 (in this case, the other result could have been a 1), the next line of the dialplan will be executed. This line will perform a Goto() and continue execution in the PublicExtensions context (presumably to dial our destination endpoint).

Assuming our CheckPublicExtensionResult variable was a 0, our dialplan will continue at the CheckEmailLookup priority label, where we use the subroutine subEmailToExtensionLookup via a GoSub().# We pass the value contained within the FilteredExtension channel variable to the subroutine, but you'll notice that we've wrapped it in the TOLOWER() dialplan function (which expects your email addresses to be stored in lowercase as opposed to mixed case).

Upon return from the subEmailToExtensionLookup subroutine, we check the GOSUB_RETVAL channel variable (which was automatically set when the subroutine returned). The result will be one of two things: the extension number that matches the name that was passed to the subroutine, or the string NoResult. Our dialplan checks ${GOSUB_RETVAL}, and if it contains NoResult, the caller is passed to the i (invalid) extension, where we inform the caller that the extension dialed is invalid. If all is well, the call will continue execution in the PublicExtensions context.

File parsing

This little trick will allow you to use the *voicemail.conf* file to look up valid usernames against their email address. This could end up being kludgy, and it requires that the email field in *voicemail.conf* is filled out and contains a username (before the @ symbol) that you will support in your dialplan, but it's simple to code in the dialplan, and if nothing else it will give you some ideas of how you might handle providing a more automated way of linking names to extension numbers for the purpose of SIP URI dialing. Note that this method will not allow you to exclude some people from name dialing. It's all or nothing.

#We explain the use of subEmailToExtensionLookup in the following section.

We've written this as a subroutine, which is invoked something like this:

```
; where 'name' is the username as found in the email address
GoSub(subEmailToExtensionLookup,start,1(name))
```

The subroutine looks like this:

```
[subEmailToExtensionLookup]
exten => start,1,Verbose(2,Checking for user in voicemail.conf)
    same => n,Set(LOCAL(FilteredExtension)=${FILTER(a-z0-9,${ARG1})})
    same => n,Set(LOCAL(Result)=${SHELL(grep "${LOCAL(FilteredExtension)}@"
/etc/asterisk/voicemail.conf)})
    same => n,GotoIf($[${ISNULL(${LOCAL(Result)})}]?no_Result,1)
    same => n,Set(LOCAL(ExtensionToDial)=${CUT(${LOCAL(Result)},=,1)})
    same => n,Set(LOCAL(ExtensionToDial)=${FILTER(0-9,${LOCAL(ExtensionToDial)})})
    same => n,Return(${LOCAL(ExtensionToDial)})

exten => no_Result,1,Verbose(2,No user ${ARG1} found in voicemail.conf)
    same => n,Return(NoResult)
```

Let's go over this code, because there are some useful actions being performed that you may be able to apply for other purposes as well.

First, a channel variable named `FilteredExtension` is created. This variable is local to the subroutine:

```
Set(LOCAL(FilteredExtension)=${FILTER(a-z0-9,${ARG1})})
```

The `FILTER()` function looks at the entire `${ARG1}` and removes any nonalphanumeric characters. This is primarily for security reasons. We are passing this string out to the shell, so it's critical to ensure it will only contain characters that we expect.

The next step is where the coolness happens:

```
Set(LOCAL(Result)=${SHELL(grep "${LOCAL(FilteredExtension)}@" /etc/asterisk/voicemail.conf)})
```

The shell is invoked in order to run the *grep* shell application, which will search through the *voicemail.conf* file, return any lines that contain name@, and assign the result to the variable ${Result}:

```
GotoIf($[${ISNULL(${LOCAL(Result)})}]?no_result,1)
```

If no lines contain the string we're looking for, we'll return from the subroutine the value NoResult (which will be found in the ${GOSUB_RETVAL} channel variable). The dialplan section that called the subroutine will need to handle this condition.

We've created an extension named no_result for this purpose:

```
exten => no_result,1,Verbose(2,No user ${ARG1} found in voicemail.conf)
    same => n,Return(NoResult)
```

If ${Result} is not null, the next steps will clean up ${Result} in order to extract the extension number* of the user with the name passed in ${ARG1}:

```
Set(LOCAL(ExtensionToDial)=${CUT(${LOCAL(Result)},=,1)})
```

The CUT() function will use the = symbol as the field delimiter and will assign the value from the first field found in ${Result} to the new variable ExtensionToDial. From there, we simply need to trim any trailing spaces by filtering all nonnumeric characters:

```
Set(LOCAL(ExtensionToDial)=${FILTER(0-9,${LOCAL(ExtensionToDial)})})
```

We can now return the extension number of the name we received:

```
Return(${LOCAL(ExtensionToDial)})
```

This example was something we whipped up for the purposes of illustrating some methods you can employ in order to easily match names to extension numbers for the purposes of SIP URI dialing. This is by no means the best way of doing this, but it is fairly simple to implement, and in many cases may be all that you need.

Database lookup

Using a database is by far the best way to handle user information on larger, more complex systems. We will discuss integrating Asterisk with databases in more detail in Chapter 16, but it is useful to introduce the concept here.

A database is ideal for handling name lookup, as it makes maintenance of user data (and integration with external systems such as web interfaces) far simpler. However, it does require a bit more effort to design and implement.

The example we will use in this chapter will work, but for a production environment it is probably too simplistic. Our goal here is simply to give you enough information to understand the concept; a tighter integration is part of what is covered in Chapter 16.

First, we'll need a table to handle our name-to-extension mapping. This could be a separate table from the main user table, or it could be handled in the main user table, provided that that table contains a field that will contain the exact strings that users will publish as their SIP URIs (as an example, some companies have rules regarding how email addresses look, so Leif might have a URI such as lmadsen@shifteight.org, or leif.madsen@shifteight.org).

 If you are serious about implementing this example in a production system, make sure you are familiar with the material in Chapter 16, as some key concepts are covered there that we omit here.

Our sample NameMapping table looks like Table 12-2.

---

* In actual fact, what we are extracting is the voicemail box number; however, this number is generally going to be the same as the user's dialable internal extension number. If it is not the same, this particular technique will not accomplish name-to-extension lookups, and another way will have to be found.

Table 12-2. NameMapping table

Name	Extension	Context
leif	100	publicExtensions
leif.madsen	100	publicExtensions
lmadsen	100	publicExtensions
jim	101	publicExtensions
reception	0	Services[a]
voicemail	*98	Services

[a] Make sure this context exists on your system.

We believe that having a separate table that only handles name-to-extension/context mapping is the most useful solution, since this table can be used to handle more than just users with telephone sets. You are encouraged to come up with other ways to handle this that may be more suitable to your environment.

In the dialplan, we would refer to this table using Asterisk's func_odbc function:

```
[subLookupNameInNameMappingTable]
exten => start,1,Verbose(2,Looking up ${ARG1})

; where 'name' is the username as found in the email address
    same => n,Set(ARRAY(CalleeExtension,CalleeContext)=${GET_NAME_LOOKUP(${ARG1})})
    same => n,GotoIf($[${ISNULL(${CalleeExtension})}]?no_result,1)
    same => n,GotoIf($[${ISNULL(${CalleeContext})}]?no_result,1)
    same => n,Return() ; You'll need to handle the new CalleeExtension and
                       ; CalleeContext variables in the code that called this
                       ; subroutine

exten => no_result,1,Verbose(2,Name was not found in the database.)
    same => n,Return(NoResult)
```

The */etc/asterisk/func_odbc.conf* file will require the following entry:

```
[NAME_LOOKUP](DB)
prefix=GET
SELECT Extension,Context FROM NameMapping WHERE Name='${ARG1}'
```

Keep in mind that there's nothing to say you can't reference more than one datastore to look up names. For example, you might have a table such as the one we've described here, but also have a secondary lookup that goes to, say, an LDAP database to try to resolve names there as well. This can get complicated to configure and maintain, but if designed right it can also mean that your Asterisk system can be tightly integrated with other systems in your enterprise.

Details on how to handle all of this in your dialplan are beyond the scope of this book. Suffice it to say that in your dialplan you will still need to handle the values that your subroutine creates or assigns.

Dialing SIP URIs from Asterisk

Asterisk can dial a SIP URI as easily as any other sort of destination, but it is the endpoint (namely, your telephone) that is ultimately going to shoulder the burden of composing the address, and there lies the difficulty.

Most SIP telephones will allow you to compose a SIP URI using the dialpad. This sounds like a great idea at first, but since there are no typewriter keys on a phone set, in order to dial something like `jim.vanmeggelen@shifteight.org` what you would need to actually input into the phone would be something along the lines of:

```
5-444-6-*-888-2-66(pause)-6-33-4(pause)-4-33-555-33-66-#-7777-44(pause)-444-333-8-33-
444-(pause)-4(pause)-44-8-*-666-777-4
```

To support this in your dialplan, you would need something similar to this[†]:

```
exten => _[0-9a-zA-Z].,1,Verbose()
    same => n,Set(FilteredExtension=${FILTER(0-9a-zA-Z@-_.,${EXTEN})})
    same => n,Dial(SIP/${FilteredExtension})
```

It's simple, it's fun, and it works! … ?

The reality is that until all phones support complex and flexible address books, as well as a QWERTY-style keyboard (perhaps via touchscreen), SIP URI dialing is not going to take off.

If you have a SIP URI that you want to dial on a regular basis (for example, during the writing of this book there were many calls made between Jim and Leif), you could add something like this to your dialplan:

```
exten => 5343,1,Dial(SIP/leif.madsen@shifteight.org)
```

With this in your dialplan, you could dial 5343 (LEIF) on your phone and the Asterisk dialplan would translate it into the appropriate SIP URI. It's not practical for a large number of URIs, but for a few here and there it can be a helpful shortcut.

Nevertheless, keep reading, because there are some very useful components of DNS that simplify the process of dialing directly between systems without the use of the PSTN.

[†] Technically, the characters ! # $ % & ' * + / = ? ^ ` { | } ~ are also valid as part of the local-part of an email address; however, they are uncommon, and we have elected not to allow them in our dialplan examples.

ENUM and E.164

Although the SIP protocol really doesn't think in terms of phone numbers, the reality is that phone numbers are not going away any time soon, and if you want to properly integrate a VoIP system with as many telephone networks as possible, you're going to need to handle the PSTN in some way.

ENUM maps telephone numbers onto the Domain Name System (DNS). In theory, ENUM is a great idea. Why not cut out the PSTN altogether, and simply route phone calls directly between endpoints using the same numbering plan? We're not sure this idea is ever going to become what the emerging telecom community would like it to be, though. The reason? Nobody really can say who owns phone numbers.

E.164 and the ITU

The International Telecommunication Union (ITU) is a United Nations agency that is actually older than the UN itself. It was founded in 1865 as the International Telegraph Union. The ITU-T sector, known for many decades as CCITT (Comité consultatif international téléphonique et télégraphique), is the standards body responsible for all of the protocols used by the PSTN, as well as many that are used in VoIP. Prior to the advent of VoIP, the workings of the ITU-T sector were of little interest to the average person, and membership was generally limited to industries and institutions that had a vested interest in telecommunications standards.

ITU standards tend to follow a letter-dot-number format. ITU-T standards you may have heard of include H.323, H.264, G.711, G.729, and so forth.

E.164 is the ITU-T standard that defines the international numbering plan for the PSTN. If you've ever used a telephone, you've used E.164 addressing.

Each country in the world has been assigned a country code,[‡] and control of addressing in those countries is handled by the local authorities.

E.164 numbers are limited to 15 digits in length (excluding the prefix).

In Asterisk, there is nothing special that needs to be done in order to handle E.164 addressing, other than to make sure your dialplan is suitable to the needs of any PSTN-compatible channels you may have.

For example, if you're operating in a NANP country, you will probably need to have the following pattern matches:

```
_NXXNXXXXXX
_1NXXNXXXXXX
_011X.
_N11
```

---

‡ With the exception of 24 countries and territories in country code 1, which are all part of the North American Numbering Plan Authority (NANPA).

In the UK, you might need something more like this:

```
_0[123789]XXXXXXXX
_0[123789]XXXXXXX
```

And in Australia, your dialplan might have these pattern matches:

```
_NXXXXXXX
_0XXXXXXXXX
```

 Please don't just copy and paste these pattern matches into your dialplan. The peculiarities of regional dialplans are tricky, and change constantly. One important item that needs to be carefully considered is the region-specific number for emergency calling, as discussed in "Emergency Dialing" on page 154. You don't want to get this stuff wrong.

The North American Numbering Plan Authority

In much of North America, the North American Numbering Plan (NANP) is in use. All countries in the NANP are assigned to country code 1. Canada and the US are the most well-known of these countries, but the NANP actually includes around 24 different countries and territories (mostly in the Caribbean).

ENUM

In order to allow the mapping of E.164 numbers onto the DNS namespace, a way of representing phone numbers as DNS names had to be devised.

This concept is defined in RFC 3761, helpfully named "The E.164 to Uniform Resource Identifiers (URI) Dynamic Delegation Discovery System (DDDS) Application (ENUM)." ENUM reportedly stands for Electronic NUmber Mapping.

According to the RFC, converting a phone number into an ENUM-compatible address requires the following algorithm:

```
1. Remove all characters with the exception of the digits.
For example, the First Well Known Rule produced the Key
"+442079460148".  This step would simply remove the
leading "+", producing "442079460148".

2. Put dots (".") between each digit. Example:
     4.4.2.0.7.9.4.6.0.1.4.8

3. Reverse the order of the digits. Example:
     8.4.1.0.6.4.9.7.0.2.4.4

4. Append the string ".e164.arpa" to the end.  Example:
     8.4.1.0.6.4.9.7.0.2.4.4.e164.arpa
```

Clear as mud?

ENUM has not taken off. The reasons appear to be mostly political in nature. The problem stems from the fact that there is no one organization that controls numbering on the PSTN the way that IANA does for the Internet. Since no one entity has a clear mandate for managing E.164 numbers globally, the challenge of maintaining an accurate and authoritative database for ENUM has proved elusive.

Some countries in Europe have done a good job of delivering reliable ENUM databases, but in country code 1 (NANP), which contains multiple countries and therefore multiple regulatory bodies, the situation has become an illogical mess. This is hardly surprising, since the carriers that control E.164 addressing can't reasonably be expected to get enthusiastic about allowing you to bypass their networks. The organizations responsible for implementing ENUM in North America have tended to work toward creating a PSTN on the Internet, which could save them money, but not you or I.

This is not at all what is wanted. Why would I want to route VoIP calls from my system to yours across a network that wants to charge me for the privilege? SIP is designed to route calls between endpoints, and has no real use for the concept of a carrier.

The advantage of all this is supposed to be that when an ENUM lookup is performed, a valid SIP URI is returned.

Asterisk and ENUM

Asterisk can perform lookups against ENUM databases using either the `ENUMLOOKUP()` function or a combination of the `ENUMQUERY()` and `ENUMRESULT()` dialplan functions. `ENUMLOOKUP()` only returns a single value back from the lookup, and is useful when you know there is likely to only be one return value (such as the SIP URI you want the system to dial), or if you simply want to get the number of records available.

Status of ENUM Around the World

In the NANP (and many other) countries, the official *e164.arpa* zone has not been formally implemented, and therefore there is no official place to go to perform ENUM lookups for NANP numbers.

A list of the statuses of various countries' implementations of ENUM can be found at *http://enumdata.org/*. For those countries fortunate enough to have ENUM in production, you can perform ENUM lookups directly to their *e164.arpa* zones of those countries fortunate enough to have ENUM in production.

For countries without *e164.arpa* zones, there are several alternative places to perform lookups, the most popular currently being *http://www.e164.org*. Note that these organizations have no formal mandate to maintain the zones they represent. They are community-based, best-effort projects, and the data contained in them will frequently be out-of-date.

An ENUM lookup in the dialplan might look like this:

```
exten => _X.,1,Set(CurrentExten=${FILTER(0-9,${EXTEN})})
   same => n,Set(LookupResult=${ENUMLOOKUP(${CurrentExten},sip,,,e164.arpa)})
   same => n,GotoIf($[${EXISTS(${LookupResult})}]?HaveLocation,1)
   same => n,Set(LookupResult=${ENUMLOOKUP(${CurrentExten},sip,,,e164.org)})
   same => n,GotoIf($[${ISNULL(${LookupResult})}]?NormalCall,1:HaveLocation,1)

exten => HaveLocation,1,Verbose(2,Handle dialing via SIP URI returned)
   exten => ...

exten => NormalCall,1,Verbose(2,Handle dialing via standard PSTN route)
   exten => ...
```

The dialplan code we just looked at will take the number dialed and pass it to the
ENUMLOOKUP() function. It requests the method type to be sip (we want the SIP URI
returned) and the lookup to be performed first against the listings in DNS found in the
e164.arpa zone, and next against the records found at *http://www.e164.org*.

Outside the countries that have implemented it, there is little uptake of ENUM. As
such, many ENUM queries will not return any results. This is not expected to change
in the near future, and ENUM will remain a curiosity until more widely implemented.

ISN, ITAD, and freenum.org

Finally we get to the cool part of this chapter.

The biggest shortcoming of ENUM is that it uses a numbering system that is not under
the control of any Internet numbering authorities.§ The *freenum.org* project solves this
problem by utilizing a numbering scheme that is managed by IANA. This means that
a formal, globally valid, nongeographic numbering system for VoIP can be immediately
and easily implemented without getting mired in the bureaucracy and politics that
burden the E.164 numbering system.

> John Todd, who manages the project, notes that *"Freenum.org is a DNS service that
> uses ENUM-like mapping methods to allow many services to be mapped to a keypad-
> friendly string. The most obvious and widely used method for this is connecting VoIP users
> together for free by creating an easily-remembered dial string that maps to SIP URIs in
> the background. However, anything that can appear in a NAPTR record (email, instant
> messenger, web addresses) can be mapped to an ISN-style freenum.org address. The goal
> of the project is to provide free numeric pointers to the billions of phones that support only
> 0–9, * and # characters and allow those devices to communicate via VoIP or other next-
> generation protocols. The project is spread out across more than thirty DNS servers
> worldwide."*

---

§ More to the point, perhaps, is that E.164 numbers are controlled by far too many organizations, each one
subjected to different regulations, and having goals that are not always compatible with the concept of global,
free VoIP calling.

Got ISN?

The heart of the *freenum.org* concept is the ITAD Subscriber Number (ISN). The ISN is a numeric string that is composed of an extension number on your system, an asterisk character separator (*),‖ and a number that is unique to your organization called an IP Telephony Administrative Domain (ITAD) number. The advantage of the ISN is that it can be dialed from any telephone. An ISN would look something like this:

    0*1273

which would represent *extension zero at ITAD 1273#* and would resolve to *sip:0@ shifteight.org*.

You control your extension numbers (everything to the left of the *). Your ITAD is assigned by IANA (the same organization that controls IP and MAC addresses).

Once your ITAD is assigned, you will be able to publish ISNs on your website, or on business cards, or wherever you would normally publish phone numbers. Any system capable of dialing ISNs will allow its users to call you by dialing your ISN. Calls will be routed directly between the two systems using the SIP URI that *freenum.org* returns.

ITAD Subscriber Numbers (ISNs)

The ISN does not replace a SIP URI, but rather complements it by allowing dialing of VoIP numbers using only characters found on a standard telephone dialpad. In order to resolve an ISN into a valid URI, the DNS system will query the ISN against the *freenum.org* domain. Any DNS lookup against your ISN will return a URI that defines how your system expects to receive calls to that ISN.*

Management of Internet Numbering

The Internet Assigned Numbers Authority (IANA) is the body responsible for managing any numbering system that exists as a result of an RFC that requires a numerical database of some kind. The most well-known responsibility of IANA is the delegation of IP addresses to the five Regional Internet Registries that control all of the public IP addresses on the planet.† These organizations are responsible for the assignment of IP addresses within their regions.

‖ This character has nothing to do with the software that is the subject of this book; it simply refers to the * that is on the dialpad of every telephone. We wonder what might have been if, instead of Asterisk, Mark Spencer had decided to call his creation Octothorpe.

#ITAD 1273 is assigned to *shifteight.org*.

* Although *freenum.org* can handle ITADs that resolve to non-SIP URIs, the handling of multiple protocols is beyond the scope of this book. For now, we recommend you restrict your ISN to handling SIP URIs.

† AfriNIC, APNIC, ARIN, LACNIC, and RIPE NCC.

There are many other numbering schemes that have been created as a result of an RFC. Other IANA-managed numbers include MAC addresses—specifically, the Organizationally Unique Identifier (OUI) portion of the MAC addressing space.

Several years ago, a protocol named TRIP (Telephony Routing over IP) was created. While this protocol never took off, and is unlikely to see any future growth, it did offer us one incredibly useful thing: the ITAD. Since ITADs are part of an RFC, the IANA is mandated to maintain a database of ITADs. This is what makes *freenum.org* possible.

IP Telephony Administrative Domains (ITADs)

Freenum.org takes advantage of IANA's responsibility to maintain a database of ITAD numbers and allows us to build simple, standards-based, globally relevant, and community-driven numbering plans for VoIP.‡ You can find the list of currently assigned ITAD numbers at *http://www.iana.org/assignments/trip-parameters/trip-parameters .xml#trip-parameters-5*.

You will want to obtain your own ITAD number by submitting the form located at *http://www.iana.org/cgi-bin/assignments.pl*.

This form should be filled out as shown in Figure 12-1.

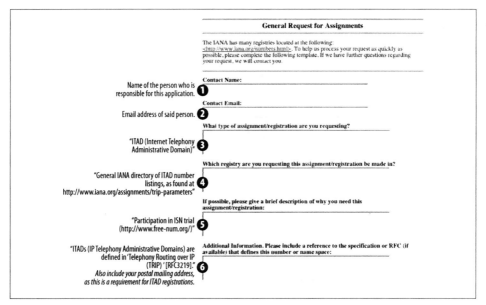

Figure 12-1. Request for Assignments form

---

‡ Note that *freenum.org* has consulted with the folks at IANA in regard to the use of ITADs with protocols other than TRIP.

Your application will be reviewed by a Real Human Being™, and within a few days you should be assigned an ITAD by IANA. A few days later, you will also receive information for your *freenum.org* account (there is currently a simple review process to ensure that bots and spammers don't abuse the system). You will then need to log onto the *freenum.org* site and define the parameters for your ITAD.

Create a DNS Entry for Your ITAD

In the top-right corner of the *freenum.org* site, you will see a *Sign in here* link. Your username is the email address you registered with IANA, and your password will have been emailed to you by the *freenum.org* system.§

You will be presented with a list of your assigned ITADs. In order for your new ITAD to work, you will need to ensure the DNS records are up-to-date.

 There are two methods of handling DNS for your ITAD. The first (and simplest) is to have a NAPTR record inserted into the *freenum.org* zone. The other way is to create a zone for your ITAD, and have *freenum.org* delegate that zone to your name servers. We will only discuss the first method here, but if you are familiar with NAPTR/ENUM administration for a DNS server, you can use the second method.

The *freenum.org* folks have created the Freenum Automated Self-Service Tool (FASST) to simplify DNS record entry for you. The essential fields will already be filled out. The only thing you need to change is under the *DNS Setting* section of the form: specify the hostname of your PBX and save the changes. The FASST tool uses a regular expression to convert an ISN lookup to a SIP URI.

In order to specify your hostname, you will need to modify the sample regular expression provided by FASST, changing the sample hostname `sip.yourdomain.com` to the hostname of your PBX. So, for example, in our case we would want to change:

```
!^\\+*([^\\*]*)!sip:\\1@sip.yourdomain.com!
```

to:

```
!^\\+*([^\\*]*)!sip:\\1@pbx.shifteight.org!
```

The other fields in the DNS entry should not be changed unless you know what you are doing. The rest of the fields in the form are optional, and can be filled out as you see fit.

---

§ This may take a few days, so if you've received your ITAD from IANA but not yet a password from *freenum.org*, give it some time.

> John Todd notes: *"For those sites which have extremely complex configurations or geographically diverse offices with different SIP servers handling different prefixes (for instance: 12xxx goes to the Asterisk server in France, 13xxx to the Asterisk server in Germany, and so on) then there are more sophisticated methods where you run your own delegated zone out of the freenum.org domain, but those are outside the scope of this book but can be learned about on the freenum.org site."*

Testing Your ITAD

As is often the case with DNS changes, it can take a few days for your changes to propagate through the system. To check, you can Google for "online dig tool" to find a web-based lookup tool, or use the *dig* tool under Linux:

```
$ dig NAPTR 4.3.2.1.1273.freenum.org
```

Once your record is updated in the system, the result will include the following:

```
;; ANSWER SECTION:
4.3.2.1.1273.freenum.org. 86400 IN NAPTR 100 10 "u" "E2U+sip"
"!^\\+*([^\\*]*)!sip:\\1@shifteight.org!" .
```

If the answer section does not include the regular expression containing your domain name, the records have not updated and you should wait a few more hours (or even leave it for a day).

Using ISNs in Your Asterisk System

So now that you've got your own ITAD (you did sign up, right?), you'll want to make it available to others, and also configure your dialplan to allow you to dial other ITADs.

Under the [globals] section of your dialplan (*/etc/asterisk/extensions.conf*), add a global variable that contains your ITAD:

```
[globals]
ITAD = 1273 ; replace '1273' with your own ITAD number
```

To allow calling to ITADs from your system, you will need something like the following dialplan code[||]:

```
[OutgoingISN]
exten => _X*X!,1,GoSub(subFreenum,start,1(${EXTEN}))
exten => _XX*X!,1,GoSub(subFreenum,start,1(${EXTEN}))
exten => _XXX*X!,1,GoSub(subFreenum,start,1(${EXTEN}))
exten => _XXXX*X!,1,GoSub(subFreenum,start,1(${EXTEN}))
exten => _XXXXX*X!,1,GoSub(subFreenum,start,1(${EXTEN}))
; you may need to add more lines here to handle XXXXXX*X, XXXXXXX*X, and so forth
```

[||] If people publish the users' full DIDs instead of their internal extension numbers, the pattern matches will need to support up to 15 digits.

```
[subFreenum]
exten => start,1,Verbose(2,Performing ISN lookup)
    same => n,Set(ISN=${FILTER(0-9*,${ARG1})})
    same => n,Set(Result=${ENUMLOOKUP(${ISN},sip,s,,freenum.org)})
    same => n,GotoIf($[${EXISTS(${Result})}]?call,1:no_result,1)

exten => call,1,Verbose(2,Placing call to ISN --${ISN}-- via ${Result})
    same => n,Dial(SIP/${Result})
    same => n,Return()

exten => no_result,1,Verbose(2,Lookup for ISN: --${ISN}-- returned no result)
    same => n,Playback(silence/1&invalid)
    same => n,Return()
```

We have added two new contexts to our dialplan: OutgoingISN and subFreenum. The OutgoingISN context controls who can dial ISN numbers from within your dialplan. If you have been following our examples throughout this book, you should have a context called LocalSets, which is the context where all your telephones enter the dialplan. Including OutgoingISN within LocalSets enables dialing of ISN numbers:

```
[LocalSets]
include => OutgoingISN  ; include the context that enables ISN dialing
include => CallPlace    ; use subroutine to determine what you can dial
```

We have placed the OutgoingISN include above the CallPlace include because Asterisk will perform extension matching in the order of the includes, and since CallPlace has a more general pattern match than our OutgoingISN pattern matches, we need to make sure OutgoingISN appears first.

The magic for dialing ISN numbers is handled in the subFreenum context. Our Outgoin gISN context will pass the requested extension (e.g., 1234*256) to the subFreenum subroutine. After the NoOp() on the first line, the subroutine will filter the request for numbers and the asterisk (*) character to make the extension safe. The result will then be assigned to the ISN channel variable:

```
exten => start,n,Set(ISN=${FILTER(0-9*,${ARG1})})
```

The subroutine will then perform a lookup for the ISN via the DNS system using the ENUMLOOKUP() dialplan function. Options passed to the ENUMLOOKUP() function include:

- The ISN number to look up
- The method type to look up and return (SIP)
- The s option, which tells Asterisk to perform an ISN-style lookup instead of a standard ENUM lookup
- The zone suffix for performing the lookups (we'll use freenum.org, but the default is e164.arpa)

Our code for performing the lookup then looks like this:

```
exten => start,n,Set(Result=${ENUMLOOKUP(${ISN},sip,s,,freenum.org)})
```

Following the lookup and storing the result in the ${Result} channel variable, our subroutine will verify whether we received a result or not:

```
exten => start,n,GotoIf($[${EXISTS(${Result})}]?call,1:no_result,1)
```

If no result is received, the call will be handled in the no_result extension. If a result is received back from our lookup, then execution will continue at the call extension where the call will be placed using the result stored in the ${Result} channel variable.

Receiving calls to your ITAD

Receiving calls to your ITAD is much simpler. If your system supports incoming SIP URIs, ISNs will already work for you.# We showed the configuration required to accept calls to your system in "Accepting Calls to Your System" on page 240.

Security and Identity

It is a sad fact of the Internet that there are a few selfish, greedy criminal types out there who think nothing of attempting to take advantage of people for their own gain. In telecom, this behavior represents several risks to you.

In this section, we will focus on security issues relating to the portions of your system that you intend to make publicly available through the Internet. While it would be simple to just refuse to allow any sort of external connections, the reality is that if you want people to be able to call you for free from the Internet (for example, if you intend to publish your company's SIP URIs on your web page), you are going to have to define a secure place within your system where those calls will arrive. Securing your incoming public VoIP connections is conceptually similar to implementing a DMZ in traditional networking.*

In Asterisk, certain contexts in your dialplan cannot be trusted. This means that you will need to carefully consider what resources are available to channels that enter the system through these contexts, and ensure that only certain services and features are available.

#If you've set up your ITAD and ISN correctly, the conversion from ISN dial string to SIP URI will take place before the call arrives on your doorstep.

* A DMZ is any portion of your network that you expose to the Internet (such as your website), and therefore cannot completely trust. It is not uncommon for organizations to place the PBX within a DMZ.

Toll Fraud

Toll fraud is by far the biggest risk to your phone system in terms of the potential for ruinous cost. It is not unheard of for fraudsters to rack up tens of thousands of dollars in stolen phone calls over the course of a few days.

Toll fraud is not a new thing, having existed prior to VoIP; however, the enabling nature of VoIP means that it is easier for fraudsters to take advantage of unsecured systems. Most carriers will not take responsibility for these costs, and thus if your system is compromised you could be stuck with a very large phone bill. While carriers are getting better and better at alerting their customers to suspicious activity, that does not absolve you of responsibility for ensuring your system is hardened against this very real and very dangerous threat.

Within your Asterisk system, it is vitally important that you know what resources on your system are exposed to the outside world and ensure that those resources are secure.

The most common form of toll fraud these days is accomplished by brute-force attack. In this scenario, the thieves will have a script that will contact your system and attempt to register as a valid user. If they are able to register as a telephone on your system, the flood of calls will commence, and you will be stuck with the bill. If you are using simple extension numbers and easy-to-guess passwords, and your system accepts registrations from outside your firewall, it is certain that you will eventually be the victim of toll fraud.

Brute-force attacks can also cause performance problems with your system, as one of these scripts can flood your router and PBX with massive numbers of registration attempts.

The following tactics have proven successful in minimizing the risk of toll fraud:

1. Do not use easy-to-guess passwords. Passwords should be at least eight characters long and contain a mix of digits, letters, and characters. 8a$j03H% is a good password.[†] 1234 is not.

2. Do not use extension numbers for your SIP registrations in *sip.conf*. Instead of [1000], use something like a MAC address (something like [0004f2123456] would be much more difficult for a brute-force script to guess).

3. Use an analysis script such as *fail2ban* to tweak your internal firewall to block IP addresses that are displaying abusive behavior, such as massive packet floods.

> The *fail2ban* daemon is emerging as a popular way to automatically respond to security threats. We'll discuss it further in Chapter 26.

---

† Actually, since it's published in this book, it is no longer a good password, but you get the idea.

Spam over Internet Telephony (SPIT)

VoIP spam has not yet taken off, but rest assured, it will. Spammers all over the world are drooling at the prospect of being able to freely assault anyone and everyone with an Internet-enabled phone system.

Like email, VoIP entails a certain level of trust, in that it assumes that every phone call is legitimate. Unfortunately, as with email spam, it only takes a few bad apples to spoil things for the rest of us.

Many organizations and persons are working on ways to address SPIT now, before it becomes a problem. Some concepts being worked on include certificates and whitelists. No one method has emerged as the definitive solution.

While it would be easy to simply lock our systems away from the world, the fact is that Internet telephony is something that every business will be expected to support in the not-too-distant future. SPIT will increasingly become a problem as more and more unsavory characters decide that this is the new road to riches.

Solving the SPIT problem will be an ongoing process: a battle between us and The Bad Guys™.

Distributed Denial of Service Attacks

SIP denial of service attacks are already happening on the Internet. Amazon's EC2 cloud has become a popular place to originate these attacks from, and other cloud-based or compromised systems will become popular for these activities as well. The actual attacks are not strictly denial of service attacks (in the sense that they are not deliberately trying to choke your system); rather, they are attack campaigns that are typically trying to use brute force to locate exploitable holes in any systems they can find. As the sheer number of these attacks increases, the effect on the network will be similar to that of email spam.

The previously mentioned *fail2ban* daemon can be useful in minimizing the effects of these attacks. Refer to Chapter 26 for more details.

Phishing

When a VoIP system has been compromised, one popular use of the compromised system is to relay fraud campaigns using the identity of the compromised system. Criminals engaging in so-called phishing expeditions will make random calls to lists of numbers, attempting to obtain credit card or other sensitive information, while posing as your organization.

Security Is an Ongoing Process

In contrast to previous editions, throughout this book we have tried to provide examples and best practices that take security into consideration at all stages. Whatever you are working on, you should be thinking about security. While implementing good security requires more design, development, and testing effort, it will save you time and money in the long run.

Most security holes happen as a result of something that was hastily implemented and wasn't locked down later. "I'll just quickly build this now, and I'll clean it up later" are words you never want to say (or hear).

Conclusion

One of the dreams of VoIP was that it was going to make phone calls free. Over a decade later, we're still paying for our phone calls. The technology has existed for some time, but the ease of use has not been there.

It costs nothing to register your ITAD and set up your system to handle ISNs. If every Asterisk system deployed had an ITAD, and people started publishing their ISNs on websites, vcards, and business cards, the weight of the Asterisk community would drive industry adoption.

Security considerations for VoIP have to be taken into consideration, but we expect that the benefits will outweigh the risks.

Our collective dream of free Internet calling may be closer than we think.

Automatic Call Distribution (ACD) Queues

*An Englishman, even if he is alone, forms an orderly
queue of one.*

—George Mikes

Automatic Call Distribution (ACD), or call queuing, provides a way for a PBX to queue up incoming calls from a group of users: it aggregates multiple calls into a holding pattern and assigns each call a rank that determines the order in which that call should be delivered to an available agent (typically, first in first out). When an agent becomes available, the highest-ranked caller in the queue is delivered to that agent, and everyone else moves up a rank.

If you have ever called an organization and heard "all of our representatives are busy," you have experienced ACD. The advantage of ACD to the callers is that they don't have to keep dialing back in an attempt to reach someone, and the advantages to the organizations are that they are able to better service their customers and to temporarily handle situations where there are more callers than there are agents.*

 There are two types of call centers: inbound and outbound. ACD refers to the technology that handles inbound call centers, whereas the term Predictive Dialer refers to the technology that handles outbound call centers. In this book we will primarily focus on inbound calling.

---

* It is a common misconception that a queue can allow you to handle more calls. This is not strictly true, in that your callers will still want to speak to a live person, and they will only be willing to wait for so long. In other words, if you are short-staffed, your queue could end up being nothing more than an obstacle to your callers. The ideal queue is invisible to the callers, since their calls get answered immediately without them having to hold.

We've all been frustrated by poorly designed and managed queues: enduring hold music from a radio that isn't in tune, mind-numbing wait times, and pointless messages that tell you every 20 seconds how important your call is, despite that fact that you've been waiting for 30 minutes and have heard the message so many times you can quote it from memory. From a customer service perspective, queue design may be one of the most important aspects of your telephone system. As with an automated attendant, what must be kept in mind above all else is that *your callers are not interested in holding in a queue*. They called because *they want to talk to you*. All your design decisions must keep this crucial fact front-and-center in your mind: people want to talk to other people; not to your phone system.[†]

The purpose of this chapter is to teach you how to create and design queues that get callers to their intended destinations as quickly and painlessly as possible.

> In this chapter, we may flip back and forth between the usage of the terms *queue members* and *agents*. Unless we are talking about agents logged in via chan_agent (using AgentLogin()), we're almost certainly talking about queue members as added via AddQueueMember() or the CLI commands (which we'll discuss in this chapter). Just know that there is a difference in Asterisk between an *agent* and a *queue member*, but that we'll use the term *agent* loosely to simply describe an endpoint as called by a Queue().

Creating a Simple ACD Queue

To start with, we're going to create a simple ACD queue. It will accept callers and attempt to deliver them to a member of the queue.

> In Asterisk, the term *member* refers to a peer assigned to a queue that can be dialed, such as SIP/0000FFFF0001. An *agent* technically refers to the Agent channel also used for dialing endpoints. Unfortunately, the Agent channel is a deprecated technology in Asterisk, as it is limited in flexibility and can cause unexpected issues that can be hard to diagnose and resolve. We will not be covering the use of chan_agent, so be aware that we will generally use the term *member* to refer to the telephone device and *agent* to refer to the person who handles the call. Since one isn't generally effective without the other, either term may refer to both.

We'll create the queue(s) in the *queues.conf* file, and manually add queue members to it through the Asterisk console. In the section "Queue Members" on page 266, we'll

---

† There are several books available that discuss call center metrics and available queuing strategies, such as James C. Abbott's *The Executive Guide to Call Center Metrics* (Robert Houston Smith).

look into how to create a dialplan that allows us to dynamically add and remove queue members (as well as pause and unpause them).

The first step is to create your *queues.conf* file in the */etc/asterisk* configuration directory:

```
$ cd /etc/asterisk/
$ touch queues.conf
```

Populate it with the following configuration, which will create two queues named [sales] and [support]. You can name them anything you want, but we will be using these names later in the book, so if you use different queue names from what we've recommended here, make note of your choices for future reference:

```
[general]
autofill=yes              ; distribute all waiting callers to available members
shared_lastcall=yes       ; respect the wrapup time for members logged into more
                          ; than one queue

[StandardQueue](!)        ; template to provide common features
musicclass=default        ; play [default] music
strategy=rrmemory         ; use the Round Robin Memory strategy
joinempty=no              ; do not join the queue when no members available
leavewhenempty=yes        ; leave the queue when no members available
ringinuse=no              ; don't ring members when already InUse (prevents
                          ; multiple calls to an agent)

[sales](StandardQueue)    ; create the sales queue using the parameters in the
                          ; StandardQueue template

[support](StandardQueue)  ; create the support queue using the parameters in the
                          ; StandardQueue template
```

The [general] section defines the default behavior and global options. We've only specified two options in the [general] section, since the built-in defaults are sufficient for our needs at this point.

The first option is autofill, which tells the queue to distribute all waiting callers to all available members immediately. Previous versions of Asterisk would only distribute one caller at a time, which meant that while Asterisk was signaling an agent, all other calls were held (even if other agents were available) until the first caller in line had been connected to an agent (which obviously led to bottlenecks in older versions of Asterisk where large, busy queues were being used). Unless you have a particular need for backward-compatibility, *this option should always be set to yes*.

The second option in the [general] section of *queues.conf* is shared_lastcall. When we enable shared_lastcall, the last call to an agent who is logged into multiple queues will be the call that is counted for wrapup time[‡] in order to avoid sending a call to an agent from another queue during the wrap period. If this option is set to no, the wrap timer will only apply to the queue the last call came from, which means an agent who

---

[‡] Wrapup time is used for agents who may need to perform some sort of logging or other function once a call is done. It gives them a grace period of several seconds in order to perform this task before taking another call.

was wrapping up a call from the support queue might still get a call from the sales queue. This option should also always be set to yes (the default).

The next section, [StandardQueue] is the template we'll apply to our sales and support queues (we declared it a template by adding (!)). We've defined the musicclass to be the default music on hold, as configured in the *musiconhold.conf* file. The strategy we'll employ is rrmemory, which stands for Round-Robin with Memory. The rrmemory strategy works by rotating through the agents in the queue in sequential order, keeping track of which agent got the last call, and presenting the next call to the next agent. When it gets to the last agent, it goes back to the top (as agents log in, they are added to the end of the list). We've set joinempty to no since it is generally bad form to put callers into a queue where there are no agents available to take their calls.

You could set this to yes for ease of testing, but we would not recommend putting it into production unless you are using the queue for some function that is not about getting your callers to your agents. Nobody wants to wait in a line that is not going anywhere.

The leavewhenempty option is used to control whether callers should fall out of the Queue() application and continue on in the dialplan if no members are available to take their calls. We've set this to yes because it makes no sense to wait in a line that's not going anywhere.

From a business perspective, you should be telling your agents to clear all calls out of the queue before logging off for the day. If you find that there are a lot of calls queued up at the end of the day, you might want to consider extending someone's shift to deal with them. Otherwise, they'll just add to your stress when they call back the next day, in a worse mood.

The alternative is to use GotoIfTime() near the end of the day to redirect callers to voicemail, or some other appropriate location in your dialplan.

Finally, we've set ringinuse to no, which tells Asterisk not to ring members when their devices are already ringing. The purpose of setting ringinuse to no is to avoid multiple calls to the same member from one or more queues.

It should be mentioned that joinempty and leavewhenempty are looking for either no members logged into the queue, or all members unavailable. Agents that are Ringing or InUse are not considered unavailable, so will not block callers from joining the queue or cause them to be kicked out when joinempty=no and/or leavewhenempty=yes.

Once you've finished configuring your *queues.conf* file, you can save it and reload the *app_queue.so* module from your Asterisk CLI:

```
$ asterisk -r
*CLI> module reload app_queue.so
    -- Reloading module 'app_queue.so' (True Call Queueing)
```

Then verify that your queues were loaded into memory:

```
localhost*CLI> queue show
support      has 0 calls (max unlimited) in 'rrmemory' strategy
(0s holdtime, 0s talktime), W:0, C:0, A:0, SL:0.0% within 0s
   No Members
   No Callers

sales        has 0 calls (max unlimited) in 'rrmemory' strategy
(0s holdtime, 0s talktime), W:0, C:0, A:0, SL:0.0% within 0s
   No Members
   No Callers
```

Now that you've created the queues, you need to configure your dialplan to allow calls to enter the queue.

Add the following dialplan logic to the *extensions.conf* file:

```
[Queues]
exten => 7001,1,Verbose(2,${CALLERID(all)} entering the support queue)
same => n,Queue(support)
same => n,Hangup()

exten => 7002,1,Verbose(2,${CALLERID(all)} entering the sales queue)
same => n,Queue(sales)
same => n,Hangup()

[LocalSets]
include => Queues      ; allow phones to call queues
```

We've included the Queues context in the LocalSets context so that our telephones can call the queues we've set up. In Chapter 15, we'll define menu items that go to these queues. Save the changes to your *extensidons.conf* file, and reload the dialplan with the *dialplan reload* CLI command.

If you dial extension 7001 or 7002 at this point, you will end up with output like the following:

```
    -- Executing [7001@LocalSets:1] Verbose("SIP/0000FFFF0003-00000001",
       "2,"Leif Madsen" <100> entering the support queue") in new stack
   == "Leif Madsen" <1--> entering the support queue
    -- Executing [7001@LocalSets:2] Queue("SIP/0000FFFF0003-00000001",
       "support") in new stack
       [2011-02-14 08:59:39] WARNING[13981]: app_queue.c:5738 queue_exec:
       Unable to join queue 'support'
    -- Executing [7001@LocalSets:3]
       Hangup("SIP/0000FFFF0003-00000001", "") in new stack
   == Spawn extension (LocalSets, 7001, 3) exited non-zero on
       'SIP/0000FFFF0003-00000001'
```

You don't join the queue at this point, as there are no agents in the queue to answer calls. Because we have `joinempty=no` and `leavewhenempty=yes` configured in *queues.conf*, callers will not be placed into the queue. (This would be a good opportunity to experiment with the `joinempty` and `leavewhenempty` options in *queues.conf* to better understand their impact on queues.)

In the next section, we'll demonstrate how to add members to your queue (as well as other member interactions with the queue, such as pause/unpause).

Queue Members

Queues aren't very useful without someone to answer the calls that come into them, so we need a method for allowing agents to be logged into the queues to answer calls. There are various ways of going about this, and we'll show you how to add members to the queue both manually (as an administrator) and dynamically (as the agent). We'll start with the Asterisk CLI method, which allows you to easily add members to the queue for testing and minimal dialplan changes. We'll then expand upon that, showing you how to add dialplan logic allowing agents to log themselves into and out of the queues and to pause and unpause themselves in queues they are logged into.

Controlling Queue Members via the CLI

We can add queue members to any available queue through the Asterisk CLI command *queue add*. The format of the *queue add* command is (all on one line):

```
*CLI> queue add member <channel> to <queue> [[[penalty <penalty>] as
<membername>] state_interface <interface>]
```

The *<channel>* is the channel we want to add to the queue, such as SIP/0000FFFF0003, and the *<queue>* name will be something like support or sales—any queue name that exists in */etc/asterisk/queues.conf*. For now we'll ignore the *<penalty>* option, but we'll discuss it in "Advanced Queues" on page 283 (penalty is used to control the rank of a member within a queue, which can be important for agents who are logged into multiple queues). We can define the *<membername>* to provide details to the queue-logging engine. The `state_interface` option is something that we should delve a bit more into at this junction. Because it is so important for all aspects of queues and their members in Asterisk, we've written a little section about it, so go ahead and read "An Introduction to Device State" on page 273. Once you've set that up, come back here and continue on. Don't worry, we'll wait.

Now that you've added `callcounter=yes` to *sip.conf* (we'll be using SIP channels throughout the rest of our examples), let's see how to add members to our queues from the Asterisk CLI.

Adding a queue member to the **support** queue can be done with the *queue add member* command:

```
*CLI> queue add member SIP/0000FFFF0001 to support
Added interface 'SIP/0000FFFF0001' to queue 'support'
```

A query of the queue will verify that our new member has been added:

```
*CLI> queue show support
support      has 0 calls (max unlimited) in 'rrmemory' strategy
(0s holdtime, 0s talktime), W:0, C:0, A:0, SL:0.0% within 0s
   Members:
      SIP/0000FFFF0001 (dynamic) (Not in use) has taken no calls yet
   No Callers
```

To remove a queue member, you would use the *queue remove member* command:

```
*CLI> queue remove member SIP/0000FFFF0001 from support
Removed interface 'SIP/0000FFFF0001' from queue 'support'
```

Of course, you can use the *queue show* command again to verify that your member has been removed from the queue.

We can also pause and unpause members in a queue from the Asterisk console, with the *queue pause member* and *queue unpause member* commands. They take a similar format to the previous commands we've been using:

```
*CLI> queue pause member SIP/0000FFFF0001 queue support reason DoingCallbacks
paused interface 'SIP/0000FFFF0001' in queue 'support' for reason 'DoingCallBacks'

*CLI> queue show support
support      has 0 calls (max unlimited) in 'rrmemory' strategy
(0s holdtime, 0s talktime), W:0, C:0, A:0, SL:0.0% within 0s
   Members:
      SIP/0000FFFF0001 (dynamic) (paused) (Not in use) has taken no calls yet
   No Callers
```

By adding a reason for pausing the queue member, such as `lunchtime`, you ensure that your queue logs will contain some additional information that may be useful. Here's how to unpause the member:

```
*CLI> queue unpause member SIP/0000FFFF0001 queue support reason off-break
unpaused interface 'SIP/0000FFFF0001' in queue 'support' for reason 'off-break'

*CLI> queue show support
support      has 0 calls (max unlimited) in 'rrmemory' strategy
(0s holdtime, 0s talktime), W:0, C:0, A:0, SL:0.0% within 0s
   Members:
      SIP/0000FFFF0001 (dynamic) (Not in use) has taken no calls yet
   No Callers
```

In a production environment, the CLI would not normally be the best way to control the state of agents in a queue. Instead, there are dialplan applications that allow agents to inform the queue as to their availability.

Controlling Queue Members with Dialplan Logic

In a call center staffed by live agents, it is most common to have the agents themselves log in and log out at the start and end of their shifts (or whenever they go for lunch, or to the bathroom, or are otherwise not available to the queue).

To enable this, we will make use of the following dialplan applications:

- `AddQueueMember()`
- `RemoveQueueMember()`

While logged into a queue, it may be that an agent needs to put herself into a state where she is temporarily unavailable to take calls. The following applications will allow this:

- `PauseQueueMember()`
- `UnpauseQueueMember()`

It may be easier to think of these applications in the following manner: the add and remove applications are used to log in and log out, and the pause/unpause pair are used for short periods of agent unavailability. The difference is simply that pause/unpause set the member as unavailable/available without actually removing them from the queue. This is mostly useful for reporting purposes (if a member is paused, the queue supervisor can see that she is logged into the queue, but simply not available to take calls at that moment). If you're not sure which one to use, we recommend that the agents use add/remove whenever they are not going to be available to take calls.

Using Pause and Unpause

The use of pause and unpause is a matter of preference. In some environments, these options may be used for all activities during the day that render an agent unavailable (such as during the lunch hour and when performing work that is not queue-related). In most call centers, however, if an agent is not beside his phone and ready to take a call at that moment, he should not be logged in at all, even if he is only going to be away from his desk for a few minutes (such as for a bathroom break).

Some supervisors like to use the add/remove and pause/unpause settings as a sort of punch clock, so that they can track when their staff arrive for work and leave at the end of the day, and how long they spend at their desks and on breaks. We do not feel this is a sound practice, as the purpose of these applications is to inform the queue as to agent availability, not to enable tracking of employees' activities.

An important thing to note here relates to the `joinempty` setting in *queues.conf*, which was discussed earlier. If an agent is paused, he is considered as logged into the queue. Let's say it is near the end of the day, and one agent put himself into pause a few hours earlier to work on a project. All the other agents have logged out and gone home. A call comes in. The queue will note that an agent is logged into the queue, and will therefore queue the call, even though the reality is that there are no people actually staffing that queue at that time. This caller may end up holding in an unstaffed queue indefinitely.

In short, agents who are not sitting at their desks and planning to be available to take calls in the next few minutes should log out. Pause/unpause should only be used for brief moments of unavailability (if at all). If you want to use your phone system as a punch clock, there are lots of great ways to do that in Asterisk, but the queue member applications are not the way we would recommend.

Let's build some simple dialplan logic that will allow our agents to indicate their availability to the queue. We are going to use the CUT() dialplan function to extract the name of our channel from our call to the system, so that the queue will know which channel to log into the queue.

We have built this dialplan to show a simple process for logging into and out of a queue, and changing the paused status of a member in a queue. We are doing this only for a single queue that we previously defined in the *queues.conf* file. The status channel variables that the AddQueueMember(), RemoveQueueMember(), PauseQueueMember(), and UnpauseQueueMember() applications set might be used to Playback() announcements to the queue members after they've performed certain functions to let them know whether they have successfully logged in/out or paused/unpaused):

```
[QueueMemberFunctions]

exten => *54,1,Verbose(2,Logging In Queue Member)
    same => n,Set(MemberChannel=${CHANNEL(channeltype)}/${CHANNEL(peername)})
    same => n,AddQueueMember(support,${MemberChannel})

; ${AQMSTATUS}
;    ADDED
;    MEMBERALREADY
;    NOSUCHQUEUE

exten => *56,1,Verbose(2,Logging Out Queue Member)
    same => n,Set(MemberChannel=${CHANNEL(channeltype)}/${CHANNEL(peername)})
    same => n,RemoveQueueMember(support,${MemberChannel})

; ${RQMSTATUS}:
;    REMOVED
;    NOTINQUEUE
;    NOSUCHQUEUE

exten => *72,1,Verbose(2,Pause Queue Member)
    same => n,Set(MemberChannel=${CHANNEL(channeltype)}/${CHANNEL(peername)})
    same => n,PauseQueueMember(support,${MemberChannel})

; ${PQMSTATUS}:
;    PAUSED
;    NOTFOUND

exten => *87,1,Verbose(2,Unpause Queue Member)
    same => n,Set(MemberChannel=${CHANNEL(channeltype)}/${CHANNEL(peername)})
    same => n,UnpauseQueueMember(support,${MemberChannel})
```

```
;  ${UPQMSTATUS}:
;      UNPAUSED
;      NOTFOUND
```

Automatically Logging Into and Out of Multiple Queues

It is quite common for an agent to be a member of more than one queue. Rather than having a separate extension for logging into each queue (or demanding information from the agents about which queues they want to log into), this code uses the Asterisk database (astdb) to store queue membership information for each agent, and then loops through each queue the agents are a member of, logging them into each one in turn.

In order to for this code to work, an entry similar to the following will need to be added to the AstDB via the Asterisk CLI. For example, the following would store the member 0000FFFF0001 as being in both the support and sales queues:

```
*CLI> database put queue_agent 0000FFFF0001/available_queues support^sales
```

You will need to do this once for each agent, regardless of how many queues they are members of.

If you then query the Asterisk database, you should get a result similar to the following:

```
pbx*CLI> database show queue_agent
/queue_agent/0000FFFF0001/available_queues     : support^sales
```

The following dialplan code is an example of how to allow this queue member to be automatically added to both the support and sales queues. We've defined a subroutine that is used to set up three channel variables (MemberChannel, MemberChanType, AvailableQueues). These channel variables are then used by the login (*54), logout (*56), pause (*72), and unpause (*87) extensions. Each of the extensions uses the subSetupAvailableQueues subroutine to set these channel variables and to verify that the AstDB contains a list of one or more queues for the device the queue member is calling from:

```
[subSetupAvailableQueues]
;
; This subroutine is used by the various login/logout/pausing/unpausing routines
; in the [ACD] context. The purpose of the subroutine is centralize the retrieval
; of information easier.
;
exten => start,1,Verbose(2,Checking for available queues)

; Get the current channel's peer name (0000FFFF0001)
    same => n,Set(MemberChannel=${CHANNEL(peername)})

; Get the current channel's technology type (SIP, IAX, etc)
    same => n,Set(MemberChanType=${CHANNEL(channeltype)})

; Get the list of queues available for this agent
    same => n,Set(AvailableQueues=${DB(queue_agent/${MemberChannel}/
    available_queues)})
; *** This should all be on a single line
```

```
; if there are no queues assigned to this agent we'll handle it in the
; no_queues_available extension
    same => n,GotoIf($[${ISNULL(${AvailableQueues})}]?no_queues_available,1)

    same => n,Return()

exten => no_queues_available,1,Verbose(2,No queues available for agent
    ${MemberChannel})
; *** This should all be on a single line

; playback a message stating the channel has not yet been assigned
    same => n,Playback(silence/1&channel&not-yet-assigned)
    same => n,Hangup()

[ACD]
;
; Used for logging agents into all configured queues per the AstDB
;
;
; Logging into multiple queues via the AstDB system
exten => *54,1,Verbose(2,Logging into multiple queues per the database values)

; get the available queues for this channel
    same => n,GoSub(subSetupAvailableQueues,start,1())
    same => n,Set(QueueCounter=1)  ; setup a counter variable

; using CUT(), get the first listed queue returned from the AstDB
    same => n,Set(WorkingQueue=${CUT(AvailableQueues,^,${QueueCounter})})

; While the WorkingQueue channel variable contains a value, loop
    same => n,While($[${EXISTS(${WorkingQueue})}])

; AddQueueMember(queuename[,interface[,penalty[,options[,membername
;  [,stateinterface]]]]])
; Add the channel to a queue, setting the interface for calling
; and the interface for monitoring of device state
;
; *** This should all be on a single line
    same => n,AddQueueMember(${WorkingQueue},${MemberChanType}/
${MemberChannel},,,${MemberChanType}/${MemberChannel})


    same => n,Set(QueueCounter=$[${QueueCounter} + 1])    ; increase our counter

; get the next available queue; if it is null our loop will end
    same => n,Set(WorkingQueue=${CUT(AvailableQueues,^,${QueueCounter})})

    same => n,EndWhile()

; let the agent know they were logged in okay
    same => n,Playback(silence/1&agent-loginok)
    same => n,Hangup()

exten => no_queues_available,1,Verbose(2,No queues available for ${MemberChannel})
```

```
        same => n,Playback(silence/1&channel&not-yet-assigned)
        same => n,Hangup()

; ------------------------

; Used for logging agents out of all configured queues per the AstDB
exten => *56,1,Verbose(2,Logging out of multiple queues)

; Because we reused some code, we've placed the duplicate code into a subroutine
        same => n,GoSub(subSetupAvailableQueues,start,1())
        same => n,Set(QueueCounter=1)
        same => n,Set(WorkingQueue=${CUT(AvailableQueues,^,${QueueCounter})})
        same => n,While($[${EXISTS(${WorkingQueue})}])
        same => n,RemoveQueueMember(${WorkingQueue},${MemberChanType}/${MemberChannel})
        same => n,Set(QueueCounter=$[${QueueCounter} + 1])
        same => n,Set(WorkingQueue=${CUT(AvailableQueues,^,${QueueCounter})})
        same => n,EndWhile()
        same => n,Playback(silence/1&agent-loggedoff)
        same => n,Hangup()

; ------------------------

; Used for pausing agents in all available queues
exten => *72,1,Verbose(2,Pausing member in all queues)
        same => n,GoSub(subSetupAvailableQueues,start,1())

        ; if we don't define a queue, the member is paused in all queues
        same => n,PauseQueueMember(,${MemberChanType}/${MemberChannel})
        same => n,GotoIf($[${PQMSTATUS} = PAUSED]?agent_paused,1:agent_not_found,1)

exten => agent_paused,1,Verbose(2,Agent paused successfully)
        same => n,Playback(silence/1&unavailable)
        same => n,Hangup()

; ------------------------

; Used for unpausing agents in all available queues
exten => *87,1,Verbose(2,UnPausing member in all queues)
        same => n,GoSub(subSetupAvailableQueues,start,1())

        ; if we don't define a queue, then the member is unpaused from all queues
        same => n,UnPauseQueueMember(,${MemberChanType}/${MemberChannel})
        same => n,GotoIf($[${PQMSTATUS} = PAUSED]?agent_unpaused,1:agent_not_found,1)

exten => agent_unpaused,1,Verbose(2,Agent paused successfully)
        same => n,Playback(silence/1&available)
        same => n,Hangup()

; ------------------------

; Used by both pausing and unpausing dialplan functionality
exten => agent_not_found,1,Verbose(2,Agent was not found)
        same => n,Playback(silence/1&cannot-complete-as-dialed)
```

You could further refine these login and logout routines to take into account that the AQMSTATUS and RQMSTATUS channel variables are set each time AddQueueMember() and RemoveQueueMember() are used. For example, you could set a flag that lets the queue member know he has not been added to a queue by setting a flag, or even add recordings or text-to-speech systems to play back the particular queue that is producing the problem. Or, if you're monitoring this via the Asterisk Manager Interface, you could have a screen pop, or use JabberSend() to inform the queue member via instant messaging. (Sorry, sometimes our brains run away with us.)

An Introduction to Device State

Device states in Asterisk are used to inform various applications as to whether your device is currently in use or not. This is especially important for queues, as we don't want to send callers to an agent who is already on the phone. Device states are controlled by the channel module, and in Asterisk only chan_sip has the appropriate handling. When the queue asks for the state of a device, it first queries the channel driver (e.g., chan_sip). If the channel cannot provide the device state directly (as is the case with chan_iax2), it asks the Asterisk core to determine it, which it does by searching through channels currently in progress.

Unfortunately, simply asking the core to search through active channels isn't accurate, so getting device state from channels other than chan_sip is less reliable when working with queues. We'll explore some methods of controlling calls to other channel types in "Advanced Queues" on page 283, but for now we'll focus on SIP channels, which do not have complex device state requirements. For more information about device states, see Chapter 14.

In order to correctly determine the state of a device in Asterisk, we need to enable call counters in *sip.conf*. By enabling call counters, we're telling Asterisk to track the active calls for a device so that this information can be reported back to the channel module and the state can be accurately reflected in our queues. First, let's see what happens to our queue without the callcounter option:

```
*CLI> queue show support
support      has 0 calls (max unlimited) in 'rrmemory' strategy
(0s holdtime, 0s talktime), W:0, C:0, A:0, SL:0.0% within 0s
   Members:
      SIP/0000FFFF0001 (dynamic) (Not in use) has taken no calls yet
   No Callers
```

Now suppose we have an extension in our dialplan, 555, that calls MusicOnHold(). If we dial that extension without having enabled call counters, a query of the support queue (of which SIP/0000FFFF0001 is a member) from the Asterisk CLI will show something similar to the following:

```
-- Executing [555@LocalSets:1] MusicOnHold("SIP/0000FFFF0001-00000000",
   "") in new stack
-- Started music on hold, class 'default', on SIP/0000FFFF0001-00000000
```

```
*CLI> queue show support
support        has 0 calls (max unlimited) in 'rrmemory' strategy
(0s holdtime, 0s talktime), W:0, C:0, A:0, SL:0.0% within 0s
   Members:
      SIP/0000FFFF0001 (dynamic) (Not in use) has taken no calls yet
   No Callers
```

Notice that even though our phone should be marked as In Use because it is on a call, it does not show up that way when we look at the queue status. This is obviously a problem since the queue will consider this device as available, even though it is already on a call.

To correct this problem, we need to add `callcounter=yes` to the [general] section of our *sip.conf* file. We can also specifically configure this for any peer (since it is a peer-level configuration option); however, this is really something you'll want to set for all peers that might ever be part of a queue, so it's normally going to be best to put this option in the [general] section (it could also be assigned to a template that would be used with all peers in the queue).

Edit your *sip.conf* file so it looks similar to the following:

```
[general]
context=unauthenticated       ; default context for incoming calls
allowguest=no                 ; disable unauthenticated calls
srvlookup=yes                 ; enabled DNS SRV record lookup on outbound calls
udpbindaddr=0.0.0.0           ; listen for UDP request on all interfaces
tcpenable=no                  ; disable TCP support
callcounter=yes               ; enable device states for SIP devices
```

Then reload the chan_sip module and perform the same test again:

```
*CLI> sip reload
  Reloading SIP
   == Parsing '/etc/asterisk/sip.conf':   == Found
```

The device should now show In use when a call is in progress from that device:

```
   == Parsing '/etc/asterisk/sip.conf':   == Found
   == Using SIP RTP CoS mark 5
      -- Executing [555@LocalSets:1] MusicOnHold("SIP/0000FFFF0001-00000001",
         "") in new stack
      -- Started music on hold, class 'default', on SIP/0000FFFF0001-00000001

*CLI> queue show support
support        has 0 calls (max unlimited) in 'rrmemory' strategy
(0s holdtime, 0s talktime), W:0, C:0, A:0, SL:0.0% within 0s
   Members:
      SIP/0000FFFF0001 (dynamic) (In use) has taken no calls yet
   No Callers
```

In short, Queue() needs to know the state of a device in order to properly manage call distribution. The callcounter option in *sip.conf* is an essential component of a properly functioning queue.

The queues.conf File

We've mentioned the *queues.conf* file already, but there are many options in this file, and we figured it would be right and proper for us to go over some of them with you.

Table 13-1 contains the options available in the [general] section of *queues.conf*.

Table 13-1. Available options for [general] section of queues.conf

Options	Available values	Description
persistentmembers	yes, no	Set this to yes to store dynamically added members to queues in the AstDB so that they can be re-added upon Asterisk restart.
autofill	yes, no	With autofill disabled, the queue application will attempt to deliver calls to agents in a serial manner. This means only one call is attempted to be distributed to agents at a time. Additional callers are not distributed to agents until that caller is connected to an agent. With autofill enabled, callers are distributed to available agents simultaneously.
monitor-type	MixMonitor, <unspecified>	If you specify the value MixMonitor the MixMonitor() application will be used for recording calls within the queue. If you do not specify a value or comment the option out, the Monitor() application will be used instead.
updatecdr	yes, no	Set this to yes to populate the dstchannel field of the CDR records with the name of a dynamically added member on answer. The value is set with the AddQueueMember() application. This option is used to mimic the behavior of chan_agent channels.
shared_lastcall	yes, no	This value is used for members logged into more than one queue to have their last call be the same across all queues, in order for the queues to respect the wrap up time of other queues.

Table 13-2 describes the options available for configuring queue contexts.

Table 13-2. Available options for defined queues in queues.conf

Options	Available values	Description
musicclass	Music class as defined by *musiconhold.conf*	Sets the music class to be used by a particular queue. You can also override this value with the CHANNEL(musicclass) channel variable.
announce	Filename of the announcement	Used for playing an announcement to the agent that answered the call, typically to let him know what queue the caller is coming from. Useful when the agent is in multiple queues, especially when set to auto-answer the queue.
strategy	ringall, leastrecent, fewestcalls, random, rrmemory, linear, wrandom	• ringall: rings all available callers (default) • leastrecent: rings the interface that least recently received a call • fewestcalls: rings the interface that has completed the fewest calls in this queue

Options	Available values	Description
		• random: rings a random interface
		• rrmemory: rings members in a round-robin fashion, remembering where we left off last for the next caller
		• linear: rings members in the order specified, always starting at the beginning of the list
		• wrandom: rings a random member, but uses the members' penalties as a weight.
servicelevel	Value in seconds	Used in statistics to determine the service level of the queue (calls answered within the service level time frame).
context	Dialplan context	Allows a caller to exit the queue by pressing a *single* DTMF digit. If a context is specified and the caller enters a number, that digit will attempt to be matched in the context specified, and dialplan execution will continue there.
penaltymemberslimit	Value of 0 or greater	Used to disregard penalty values if the number of members in the queue is lower than the value specified.
timeout	Value in seconds	Specifies the number of seconds to ring a member's device. Also see timeoutpriority.
retry	Value in seconds	Specifies the number of seconds to wait before attempting the next member in the queue if the timeout value is exhausted while attempting to ring a member of the queue.
timeoutpriority	app, conf	Used to control the priority of the two possible timeout options specified for a queue. The Queue() application has a timeout value that can be specified to control the absolute time a caller can be in the queue. The timeout value in *queues.conf* controls the amount of time (along with retry) to ring a member for. Sometime these values conflict, so you can control which value takes precedence. The default is app, as this is the way it works in previous versions.
weight	Value of 0 or higher	Defines the weight of a queue. A queue with a higher weight defined will get first priority when members are associated with multiple queues.
wrapuptime	Value in seconds	The number of seconds to keep a member unavailable in a queue after completing a call.
autofill	yes, no	Same as defined in the [general] section. This value can be defined per queue.
autopause	yes, no, all	Enables/disables the automatic pausing of members who fail to answer a call. A value of all causes this member to be paused in all queues she is a member of.
maxlen	Value of 0 or higher	Specifies the maximum number of callers allowed to be waiting in a queue. A value of zero means an unlimited number of callers are allowed in the queue.

Options	Available values	Description
setinterfacevar	yes, no	If set to yes, the following channel variables will be set just prior to connecting the caller with the queue member: • MEMBERINTERFACE: the member's interface, such as Agent/1234 • MEMBERNAME: the name of the member • MEMBERCALLS: the number of calls the interface has taken • MEMBERLASTCALL: the last time the member took a call • MEMBERPENALTY: the penalty value of the member • MEMBERDYNAMIC: indicates whether the member was dynamically added to the queue or not • MEMBERREALTIME: indicates whether the member is included from real time or not
setqueueentryvar	yes, no	If set to yes, the following channel variables will be set just prior to the call being bridged: • QEHOLDTIME: the amount of time the caller was held in the queue • QEORIGINALPOS: the position the caller originally entered the queue at
setqueuevar	yes, no	If set to yes, the following channel variables will be set just prior to the call being bridged: • QUEUENAME: the name of the queue • QUEUEMAX: the maximum number of calls allowed in this queue • QUEUESTRATEGY: the strategy method defined for the queue • QUEUECALLS: the number of calls currently in the queue • QUEUEHOLDTIME: the current average hold time of callers in the queue • QUEUECOMPLETED: the number of completed calls in this queue • QUEUEABANDONED: the number of abandoned calls • QUEUESRVLEVEL: the queue service level • QUEUESRVLEVELPERF: the queue's service level performance

Options	Available values	Description
membermacro	Name of a macro defined in the dialplan	Defines a macro to be executed just prior to bridging the caller and the queue member.
announce-frequency	Value in seconds	Defines how often we should announce the caller's position and/or estimated hold time in the queue. Set this value to zero to disable.
min-announce-frequency	Value in seconds	Specifies the minimum amount of time that must pass before we announce the caller's position in the queue again. This is used when the caller's position may change frequently, to prevent the caller hearing multiple updates in a short period of time.
periodic-announce-frequency	Value in seconds	Indicates how often we should make periodic announcements to the caller.
random-periodic-announce	yes, no	If set to yes, will play the defined periodic announcements in a random order. See periodic-announce.
relative-periodic-announce	yes, no	If set to yes, the periodic-announce-frequency timer will start from when the end of the file being played back is reached, instead of from the beginning. Defaults to no.
announce-holdtime	yes, no, once	Defines whether the estimated hold time should be played along with the periodic announcements. Can be set to yes, no, or only once.
announce-position	yes, no, limit, more	Defines whether the caller's position in the queue should be announced to her. If set to no, the position will never be announced. If set to yes, the caller's position will always be announced. If the value is set to limit, the caller will hear her position in the queue only if it is within the limit defined by announce-position-limit. If the value is set to more, the caller will hear her position if it is beyond the number defined by announce-position-limit.
announce-position-limit	Number of zero or greater	Used if you've defined announce-position as either limit or more.
announce-round-seconds	Value in seconds	If this value is nonzero, we'll announce the number of seconds as well, and round them to the value defined.
queue-thankyou	Filename of prompt to play	If not defined, will play the default value ("Thank you for your patience"). If set to an empty value, the prompt will not be played at all.
queue-youarenext	Filename of prompt to play	If not defined, will play the default value ("You are now first in line"). If set to an empty value, the prompt will not be played at all.
queue-thereare	Filename of prompt to play	If not defined, will play the default value ("There are"). If set to an empty value, the prompt will not be played at all.

Options	Available values	Description
queue-callswaiting	Filename of prompt to play	If not defined, will play the default value ("calls waiting"). If set to an empty value, the prompt will not be played at all.
queue-holdtime	Filename of prompt to play	If not defined, will play the default value ("The current estimated hold time is"). If set to an empty value, the prompt will not be played at all.
queue-minutes	Filename of prompt to play	If not defined, will play the default value ("minutes"). If set to an empty value, the prompt will not be played at all.
queue-seconds	Filename of prompt to play	If not defined, will play the default value ("seconds"). If set to an empty value, the prompt will not be played at all.
queue-reporthold	Filename of prompt to play	If not defined, will play the default value ("Hold time"). If set to an empty value, the prompt will not be played at all.
periodic-announce	A set of periodic announcements to be played, separated by commas	Prompts are played in the order they are defined. Defaults to queue-periodic-announce ("All representatives are currently busy assisting other callers. Please wait for the next available representative").
monitor-format	gsm, wav, wav49, <any valid file format>	Specifies the file format to use when recording. If monitor-format is commented out, calls will not be recorded.
monitor-type	MixMonitor, <unspecified>	Same as monitor-type as defined in the [general] section, but on a per-queue basis.
joinempty	paused, penalty, inuse, ringing, unavailable, invalid, unknown, wrapup	Controls whether a caller is added to the queue when no members are available. Comma-separated options can be included to define how this option determines whether members are available. The definitions for the values are:<ul><li>paused: members are considered unavailable if they are paused.</li><li>penalty: members are considered unavailable if their penalties are less than QUEUE_MAX_PENALTY.</li><li>inuse: members are considered unavailable if their device status is In Use.</li><li>ringing: members are considered unavailable if their device status is Ringing.</li><li>unavailable: applies primarily to agent channels; if the agent is not logged in but is a member of the queue, it is considered unavailable.</li><li>invalid: members are considered unavailable if their device status is Invalid. This is typically an error condition.</li><li>unknown: members are considered unavailable if device status is unknown.</li></ul>

Options	Available values	Description
		• wrapup: members are considered unavailable if they are currently in the wrapup time after the completion of a call.
leavewhenempty	paused, penalty, inuse, ringing, unavailable, invalid, unknown, wrapup	Used to control whether callers are kicked out of the queue when members are no longer available to take calls. See joinempty for more information on the assignable values.
eventwhencalled	yes, no, vars	If set to yes, the following manager events will be sent to the Asterisk Manager Interface (AMI): • AgentCalled • AgentDump • AgentConnect • AgentComplete If set to vars, all channel variables associated with the agent will also be sent to the AMI.
eventmemberstatus	yes, no	If set to yes, the QueueMemberStatus event will be sent to AMI. Note that this may generate a lot of manager events.
reportholdtime	yes, no	Enables reporting of the caller's hold time to the queue member prior to bridging.
ringinuse	yes, no	Used to avoid sending calls to members whose status is In Use. Recall from our discussion in the preceding section that only the SIP channel driver is currently able to accurately report this status.
memberdelay	Value in seconds	Used if you want there to be a delay prior to the caller and queue member being connected to each other.
timeoutrestart	yes, no	If set to yes, resets the timeout for an agent to answer if either a BUSY or CONGESTION status is received from the channel. This can be useful if the agent is allowed to reject or cancel a call.
defaultrule	Rule as defined in *queuerules.conf*	Associates a queue rule as defined in *queuerules.conf* to this queue, which is used to dynamically change the minimum and maximum penalties, which are then used to select an available agent. See "Changing Penalties Dynamically (queuerules.conf)" on page 285.
member	Device	Used to define static members in a queue. To define a static member, you supply its *Technology/Device_ID* (e.g., Agent/1234, SIP/0000FFFF0001, DAHDI/g0/14165551212).

The agents.conf File

If you've browsed through the samples in the *~/src/asterisk-complete/1.8/configs/* directory, you may have noticed the *agents.conf* file. It may seem tempting, and it has its places, but overall the best way to implement queues is through the use of SIP channels. There are two reasons for this. The first is that SIP channels are the only type that provide true device state information. The other reason is that agents are always logged in when using the agent channel, and if you're using remote agents, the bandwidth requirements may be greater than you wish. However, in busy call centers it may be desirable to force agents to answer calls immediately rather than having them press the answer button on the phone.

The *agents.conf* file is use to define agents for queues using the agents channel. This channel is similar in nature to the other channel types in Asterisk (local, SIP, IAX2, etc.), but it is more of a pseudo-channel in that it is used to connect callers to agents who have logged into the system using other types of transport channel. For example, suppose we use our SIP-enabled phone to log in to Asterisk using the `AgentLogin()` dialplan application. Once we're logged in, the channel remains online the entire time it is available (logged on), and calls are then passed to it through the agent channel.

Let's take a look at the various options available to us in the *agents.conf* file to get a better idea of what it provides us. Table 13-3 shows the single option available in the `[general]` section of *agents.conf*. Table 13-4 shows the available options under the `[agents]` header.

Table 13-3. Options available under the [general] header in agents.conf

Options	Available values	Description
multiplelogin	yes, no	If set to yes, a single line on a device can log in as multiple agents. Defaults to yes.

Table 13-4. Options available under the [agents] header in agents.conf

Options	Available values	Description
maxloginretries	Integer value	Specifies the maximum number of tries an agent has to log in before the system considers it a failed attempt and ends the call. Defaults to 3.
autologoff	Value in seconds	Specifies the number of seconds for which an agent's device should ring before the agent is automatically logged off.
autologoffunavail	yes, no	If set to yes, the agent is automatically logged off when the device being called returns a status of CHANUNAVAIL.
ackcall	yes, no	If set to yes, the agent must enter a single DTMF digit to accept the call. To be used in conjunction with acceptdtmf. Defaults to no.
acceptdtmf	Single DTMF character	Used in conjunction with ackcall, this option defines the DTMF character to be used to accept a call. Defaults to #.
endcall	yes, no	If set to yes, allows an agent to end a call with a single DTMF digit. To be used in conjunction with enddtmf. Defaults to yes.

Options	Available values	Description
enddtmf	Single DTMF character	Used in conjunction with endcall, this option defines the DTMF character to be used to end a call. Defaults to *.
wrapuptime	Value in milliseconds	Specifies the amount of time after disconnection of a caller from an agent for which the agent will not be available to accept another call. Used in situations where agents must perform a function after each call (such as entering call details into a log).
musiconhold	Music class as defined in *musiconhold.conf*	Defines the default music class agents listen to when logged in.
goodbye	Name of file (relative to */var/lib/asterisk/sounds/<lang>/*)	Defines the default goodbye sound played to agents. Defaults to *vm-goodbye*.
updatecdr	yes, no	Used in call detail records to change the source channel field to the agent/*agent_id*.
group	Integer value	Allows you to define groups for sets of agents. *The use of agent groups is essentially deprecated functionality that we do not recommend you use.* If you define group1, you can use Agent/@1 in *queues.conf* to call that group of agents. The call will be connected arbitrarily to one of those agents. If no agents are available, it will return back to the queue like any other unanswered call. If you use Agent/:1, it will wait for a member of the group to become available. *The use of strategies has no effect on agent groups. Do not use these.*
recordagentcalls	yes, no	Enables/disables the recording of agent calls. Disabled by default.
recordformat	File format (gsm, wav, etc.)	Defines the format to be used when recording agent calls. Default is wav.
urlprefix	String (URL)	Accepts a string as its argument. The string can be formed as a URL and is appended to the start of the text to be added to the name of the recording.
savecallsin	Filesystem path (e.g., */var/calls/*)	Accepts a filesystem path as its argument. Allows you to override the default path of */var/spool/asterisk/monitor/* with one of your choosing.[a]
custom_beep	Name of file (relative to */var/lib/asterisk/sounds/<lang>/*)	Accepts a filename as its argument. Can be used to define a custom notification tone to signal to an always-connected agent that there is an incoming call.

agent	Agent definition (see description)	Defines an agent for use by Queue() and AgentLogin(). These are agents that will log in and stay connected to the system, waiting for calls to be delivered by the Queue() dialplan application. Agents are defined like so:
		`agent => agent_id,agent_password,name`
		
		An example of a defined agent would be:
		`agent => 1000,1234,Danielle Roberts`

[a] Since the storage of calls will require a large amount of hard drive space, you will want to define a strategy to handle storing and managing these recordings. This location should probably reside on a separate volume, one with very high performance characteristics.

Advanced Queues

In this section we'll take a look at some of the finer-grained queue controls, such as options for controlling announcements and when callers should be placed into (or removed from) the queue. We'll also look at penalties and priorities, exploring how we can control the agents in our queue by giving preference to a pool of agents to answer the call and increase that pool dynamically based on the wait times in the queue. Finally, we'll look at using Local channels as queue members, which gives us the ability to perform dialplan functionality prior to connecting the caller to an agent.

Priority Queue (Queue Weighting)

Sometimes you need to add people to a queue at a higher priority than that given to other callers. Perhaps the caller has already spent time waiting in a queue, and an agent has taken some information but realized the caller needed to be transferred to another queue. In this case, to minimize the caller's overall wait time, it might be desirable to transfer the call to a priority queue that has a higher weight (and thus a higher preference), so it will be answered quickly.

Setting a higher priority on a queue is done with the `weight` option. If you have two queues with differing weights (e.g., support and support-priority), agents assigned to both queues will be passed calls from the higher-priority queue in preference to calls from the lower-priority queue. Those agents will not take any calls from the lower-priority queue until the higher-priority queue is cleared. (Normally, there will be some agents who are assigned only to the lower-priority queue, to ensure that those calls are dealt with in a timely manner.) For example, if we place queue member James Shaw into both the support and support-priority queues, callers in the support-priority queue will have a preferred standing with James over callers in the support queue.

Let's take a look at how we could make this work. First, we need to create two queues that are identical except for the `weight` option. We can use a template for this to ensure that the two queues remain identical if anything should need to change in the future:

```
[support_template](!)
musicclass=default
strategy=rrmemory
joinempty=no
leavewhenempty=yes
ringinuse=no

[support](support_template)
weight=0

[support-priority](support_template)
weight=10
```

With our queues configured (and subsequently reloaded using *module reload app_queue.so* from the Asterisk console), we can now create two extensions to transfer callers to. This can be done wherever you would normally place your dialplan logic to perform transfers. We're going to use the LocalSets context, which we've previously enabled as the starting context for our devices:

```
[LocalSets]
include => Queue  ; allow direct transfer of calls to queues

[Queues]
exten => 7000,1,Verbose(2,Entering the support queue)
    same => n,Queue(support)              ; standard support queue available
                                          ; at extension 7000
    same => n,VoiceMail(7000@queues,u)    ; if there are no members in the queue,
                                          ; we exit and send the caller to voicemail
    same => n,Hangup()

exten => 8000,1,Verbose(2,Entering the priority support queue)
    same => n,Queue(support-priority)     ; priority queue available at
                                          ; extension 8000
    same => n,VoiceMail(7000@queues,u)    ; if there are no members in the queue,
                                          ; we exit and send the caller to voicemail
    same => n,Hangup()
```

There you have it: two queues defined with different weights. We've configured our standard queues to start at extension 7000, and our priority queues to start at 8000. We can mirror this for several queues by simply matching between the 7XXX and 8XXX ranges. So, for example, if we have our sales queue at extension 7004, our priority-sales queue (for returning customers, perhaps?) could be placed in the mirrored queue at 8004, which has a higher weight.

The only other configuration left to do is to make sure some or all of your queue members are placed in both queues. If you have more callers in your 7XXXX queues, you may want to have more queue members logged into that queue, with a percentage of your queue members logged into both queues. Exactly how you wish to configure your queues will depend on your local policy and circumstances.

Queue Member Priority

Within a queue, we can *penalize* members in order to lower their preference for being called when there are people waiting in a particular queue. For example, we may penalize queue members when we want them to be a member of a queue, but to be used only when the queue gets full enough that all our preferred agents are unavailable. This means we can have three queues (say, support, sales, and billing), each containing the same three queue members: James Shaw, Kay Madsen, and Danielle Roberts.

Suppose, however, that we want James Shaw to be the preferred contact in the support queue, Kay Madsen preferred in sales, and Danielle Roberts preferred in billing. By penalizing Kay Madsen and Danielle Roberts in support, we ensure that James Shaw will be the preferred queue member called. Similarly, we can penalize James

Shaw and Danielle Roberts in the sales queue so Kay Madsen is preferred, and penalize James Shaw and Kay Madsen in the billing queue so Danielle Roberts is preferred.

Penalizing queue members can be done either in the *queues.conf* file, if you're specifying queue members statically, or through the AddQueueMember() dialplan application. Let's look at how our queues would be set up with static members in *queues.conf*. We'll be using the StandardQueue template we defined earlier in this chapter:

```
[support](StandardQueue)
member => SIP/0000FFFF0001,0,James Shaw         ; preferred
member => SIP/0000FFFF0002,10,Kay Madsen        ; second preferred
member => SIP/0000FFFF0003,20,Danielle Roberts  ; least preferred

[sales](StandardQueue)
member => SIP/0000FFFF0002,0,Kay Madsen
member => SIP/0000FFFF0003,10,Danielle Roberts
member => SIP/0000FFFF0001,20,James Shaw

[billing](StandardQueue)
member => SIP/0000FFFF0003,0,Danielle Roberts
member => SIP/0000FFFF0001,10,James Shaw
member => SIP/0000FFFF0002,20,Kay Madsen
```

By defining different penalties for each member of the queue, we can help control the preference for where callers are delivered, but still ensure that other queue members will be available to answer calls if the preferred member is unavailable. Penalties can also be defined using AddQueueMember(), as the following example demonstrates:

```
exten => *54,1,Verbose(2,Logging In Queue Member)
    same => n,Set(MemberChannel=${CHANNEL(channeltype)}/${CHANNEL(peername)})

; *CLI> database put queue support/0000FFFF0001/penalty 0
    same => n,Set(QueuePenalty=${DB(queue/support/${CHANNEL(peername)}/penalty)})

; *CLI> database put queue support/0000FFFF0001/membername "James Shaw"
    same => n,Set(MemberName=${DB(queue/support/${CHANNEL(peername)}/membername)})

; AddQueueMember(queuename[,interface[,penalty[,options[,membername
; [,stateinterface]]]]])
    same => n,AddQueueMember(support,${MemberChannel},${QueuePenalty},,${MemberName})
```

Using AddQueueMember(), we've shown how you could retrieve the penalty associated with a given member name for a particular queue and assign that value to the member when she logs into the queue. Some additional abstraction would need to be done to make this work for multiple queues; for more information see "Automatically Logging Into and Out of Multiple Queues" on page 270.

Changing Penalties Dynamically (queuerules.conf)

Using the *queuerules.conf* file, it is possible to specify rules to change the values of the QUEUE_MIN_PENALTY and QUEUE_MAX_PENALTY channel variables. The QUEUE_MIN_PENALTY and QUEUE_MAX_PENALTY channel variables are used to control which members of a queue are to be used for servicing callers. Let's say we have a queue called support, and we have five queue members with various penalties ranging from 1 through 5. If prior to a caller entering the queue the QUEUE_MIN_PENALTY channel variable is set to a value of 2 and the QUEUE_MAX_PENALTY is set to a value of 4, only queue members whose penalties are set to values ranging from 2 through 4 will be considered available to answer that call:

```
[Queues]
exten => 7000,1,Verbose(2,Entering the support queue)
    same => n,Set(QUEUE_MIN_PENALTY=2)   ; set minimum queue member penalty to be used
    same => n,Set(QUEUE_MAX_PENALTY=4)   ; set maximum queue member penalty we'll use
    same => n,Queue(support)             ; entering the queue with minimum and maximum
                                         ; member penalties to be used
```

What's more, during the caller's stay in the queue, we can dynamically change the values of QUEUE_MIN_PENALTY and QUEUE_MAX_PENALTY for that caller. This allows either more or a different set of queue members to be used, depending on how long the caller waits in the queue. For instance, in the previous example, we could modify the minimum penalty to 1 and the maximum penalty to 5 if the caller has to wait more than 60 seconds in the queue.

The rules are defined using the *queuerules.conf* file. Multiple rules can be created in order to facilitate different penalty changes throughout the call. Let's take a look at how we'd define the changes described in the previous paragraph:

```
[more_members]
penaltychange => 60,5,1
```

 If you make changes to the *queuerules.conf* file and reload *app_queue.so*, the new rules will affect only new callers in the queue, not existing callers.

We've defined the rule more_members in *queuerules.conf* and passed the following values to penaltychange: 60 is the number of seconds to wait before changing the penalty values, 5 is the new QUEUE_MAX_PENALTY, and 1 is the new QUEUE_MIN_PENALTY. With our new rule defined, we must reload *app_queue.so* to make it available to us for use:

```
*CLI> module reload app_queue.so
    -- Reloading module 'app_queue.so' (True Call Queueing)
   == Parsing '/etc/asterisk/queuerules.conf':   == Found
```

We can also verify our rules at the console with *queue show rules*:

```
*CLI> queue show rules
Rule: more_members
    After 60 seconds, adjust QUEUE_MAX_PENALTY to 5 and adjust QUEUE_MIN_PENALTY to 1
```

With our rule now loaded into memory, we can modify our dialplan to make use of it. Just modify the Queue() line to include the new rule, like so:

```
[Queues]
exten => 7000,1,Verbose(2,Entering the support queue)
    same => n,Set(QUEUE_MIN_PENALTY=2)    ; set minimum queue member penalty
    same => n,Set(QUEUE_MAX_PENALTY=4)    ; set maximum queue member penalty

; Queue(queuename[,options[,URL[,announceoverride[,timeout[,AGI[,macro
; [,gosub[,rule[,position]]]]]]]]])
    same => n,Queue(support,,,,,,,,more_members)   ; entering queue with minimum and
                                                   ; maximum member penalties
```

The *queuerules.conf* file is quite flexible. We can define our rule using relative instead of absolute penalty values, and we can define multiple rules:

```
[more_members]
penaltychange => 30,+1
penaltychange => 45,,-1
penaltychange => 60,+1
penaltychange => 120,+2
```

Here, we've modified our more_members rule to use relative values. After 30 seconds, we increase the maximum penalty by 1 (which would take us to 5 using our sample dialplan). After 45 seconds, we decrease the minimum penalty by 1, and so on. We can verify our new rule changes after a *module reload app_queue.so* at the Asterisk console:

```
*CLI> queue show rules
Rule: more_members
   After 30 seconds, adjust QUEUE_MAX_PENALTY by 1 and adjust QUEUE_MIN_PENALTY by 0
   After 45 seconds, adjust QUEUE_MAX_PENALTY by 0 and adjust QUEUE_MIN_PENALTY by -1
   After 60 seconds, adjust QUEUE_MAX_PENALTY by 1 and adjust QUEUE_MIN_PENALTY by 0
   After 120 seconds, adjust QUEUE_MAX_PENALTY by 2 and adjust QUEUE_MIN_PENALTY by 0
```

Announcement Control

Asterisk has the ability to play several announcements to callers waiting in the queue. For example, you might want to announce the caller's position in the queue, the average wait time, or make periodic announcements thanking your callers for waiting (or whatever your audio files say). It's important to tune the values that control when these announcements are played to the callers, because announcing their position, thanking them for waiting, and telling them the average hold time too often may annoy them, causing them to either hang up or take it out on your agents.

There are several options in the *queues.conf* file that you can use to fine-tune what and when announcements are played to your callers. The full list of queue options is available in "The queues.conf File" on page 275, but we'll review the relevant ones here.

Table 13-5 lists the options you can use to control when announcements are played to the caller.

Table 13-5. Options related to prompt control timing within a queue

Options	Available values	Description
announce-frequency	Value in seconds	Defines how often we should announce the caller's position and/or estimated hold time in the queue. Set this value to zero to disable.
min-announce-frequency	Value in seconds	Indicates the minimum amount of time that must pass before we announce the caller's position in the queue again. This is used when the caller's position may change frequently, to prevent the caller hearing multiple updates in a short period of time.
periodic-announce-frequency	Value in seconds	Specifies how often we should make periodic announcements to the caller.
random-periodic-announce	yes, no	If set to yes, will play the defined periodic announcements in a random order. See periodic-announce.
relative-periodic-announce	yes, no	If set to yes, the periodic-announce-frequency timer will start from when the end of the file being played back is reached, instead of from the beginning. Defaults to no.
announce-holdtime	yes, no, once	Defines whether the estimated hold time should be played along with the periodic announcements. Can be set to yes, no, or only once.
announce-position	yes, no, limit, more	Defines whether the caller's position in the queue should be announced to her. If set to no, the position will never be announced. If set to yes, the caller's position will always be announced. If the value is set to limit, the caller will hear her position in the queue only if it is within the limit defined by announce-position-limit. If the value is set to more, the caller will hear her position only if it is beyond the number defined by announce-position-limit.
announce-position-limit	Number of zero or greater	Used if you've defined announce-position as either limit or more.
announce-round-seconds	Value in seconds	If this value is nonzero, we'll announce the number of seconds as well, and round them to the value defined.

Table 13-6 shows what files will be used when announcements are played to the caller.

Table 13-6. Options for controlling the playback of prompts within a queue

Options	Available values	Description
musicclass	Music class as defined by *musiconhold.conf*	Sets the music class to be used by a particular queue. You can also override this value with the CHANNEL(musicclass) channel variable.
queue-thankyou	Filename of prompt to play	If not defined, will play the default value ("Thank you for your patience"). If set to an empty value, the prompt will not be played at all.

Options	Available values	Description
queue-youarenext	Filename of prompt to play	If not defined, will play the default value ("You are now first in line"). If set to an empty value, the prompt will not be played at all.
queue-thereare	Filename of prompt to play	If not defined, will play the default value ("There are"). If set to an empty value, the prompt will not be played at all.
queue-callswaiting	Filename of prompt to play	If not defined, will play the default value ("calls waiting"). If set to an empty value, the prompt will not be played at all.
queue-holdtime	Filename of prompt to play	If not defined, will play the default value ("The current estimated hold time is"). If set to an empty value, the prompt will not be played at all.
queue-minutes	Filename of prompt to play	If not defined, will play the default value ("minutes"). If set to an empty value, the prompt will not be played at all.
queue-seconds	Filename of prompt to play	If not defined, will play the default value ("seconds"). If set to an empty value, the prompt will not be played at all.
queue-reporthold	Filename of prompt to play	If not defined, will play the default value ("Hold time"). If set to an empty value, the prompt will not be played at all.
periodic-announce	A set of periodic announcements to be played, separated by commas	Prompts are played in the order they are defined. Defaults to queue-periodic-announce ("All representatives are currently busy assisting other callers. Please wait for the next available representative").

If the number of options devoted to playing announcements to callers is any indication of their importance, it's probably in our best interest to use them to their fullest potential. The options in Table 13-5 help us define when we'll play announcements to callers, and the options in Table 13-6 help us control what we play to our callers. With those tables in hand, let's take a look at an example queue where we've defined some values. We'll use our basic queue template as a starting point:

```
[general]
autofill=yes            ; distribute all waiting callers to available members
shared_lastcall=yes     ; respect the wrapup time for members logged into more
                        ; than one queue

[StandardQueue](!)      ; template to provide common features
musicclass=default      ; play [default] music
strategy=rrmemory       ; use the Round Robin Memory strategy
joinempty=yes           ; do not join the queue when no members available
leavewhenempty=no       ; leave the queue when no members available
ringinuse=no            ; don't ring members when already InUse (prevents
                        ; multiple calls to an agent)

[sales](StandardQueue)  ; create the sales queue using the parameters in the
                        ; StandardQueue template

[support](StandardQueue) ; create the support queue using the parameters in the
                        ; StandardQueue template
```

We'll now modify the `StardardQueue` template to control our announcements:

```
[StandardQueue](!)        ; template to provide common features
musicclass=default        ; play [default] music
strategy=rrmemory         ; use the Round Robin Memory strategy
joinempty=yes             ; do not join the queue when no members available
leavewhenempty=no         ; leave the queue when no members available
ringinuse=no              ; don't ring members when already InUse (prevents
                          ; multiple calls to an agent)

; -------- Announcement Control --------
announce-frequency=30             ; announces caller's hold time and position every 30
                                  ; seconds
min-announce-frequency=30         ; minimum amount of time that must pass before the
                                  ; caller's position is announced
periodic-announce-frequency=45    ; defines how often to play a periodic announcement to
                                  ; caller
random-periodic-announce=no       ; defines whether to play periodic announcements in
                                  ; a random order, or serially
relative-periodic-announce=yes    ; defines whether the timer starts at the end of
                                  ; file playback (yes) or the beginning (no)
announce-holdtime=once            ; defines whether the estimated hold time should be
                                  ; played along with the periodic announcement
announce-position=limit           ; defines if we should announce the caller's position
                                  ; in the queue
announce-position-limit=10        ; defines the limit value where we announce the
                                  ; caller's position (when announce-position is set to
                                  ; limit or more)
announce-round-seconds=30         ; rounds the hold time announcement to the nearest
                                  ; 30-second value
```

Let's describe what we've just set in our `StandardQueue` template.

We'll announce the caller's hold time and position every 30 seconds (announce-frequency),[§] and make sure the minimum amount of time that passes before we announce it again is at least 30 seconds (min-announce-frequency). We do this to limit how often our announcements are played to the callers, in order to avoid the updates becoming annoying. Periodically, we'll play an announcement to the callers that thanks them for holding and assures them that an agent will be with them shortly. (The announcement is defined by the periodic-announcement setting. We're using the default announcement, but you can define one or more announcements yourself using periodic-announce.)

These periodic announcements will be played every 45 seconds (periodic-announce-frequency), in the order they were defined (random-period-announce). To determine when the periodic-announce-frequency timer should start, we use relative-periodic-announce. The yes setting means the timer will start after the announcement has finished playing, rather than when it starts to play. The problem you could run into if you set this to no is that if your periodic announcement runs for any significant length of time

---

[§] Callers' positions and hold times are only announced if more than one person is holding in the queue.

(lets say 30 seconds), it will appear as if it is being played every 15 seconds, rather than every 45 seconds as may be intended.

How many times we announce the hold time to the caller is controlled via the announce-holdtime option, which we've set to once. Setting the value to yes will announce it every time, and setting to no will disable it.

We configure how and when we announce the caller's estimated remaining hold time via announce-position, which we've set to limit. Using the value of limit for announce-position lets us announce the caller's position only if it is within the limit defined by announce-position-limit. So, in this case we're only announcing the callers' positions if they are in the first 10 positions of the queue. We could also use yes to announce the position every time the periodic announcement is played, set it to no to never announce it, or use the value more if we want to announce the position only when it is greater than the value set for announce-position-limit.

Our last option, announce-round-seconds, controls the value to round to when we announce the caller's hold time. In this case, instead of saying "1 minute and 23 seconds," the value would be rounded to the nearest 30-second value, which would result in a prompt of "1 minute and 30 seconds."

Overflow

Overflowing out of the queue is done either with a timeout value, or when no queue members are available (as defined by joinempty or leavewhenempty). In this section we'll discuss how to control when overflow happens.

Controlling timeouts

The Queue() application supports two kinds of timeout: one is for the maximum period of time a caller stays in the queue, and the other is how long to ring a device when attempting to connect a caller to a queue member. We'll be talking about the maximum period of time a caller stays in the queue before the call overflows to another location, such as VoiceMail(). Once the call has fallen out of the queue, it can go anywhere that a call could normally go when controlled by the dialplan.

The timeouts are specified in two locations. The timeout that indicates how long to ring queue members for is specified in the *queues.conf* file. The absolute timeout (how long the caller stays in the queue) is controlled via the Queue() application. To set a maximum amount of time for callers to stay in a queue, simply specify it after the queue name in the Queue() application:

```
[Queues]
exten => 7000,1,Verbose(2,Joining the support queue for a maximum of 2 minutes)
    same => n,Queue(support,120)
    same => n,VoiceMail(support@queues,u)
    same => n,Hangup()
```

Of course, we could define a different destination, but the `VoiceMail()` application is as good as any. Just make sure that if you're going to send callers to voicemail someone checks it regularly and calls your customers back.

Now let's say we have the scenario where we have set our absolute timeout to 10 seconds, our timeout value for ringing queue members to 5 seconds, and our retry timeout value to 4 seconds. In this scenario, we would ring the queue member for 5 seconds, then wait 4 seconds before attempting another queue member. That brings us up to 9 seconds of our absolute timeout of 10 seconds. At this point, should we ring the second queue member for 1 second and then exit the queue, or should we ring this member for the full 5 seconds before exiting?

We control which timeout value has priority with the `timeoutpriority` option in *queues.conf*. The available values are `app` and `conf`. If we want the application timeout (the absolute timeout) to take priority, which would cause our caller to be kicked out after exactly 10 seconds, we should set the `timeoutpriority` value to `app`. If we want the configuration file timeout to take priority and finish ringing the queue member, which will cause the caller to stay in the queue a little longer, we should set `timeout priority` to `conf`. The default value is `app` (which is the default behavior in previous versions of Asterisk).

Controlling when to join and leave a queue

Asterisk provides two options that control when callers can join and are forced to leave queues, based on the statuses of the queue members. The first option, `joinempty`, is used to control whether callers can enter a queue. The `leavewhenempty` option is used to control when callers already in a queue should be removed from that queue (i.e., if all of the queue members become unavailable). Both options take a comma-separated list of values that control this behavior. The factors are listed in Table 13-7.

Table 13-7. Options that can be set for joinempty or leavewhenempty

Value	Description
paused	Members are considered unavailable if they are paused.
penalty	Members are considered unavailable if their penalties are less than QUEUE_MAX_PENALTY.
inuse	Members are considered unavailable if their device status is In Use.
ringing	Members are considered unavailable if their device status is Ringing.
unavailable	Applies primarily to agent channels; if the agent is not logged in but is a member of the queue it is considered unavailable.
invalid	Members are considered unavailable if their device status is Invalid. This is typically an error condition.
unknown	Members are considered unavailable if device status is unknown.
wrapup	Members are considered unavailable if they are currently in the wrapup time after the completion of a call.

For `joinempty`, prior to placing a caller into the queue, all the members are checked for availability using the factors you list as criteria. If all members are deemed to be unavailable, the caller will not be permitted to enter the queue, and dialplan execution will continue at the next priority.‖ For the `leavewhempty` option, the members' statuses are checked periodically against the listed conditions; if it is determined that no members are available to take calls, the caller is removed from the queue, with dialplan execution continuing at the next priority.

An example use of `joinempty` could be:

```
joinempty=paused,inuse,invalid
```

With this configuration, prior to a caller entering the queue the statuses of all queue members will be checked, and the caller will not be permitted to enter the queue unless at least one queue member is found to have a status that is not `paused`, `inuse`, or `invalid`.

The `leavewhenempty` example could be something like:

```
leavewhenempty=inuse,ringing
```

In this case, the queue members' statuses will be checked periodically, and callers will be removed from the queue if no queue members can be found who do not have a status of either `inuse` or `ringing`.

Previous versions of Asterisk used the values `yes`, `no`, `strict`, and `loose` as the available values to be assigned. The mapping of those values is shown in Table 13-8.

Table 13-8. Mapping between old and new values for controlling when callers join and leave queues

Value	Mapping (joinempty)	Mapping (leavewhenempty)
yes	(empty)	penalty,paused,invalid
no	penalty,paused,invalid	(empty)
strict	penalty,paused,invalid,unavailable	penalty,paused,invalid,unavailable
loose	penalty,invalid	penalty,invalid

Using Local Channels

The use of Local channels as queue members is a popular way of executing parts of the dialplan and performing checks prior to dialing the actual agent's device. For example, it allows us to do things like start recording the call, set up channel variables, write to a log file, set a limit on the call length (e.g., if it is a paid service), or do any of the other things we might need to do once we know which location we're going to call.

When using Local channels for queues, they are added just like any other channels. In the *queues.conf* file, adding a Local channel would look like this:

‖ If the priority n+1 from where the Queue() application was called is not defined, the call will be hung up.

```
; queues.conf
[support](StandardQueue)
member => Local/SIP-0000FFFF0001@MemberConnector   ; pass the technology to dial over
                                                   ; and the device identifier,
                                                   ; separated by a hyphen. We'll
                                                   ; break it apart inside the
                                                   ; MemberConnector context.
```

 Notice how we passed the type of technology we want to call along with the device identifier to the MemberConnector context. We've simply used a hyphen (although we could have used nearly anything as a separator argument) as the field marker. We'll use the CUT() function inside the MemberConnector context and assign the first field (SIP) to one channel variable and the second field (0000FFFF0001) to another channel variable, which will then be used to call the endpoint.

Passing information to be later "exploded" in the context used by the Local channel is a common and useful technique (kind of like the explode() function in PHP).

Of course, we'll need the MemberConnector context to actually connect the caller to the agent:

```
[MemberConnector]
exten => _[A-Za-z0-9].,1,Verbose(2,Connecting ${CALLERID(all)} to Agent at ${EXTEN})

    ; filter out any bad characters, allowing alphanumeric characters and the hyphen
    same => n,Set(QueueMember=${FILTER(A-Za-z0-9\-,${EXTEN})})

    ; assign the first field of QueueMember to Technology using the hyphen separator
    same => n,Set(Technology=${CUT(QueueMember,-,1)})

    ; assign the second field of QueueMember to Device using the hyphen separator
    same => n,Set(Device=${CUT(QueueMember,-,2)})

    ; dial the agent
    same => n,Dial(${Technology}/${Device})
    same => n,Hangup()
```

So, now we've passed our queue member to the context, and we can dial the device. However, because we're using the Local channel as the queue member, the Queue() won't necessarily know the state the call is in, especially when the Local channel is optimized out of the path (see *https://wiki.asterisk.org/wiki/display/AST/Local+Channel +Modifiers* for information about the /n modifier, which causes the Local channel to not be optimized out of the path). The queue will be monitoring the state of the Local channel, and not that of the device we really want to monitor.

Luckily, we can give the Queue() the actual device to monitor and associate that with the Local channel, so that the Local channel's state is always that of the device we'll end up calling. Our queue member would be modified in the *queues.conf* file like so:

```
; queues.conf
[support](StandardQueue)
member => Local/SIP-0000FFFF0001@MemberConnector,,,SIP/0000FFFF0001
```

 Only SIP channels are capable of sending back reliable device state information, so it is highly recommended that you use only these channels when using Local channels as queue members.

You can also use the `AddQueueMember()` and `RemoveQueueMember()` applications to add members to and remove members from a queue, just like with any other channel. `AddQueueMember()` also has the ability to set the state interface, which we defined statically in the *queues.conf* file. An example of how you might do this follows:

```
[QueueMemberLogin]
exten => 500,1,Verbose(2,Logging in device ${CHANNEL(peername)} into the support queue)

    ; Save the device's technology to the MemberTech channel variable
    same => n,Set(MemberTech=${CHANNEL(channeltype)})

    ; Save the device's identifier to the MemberIdent channel variable
    same => n,Set(MemberIdent=${CHANNEL(peername)})

    ; Build up the interface name and assign it to the Interface channel variable
    same => n,Set(Interface=${MemberTech}/${MemberIdent})

    ; Add the member to the support queue using a Local channel. We're using the same
    ; format as before, separating the technology and the device indentifier with
    ; a hyphen and passing that information to the MemberConnector context. We then
    ; use the IF() function to determine if the member's technology is SIP and, if so,
    ; to pass back the contents of the Interface channel variable as the value to the
    ; state interface field of the AddQueueMember() application.
    ;
    ; *** This line should not have any line breaks
    same => n,AddQueueMember(support,Local/${MemberTech}-${MemberIdent}
@MemberConnector,,,${IF($[${MemberTech} = SIP]?${Interface})})
    same => n,Playback(silence/1)

    ; Play back either the agent-loginok or agent-incorrect file, depending on what
    ; the AQMSTATUS variable is set to.
    same => n,Playback(${IF($[${AQMSTATUS} = ADDED]?agent-loginok:agent-incorrect)})
    same => n,Hangup()
```

Now that we can add devices to the queue using Local channels, let's look at how we might control the number of calls to either non-SIP channels or devices with more than one line on them. We can make use of the `GROUP()` and `GROUP_COUNT()` functions to track call counts to an endpoint. We'll modify our `MemberConnector` context to take this into account:

```
[MemberConnector]
exten => _[A-Za-z0-9].,1,Verbose(2,Connecting ${CALLERID(all)} to Agent at ${EXTEN})

    ; filter out any bad characters, allowing alphanumeric characters and the hyphen
    same => n,Set(QueueMember=${FILTER(A-Za-z0-9\-,${EXTEN})})
```

```
; assign the first field of QueueMember to Technology using the hyphen separator
same => n,Set(Technology=${CUT(QueueMember,-,1)})

; assign the second field of QueueMember to Device using the hyphen separator
same => n,Set(Device=${CUT(QueueMember,-,2)})

; Increase the value of the group inside the queue_members category by one
same => n,Set(GROUP(queue_members)=${Technology}-${Device})

; Check if the group@category is greater than 1, and, if so, return Congestion()
; (too many channels)
;
; *** This line should not have any line breaks
same => n,ExecIf($[${GROUP_COUNT(${Technology}-${Device}@queue_members)} > 1]
?Congestion())

; dial the agent
same => n,Dial(${Technology}/${Device})
same => n,Hangup()
```

The passing back of Congestion() will cause the caller to be returned to the queue (while this is happening, the caller gets no indication that anything is amiss and keeps hearing music until we actually connect to the device). While this is not an ideal situation because the queue will keep trying the member over and over again (or at least include it in the cycle of agents, depending on how many members you have and their current statuses), it is better than an agent getting multiple calls at the same time.

We've also used this same method to create a type of reservation process. If you want to call an agent directly (for example, if the caller needs to follow up with a particular agent), you could reserve that agent by using the GROUP() and GROUP_COUNT() functions to essentially pause the agent in the queue until the caller can be connected. This is particularly useful in situations where you need to play some announcements to the caller prior to connecting her with the agent, but you don't want the agent to get connected to another caller while the announcements are being played.

Queue Statistics: The queue_log File

The *queue_log* file located in */var/log/asterisk/* contains information about the queues defined in your system (when a queue is reloaded, when queue members are added or removed, etc.) and about calls into the queues (e.g., their status and what channels the callers were connected to). The queue log is enabled by default, but can be controlled via the *logger.conf* file. There are three options related to the *queue_log* file specifically:

queue_log
> Controls whether the queue log is enabled or not. Valid values are yes or no (defaults to yes).

`queue_log_to_file`
> Controls whether the queue log should be written to a file even when a real time backend is present. Valid values are **yes** or **no** (defaults to **no**).

`queue_log_name`
> Controls the name of the queue log. The default is `queue_log`.

The queue log is a pipe-separated list of events. The fields in the *queue_log* file are as follows:

- Epoch timestamp of the event
- Unique ID of the call
- Name of the queue
- Name of bridged channel
- Type of event
- Zero or more event parameters

The information contained in the event parameters depends on the type of event. A sample *queue_log* file might look something like the following:

```
1292281046|psy1-1292281041.87|7100|NONE|ENTERQUEUE||4165551212|1
1292281046|psy1-1292281041.87|7100|Local/9996@MemberConnector|RINGNOANSWER|0
1292281048|psy1-1292281041.87|7100|Local/9990@MemberConnector|CONNECT|2
|psy1-1292281046.90|0

1292284121|psy1-1292281041.87|7100|Local/9990@MemberConnector|COMPLETECALLER|2|3073|1
1292284222|MANAGER|7100|Local/9990@MemberConnector|REMOVEMEMBER|
1292284222|MANAGER|7200|Local/9990@MemberConnector|REMOVEMEMBER|
1292284491|MANAGER|7100|Local/9990@MemberConnector|ADDMEMBER|
1292284491|MANAGER|7200|Local/9990@MemberConnector|ADDMEMBER|
1292284519|psy1-1292284515.93|7100|NONE|ENTERQUEUE||4165551212|1
1292284519|psy1-1292284515.93|7100|Local/9996@MemberConnector|RINGNOANSWER|0
1292284521|psy1-1292284515.93|7100|Local/9990@MemberConnector|CONNECT|2
|psy1-1292284519.96|0

1292284552|MANAGER|7100|Local/9990@MemberConnector|REMOVEMEMBER|
1292284552|MANAGER|7200|Local/9990@MemberConnector|REMOVEMEMBER|
1292284562|psy1-1292284515.93|7100|Local/9990@MemberConnector|COMPLETECALLER|2|41|1
```

As you can see from this example, there might not always be a unique ID for the event. In some cases external services, such as the Asterisk Manager Interface (AMI), perform actions on the queue; in this case you'll see something like `MANAGER` in the Unique ID field.

The available events and the information they provide are described in Table 13-9.

Table 13-9. Events in the Asterisk queue log

Event	Information provided
ABANDON	Written when a caller in a queue hangs up before his call is answered by an agent. Three parameters are provided for ABANDON: the position of the caller at hangup, the original position of the caller when entering the queue, and the amount of time the caller waited prior to hanging up.
ADDMEMBER	Written when a member is added to the queue. The bridged channel name will be populated with the name of the channel added to the queue.
AGENTDUMP	Indicates that the agent hung up on the caller while the queue announcement was being played, prior to them being bridged together.
AGENTLOGIN	Recorded when an agent logs in. The bridged channel field will contain something like Agent/9994 if logging in with chan_agent, and the first parameter field will contain the channel logging in (e.g., SIP/0000FFFF0001).
AGENTLOGOFF	Logged when an agent logs off, along with a parameter indicating how long the agent was logged in for.
COMPLETEAGENT	Recorded when a call is bridged to an agent and the agent hangs up, along with parameters indicating the amount of time the caller was held in the queue, the length of the call with the agent, and the original position at which the caller entered the queue.
COMPLETECALLER	Same as COMPLETEAGENT, except the caller hung up and not the agent.
CONFIGRELOAD	Indicates that the queue configuration was reloaded (e.g., via *module reload app_queue.so*).
CONNECT	Written when the caller and the agent are bridged together. Three parameters are also written: the amount of time the caller waited in the queue, the unique ID of the queue member's channel to which the caller was bridged, and the amount of time the queue member's phone rang prior to being answered.
ENTERQUEUE	Written when a caller enters the queue. Two parameters are also written: the URL (if specified) and the caller ID of the caller.
EXITEMPTY	Written when the caller is removed from the queue due to a lack of agents available to answer the call (as specified by the leavewhenempty parameter). Three parameters are also written: the position of the caller in the queue, the original position at which the caller entered the queue, and the amount of time the caller was held in the queue.
EXITWITHKEY	Written when the caller exits the queue by pressing a single DTMF key on his phone to exit the queue and continue in the dialplan (as enabled by the context parameter in *queues.conf*). Four parameters are recorded: the key used to exit the queue, the position of the caller in the queue upon exit, the original position the caller entered the queue at, and the amount of time the caller was waiting in the queue.
EXITWITHTIMEOUT	Written when the caller is removed from the queue due to timeout (as specified by the timeout parameter to Queue()). Three parameters are also recorded: the position the caller was in when exiting the queue, the original position of the caller when entering the queue, and the amount of time the caller waited in the queue.
PAUSE	Written when a queue member is paused.
PAUSEALL	Written when all members of a queue are paused.
UNPAUSE	Written when a queue member is unpaused.
UNPAUSEALL	Written when all members of a queue are unpaused.

Event	Information provided
PENALTY	Written when a member's penalty is modified. The penalty can be changed through several means, such as the QUEUE_MEMBER_PENALTY() function, through using Asterisk Manager Interface, or the Asterisk CLI commands.
REMOVEMEMBER	Written when a queue member is removed from the queue. The bridge channel field will contain the name of the member removed from the queue.
RINGNOANSWER	Logged when a queue member is rung for a period of time, and the timeout value for ringing the queue member is exceeded. A single parameter will also be written indicating the amount of time the member's extension rang.
TRANSFER	Written when a caller is transferred to another extension. Additional parameters are also written, which include: the extension and context the caller was transferred to, the hold time of the caller in the queue, the amount of time the caller was speaking to a member of the queue, and the original position of the caller when he entered the queue.[a]
SYSCOMPAT	Recorded if an agent attempts to answer a call, but the call cannot be set up due to incompatibilities in the media setup.

[a] Please note that when the caller is transferred using SIP transfers (rather than the built-in transfers triggered by DTMF and configured in *features.conf*), the TRANSFER event may not be reliable.

Conclusion

We started this chapter with a look at basic call queues, discussing what they are, how they work, and when you might want to use one. After building a simple queue, we explored how to control queue members through various means (including the use of Local channels, which provide the ability to perform some dialplan logic just prior to connecting to a queue member). We also explored all the options available to us in the *queues.conf*, *agents.conf*, and *queuerules.conf* files, which offer us fine-grained control over any queues we configure. Of course, we need the ability to monitor what our queues are doing, so we looked finally at the queue log and the myriad of events and event parameters written when various things happen in our queues.

With the knowledge provided in this chapter, you should be well on your way to implementing a successful set of queues for your company.

Device States

Out of clutter, find simplicity.

—Albert Einstein

It is often useful to be able to determine the state of the devices that are attached to a telephone system. For example, a receptionist might require the ability to see the statuses of all the people in the office in order to determine whether somebody can take a phone call. Asterisk itself needs this same information. As another example, if you were building a call queue, as discussed in Chapter 13, Asterisk needs to know when an agent is available so that another call can be delivered. This chapter discusses device state concepts in Asterisk, as well as how devices and applications use and access this information.

Device States

There are two types of devices that device states refer to: real devices and virtual devices. Real devices are telephony endpoints that can make or receive calls, such as SIP phones. Virtual devices include things that are inside of Asterisk, but provide useful state information. Table 14-1 lists the available virtual devices in Asterisk.

Table 14-1. Virtual devices in Asterisk

Virtual device	Description
`MeetMe:` `<conference bridge>`	The state of a MeetMe conference bridge. The state will reflect whether or not the conference bridge currently has participants called in. More information on using `MeetMe()` for call conferencing can be found in "Conferencing with MeetMe()" on page 218.
`SLA:<shared line>`	Shared Line Appearance state information. This state is manipulated by the `SLATrunk()` and `SLAStation()` applications. More detail can be found in "Shared Line Appearances" on page 318.
`Custom:<custom name>`	Custom device states. These states have custom names and are modified using the `DEVICE_STATE()` function. Example usage can be found in "Using Custom Device States" on page 307.

Virtual device	Description
`Park:<exten@context>`	The state of a spot in a call parking lot. The state information will reflect whether or not a caller is currently parked at that extension. More information about call parking in Asterisk can be found in "Parking Lots" on page 228.
`Calendar:<calendar name>`	Calendar state. Asterisk will use the contents of the named calendar to set the state to `available` or `busy`. More information about calendar integration in Asterisk can be found in Chapter 18.

A device state is a simple one-to-one mapping to a device. Figure 14-1 shows this mapping.

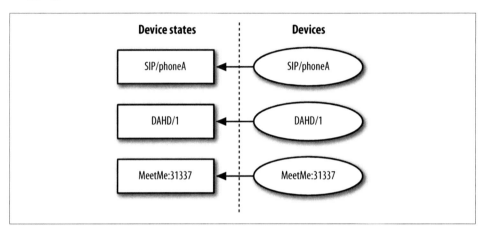

Figure 14-1. Device state mappings

Checking Device States

The `DEVICE_STATE()` dialplan function can be used to read the current state of a device. Here is a simple example of it being used in the dialplan:

```
exten => 7012,1,Answer()

; *** This line should not have any line breaks
    same => n,Verbose(3,The state of SIP/0004F2060EB4 is
${DEVICE_STATE(SIP/0004F2060EB4)})
    same => n,Hangup()
```

If we call extension 7012 from the same device that we are checking the state of, the following verbose message comes up on the Asterisk console:

```
    -- The state of SIP/0004F2060EB4 is INUSE
```

Chapter 20 discusses the Asterisk Manager Interface (AMI). The Get Var manager action can be used to retrieve device state values in an external program. You can use it to get the value of either a normal variable or a dialplan function, such as DEVICE_STATE().

The following list includes the possible values that will come back from the DEVICE_STATE() function:

- UNKNOWN
- NOT_INUSE
- INUSE
- BUSY
- INVALID
- UNAVAILABLE
- RINGING
- RINGINUSE
- ONHOLD

Extension States

Extension states are another important concept in Asterisk. Extension states are what SIP devices subscribe to for presence information. (SIP presence is discussed in more detail in "SIP Presence" on page 306). The state of an extension is determined by checking the state of one or more devices. The list of devices that map to extension states is defined in the Asterisk dialplan, */etc/asterisk/extensions.conf*, using a special hint directive. Figure 14-2 shows the mapping between devices, device states, and extension states.

Hints

To define an extension state hint in the dialplan, the keyword hint is used in place of a priority. Here is a simple example dialplan that relates to Figure 14-2:

```
[default]

exten => 1234,hint,SIP/phoneA&SIP/phoneB&SIP/phoneC

exten => 5555,hint,DAHDI/1

exten => 31337,hint,MeetMe:31337
```

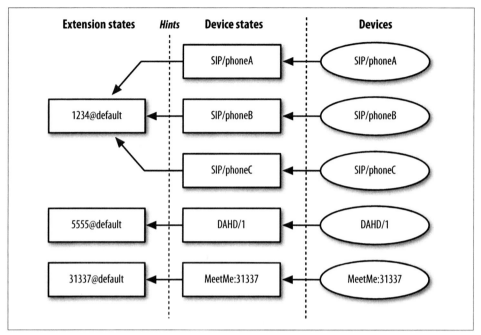

Figure 14-2. Extension state mappings

Typically, hints are simply defined along with the rest of the extension. This next example adds simple extension entries for what would happen if each of these extensions were called:

```
[default]

exten => 1234,hint,SIP/phoneA&SIP/phoneB&SIP/phoneC
exten => 1234,1,Dial(SIP/phoneA&SIP/phoneB&SIP/phoneC)

exten => 5555,hint,DAHDI/1
exten => 5555,1,Dial(DAHDI/1)

exten => 31337,hint,MeetMe:31337
exten => 31337,1,MeetMe(31337,dM)
```

In our example we've made a direct correlation between the hint's extension number and the extension number being dialed, although there is no requirement that that be the case.

Checking Extension States

The easiest way to check the current state of an extension is at the Asterisk CLI. The *core show hints* command will show you all currently configured hints. Consider the following hint definition:

```
[phones]

exten => 7001,hint,SIP/0004F2060EB4
```

When *core show hints* is executed at the Asterisk CLI, the following output is presented when the device is currently in use:

```
*CLI> core show hints

    -= Registered Asterisk Dial Plan Hints =-
        7001@phones   : SIP/0004F2060EB4    State:InUse    Watchers  0
----------------
- 1 hints registered
```

In addition to showing you the state of the extension, the output of *core show hints* also provides a count of watchers. A *watcher* is something in Asterisk that has subscribed to receive updates on the state of this extension. If a SIP phone subscribes to the state of an extension, the watcher count will be increased.

Extension state can also be retrieved with a dialplan function, EXTENSION_STATE(). This function operates similarly to the DEVICE_STATE() function described in the preceding section. The following example shows an extension that will print the current state of another extension to the Asterisk console:

```
exten => 7013,1,Answer()
    same => n,Verbose(3,The state of 7001@phones is ${EXTENSION_STATE(7001@phones)})
    same => n,Hangup()
```

When this extension is called, this is the verbose message that shows up on the Asterisk console:

```
    -- The state of 7001@phones is INUSE
```

The following list includes the possible values that may be returned back from the EXTENSION_STATE() function:

- UNKNOWN
- NOT_INUSE
- INUSE
- BUSY
- UNAVAILABLE
- RINGING
- RINGINUSE
- HOLDINUSE
- ONHOLD

SIP Presence

Asterisk provides the ability for devices to subscribe to extension state using the SIP protocol. This functionality is often referred to as BLF (Busy Lamp Field).[*]

Asterisk Configuration

To get this working, hints must be defined in */etc/asterisk/extensions.conf* (see "Hints" on page 303 for more information on configuring hints in the dialplan). Additionally, there are some important options that must be set in the configuration file for the SIP channel driver, which is */etc/asterisk/sip.conf*. The following list discusses these options:

callcounter
> Enables/disables call counters. This must be enabled for Asterisk to be able to provide state information for SIP devices. This option may be set either in the [general] section or in peer-specific sections of *sip.conf*.

> If you would like device states to work for SIP devices, you must at least set the callcounter option to yes. Otherwise, the SIP channel driver will not bother tracking calls to and from devices and will provide no state information about them.

busylevel
> Sets the number of calls that must be in progress for Asterisk to report that a device is busy. This option may only be set in peer-specific sections of *sip.conf*. By default, this option is not set. This means that Asterisk will report that a device is in use, but never busy.

call-limit
> This option has been deprecated in favor of using the GROUP() and GROUP_COUNT() functions in the Asterisk dialplan. You may find older documentation that suggests that this option is required for SIP presence to work. That used to be the case, but this option has been replaced by the callcounter option for that purpose.

allowsubscribe
> Allows you to disable support for subscriptions. If this option has not been set, subscriptions will be enabled. To disable subscription support completely, set allowsubscribe to no in the [general] section of *sip.conf*.

---

[*] Some also like to call these "blinky lamps" or "blinky lights" for their phones. Geeks and their LEDs…

`subscribecontext`

Allows you to set a specific context for subscriptions. Without this set, the context defined by the `context` option will be used. This option may be set either in the [general] section or in peer-specific sections of *sip.conf*.

`notifyringing`

Controls whether or not a notification will be sent when an extension goes into a ringing state. This option is set to **yes** by default. It only has an effect on subscriptions that use the *dialog-info* event package. This option can only be set globally in the [general] section of *sip.conf*.

`notifyhold`

Allows `chan_sip` to set SIP devices' states to `ONHOLD`. This is set to **yes** by default. This option can only be set globally in the [general] section of *sip.conf*.

`notifycid`

Enables/disables sending of an inbound call's caller ID information to an extension. This option applies to devices that subscribe to `dialog-info+xml`-based extension state notifications, such as Snom phones. Displaying caller ID information can be useful to help an agent decide whether to execute a pickup on an incoming call. This option is set to **no** by default.

 This magic pickup only works if the extension and context of the hint are the same as the extension and context of the incoming call. Notably, the usage of the `subscribecontext` option usually breaks this option. This option can also be set to the value `ignore-con text`. This will bypass the context issue, but should only be used in an environment where there is only a single instance of the extension that has been subscribed to. Otherwise, you might accidentally pick up calls that you did not mean to pick up.

Using Custom Device States

Asterisk provides the ability to create custom device states. This lends itself to the development of some interesting custom applications. We'll start by showing the basic syntax for controlling custom device states, and then we'll build an example that uses them.

Custom device states all start with a prefix of `Custom:`. The text that comes after the prefix can be anything you want. To set or read the value of a custom device state, use the `DEVICE_STATE()` dialplan function. For example, to set a custom device state:

```
exten => example,1,Set(DEVICE_STATE(Custom:example)=BUSY)
```

Similarly, to read the current value of a custom device state:

```
exten => Verbose(1,The state of Custom:example is ${DEVICE_STATE(Custom:example)})
```

Custom device states can be used as a way to directly control the state shown on a device that has subscribed to the state of an extension. Just map an extension to a custom device state using a hint in the dialplan:

```
exten => example,hint,Custom:example
```

An Example

There are a number of interesting use cases for custom device states. In this section we will build an example that implements a custom "do not disturb" (DND) button on a SIP phone. This same approach could be applied to many other things that you might like to be able to toggle at the touch of a button. For example, this approach could be used to let members know if they are currently logged into a queue or not.

The first piece of the example is the hint in the dialplan. This is required so BLF can be configured on a SIP phone to subscribe to this extension. In this case, the phone must be configured to subscribe to the state of DND_7015:

```
exten => DND_7015,hint,Custom:DND_7015
```

Next, we will create an extension that will be called when the user presses the key associated with the custom DND feature. It is interesting to note that this extension does nothing with audio. In fact, the user of the phone most likely will not even know that a call is placed when he presses the button. As far as the user is concerned, pressing that key simply turns on or off the light next to the button that reflects whether or not DND is enabled. The extension should look like this:

```
exten => DND_7015,1,Answer()
    same => n,GotoIf($["${DEVICE_STATE(Custom:DND_7015)}"="BUSY"]?turn_off:turn_on)

    same => n(turn_off),Set(DEVICE_STATE(Custom:DND_7015)=NOT_INUSE)
    same => n,Hangup()

    same => n(turn_on),Set(DEVICE_STATE(Custom:DND_7015)=BUSY)
    same => n,Hangup()
```

The final part of this example shows how the DND state is used in the dialplan. If DND is enabled, a message is played to the caller saying that the agent is unavailable. If it is disabled, a call will be made to a SIP device:

```
exten => 7015,1,GotoIf($["${DEVICE_STATE(Custom:DND_7015)}"="BUSY"]?busy:available)
    same => n(available),Verbose(3,DND is currently off for 7015.)
    same => n,Dial(SIP/exampledevice)
    same => n,Hangup()

    same => n(busy),Verbose(3,DND is on for 7015.)
    same => n,Playback(vm-theperson)
    same => n,Playback(digits/7&digits/0&digits/1&digits/5)
    same => n,Playback(vm-isunavail)
    same => n,Playback(vm-goodbye)
    same => n,Hangup()
```

Example 14-1 shows the full example as it would appear in /etc/asterisk/extensions.conf.

Example 14-1. Custom "do not disturb" functionality using custom device states

```
;
; A hint so a phone can use BLF to signal the DND state.
;
exten => DND_7015,hint,Custom:DND_7015


;
; An extension to dial when the user presses the custom DND
; key on his phone.  This will toggle the state and will result
; in the light on the phone turning on or off.
;
exten => DND_7015,1,Answer()
    same => n,GotoIf($["${DEVICE_STATE(Custom:DND_7015)}"="BUSY"]?turn_off:turn_on)

    same => n(turn_off),Set(DEVICE_STATE(Custom:DND_7015)=NOT_INUSE)
    same => n,Hangup()

    same => n(turn_on),Set(DEVICE_STATE(Custom:DND_7015)=BUSY)
    same => n,Hangup()



;
; Example usage of the DND state.
;
exten => 7015,1,GotoIf($["${DEVICE_STATE(Custom:DND_7015)}"="BUSY"]?busy:available)
    same => n(available),Verbose(3,DND is currently off for 7015.)
    same => n,Dial(SIP/exampledevice)
    same => n,Hangup()

    same => n(busy),Verbose(3,DND is on for 7015.)
    same => n,Playback(vm-theperson)
    same => n,Playback(digits/7&digits/0&digits/1&digits/5)
    same => n,Playback(vm-isunavail)
    same => n,Playback(vm-goodbye)
    same => n,Hangup()
```

Distributed Device States

Asterisk is primarily designed to run on a single system. However, as requirements for scalability increase, it is common for deployments to require multiple Asterisk servers. Since that has become increasingly common, some features have been added to make it easier to coordinate multiple Asterisk servers. One of those features is *distributed device state* support.

What this means is that if a device is on a call on one Asterisk server, the state of that device on all servers reflects that. To be more specific, the way this works is that every server knows the state of each device from the perspective of each server. Using this collection of states, each server will calculate what the overall device state value is to report to the rest of Asterisk.

To accomplish distributed device state, some sort of messaging mechanism must be used for the servers to communicate with each other. Two such mechanisms are supported as of Asterisk 1.8: AIS and XMPP.

Using OpenAIS

The Application Interface Specification (AIS) is a standardized set of messaging middleware APIs. The definition for the APIs is provided by the Service Availability Forum (*http://www.saforum.org*). The open source implementation of AIS that was used for the development and testing of this functionality is OpenAIS (*http://www.openais.org*), which is built on Corosync (*http://www.corosync.org*).

Corosync, and thus OpenAIS, is built in such a way that nodes must be located on the same high-speed, low-latency LAN. If your deployment is geographically distributed, you should use the XMPP-based distributed device state support, which is discussed in "Using XMPP" on page 314.

Installation

The first step to getting the necessary components installed is to install Corosync and OpenAIS. Corosync depends on the NSS library. Install the *libnss3-dev* package on Ubuntu or the *nss-devel* package on CentOS.

Next, install Corosync and OpenAIS. There may be packages available, but they are also fairly straightforward to install from source. Download the latest releases from the Corosync and OpenAIS home pages. Then, execute the following commands to compile and install each package:

```
$ tar xvzf corosync-1.2.8.tar.gz
$ cd corosync-1.2.8
$ ./configure
$ make
$ sudo make install

$ tar xvzf openais-1.1.4.tar.gz
$ cd openais-1.1.4
$ ./configure
$ make
$ sudo make install
```

If you installed Asterisk prior to installing Corosync and OpenAIS, you will need to recompile and reinstall Asterisk to get AIS support. Start by running the Asterisk *configure* script. The *configure* script is responsible for inspecting the system to find out which optional dependencies can be found so that the build system knows which modules can be built:

```
$ cd /path/to/asterisk
$ ./configure
```

After running the *configure* script, run the *menuselect* tool to ensure that Asterisk has been told to build the `res_ais` module (this module can be found in the *Resource Modules* section of *menuselect*):

```
$ make menuselect
```

Finally, compile and install Asterisk:

```
$ make
$ sudo make install
```

 This is a pretty quick and crude set of instructions for compiling and installing Asterisk. For a much more complete set of instructions, please see Chapter 3.

OpenAIS configuration

Now that OpenAIS has been installed, it needs to be configured. There is a configuration file for both OpenAIS and Corosync that must be put in place. Check to see if */etc/ais/openais.conf* and */etc/corosync/corosync.conf* exist. If they do not exist, copy in the sample configuration files:

```
$ sudo mkdir -p /etc/ais
$ cd openais-1.1.4
$ sudo cp conf/openais.conf.sample /etc/ais/openais.conf

$ sudo mkdir -p /etc/corosync
$ cd corosync-1.2.8
$ sudo cp conf/corosync.conf.sample /etc/corosync/corosync.conf
```

Next, you will need to edit both the *openais.conf* and *corosync.conf* files. There are a number of options here, but the most important one that must be changed is the `bindnetaddr` option in the `totem-interface` section. This must be set to the IP address of the network interface that this node will use to communicate with the rest of the cluster:

```
totem {
    ...
    interface {
        ringnumber: 0
        bindnetaddr: 10.24.22.144
        mcastaddr: 226.94.1.1
        mcastport: 5405
    }
}
```

For detailed documentation on the rest of the options in these configuration files, see the associated manpages:

```
$ man openais.conf
$ man corosync.conf
```

To get started with testing out basic OpenAIS connectivity, try starting the *aisexec* application in the foreground and watching the output:

```
$ sudo aisexec -f
```

For example, if you watch the output of *aisexec* on the first node while you bring up the second node, you should see output that reflects that the cluster now has two connected nodes:

```
Nov 13 06:55:30 corosync [CLM   ] CLM CONFIGURATION CHANGE
Nov 13 06:55:30 corosync [CLM   ] New Configuration:
Nov 13 06:55:30 corosync [CLM   ]       r(0) ip(10.24.22.144)
Nov 13 06:55:30 corosync [CLM   ]       r(0) ip(10.24.22.242)
Nov 13 06:55:30 corosync [CLM   ] Members Left:
Nov 13 06:55:30 corosync [CLM   ] Members Joined:
Nov 13 06:55:30 corosync [CLM   ]       r(0) ip(10.24.22.242)
Nov 13 06:55:30 corosync [TOTEM ] A processor joined or left the membership and a new
membership was formed.
Nov 13 06:55:30 corosync [MAIN  ] Completed service synchronization, ready to provide
service.
```

 If you have any trouble getting the nodes to sync up with each other, one thing to check is that there are no firewall rules on the nodes that are blocking the multicast traffic that is used for the nodes to communicate with each other.

Asterisk configuration

The **res_ais** module for Asterisk has a single configuration file, */etc/asterisk/ais.conf*. One short section is required in this file to enable distributed device state in an AIS cluster. Place the following contents in the */etc/asterisk/ais.conf* file:

```
[device_state]

type = event_channel
publish_event = device_state
subscribe_event = device_state
```

There is an Asterisk CLI command that can be used to ensure that this configuration has been loaded properly:

```
*CLI> ais evt show event channels

=============================================================
=== Event Channels ==========================================
=============================================================
===
=== -----------------------------------------------------------
=== Event Channel Name: device_state
=== ==> Publishing Event Type: device_state
=== ==> Subscribing to Event Type: device_state
```

```
=== --------------------------------------------------------
===
============================================================
```

Another useful Asterisk CLI command provided by the `res_ais` module is used to list the members of the AIS cluster:

```
*CLI> ais clm show members

============================================================
=== Cluster Members ========================================
============================================================
===
=== --------------------------------------------------------
=== Node Name: 10.24.22.144
=== ==> ID: 0x9016180a
=== ==> Address: 10.24.22.144
=== ==> Member: Yes
===
=== --------------------------------------------------------
===
=== --------------------------------------------------------
=== Node Name: 10.24.22.242
=== ==> ID: 0xf216180a
=== ==> Address: 10.24.22.242
=== ==> Member: Yes
=== --------------------------------------------------------
===
============================================================
```

Testing device state changes

Now that you've set up and configured distributed device state using OpenAIS, there are some simple tests that can be done using custom device states to ensure that device states are being communicated between the servers. Start by creating a test hint in the Asterisk dialplan, */etc/asterisk/extensions.conf*:

```
[devstate_test]

exten => foo,hint,Custom:abc
```

Now, you can adjust the custom device state from the Asterisk CLI using the *dialplan set global* CLI command and then check the state on each server using the *core show hints* command. For example, we can use this command to set the state on one server:

```
pbx1*CLI> dialplan set global DEVICE_STATE(Custom:abc) INUSE

    -- Global variable 'DEVICE_STATE(Custom:abc)' set to 'INUSE'
```

and then, check the state on another server using this command:

```
*CLI> core show hints

-= Registered Asterisk Dial Plan Hints =-
    foo@devstatetest        : Custom:abc        State:InUse      Watchers  0
```

If you would like to dive deeper into the processing of distributed device state changes, there are some useful debug messages that can be enabled. First, enable debug on the Asterisk console in */etc/asterisk/logger.conf*. Then, enable debugging at the Asterisk CLI:

```
*CLI> core set debug 1
```

With the debug output enabled, you will see some messages that show how Asterisk is processing each state change. When the state of a device changes on one server, Asterisk checks the state information it has for that device on all servers and determines the overall device state. The following examples illustrate:

```
*CLI> dialplan set global DEVICE_STATE(Custom:abc) NOT_INUSE

    -- Global variable 'DEVICE_STATE(Custom:abc)' set to 'NOT_INUSE'

[Nov 13 13:27:12] DEBUG[14801]: devicestate.c:652
handle_devstate_change: Processing device state change for 'Custom:abc'
[Nov 13 13:27:12] DEBUG[14801]: devicestate.c:602
process_collection: Adding per-server state of 'Not in use' for 'Custom:abc'
[Nov 13 13:27:12] DEBUG[14801]: devicestate.c:602
process_collection: Adding per-server state of 'Not in use' for 'Custom:abc'
[Nov 13 13:27:12] DEBUG[14801]: devicestate.c:609
process_collection: Aggregate devstate result is 'Not in use' for 'Custom:abc'
[Nov 13 13:27:12] DEBUG[14801]: devicestate.c:631
process_collection: Aggregate state for device 'Custom:abc' has changed to
'Not in use'

*CLI> dialplan set global DEVICE_STATE(Custom:abc) INUSE

    -- Global variable 'DEVICE_STATE(Custom:abc)' set to 'INUSE'

[Nov 13 13:29:30] DEBUG[14801]: devicestate.c:652 handle_devstate_change:
Processing device state change for 'Custom:abc'
[Nov 13 13:29:30] DEBUG[14801]: devicestate.c:602 process_collection:
Adding per-server state of 'Not in use' for 'Custom:abc'
[Nov 13 13:29:30] DEBUG[14801]: devicestate.c:602 process_collection:
Adding per-server state of 'In use' for 'Custom:abc'
[Nov 13 13:29:30] DEBUG[14801]: devicestate.c:609 process_collection:
Aggregate devstate result is 'In use' for 'Custom:abc'
[Nov 13 13:29:30] DEBUG[14801]: devicestate.c:631 process_collection:
Aggregate state for device 'Custom:abc' has changed to 'In use'
```

Using XMPP

The eXtensible Messaging and Presence Protocol (XMPP) (*http://www.xmpp.org*), formerly (and still commonly) known as *Jabber*, is an IETF standardized communications protocol. It is most commonly known as an IM protocol, but it can be used for a number of other interesting applications as well. The XMPP Standards Foundation (XSF) works to standardize extensions to the XMPP protocol. One such extension, referred to as PubSub, provides a publish/subscribe mechanism.

Asterisk has the ability to use XMPP PubSub to distribute device state information. One of the nice things about using XMPP to accomplish this is that it works very well for geographically distributed Asterisk servers.

Installation

To distribute device states using XMPP, you will need an XMPP server that supports PubSub. One such server that has been successfully tested against Asterisk is Tigase (*http://www.tigase.org*).

The Tigase website has instructions for installing and configuring the Tigase server. We suggest that you follow those instructions (or the instructions provided for whatever other server you may choose to use) and come back to this book when you're ready to work on the Asterisk-specific parts.

On the Asterisk side of things, you will need to ensure that you have installed the res_jabber module. You can check to see if it is already loaded at the Asterisk CLI:

```
*CLI> module show like jabber

Module                      Description                     Use Count
res_jabber.so               AJI - Asterisk Jabber Interface 0
1 modules loaded
```

If you are using a custom */etc/asterisk/modules.conf* file that lists only specific modules to be loaded, you can also check the filesystem to see if the module was compiled and installed:

```
$ ls -l /usr/lib/asterisk/modules/res_jabber.so

-rwxr-xr-x 1 root root 837436 2010-11-12 15:33 /usr/lib/asterisk/modules/res_jabber.so
```

If you do not yet have res_jabber installed, you will need to install the iksemel and OpenSSL libraries. Then, you will need to recompile and reinstall Asterisk. Start by running the Asterisk *configure* script, which is responsible for inspecting the system and locating optional dependencies, so that the build system knows which modules can be built:

```
$ cd /path/to/asterisk
$ ./configure
```

After running the *configure* script, run the *menuselect* tool to ensure that Asterisk has been told to build the res_jabber module. This module can be found in the *Resource Modules* section of *menuselect*:

```
$ make menuselect
```

Finally, compile and install Asterisk:

```
$ make
$ sudo make install
```

This is a pretty quick and crude set of instructions for compiling and installing Asterisk. For a much more complete set of instructions, please see Chapter 3.

Creating XMPP accounts

Unfortunately, Asterisk is currently not able to register new accounts on an XMPP server. You will have to create an account for each server via some other mechanism. The method we used while testing was to use an XMPP client such as Pidgin (*http:// www.pidgin.im*) to complete the account registration process. After account registration is complete, the XMPP client is no longer needed. For the rest of the examples, we will use the following two buddies, both of which are on the server *jabber.shifteight.org*:

- `server1@jabber.shifteight.org/astvoip1`
- `server2@jabber.shifteight.org/astvoip2`

Asterisk configuration

The */etc/asterisk/jabber.conf* file will need to be configured on each server. We will show the configuration for a two-server setup here, but the configuration can easily be expanded to more servers as needed. Example 14-2 shows the contents of the configuration file for server 1 and Example 14-3 shows the contents of the configuration file for server 2. For additional information on the *jabber.conf* options associated with distributed device states, see the *configs/jabber.conf.sample* file that is included in the Asterisk source tree.

Example 14-2. jabber.conf for server1

```
[general]
autoregister = yes

[asterisk]
type = client
serverhost = jabber.shifteight.org
pubsub_node = pubsub.jabber.shifteight.org
username = server1@jabber.shifteight.org/astvoip1
secret = mypassword
distribute_events = yes
status = available
usetls = no
usesasl = yes
buddy = server2@jabber.shifteight.org/astvoip2
```

Example 14-3. jabber.conf for server2

```
[general]
autoregister = yes

[asterisk]
type = client
```

```
serverhost = jabber.shifteight.org
pubsub_node = pubsub.jabber.shifteight.org
username = server2@jabber.shifteight.org/astvoip2
secret = mypassword
distribute_events = yes
status = available
usetls = no
usesasl = yes
buddy = server1@jabber.shifteight.org/astvoip1
```

Testing

To ensure that everything is working properly, start by doing some verification of the *jabber.conf* settings on each server. There are a couple of relevant Asterisk CLI commands that can be used here. The first is the *jabber show connected* command, which will verify that Asterisk has successfully logged in with an account on the *jabber* server. The output of this command on the first server shows:

```
*CLI> jabber show connected

Jabber Users and their status:
      User: server1@jabber.shifteight.org/astvoip1    - Connected
----
    Number of users: 1
```

Meanwhile, if *jabber show connected* is executed on the second server, it shows:

```
*CLI> jabber show connected

Jabber Users and their status:
      User: server2@jabber.shifteight.org/astvoip2    - Connected
----
    Number of users: 1
```

The next useful command for verifying the setup is *jabber show buddies*. This command allows you to verify that the other server is correctly listed on your buddy list. It also lets you see if the other server is seen as currently connected. If you were to run this command on the first server without Asterisk currently running on the second server, the output would look like this:

```
*CLI> jabber show buddies

Jabber buddy lists
Client: server1@jabber.shifteight.org/astvoip1
    Buddy: server2@jabber.shifteight.org
          Resource: None
    Buddy: server2@jabber.shifteight.org/astvoip2
          Resource: None
```

Next, start Asterisk on the second server and run *jabber show buddies* on that server. The output will contain more information, since the second server will see the first server online:

```
*CLI> jabber show buddies

Jabber buddy lists
Client: server2@jabber.shifteight.org/astvoip2
    Buddy: server1@jabber.shifteight.org
            Resource: astvoip1
                node: http://www.asterisk.org/xmpp/client/caps
                version: asterisk-xmpp
                Jingle capable: yes
            Status: 1
            Priority: 0
    Buddy: server1@jabber.shifteight.org/astvoip1
            Resource: None
```

At this point, you should be ready to test out the distribution of device states. The procedure is the same as that for testing device states over AIS, which can be found in "Testing device state changes" on page 313.

Shared Line Appearances

In Asterisk, Shared Line Appearances (SLA)—sometimes also referred to in the industry as Bridged Line Appearances (BLA)—can be used. This functionality can be used to satisfy two primary use cases, which include emulating a simple key system and creating shared extensions on a PBX.

Building key system emulation is the use case for which these applications were primarily designed. In this environment, you have some small number of trunks coming into the PBX, such as analog phone lines, and each phone has a dedicated button for calls on that trunk. You may refer to these trunks as line 1, line 2, and line 3, for example.

The second primary use case is for creating shared extensions on your PBX. This use case seems to be the most common these days. There are many reasons you might want to do this. One example is that you may want an extension to appear on both the phones of an executive and her administrative assistant. Another example would be if you want the same extension to appear on all of the phones in the same lab.

While these use cases are supported to an extent, there are limitations. There is still more work to be done in Asterisk to make these features work really well for what people want to do with them. These limitations are discussed in "Limitations" on page 328.

Installing the SLA Applications

The SLA applications are built on two key technologies in Asterisk. The first is device state processing, and the second is conferencing. Specifically, the conferencing used by these applications is the `MeetMe()` application. The SLA applications come with the same module as the `MeetMe()` application, so you must install the `app_meetme` module.

You can check at the Asterisk CLI to see if you already have the module:

```
pbx*CLI> module show like app_meetme.so

Module                        Description                    Use Count
0 modules loaded
```

In this case, the module is not present. The most common reason that an Asterisk system does not have the `app_meetme` module is because DAHDI has not been installed. The `MeetMe()` application uses DAHDI to perform conference mixing. Once DAHDI is installed (refer to Chapter 3 for installation information), rerun the Asterisk *configure* script, recompile, and reinstall. Once the module has been properly installed, you should be able to see it at the CLI:

```
*CLI> module show like app_meetme.so

Module                        Description                    Use Count
app_meetme.so                 MeetMe conference bridge       0
1 modules loaded
```

Once the `app_meetme` module is loaded, you should have both the `SLAStation()` and `SLATrunk()` applications available:

```
*CLI> core show applications like SLA

    -= Matching Asterisk Applications =-
            SLAStation: Shared Line Appearance Station.
             SLATrunk: Shared Line Appearance Trunk.
    -= 2 Applications Matching =-
```

Configuration Overview

The two main configuration files that must be edited to set up SLA are */etc/asterisk/extensions.conf* and */etc/asterisk/sla.conf*. The *sla.conf* file is used for defining trunks and stations. A station is any SIP phone that will be using SLA. Trunks are the literal trunks or shared extensions that will be appearing on two or more stations. The Asterisk dialplan, *extensions.conf*, provides some important glue that pulls an SLA configuration together. The dialplan includes some extension state hints and extensions that define how calls get into and out of an SLA setup. The next few sections provide detailed examples of the configuration for a few different use cases.

Key System Example with Analog Trunks

This usage of SLA comes with the simplest configuration.[†] This scenario would typically be used for a fairly small installation, where you have a few analog lines and SIP

---

† Admittedly, none of the configuration for SLA is simple.

phones that all have line keys directly associated with the analog lines. For the purposes of this example, we will say we have two analog lines and four SIP phones. Each SIP phone will have a button for line1 and a button for line2. This section will assume that you have done some configuration up front, including:

- Configuring the four SIP phones. For more information on setting up SIP phones, see Chapter 5.
- Configuring the two analog lines. Fore more information on setting up analog lines with Asterisk, see Chapter 7.

For this example, we will use the following device names for the SIP phones and analog lines. Be sure to adapt the examples to match your own configuration:

- SIP/station1
- SIP/station2
- SIP/station3
- SIP/station4

- DAHDI/1
- DAHDI/2

sla.conf

As mentioned previously, *sla.conf* contains configuration that maps devices to trunks and stations. For this example, we will start by defining the two trunks:

```
[line1]
type = trunk
device = DAHDI/1

[line2]
type = trunk
device = DAHDI/2
```

Next, we will set up the station definitions. We have four SIP phones, which will each use both trunks. Note that the section names in *sla.conf* for stations do not need to match the SIP device names, but it is done that way here for convenience:

```
[station1]
type = station
device = SIP/station1
trunk = line1
trunk = line2

[station2]
type = station
device = SIP/station2
trunk = line1
trunk = line2
```

```
[station3]
type = station
device = SIP/station3
trunk = line1
trunk = line2

[station4]
type = station
device = SIP/station4
trunk = line1
trunk = line2
```

The station configuration is a bit repetitive. Asterisk configuration file template sections come in handy here to collapse the configuration down a bit. Here is the station configuration again, but this time using a template:

```
[station](!)
type = trunk
trunk = line1
trunk = line2

[station1](station)
device = SIP/station1

[station2](station)
device = SIP/station2

[station3](station)
device = SIP/station3

[station4](station)
device = SIP/station4
```

extensions.conf

The next configuration file required for this example is */etc/asterisk/extensions.conf*. There are three contexts. First, we have the line1 and line2 contexts. When a call comes in on one of the analog lines, it will come in to one of these contexts in the dialplan and execute the SLATrunk() application. This application will take care of ringing all of the appropriate stations:

```
[line1]

exten => s,1,SLATrunk(line1)

[line2]

exten => s,1,SLATrunk(line2)
```

The next section of the dialplan is the sla_stations context. All calls from the SIP phones should be sent to this context. Further, the SIP phones should be configured such that as soon as they go off-hook, they immediately make a call to the station1 extension (or station2, station3, etc., as appropriate). If the line1 key on the phone is pressed, a call should be sent to the station1_line1 extension (or station2_line1, etc.).

Any time that a phone goes off-hook or a line key is pressed, the call that is made will immediately connect it to one of the analog lines. For a line that is not already in use, the analog line will be providing a dialtone, and the user will be able to send digits to make a call. If a user presses a line key for a line that is already in use, that user will be bridged into the existing call on that line. The sla_stations context looks like this:

```
[sla_stations]

exten => station1,1,SLAStation(station1)
exten => station1_line1,hint,SLA:station1_line1
exten => station1_line1,1,SLAStation(station1_line1)
exten => station1_line2,hint,SLA:station1_line2
exten => station1_line2,1,SLAStation(station1_line2)

exten => station2,1,SLAStation(station2)
exten => station2_line1,hint,SLA:station2_line1
exten => station2_line1,1,SLAStation(station2_line1)
exten => station2_line2,hint,SLA:station2_line2
exten => station2_line2,1,SLAStation(station2_line2)

exten => station3,1,SLAStation(station3)
exten => station3_line1,hint,SLA:station3_line1
exten => station3_line1,1,SLAStation(station3_line1)
exten => station3_line2,hint,SLA:station3_line2
exten => station3_line2,1,SLAStation(station3_line2)

exten => station4,1,SLAStation(station4)
exten => station4_line1,hint,SLA:station4_line1
exten => station4_line1,1,SLAStation(station4_line1)
exten => station4_line2,hint,SLA:station4_line2
exten => station4_line2,1,SLAStation(station4_line2)
```

Additional phone configuration tasks

The previous section covered the dialplan for trunks and stations. There are some specific things to keep in mind when setting up phones for use with this setup. First, each phone should be configured to send a call as soon as it is taken off-hook.

The other important item is the configuration of the line keys. Asterisk uses extension state subscriptions to control the LEDs next to the line buttons. Beyond that, each line key should be configured as a speed-dial. Use the following checklist for your line key configuration (how you accomplish these tasks will depend on the specific phones you are using):

- Set the label of the key to be *Line 1* (etc.), or whatever you deem appropriate.
- Set up the keys such that the *Line 1* key on station1 subscribes to the state of station1_line1, and so on. This is required so Asterisk can make the LEDs reflect the state of the lines.

- Ensure that if the *Line 1* key on `station1` is pressed a call is sent to the `station1_line1` extension, and so on.

Key System Example with SIP Trunks

This example is intended to be identical in functionality to the previous example. The difference is that instead of using analog lines as trunks, we will use a connection to a SIP provider that will terminate the calls to the PSTN. For more information on setting up Asterisk to connect to a SIP provider, see Chapter 7.

sla.conf

The *sla.conf* file for this scenario is a bit tricky.[‡] You might expect to see the device line in the trunk configuration have a SIP channel listed, but instead we're going to use a Local channel. This will allow us to use some additional dialplan logic for call processing. The purpose of the Local channel will become clearer in the next section, when the dialplan example is discussed. Here are the trunk configurations:

```
[line1]
type = trunk
device = Local/disa@line1_outbound

[line2]
type = trunk
device = Local/disa@line2_outbound
```

The station configuration is identical to the last example, so let's get right to it:

```
[station](!)
type = trunk
trunk = line1
trunk = line2

[station1](station)
device = SIP/station1

[station2](station)
device = SIP/station2

[station3](station)
device = SIP/station3

[station4](station)
device = SIP/station4
```

‡ Read: a hack.

extensions.conf

As in the last example, you will need `line1` and `line2` contexts to process incoming calls on these trunks:

```
[line1]

exten => s,1,SLATrunk(line1)


;
; If the provider specifies your phone number when sending you
; a call, you will need another rule in the dialplan to match that.
;
exten => _X.,1,Goto(s,1)

[line2]

exten => s,1,SLATrunk(line2)

exten => _X.,1,Goto(s,1)
```

This example requires an `sla_stations` context, as well. This is for all calls coming from the phones. It's the same as it was in the last example:

```
[sla_stations]

exten => station1,1,SLAStation(station1)
exten => station1_line1,hint,SLA:station1_line1
exten => station1_line1,1,SLAStation(station1_line1)
exten => station1_line2,hint,SLA:station1_line2
exten => station1_line2,1,SLAStation(station1_line2)

exten => station2,1,SLAStation(station2)
exten => station2_line1,hint,SLA:station2_line1
exten => station2_line1,1,SLAStation(station2_line1)
exten => station2_line2,hint,SLA:station2_line2
exten => station2_line2,1,SLAStation(station2_line2)

exten => station3,1,SLAStation(station3)
exten => station3_line1,hint,SLA:station3_line1
exten => station3_line1,1,SLAStation(station3_line1)
exten => station3_line2,hint,SLA:station3_line2
exten => station3_line2,1,SLAStation(station3_line2)

exten => station4,1,SLAStation(station4)
exten => station4_line1,hint,SLA:station4_line1
exten => station4_line1,1,SLAStation(station4_line1)
exten => station4_line2,hint,SLA:station4_line2
exten => station4_line2,1,SLAStation(station4_line2)
```

The last piece of the dialplan that is required is the implementation of the `line1_out bound` and `line2_outbound` contexts. This is what the SLA applications use when they want to send calls out to a SIP provider. The key to this setup is the usage of the `DISA()` application. In the last example, phones were directly connected to an analog line. This allowed the upstream switch to provide a dialtone, collect digits, and then complete the call. In this example, we use the DISA() application locally to provide a dialtone and collect digits. Once a complete number has been dialed, the call will proceed to go out to a SIP provider:

```
[line1_outbound]

exten => disa,1,DISA(no-password,line1_outbound)

;
; Add extensions for whatever numbers you would like to
; allow to be dialed.
;
exten => _1NXXNXXXXXX,1,Dial(SIP/${EXTEN}@myprovider)

[line2_outbound]

exten => disa,1,DISA(no-password,line2_outbound)

exten => _1NXXNXXXXXX,1,Dial(SIP/${EXTEN}@myprovider)
```

Shared Extension Example

The previous two examples were for small key system emulation. For this example, we'll try something quite different. Many PBX vendors offer the ability to have the same extension shared across multiple phones. This is not simply a matter of having multiple phones ring when an extension is called: it is deeper integration than that. The behavior of the line key for a shared extension is similar to that of a line key on a key system. For example, you can simply put a call on hold from one phone and pick it up from another. Also, if multiple phones press the key for the shared extension, they will all be bridged into the same call. That is why this functionality is often also referred to as Bridged Line Appearances (BLA).

In the previous two examples, we had two trunks and four stations. For this example, we're going to set up a single shared extension on two phones. The shared extension will be referred to as extension 5001.

sla.conf

Every usage of the SLA applications requires trunk and station definitions. This example, like the previous ones, will be making use of the DISA() application and the *sla.conf* file will look very similar:

```
[5001]
type = trunk
device = Local/disa@5001_outbound

[5001_phone1]
device = SIP/5001_phone1
trunk = 5001

[5001_phone2]
```

```
device = SIP/5001_phone2
trunk = 5001
```

extensions.conf

The first part of the dialplan that is required is what will be executed when extension 5001 is dialed on the PBX. Normally, to call a phone you would use the `Dial()` application. In this case, we're going to use the `SLATrunk()` application. This will take care of ringing both phones and keeping them bridged together:

```
exten => 5001,1,SLATrunk(5001)
```

Next, we will need a context that will be used for making outbound calls from this shared extension. This assumes that 5001_phone1 and 5001_phone2 have been configured with their **context** options set to 5001 in *sip.conf*:

```
[5001]

;
; This extension is needed if you want the shared extension to
; be used by default. In that case, have this extension dialed
; when the phone goes off-hook.
;
exten => 5001_phone1,1,SLAStation(5001_phone1)
;
; This is the extension that should be dialed when the 5001 key is
; pressed on 5001_phone1.
;
exten => 5001_phone1_5001,hint,SLA:5001_phone1_5001
exten => 5001_phone1_5001,1,SLAStation(5001_phone1_5001)

exten => 5001_phone2,1,SLAStation(5001_phone2)
exten => 5001_phone2_5001,hint,SLA:5001_phone2_5001
exten => 5001_phone2_5001,1,SLAStation(5001_phone2_5001)
```

Finally, we need an implementation of the **5001_outbound** context. This will be used to provide a dialtone and collect digits on the bridged line:

```
[5001_outbound]

exten => disa,1,DISA(no-password,5001_outbound)

;
; This context will also need to be able to see whatever
; extensions you would like to be reachable from this extension.
;
include => pbx_extensions
```

Additional Configuration

The */etc/asterisk/sla.conf* file has some optional configuration parameters that were not used in any of the examples in this chapter. To give you an idea of what other behavior can be configured, the options are covered here. This file has a [general] section that is reserved for global configuration options. Currently, there is only a single option that can be specified in this section:

attemptcallerid = *yes*
> This option specifies whether or not the SLA applications should attempt to pass caller ID information. It is set to no by default. If this is enabled, the display of the phones may not be what you would expect in some situations.

The trunk definitions in the previous examples only specified the type and device. Here are some additional options that can be specified for a trunk:

autocontext = line1
> If this option is set, Asterisk will automatically create a dialplan context using this name. The context will contain an s extension that executes the SLATrunk() application with the appropriate argument for this trunk. By default, all dialplan entries must be created manually.

ringtimeout = 20
> This option allows you to specify the number of seconds to allow an inbound call on this trunk to ring before the SLATrunk() application will exit and consider it an unanswered call. By default, this option is not set.

barge = no
> The barge option specifies whether or not other stations are allowed to join a call that is in progress on this trunk by pressing the same line button. Barging into a call on a trunk is allowed by default.

hold = private
> The hold option specifies hold permissions for this trunk. If this option is set to open, any station can place this trunk on hold and any other station is allowed to take it back off of hold. If this option is set to private, only the station that placed the trunk on hold is allowed to take it back off of hold. This option is set to open by default.

When we defined the stations in the previous examples, we only supplied the type, device, and a list of trunks. However, station definitions accept some additional configuration options, as well. They are listed here:

autocontext = sla_stations
> If this option is specified, Asterisk will automatically create the extensions required for calls coming from this station in the context specified. This is off by default, which means that all extensions must be specified manually.

```
ringtimeout = 20
```
A timeout may be specified in seconds for how long this station will ring before the call is considered unanswered. There is no timeout set by default.

```
ringdelay = 5
```
A ring delay in seconds can be specified for a station. If a delay is specified, this station will not start ringing until this number of seconds after the call first came in on this shared line. There is no delay set by default.

```
hold = private
```
Hold permissions can be specified for a specific station as well. If this option is set to **private**, any trunks put on hold by this station can only be picked back up by this station. By default, this is set to **open**.

```
trunk = line1,ringtimeout=20
```
A **ringtimeout** can be applied to calls coming from only a specific trunk.

```
trunk = line1,ringdelay=5
```
A **ringdelay** can also be applied to calls from a specific trunk.

Limitations

While Asterisk makes many things easy, SLA is not one of them. Despite this functionality being intended to emulate simple features, the configuration required to make it work is fairly complex. Someone who is new to Asterisk and only wants a simple key system setup will have to learn a lot of complex Asterisk and SIP phone concepts to get it working.

Another feature that still needs some development work before it will work seamlessly with SLA is caller ID. At the time that this functionality was written, Asterisk did not have the appropriate infrastructure in place to be able to update caller ID information throughout the duration of the call. Based on how this functionality is implemented, this infrastructure is required to make the display on the phones useful. It does exist as of Asterisk 1.8 but the SLA applications have not yet been updated to use it. The end result is that you can either have no caller ID information at all, or you can enable it and understand that the phone displays are not always going to display correctly as changes happen throughout the duration of a call.

Another limitation, most relevant to usage of shared extensions, is that transfers do not work. The main reason is that transfers generally involve putting a call on hold for a short time. Call hold is processed in a special way with SLA, in that the held call is not controlled by the phone that initiated the hold. This breaks transfer processing.

In summary, SLA is not necessarily simple to set up, and it comes with some significant limitations. With that said, if what does exist suits your needs, by all means go for it.

Conclusion

This chapter discussed many aspects of device state handling in Asterisk. We started by discussing the core concepts of device states and extension states, and built up from there. We covered how SIP phones can subscribe to states, tools for creating custom states, and two mechanisms that can be used for distributing states among many servers. Finally, we covered one of the features in Asterisk, Shared Line Appearances, that relies heavily on the device state infrastructure in Asterisk to operate.

The Automated Attendant

*I don't answer the phone. I get the feeling whenever I do
that there will be someone on the other end.*

—Fred Couples

In many PBXs it is common to have a menuing system in place to answer incoming calls automatically, and allow the callers to direct themselves to various extensions and resources in the system through menu choices. This is known in the telecom industry as an automated attendant (AA). An auto attendant normally provides the following features:

- Transfer to extension
- Transfer to voicemail
- Transfer to a queue
- Play message (e.g., "our address is...")
- Connect to a submenu (e.g., "for a listing of our departments...")
- Connect to reception
- Repeat choices

For anything else—especially if there is external integration required, such as a database lookup—an Interactive Voice Response (IVR) would normally be needed.

An Auto Attendant Is Not an IVR

In the open source telecom community, you will often hear the term IVR used to describe an automated attendant. However, in the telecom industry, an IVR is distinct from an auto attendant. For this reason, when you are talking to somebody about any sort of telecom menu, you should ensure that you are talking about the same thing. To a telecom professional, the term IVR implies a relatively complex and involved development effort (and subsequent costs), whereas an automated attendant is a simple and inexpensive thing that is common to most PBXs.

In this chapter we talk about building an automated attendant. In Chapter 17 we will discuss IVR.[*]

Designing Your Auto Attendant

The most common mistake beginners make when designing an AA is needless complexity. While there can be much joy and sense of accomplishment in the creation of a multilevel AA with dozens of nifty options and oodles of really cool prompts, your callers have a different agenda. The reason people make phone calls is primarily because they want to talk to someone. While people have become used to the reality of auto attendants, and in some cases they can speed things up, for the most part people would prefer to speak to somebody live. This means that there are two fundamental rules that every auto attendant should adhere to:

1. Keep it simple.
2. Make sure you always include a handler for the folks who are going to press 0 whenever they hear an auto attendant. If you do not want to have a 0 option, be aware that many people will be insulted by this, and they will hang up and not call back. In business, this is generally a bad thing.

Before you start to code your AA, it is wise to design it. You will need to define a call flow, and you will need to specify the prompts that will play at each step. Software diagramming tools can be useful for this, but there's no need to get fancy. Table 15-1 provides a good template for a basic auto attendant that will do what you need.

Table 15-1. A basic automated attendant

Step or choice	Sample prompt	Notes	Filename
Greeting—business hours	Thank you for calling ABC company.	Day greeting.	*daygreeting.wav*
Greeting—non-business hours	Thank you for calling ABC company. Our office is now closed.	Night greeting.	*nightgreeting.wav*
Main menu	If you know the extension of the person you wish to reach, please enter it now. For sales, please press 1, for service, press 2, for our company directory, press #. For our	Main menu prompt.	*mainmenu.wav*

---

[*] It should be noted that Asterisk is an excellent IVR-creation tool. It's not bad for building automated attendants either.

Step or choice	Sample prompt	Notes	Filename
	address and fax information, please press 3. To repeat these choices press 9, or you can remain on the line or press 0 to be connected to our operator.		
1	Please hold while we connect your calls.	Transfer to sales queues.	*holdwhileweconnect.wav*
2	Please hold while we connect your call.	Transfer to support queue.	*holdwhileweconnect.wav*
#	n/a	Run `Directory()` application	n/a
3	Our address is [address]. Our fax number is [fax number]. etc.	Play a recording containing address and fax information. Return caller to menu prompt when done.	*faxandaddress.wav*
0	Transferring to our attendant. Please hold.	Transfer to reception/operator.	*transfertoreception.wav*
9	n/a	Repeat. Replay menu prompt (but not greeting).	n/a
t	n/a	Timeout. If the caller does not make a choice, treat the call as if caller has dialed 0.	
i	You have made an invalid selection. Please try again.	Caller pressed an invalid digit: replay menu prompt (but not greeting).	*invalid.wav*
_XXX[a]	n/a	Transfer call to dialed extension.	*holdwhileweconnect.wav*

[a] This pattern match must be relevant to your extension range.

Let's go over the various components of this template. Then we'll show you the dialplan code required to implement it, as well as how to create prompts for it.

The Greeting

The first thing the caller hears is actually two prompts.

The first prompt is the greeting. The only thing the greeting should do is greet the caller. Examples of a greeting might be "Thank you for calling Van Meggelen and Associates," "Welcome to Leif's School of Wisdom and T-Shirt Design," or "You have reached the offices of Dewey, Cheetum, and Howe, Attorneys." That's it—the choices for the caller will come later. This allows you to record different greetings without having to record a whole new menu. For example, for a few weeks each year you might want your greeting to say "Season's greetings" or whatever, but your menu will not need to change. Also, if you want to play a different recording after hours ("Thank you for calling. Our office is now closed."), you can use different greetings, but the heart of the menu can stay the same. Finally, if you want to be able to return callers to the menu from a different part of the system, you will normally not want them to hear the greeting again.

The Main Menu

The main menu prompt is where you inform your callers of the choices available to them. You should speak this as quickly as possible (without sounding rushed).[†] When you record a choice, *always tell the users the action that will be taken before giving them the digit option to take that action*. So, don't say "press 1 for sales," but rather say "for sales, press 1." The reason for this is that most people will not pay full attention to the prompt until they hear the choice that is of interest to them. Once they hear their choice, you will have their full attention and can tell them what button to press to get them to where they want to go.

Another point to consider is what order to put the choices in. A typical business, for example, will want sales to be the first menu choice, and most callers will expect this as well. The important thing is to think of your callers. For example, most people will not be interested in address and fax information, so don't make that the first choice.[‡] Think about the goal of getting the callers to their intended destinations as quickly as possible when you make your design choices. Ruthlessly cut anything that is not absolutely essential.

Selection 1

Option 1 in our example will be a simple transfer. Normally this would be to a resource located in another context, and it would typically have an internal extension number so that internal users could also transfer calls to it. In this example, we are going to use this option to send callers to the queue called `sales` that was created in Chapter 13.

Selection 2

Option 2 will be technically identical to option 1. Only the destination will be different. This selection will transfer callers to the `support` queue.

[†] If necessary, you can use an audio editing program such as Audacity to remove silence, and even to speed up the recording a bit.

[‡] In fact, we don't normally recommend this in an AA because it adds to what the caller has to listen to, and most people will go to a website for this sort of information.

Selection

It's good to have the option for the directory as close to the beginning of the recording as possible. Many people will use a directory if they know it is there, but can't be bothered to listen to the whole menu prompt to find out about it. Impatient people will press 0, so the sooner you tell them about the directory, the more chance you'll have that they'll use it, and thus reduce the workload on your receptionist.

Selection 3

When you have an option that does nothing but play a recording back to the caller (such as address and fax information), you can leave all the code for that in the same context as the menu, and simply return the caller to the main menu prompt at the end of the recording. In general, these sorts of options are not as useful as we would like to think they are, so in most cases you'll probably want to leave this out.

Selection 9

It is very important to give the caller the option to hear the choices again. Many people will not be paying attention throughout the whole menu, and if you don't give them the option to hear the choices again, they will most likely press 0.

Note that you do not have to play the greeting again, only the main menu prompt.

Selection 0

As stated before, and whether you like it or not, this is the choice that many (possibly the majority) of your callers will select. If you really don't want to have somebody handle these calls, you can send this extension to a mailbox, but we don't recommend it. If you are a business, many of your callers will be your customers. You want to make it easy for them to get in touch with you. Trust us.

Timeout

Many people will call a number, and not pay too much attention to what is happening. They know that if they just wait on the line, they will eventually be transferred to the operator. Or perhaps they are in their cars, and really shouldn't be pressing buttons on their phones. Either way, oblige them. If they don't make any selection, don't harass them and force them to do so. Connect them to the operator.

Invalid

People make mistakes. That's OK. The invalid handler will let them know that whatever they have chosen is not a valid option and will return them to the menu prompt so that they can try again. Note that you should not play the greeting again, only the main menu prompt.

Dial by Extension

If somebody calls your system and knows the extension she wants to reach, your automated attendant should have code in place to handle this.

 Although Asterisk can handle an overlap between menu choices and extension numbers (i.e., you can have a menu choice 1 and extensions from 100–199), it is generally best to avoid this overlap. Otherwise, the dialplan will always have to wait for the interdigit timeout whenever somebody presses 1, because it won't know if they are planning to dial extension 123. The interdigit timeout is the delay the system will allow between digits before it assumes the entire number has been input. This timer ensures callers have enough time to dial a multidigit extension, but it also causes a delay in the processing of single-digit inputs.

Building Your Auto Attendant

After you have designed your auto attendant, there are three things you need to do to make it work properly:

- Record prompts.
- Build the dialplan for the menu.
- Direct the incoming channels to the auto attendant context.

We will start by talking about recordings.

Recording Prompts

Recording prompts for a telephone system is a critical task. This is what your callers will hear when they interact with your system, and the quality and professionalism of these prompts will reflect on your organization.

Asterisk is very flexible in this regard and can work with many different audio formats. We have found that, in general, the most useful format to use is WAV. Files saved in this format can be of many different kinds, but only one type of WAV file will work with Asterisk: files must be encoded in 16-bit, 8000 Hz, mono format.

Recommended Prompt File Format

The WAV file format we have recommended is useful for system prompts because it is a format that can easily be converted to any other format that your phones might use without distortion, and one that almost any computer can play without any special software. Thus, not only can Asterisk handle the file easily, but it is also easy to work with it on a PC (which can be useful). Asterisk can handle other file formats as well, and in some cases these may be more suitable to your needs, but in general we find 16-

---

bit 8-kHz WAV files to be the easiest to work with and, most of the time, the best-possible quality.

There are essentially two ways to get prompts into a system. One is to record sound files in a studio or on a PC, and then move those files into the system. A second way is to record the prompts directly onto the system using a telephone set. We prefer the second method.

Our advice is this: don't get hung up on the complexities of recording audio through a PC or in a studio.[§] It is generally not necessary. A telephone set will produce excellent-quality recordings, and the reasons are simple: the microphone and electronics in a telephone are carefully designed to capture the human voice in a format that is ideal for transmission on telephone networks, and therefore a phone set is also ideal for doing prompts. The set will capture the audio in the correct format, and will filter out background noise and normalize the decibel level.

 Yes, a properly produced studio prompt will be superior to a prompt recorded over a telephone, but if you don't have the equipment or experience, take our advice and use a telephone to do your recordings, because a poorly produced studio prompt will be much worse.

Using the dialplan to create recordings

The simplest method of recording prompts is to use the Record() application. For example:

```
[context_for_my_handset]
exten => 101,1,Playback(vm-intro)
exten => 101,n,Record(maingreeting.wav)
exten => 101,n,Wait(2)
exten => 101,n,Playback(maingreeting)
exten => 101,n,Hangup
```

This extension plays a prompt, issues a beep, makes a recording, and plays that recording back.[‖] It's notable that the Record() application takes the entire filename as its argument, while the Playback() application excludes the filetype extension (*.wav*, *.gsm*, etc.). This is because the Record() application needs to know which format the recording should be made in, while the Playback() application does not. Instead, Playback() automatically selects the best audio format available, based upon the codec your handset is using and the formats available in the *sounds* folder (for example, if you

---

§ Unless you are an expert in these areas, in which case go for it!

‖ The *vm-intro* prompt isn't perfect (it asks you to leave a message), but it's close enough for our purposes. The usage instructions at least are correct: press pound to end the recording. Once you've gotten the hang of recording prompts, you can go back, record a custom prompt, and change priority 1 to reflect more appropriate instructions for recording your own prompts.

have a *maingreeting.wav* and a *maingreeting.gsm* file in your *sounds* folder, Play back() will select the one that requires the least CPU to play back to the caller).

You'll probably want a separate extension for recording each of the prompts, possibly hidden away from your normal set of extensions, to avoid a mistyped extension from wiping out any of your current menu prompts. If the number of prompts that you have is large, repeating this extension with slight modifications for each will get tedious, but there are ways around that. We'll show you how to make your prompt recording more intelligent in Chapter 17, but for now, this method will suffice.

The Dialplan

Here is the code required to create the auto attendant that we designed earlier. We will often use blank lines before labels within an extension in order to make the dialplan easier to read, but note that just because there is a blank line does not mean there is a different extension:

```
[main_menu]

exten => s,1,Verbose(1, Caller ${CALLERID(all)} has entered the auto attendant)
    same => n,Answer()

; this sets the inter-digit timer
    same => n,Set(TIMEOUT(digit)=2)

; wait one second to establish audio
    same => n,Wait(1)

; If Mon-Fri 9-5 goto label daygreeting
    same => n,GotoIfTime(9:00-17:00,mon-fri,*,*?daygreeting:afterhoursgreeting)

    same => n(afterhoursgreeting),Background(after-hours) ; AFTER HOURS GREETING
    same => n,Goto(menuprompt)

    same => n(daygreeting),Background(daytime)    ; DAY GREETING
    same => n,Goto(menuprompt)

    same => n(menuprompt),Background(main-menu) ; MAIN MENU PROMPT
    same => n,WaitExten(4)                      ; more than 4 seconds is probably
                                                ; too much
    same => n,Goto(0,1)                         ; Treat as if caller has pressed '0'

exten => 1,1,Verbose(1,
    same => n,Goto(Queues,7002,1)     ; Sales Queue - see Chapter 13 for details

exten => 2,1,Verbose(1,
    same => n,Goto(Queues,7001,1)     ; Service Queue - see Chapter 13 for details

exten => 3,1,Verbose(1,
    same => n,Background()            ; Address and fax info
    same => n,Goto(s,menuprompt)      ; Take caller back to main menu prompt
```

```
exten => #,1,Verbose(1,
    same => n,Directory() ;

exten => 0,1,Verbose(1,
    same => n,Dial(SIP/operator)        ; Operator extension/queue

exten => i,1,Verbose(1,
    same => n,Playback(invalid)
    same => n,Goto(s,menuprompt)

exten => t,1,Verbose(1,
    same => n,Goto(0,1)


; You will want to have a pattern match for the various extensions
; that you'll allow external callers to dial
; BUT DON'T JUST INCLUDE THE LocalSets CONTEXT
; OR EXTERNAL CALLERS WILL BE ABLE TO MAKE CALLS OUT OF YOUR SYSTEM

; WHATEVER YOU DO HERE, TEST IT CAREFULLY TO ENSURE EXTERNAL CALLERS
; WILL NOT BE ABLE TO DO ANYTHING BUT DIAL INTERNAL EXTENSIONS

exten => _1XX,1,Verbose(1,Call to an extension starting with '1'
    same => n,Goto(InternalSets,${EXTEN},1)
```

Delivering Incoming Calls to the Auto Attendant

Any call coming into the system will enter the dialplan in the context defined for whatever channel the call arrives on. In many cases this will be a context named incoming, or from-pstn, or something similar. The calls will arrive either with an extension (as would be the case with a DID) or without one (which would be the case with a traditional analog line).

Whatever the name of the context, and whatever the name of the extension, you will want to send each incoming call to the menu. Here are a few examples:

```
[from-pstn] ; an analog line that has context=from-pstn (typically a DAHDI channel)
exten => s,1,Goto(main_menu,s,1)

[incoming] ; a DID coming in on a channel with context=incoming (PRI, SIP, or IAX)
exten => 4169671111,1,Goto(main_menu,s,1)
```

Depending on how you configure your incoming channels, you will generally want to use the Goto() application if you want to send the call to an auto attendant. This is far neater than just coding everything in the incoming context.

IVR

We'll cover Interactive Voice Response (IVR) in more depth in Chapter 17 but before we do that, we're going to talk about something that is essential to any IVR: database integration is the subject of the next chapter.

Conclusion

An automated attendant can provide a very useful service to callers. However, if it is not designed and implemented well, it can also be a barrier to your callers that may well drive them away. Take the time to carefully plan out your auto attendant, and keep it simple.

Relational Database Integration

*Few things are harder to put up with than the annoyance
of a good example.*

—Mark Twain

In this chapter we are going to explore integrating some Asterisk features and functions into a database. There are several databases available for Linux, but we have chosen to limit our discussion to the two most popular: PostgreSQL and MySQL.

We will also explain how to configure Linux to connect to a Microsoft SQL database via ODBC; however, configuration of the Windows/Microsoft portion is beyond the scope of this book.

Regardless of which database you use, this chapter focuses primarily on the ODBC connector, so as long as you have some familiarity with getting your favorite database ODBC-ready, you shouldn't have any problems with this chapter.

Integrating Asterisk with databases is one of the fundamental aspects of building a large clustered or distributed system. The power of the database will enable you to use dynamically changing data in your dialplans, for tasks such as sharing information across an array of Asterisk systems or integrating with web-based services. Our favorite dialplan function, which we will cover later in this chapter, is func_odbc.

While not all Asterisk deployments will require relational databases, understanding how to harness them opens a treasure chest full of new ways to design your telecom solution.

Installing and Configuring PostgreSQL and MySQL

In the following sections we will show how to install and configure PostgreSQL and MySQL on both CentOS and Ubuntu.* It is recommended that you only install one database at a time while working through this section. Pick the database you are most comfortable with, as there is no wrong choice.

Installing PostgreSQL for CentOS

The following command can be used to install the PostgreSQL server and its dependencies from the console:

```
$ sudo yum install -y postgresql-server
Install      3 Package(s)
Upgrade      0 Package(s)

Total download size: 6.9 M
Is this ok [y/N]: y
```

Then start the database, which will take a few seconds to initialize for the first time:

```
$ sudo service postgresql start
```

Now head to "Configuring PostgreSQL" on page 343 for instructions on how to perform the initial configuration.

Installing PostgreSQL for Ubuntu

To install PostgreSQL on Ubuntu, run the following command. You will be prompted to also install any additional packages that are dependencies of the application. Press ⎡Enter⎤ to accept the list of dependencies, at which point the packages will be installed and PostgreSQL will be automatically started and initialized:

```
$ sudo apt-get install postgresql
...
After this operation, 19.1MB of additional disk space will be used.
Do you want to continue [Y/n]? y
```

Now head to "Configuring PostgreSQL" on page 343 for instructions on how to perform the initial configuration.

---

* On a large, busy system you will want to install the database on a completely separate box from your Asterisk system.

Installing MySQL for CentOS

To install MySQL on CentOS, run the following command. You will be prompted to install several dependencies. Press Enter to accept, and the MySQL server and dependency packages will be installed:

```
$ sudo yum install mysql-server
Install      5 Package(s)
Upgrade      0 Package(s)

Total download size: 27 M
Is this ok [y/N]: y
```

Then start the MySQL database by running:

```
$ sudo service mysqld start
```

Now head to "Configuring MySQL" on page 345 to perform the initial configuration.

Installing MySQL for Ubuntu

To install MySQL on Ubuntu, run the following command. You will be prompted to install several dependencies. Press Enter to accept, and the MySQL server and its dependency packages will be installed:

```
$ sudo apt-get install mysql-server
Need to get 24.0MB of archives.
After this operation, 60.6MB of additional disk space will be used.
Do you want to continue [Y/n]? y
```

During the installation, you will be placed into a configuration wizard to help you through the initial configuration of the database. You will be prompted to enter a new password for the *root* user. Type in a strong password and press Enter. You will then be asked to confirm the password. Type your strong password again, followed by Enter. You will then be returned to the console, where the installation will complete. The MySQL service will now be running.

Now head to "Configuring MySQL" on page 345 to perform the initial configuration.

Configuring PostgreSQL

Next, create a user called *asterisk*, which you will use to connect to and manage the database. You can switch to the *postgres* user by using the following command:

```
$ sudo su - postgres
```

At the time of this writing, PostgreSQL version 8.1.*x* is utilized on CentOS, and 8.4.x on Ubuntu.

Then run the following commands to create the *asterisk* user in the database and set up permissions:

```
$ createuser -P
Enter name of user to add: asterisk
Enter password for new user:
Enter it again:
Shall the new role be a superuser? (y/n) n
Shall the new user be allowed to create databases? (y/n) y
Shall the new user be allowed to create more new users? (y/n) n
CREATE ROLE
```

Now, edit the *pg_hba.conf* file in order to allow the *asterisk* user you just created to connect to the PostgreSQL server over the TCP/IP socket.

On CentOS, this file will be located at */var/lib/pgsql/data/pg_hba.conf*. On Ubuntu, you will find it at */etc/postgresql/8.4/main/pg_hba.conf*.

At the end of the file, replace everything below this line:

```
# TYPE DATABASE USER CIDR-ADDRESS METHOD
```

with the following:

```
# TYPE   DATABASE USER      CIDR-ADDRESS  METHOD
host     all      asterisk  127.0.0.1/32  md5
local    all      asterisk                trust
```

Configuring PostgreSQL Database Access via IPv6 localhost

Also, on Ubuntu you will likely need to add the following line:

```
host    all     asterisk        ::1/128                         md5
```

Without it, when you get to "Validating the ODBC Connector" on page 351 you may end up with the following error when connecting:

```
[28000][unixODBC]FATAL:  no pg_hba.conf entry for host "::1", user "asterisk",
database "asterisk", SSL off
[ISQL]ERROR: Could not SQLConnect
```

Now you can create the database that we will use throughout this chapter. Call the database *asterisk* and set the owner to your *asterisk* user:

```
$ createdb --owner=asterisk asterisk
CREATE DATABASE
```

You can set the password for the *asterisk* user like so:

```
$ psql -d template1
template1=# "ALTER USER asterisk WITH PASSWORD 'password'"
template1=# \q
```

Exit from the *postgres* user:

```
$ exit
```

Then restart the PostgreSQL server. On CentOS:

```
$ sudo service postgresql restart
```

 You need to restart the PostgreSQL service because you made changes to *pg_hba.conf*, not because you added a new user or changed the password.

On Ubuntu:

```
$ sudo /etc/init.d/postgresql-8.4 restart
```

 On Ubuntu 10.10 and newer the version number seems to be dropped, so it may just be */etc/init.d/postgresql restart*.

You can verify your connection to the PostgreSQL server via TCP/IP, like so:

```
$ psql -h 127.0.0.1 -U asterisk
Password for user asterisk:

Welcome to psql 8.1.21, the PostgreSQL interactive terminal.

Type:  \copyright for distribution terms
       \h for help with SQL commands
       \? for help with psql commands
       \g or terminate with semicolon to execute query
       \q to quit

asterisk=>
```

You're now ready to move on to "Installing and Configuring ODBC" on page 346.

Configuring MySQL

With the MySQL database now running, you should secure your installation. Conveniently, there is a script you can execute that will allow you to enter a new password[†] for the *root* user, along with some additional options. The script is pretty straightforward, and after entering and confirming your root password you can continue to select the defaults unless you have a specific reason not to.

Execute the following script:

```
$ sudo /usr/bin/mysql_secure_installation
```

[†] If you installed on Ubuntu, you will have already set the root password. You will have to enter that password while executing the script, at which point it will say you've already set a root password, so you don't need to change it.

Then connect to the database console so you can create your *asterisk* user and set up permissions:

```
$ mysql -u root -p
Enter password:
```

After entering the password, you will be presented with the mysql console prompt. You can now create your *asterisk* user by executing the CREATE USER command. The % is a wildcard indicating the *asterisk* user can connect from any host and is IDENTIFIED BY the password *some_secret_password* (which you should obviously change). Note the trailing semicolon:

```
mysql> CREATE USER 'asterisk'@'%' IDENTIFIED BY 'some_secret_password';
Query OK, 0 rows affected (0.00 sec)
```

Let's also create the initial database you'll use throughout this chapter:

```
mysql> CREATE DATABASE asterisk;
Query OK, 1 rows affected (0.00 sec)
```

Now that you've created your user and database, you need to assign permissions for the *asterisk* user to access the *asterisk* database:

```
mysql> GRANT ALL PRIVILEGES ON asterisk.* TO 'asterisk'@'%';
Query OK, 0 rows affected (0.00 sec)
```

Finally, exit from the console and verify that your permissions are correct by logging back into the *asterisk* database as the *asterisk* user:

```
mysql> exit
Bye
# mysql -u asterisk -p asterisk
Enter password:

mysql>
```

You're now ready to move on to "Installing and Configuring ODBC" on page 346.

Installing and Configuring ODBC

The ODBC connector is a database abstraction layer that makes it possible for Asterisk to communicate with a wide range of databases without requiring the developers to create a separate database connector for every database Asterisk wants to support. This saves a lot of development effort and code maintenance. There is a slight performance cost, because we are adding another application layer between Asterisk and the database, but this can be mitigated with proper design and is well worth it when you need powerful, flexible database capabilities in your Asterisk system.

Before you install the connector in Asterisk, you have to install ODBC into Linux itself. To install the ODBC drivers, use one of the following commands.

On CentOS:

```
$ sudo yum install unixODBC unixODBC-devel libtool-ltdl libtool-ltdl-devel
```

 If you're using a 64-bit installation, remember to add *.x86_64* to the end of your development packages to make sure the i386 packages are not also installed, as stability problems can result if Asterisk links against the wrong libraries.

On Ubuntu:

```
$ sudo apt-get install unixODBC unixODBC-dev
```

 See Chapter 3 for the matrix of packages you should have installed.

You'll also need to install the *unixODBC* development package, because Asterisk uses it to build the ODBC modules we will be using throughout this chapter.

 The *unixODBC* drivers shipped with distributions are often a few versions behind the officially released versions on the *http://www.unixodbc .org* website. If you have stability issues while using *unixODBC*, you may need to install from source. Just be sure to remove the *unixODBC* drivers via your package manager first, and then update the paths in your */etc/odbcinst.ini* file.

By default, CentOS will install the drivers for connecting to PostgreSQL databases via ODBC. To install the drivers for MySQL, execute the following command:

```
$ sudo yum install mysql-connector-odbc
```

To install the PostgreSQL ODBC connector on Ubuntu:

```
$ sudo apt-get install odbc-postgresql
```

Or to install the MySQL ODBC connector on Ubuntu:

```
$ sudo apt-get install libmyodbc
```

Configuring ODBC for PostgreSQL

Configuration for the PostgreSQL ODBC driver is done in the */etc/odbcinst.ini* file.

On CentOS the default file already contains some data, including that for PostgreSQL, so just verify that the data exists. The file will look like the following:

```
[PostgreSQL]
Description     = ODBC for PostgreSQL
```

```
Driver          = /usr/lib/libodbcpsql.so
Setup           = /usr/lib/libodbcpsqlS.so
FileUsage       = 1
```

On Ubuntu, the */etc/odbcinst.ini* file will be blank, so you'll need to add the data to that configuration file. Add the following to the *odbcinst.ini* file:

```
[PostgreSQL]
Description     = ODBC for PostgreSQL
Driver          = /usr/lib/odbc/psqlodbca.so
Setup           = /usr/lib/odbc/libodbcpsqlS.so
FileUsage       = 1
```

 On 64-bit systems, you will need to change the path of the libraries from */usr/lib/* to */usr/lib64/* in order to access the correct library files.

In either case, you can use *cat > /etc/odbcinst.ini* to write a clean configuration file, as we've done in other chapters. Just use $\boxed{\text{Ctrl}}$+$\boxed{\text{D}}$ to save the file once you're done.

Verify that the system is able to see the driver by running the following command. It should return the label name PostgreSQL if all is well:

```
$ odbcinst -q -d
[PostgreSQL]
```

Next, configure the */etc/odbc.ini* file, which is used to create an identifier that Asterisk will use to reference this configuration. If at any point in the future you need to change the database to something else, you simply need to reconfigure this file, allowing Asterisk to continue to point to the same place[‡]:

```
[asterisk-connector]
Description     = PostgreSQL connection to 'asterisk' database
Driver          = PostgreSQL
Database        = asterisk
Servername      = localhost
UserName        = asterisk
Password        = welcome
Port            = 5432
Protocol        = 8.1
ReadOnly        = No
RowVersioning   = No
ShowSystemTables = No
ShowOidColumn   = No
FakeOidIndex    = No
ConnSettings    =
```

[‡] Yes, this is excessively verbose. The only entries you really need are Driver, Database, and Servername. Even the UserName and Password are specified elsewhere, as you'll see later (although these are required when testing, as in "Validating the ODBC Connector" on page 351).

Configuring ODBC for MySQL

Configuration for the MySQL ODBC driver is done in the */etc/odbcinst.ini* file.

On CentOS the default file already contains some data, including that for MySQL, but it needs to be uncommented and requires a couple of changes. Replace the existing text with the following:

```
[MySQL]
Description = ODBC for MySQL
Driver = /usr/lib/libmyodbc3.so
Setup = /usr/lib/libodbcmyS.so
FileUsage = 1
```

On Ubuntu, the */etc/odbcinst.ini* file will be blank, so you'll need to add the data to that configuration file. Add the following to the *odbcinst.ini* file:

```
[MySQL]
Description = ODBC for MySQL
Driver = /usr/lib/odbc/libmyodbc.so
Setup = /usr/lib/odbc/libodbcmyS.so
FileUsage = 1
```

 On 64-bit systems, you will need to change the path of the libraries from */usr/lib/* to */usr/lib64/* in order to access the correct library files.

In either case, you can use *cat > /etc/odbcinst.ini* to write a clean configuration file, as we've done in other chapters. Just use ⎡Ctrl⎤+⎡D⎤ to save the file once you're done.

Verify that the system is able to see the driver by running the following command. It should return the label name MySQL if all is well:

```
# odbcinst -q -d
[MySQL]
```

Next, configure the */etc/odbc.ini* file, which is used to create an identifier that Asterisk will use to reference this configuration. If at any point in the future you need to change the database to something else, you simply need to reconfigure this file, allowing Asterisk to continue to point to the same place:

```
[asterisk-connector]
Description     = MySQL connection to 'asterisk' database
Driver          = MySQL
Database        = asterisk
Server          = localhost
UserName        = asterisk
Password        = welcome
Port            = 3306
Socket          = /var/lib/mysql/mysql.sock
```

On Ubuntu 10.10, the socket location is */var/run/mysqld/mysqld.sock*.

Configuring ODBC for Microsoft SQL

Connecting to Microsoft SQL (MS SQL) is similar to connecting to either MySQL or PostgreSQL, as we've previously discussed. The configuration of MS SQL is beyond the scope of this book, but the following information will get your Asterisk box configured to connect to your MS SQL database once you've enabled the appropriate permissions on your database.

To connect to MS SQL, you need to install the FreeTDS drivers using the package manager (or by compiling via the source files available at *http://www.freetds.org*).

On CentOS:

```
$ sudo yum install freetds
```

On Ubuntu:

```
$ sudo apt-get install freetds
```

After installing the drivers, you need to configure the */etc/odbcinst.ini* file, which tells the system where the driver files are located.

Insert the following text into the */etc/odbcinst.ini* file with your favorite text editor or with the following command:

```
$ sudo cat > /etc/odbcinst.ini
[FreeTDS]
Description = ODBC for Microsoft SQL
Driver      = /usr/lib/libtdsodbc.so
UsageCount  = 1
Threading   = 2
Ctrl+D
```

 If you compiled via source, the files may be located in */usr/local/lib/* or (if you compiled on a 64-bit system) */usr/local/lib64/*.

Verify that the system is able to see the driver by running the following command. It should return the label name FreeTDS if all is well:

```
$ odbcinst -q -d
[FreeTDS]
```

Once you've configured the drivers, you need to modify the */etc/odbc.ini* file to control how to connect to the database:

```
[asterisk-connector]
Description     = MS SQL connection to 'asterisk' database
Driver          = FreeTDS
Database        = asterisk
Server          = 192.168.100.1
UserName        = asterisk
Password        = welcome
Trace           = No
TDS_Version     = 7.0
Port            = 1433
```

In the next section, you will be able to validate your connection to the MS SQL server.

Validating the ODBC Connector

Now, verify that you can connect to your database using the *isql* application. *echo* the select 1 statement and pipe it into *isql*, which will then connect using the asterisk-connector section you added to */etc/odbc.ini*. You should get the following output (or at least something similar; we're looking for a result of 1 rows fetched):

```
$ echo "select 1" | isql -v asterisk-connector
+---------------------------------------+
| Connected!                            |
|                                       |
| sql-statement                         |
| help [tablename]                      |
| quit                                  |
|                                       |
+---------------------------------------+
SQL> +------------+
| ?column?   |
+------------+
| 1          |
+------------+
SQLRowCount returns 1
1 rows fetched
$ exit
```

With *unixODBC* installed, configured, and verified to work, you need to recompile Asterisk so that the ODBC modules are created and installed. Change back to your Asterisk source directory and run the *./configure* script so it knows you have installed *unixODBC*:

```
$ cd ~/src/asterisk-complete/asterisk/1.8
$ ./configure
$ make menuselect
$ make install
```

Almost everything in this chapter is turned on by default. You will want to run *make menuselect* to verify that the ODBC-related modules are enabled. These include cdr_odbc, cdr_adaptive_odbc, func_odbc, func_realtime, pbx_realtime, res_config_odbc, and res_odbc. For voicemail stored in an ODBC database, be sure to select ODBC_STORAGE from the *Voicemail Build Options* menu. You can verify that the modules exist in the */usr/lib/asterisk/modules/* directory.

Configuring res_odbc to Allow Asterisk to Connect Through ODBC

Asterisk ODBC connections are configured in the *res_odbc.conf* file located in */etc/asterisk*. The *res_odbc.conf* file sets the parameters that various Asterisk modules will use to connect to the database.

 The pooling and limit options are quite useful for MS SQL and Sybase databases. These permit you to establish multiple connections (up to limit connections) to a database while ensuring that each connection has only one statement executing at once (this is due to a limitation in the protocol used by these database servers).

Modify the *res_odbc.conf* file so it looks like the following:

```
[asterisk]
enabled => yes
dsn => asterisk-connector
username => asterisk
password => welcome
pooling => no
limit => 0
pre-connect => yes
```

The dsn option points at the database connection you configured in */etc/odbc.ini*, and the pre-connect option tells Asterisk to open up and maintain a connection to the database when loading the *res_odbc.so* module. This lowers some of the overhead that would come from repeatedly setting up and tearing down the connection to the database.

Once you've configured *res_odbc.conf*, start Asterisk and verify the database connection with the *odbc show* CLI command:

```
*CLI> odbc show

ODBC DSN Settings
-----------------

    Name:   asterisk
    DSN:    asterisk-connector
```

```
Last connection attempt: 1969-12-31 19:00:00
Pooled: No
Connected: Yes
```

Managing Databases

While it isn't within the scope of this book to teach you about how to manage your databases, it is worth at least noting briefly some of the applications you could use to help with database management. Several exist, some of which are local client applications running from your computer and connecting to the database, and others of which are web-based applications that could be served from the same computer running the database itself, thereby allowing you to connect remotely.

Some of the ones we've used include:

- phpMyAdmin (*http://www.phpmyadmin.net*)
- MySQL Workbench (*http://wb.mysql.com*)
- pgAdmin (*http://www.pgadmin.org*)
- Navicat (commercial) (*http://www.navicat.com*)

Troubleshooting Database Issues

When working with ODBC database connections and Asterisk, it is important to remember that the ODBC connection abstracts some of the information passed between Asterisk and the database. In cases where things are not working as expected, you may need to enable logging on your database platform to see what Asterisk is sending to the database (e.g., what SELECT, INSERT, or UPDATE statements are being triggered from Asterisk), what the database is seeing, and why the database may be rejecting the statements.

For example, one of the most common problems found with ODBC database integration is an incorrectly defined table, or a missing column that Asterisk expects to exist. While great strides have been made in the form of adaptive modules, not all parts of Asterisk are adaptive. In the case of ODBC voicemail storage, you may have missed a column such as flag, which is a new column not previously found in versions of Asterisk prior to 1.8.[§] In order to debug why your data is not being written to the database as expected, you should enable statement logging on the database side, and then determine what statement is being executed and why the database is rejecting it.

[§] This was actually an issue one of the authors had while working on this book, and the flag column was found by looking at the statement logging during PostgreSQL testing.

A Gentle Introduction to func_odbc

The very first use of func_odbc, which occurred while its author was still writing it, is also a good introduction to its use. A customer of one of the module's authors noted that some people calling into his switch had figured out a way to make free calls with his system. While his eventual intent was to change his dialplan to avoid those problems, he needed to blacklist certain caller IDs in the meantime, and the database he wanted to use for this was a Microsoft SQL Server database. With a few exceptions, this is the actual dialplan[||]:

```
[span3pri]
exten => _50054XX,1,NoOp()
    same => n,Set(CDR(accountcode)=pricall)
    same => n,GotoIf($[${ODBC_ANIBLOCK(${CALLERID(number)})}]?busy)
    same => n(dial),Dial(DAHDI/G1/${EXTEN})
    same => n(busy),Busy(10)
    same => n,Hangup
```

This dialplan, in a nutshell, passes through all calls to another system for routing purposes, except those calls whose caller IDs are in a blacklist. The calls coming into this system used a block of 100 7-digit DIDs. There is a mystery function in this dialplan, though: ODBC_ANIBLOCK(). This function is defined in another configuration file, *func_odbc.conf*, at runtime:

```
[ANIBLOCK]
dsn=telesys
readsql=SELECT IF(COUNT(1)>0, 1, 0) FROM Aniblock WHERE NUMBER='${ARG1}'
```

So, your ODBC_ANIBLOCK()[#] connects to a listing in *res_odbc.conf* named telesys and selects a count of records that have the NUMBER specified by the argument, which is, referring to our dialplan above, the caller ID. Nominally, this function should return either a 1 (indicating the caller ID exists in the Aniblock table) or a 0 (if it does not). This value also evaluates directly to true or false, which means we don't need to use an expression in our dialplan to complicate the logic.

Getting Funky with func_odbc: Hot-Desking

The func_odbc dialplan function is arguably the coolest and most powerful dialplan function in Asterisk. It allows you to create and use fairly simple dialplan functions that retrieve and use information from databases directly in the dialplan. There are all kinds

---

[||] This system is unfortunately no longer in service. Thus, any changes have been made for the sake of simplicity, not to conceal the business for which it was designed.

[#] We're using the IF() SQL function to make sure we return a value of 0 or 1. This works on MySQL 5.1 or later. If it does not work on your SQL installation, you could also check the returned result in the dialplan using the IF() function there.

of ways in which this might be used, such as for managing users or allowing the sharing of dynamic information within a clustered set of Asterisk machines.

What func_odbc allows you to do is define SQL queries to which you assign function names. In effect, you are creating custom functions that obtain their results by executing queries against a database. The *func_odbc.conf* file is where you specify the relationships between the function names you create and the SQL statements you wish them to perform. By referring to the named functions in the dialplan, you can retrieve and update values in the database.

> While using an external script to interact with a database (from which a flat file is created that Asterisk will read) has advantages (if the database goes down, your system will continue to function and the script will simply not update any files until connectivity to the database is restored), it also has disadvantages. A major disadvantage is that any changes you make to a user will not be available until you run the update script. This is probably not a big issue on small systems, but on large systems waiting for changes to take effect can cause issues, such as pausing a live call while a large file is loaded and parsed.
>
> You can relieve some of this by utilizing a replicated database system. Asterisk 1.6.0 and newer provide the ability to fail over to another database system. This way, you can cluster the database backend utilizing a master-master relationship (for PostgreSQL, pgcluster (*http://pgcluster .projects.postgresql.org/*) or Postgres-R (*http://postgres-r.org/*);[*] for MySQL it's native[†]), or a master-slave (for PostgreSQL, Slony-I (*http:// www.slony.info/*), for MySQL it's native) replication system.

In order to get you into the right frame of mind for what follows, we want you to picture a Dagwood sandwich.[‡]

Can you relay the total experience of such a thing by showing someone a picture of a tomato, or by waving a slice of cheese about? Not hardly. That is the conundrum we faced when trying to give a useful example of why func_odbc is so powerful. So, we decided to build the whole sandwich for you. It's quite a mouthful, but after a few bites of this, peanut butter and jelly is never going to be the same.

For our example, we decided to implement something we think could have some practical uses. Picture a small company with a sales force of five people who have to share two desks. This is not as cruel as it seems, because these folks spend most of their time on the road, and they are each only in the office for at most one day each week.

---

[*] *pgcluster* appears to be a dead project, and Postgres-R appears to be in its infancy, so there may currently be no good solution for master-master replication using PostgreSQL.

[†] There are several tutorials on the Web describing how to set up replication with MySQL.

[‡] And if you don't know what a Dagwood is, that's what Wikipedia is for. I am not that old.

Still, when they do get into the office, they'd like the system to know which desks they are sitting at, so that their calls can be directed there. Also, the boss wants to be able to track when they are in the office and control calling privileges from those phones when no one is there.

This need is typically solved by what is called a *hot-desking* feature, so we have built one for you in order to show you the power of func_odbc.

Lets start with the easy stuff, and create two desktop phones in the *sip.conf* file:

```
; sip.conf
; HOT DESK USERS
[0000FFFF0001]
type=friend
host=dynamic
secret=my_special_secret
context=hotdesk
qualify=yes

[0000FFFF0002]
type=friend
host=dynamic
secret=my_special_secret
context=hotdesk
qualify=yes


; END HOT DESK USERS
```

These two desk phones both enter the dialplan at the hotdesk context in *extensions.conf*. If you want to have these devices actually work, you will of course need to set the appropriate parameters in the devices themselves, but we covered all that in Chapter 5.

That's all for *sip.conf*. We've got two slices of bread, which is hardly a sandwich yet.

Now let's get the database part of it set up (we are assuming that you have an ODBC database created and working, as outlined in the earlier parts of this chapter). First, connect to the database console.

For PostgreSQL:

```
$ sudo su - postgres
$ psql -U asterisk -h localhost asterisk
Password:
```

Then create the table with the following bit of SQL:

```
CREATE TABLE ast_hotdesk
(
    id serial NOT NULL,
    extension int8,
    first_name text,
    last_name text,
    cid_name text,
    cid_number varchar(10),
```

```
    pin int4,
    context text,
    status bool DEFAULT false,
    "location" text,
    CONSTRAINT ast_hotdesk_id_pk PRIMARY KEY (id)
)
WITHOUT OIDS;
```

For MySQL:

```
$ mysql -u asterisk -p asterisk
Enter password:
```

Then create the table with the following bit of SQL:

```
CREATE TABLE ast_hotdesk
(
    id serial NOT NULL,
    extension int8,
    first_name text,
    last_name text,
    cid_name text,
    cid_number varchar(10),
    pin int4,
    context text,
    status bool DEFAULT false,
    location text,
    CONSTRAINT ast_hotdesk_id_pk PRIMARY KEY (id)
);
```

The table information is summarized in Table 16-1.

Table 16-1. Summary of ast_hotdesk table

Column name	Column type
id	Serial, auto-incrementing
extension	Integer
first_name	Text
last_name	Text
cid_name	Text
cid_number	Varchar 10
pin	Integer
context	Text
status	Boolean, default false
location	Text

After that, populate the database with the following information (some of the values that you see actually will change only after the dialplan work is done, but we include it here by way of example).

At the PostgreSQL console, run the following commands:

```
asterisk=> INSERT INTO ast_hotdesk ('extension', 'first_name', 'last_name',\
'cid_name','cid_number', 'pin', 'context', 'location') \
VALUES (1101, 'Leif', 'Madsen', 'Leif Madsen', '4165551101', '555',\
'longdistance','0000FFFF0001');
```

At the MySQL console, run the following commands:

```
mysql> INSERT INTO ast_hotdesk (extension, first_name, last_name, cid_name,
cid_number, pin, context, location)
VALUES (1101, 'Leif', 'Madsen', 'Leif Madsen',
'4165551101', '555', 'longdistance', '0000FFFF0001');
```

Repeat these commands, changing the VALUES as needed, for all entries you wish to have in the database. You can view the data in the ast_hotdesk table by running a simple SELECT statement from the database console:

```
mysql> SELECT * FROM ast_hotdesk;
```

which would give you something like the following output:

id	extension	first_name	last_name	cid_name	cid_number
1	1101	"Leif"	"Madsen"	"Leif Madsen"	"4165551101"
2	1102	"Jim"	"Van Meggelen"	"Jim Van Meggelen"	"4165551102"
3	1103	"Russell"	"Bryant"	"Russell Bryant"	"4165551103"
4	1104	"Mark"	"Spencer"	"Mark Spencer"	"4165551104"
5	1105	"Kevin"	"Fleming"	"Kevin Fleming"	"4165551105"

pin	context	status	location	$
"555"	"longdistance"	"TRUE"	"0000FFFF0001"	
"556"	"longdistance"	"FALSE"	""	
"557"	"local"	"FALSE"	""	
"558"	"international"	"FALSE"	""	
"559"	"local"	"FALSE"	""	

We've got the condiments now, so let's get to our dialplan. This is where the magic is going to happen.

Somewhere in *extensions.conf* we are going to have to create the hotdesk context. To start, let's define a pattern-match extension that will allow the users to log in:

```
; extensions.conf
; Hot-Desking Feature
[hotdesk]
; Hot Desk Login
exten => _#110[1-5],1,NoOp()
   same => n,Set(E=${EXTEN:1})  ; strip off the leading hash (#) symbol
   same => n,Verbose(1,Hot Desk Extension ${E} is changing status)
   same => n,Verbose(1,Checking current status of extension ${E})
   same => n,Set(${E}_STATUS=${HOTDESK_INFO(status,${E})})
   same => n,Set(${E}_PIN=${HOTDESK_INFO(pin,${E})})
```

We're not done writing this extension yet, but let's pause for a moment and see where we're at so far.

When a sales agent sits down at a desk, he logs in by dialing hash (#) plus his own extension number. In this case we have allowed the 1101 through 1105 extensions to log in with our pattern match of _#110[1-5]. You could just as easily make this less restrictive by using _#11XX (allowing 1100 through 1199). This extension uses func_odbc to perform a lookup with the HOTDESK_INFO() dialplan function. This custom function (which we will define in the *func_odbc.conf* file) performs an SQL statement and returns whatever is retrieved from the database.

We would define the new function HOTDESK_INFO() in *func_odbc.conf* like so:

```
[INFO]
prefix=HOTDESK
dsn=asterisk
readsql=SELECT ${ARG1} FROM ast_hotdesk WHERE extension = '${ARG2}'
```

That's a lot of stuff in just a few lines. Let's quickly cover them before we move on.

First of all, the prefix is optional. If you don't configure the prefix, then Asterisk adds "ODBC" to the name of the function (in this case, INFO), which means this function would become ODBC_INFO(). This is not very descriptive of what the function is doing, so it can be helpful to assign a prefix that helps to relate your ODBC functions to the tasks they are performing. We chose HOTDESK, which means that this custom function will be named HOTDESK_INFO().

The dsn attribute tells Asterisk which connection to use from *res_odbc.conf*. Since several database connections could be configured in *res_odbc.conf*, we specify which one to use here. In Figure 16-1, we show the relationship between the various file configurations and how they reference down the chain to connect to the database.

 The *func_odbc.conf.sample* file in the Asterisk source contains additional information about how to handle multiple databases and control the reading and writing of information to different DSN connections. Specifically, the readhandle, writehandle, readsql, and writesql arguments will provide you with great flexibility for database integration and control.

Finally, we define our SQL statement with the readsql attribute. Dialplan functions have two different formats that they can be called with: one for retrieving information, and one for setting information. The readsql attribute is used when we call the HOT DESK_INFO() function with the retrieve format (we could execute a separate SQL statement with the writesql attribute; we'll discuss the format for that attribute a little bit later in this chapter).

Reading values from this function would take this format in the dialplan:

```
exten => s,n,Set(RETURNED_VALUE=${HOTDESK_INFO(status,1101)})
```

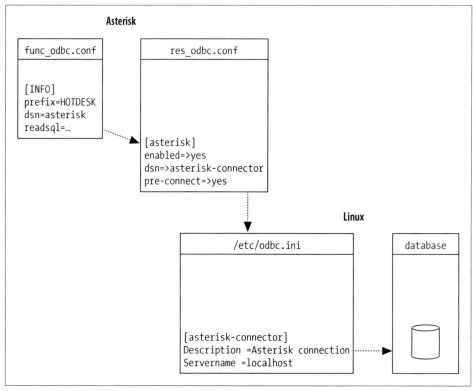

Figure 16-1. Relationships between func_odbc.conf, res_odbc.conf, /etc/odbc.ini (unixODBC), and the database connection

This would return the value located in the database within the `status` column where the `extension` column equals `1101`. The `status` and `1101` we pass to the `HOTDESK_INFO()` function are then placed into the SQL statement we assigned to the `readsql` attribute, available as `${ARG1}` and `${ARG2}`, respectively. If we had passed a third option, this would have been available as `${ARG3}`.

Multirow Functionality with func_odbc

As of Asterisk branch 1.6.0, a mode exists that allows Asterisk to handle multiple rows of data returned from the database. For example, if we were to create a dialplan function in *func_odbc.conf* that returned all available extensions, we would need to enable multirow mode for the function. This would cause the function to work a little differently, returning an ID number that could then be passed to the `ODBC_FETCH()` function to return each row in turn.

Prior to the 1.6.0 branch, we needed to use the SQL functions `LIMIT` and `OFFSET` in order to control data being returned to Asterisk for iteration. This was resource-intensive (at least in relation to multirow mode), as it required multiple queries to the database for each row.

A simple example follows. Suppose we have the following *func_odbc.conf*:

```
[ALL_AVAIL_EXTENS]
prefix=GET
dsn=asterisk-connector
mode=multirow
readsql=SELECT extension FROM ast_hotdesk WHERE status = '${ARG1}'
```

and a dialplan in *extensions.conf* that looks something like this:

```
[multirow_example]
exten => start,1,Verbose(1,Looping example)
    same => n,Set(ODBC_ID=${GET_ALL_AVAIL_EXTENS(1)})
    same => n,GotoIf($[${ODBCROWS} < 1]?no_rows,1)
    same => n,Set(COUNTER=1)
    same => n,While($[${COUNTER} <= ${ODBCROWS}])
    same => n,Set(AVAIL_EXTEN_${COUNTER}=${ODBC_FETCH(${ODBC_ID})})
    same => n,Set(COUNTER=$[${COUNTER + 1}])
    same => n,EndWhile()
    same => n,ODBCFinish()

exten => no_rows,1,Verbose(1,No rows returned)
    same => n,Playback(silence/1&invalid)
    same => n,Hangup()
```

The `ODBC_FETCH()` function will essentially treat the information as a stack, and each call to it with the passed `ODBC_ID` will pop the next row of information off the stack. We also have the option of using the `ODBC_FETCH_STATUS` channel variable, which is set once the `ODBC_FETCH()` function (which returns `SUCCESS` if additional rows are available or `FAILURE` if no additional rows are available) is called. This permits us to write a dialplan like the following, which does not use a counter, but still loops through the data. This may be useful if we're looking for something specific and don't need to look at all the data. Once we're done, the `ODBCFinish()` dialplan application should be called to clean up any remaining data.

Here's another *extensions.conf* example:

```
[multirow_example_2]
exten => start,1,Verbose(1,Looping example with break)
    same => n,Set(ODBC_ID=${GET_ALL_AVAIL_EXTENS(1)})
    same => n(loop_start),NoOp()
    same => n,Set(ROW_RESULT=${ODBC_FETCH(${ODBC_ID})})
    same => n,GotoIf($["${ODBC_FETCH_RESULT}" = "FAILURE"]?cleanup,1)
    same => n,GotoIf($["${ROW_RESULT}" = "1104"]?good_exten,1)
    same => n,Goto(loop_start)

exten => cleanup,1,Verbose(1,Cleaning up after all iterations)
    same => n,Verbose(1,We did not find the extension we wanted)
    same => n,ODBCFinish(${ODBC_ID})
    same => n,Hangup()

exten => good_exten,1,Verbose(1,Extension we want is available)
    same => n,ODBCFinish(${ODBC_ID})
    same => n,Verbose(1,Perform some action we wanted)
    same => n,Hangup()
```

We'll be using multirow mode for one of our functions later in this chapter.

After the SQL statement is executed, the value returned (if any) is assigned to the `RETURNED_VALUE` channel variable.

Using the ARRAY() Function

In our example, we are utilizing two separate database calls and assigning those values to a pair of channel variables, `${E}_STATUS` and `${E}_PIN`. This was done to simplify the example:

```
exten => _110[1-5],n,Set(${E}_STATUS=${HOTDESK_INFO(status,${E})})
    same => n,Set(${E}_PIN=${HOTDESK_INFO(pin,${E})})
```

As an alternative, we could have returned multiple columns and saved them to separate variables utilizing the `ARRAY()` dialplan function. If we had defined our SQL statement in the *func_odbc.conf* file like so:

```
readsql=SELECT pin,status FROM ast_hotdesk WHERE extension = '${E}'
```

we could have used the `ARRAY()` function to save each column of information for the row to its own variable with a single call to the database:

```
exten => _110[1-5],n,Set(ARRAY(${E}_PIN,${E}_STATUS)=${HOTDESK_INFO(${E})})
```

Using `ARRAY()` is handy any time you might get comma-separated values back and want to assign the values to separate variables, such as with `CURL()`.

So, in the first two lines of the following block of code, we are passing the value `status` and the value contained in the `${E}` variable (e.g., 1101) to the `HOTDESK_INFO()` function. The two values are then replaced in the SQL statement with `${ARG1}` and `${ARG2}`, respectively, and the SQL statement is executed. Finally the value returned is assigned to the `${E}_STATUS` channel variable.

OK, let's finish writing the pattern-match extension now:

```
        same => n,Set(${E}_STATUS=${HOTDESK_INFO(status,${E})})
        same => n,Set(${E}_PIN=${HOTDESK_INFO(pin,${E})})
        same => n,GotoIf($[${ODBCROWS} < 0]?invalid_user,1)
    ; check if ${E}_STATUS is NULL
        same => n,GotoIf($[${${E}_STATUS} = 1]?logout,1:login,1)
```

After assigning the value of the `status` column to the `${E}_STATUS` variable (if the user dials extension 1101, the variable name will be 1101_STATUS), we check if we've received a value back from the database (error checking) using the `${ODBCROWS}` channel variable.

The last row in the block checks the status of the phone and, if the agent is currently logged in, logs him off. If the agent is not already logged in, it will go to extension `login`, priority 1 within the same context.

Remember that in a traditional phone system all extensions must be numbers, but in Asterisk, extensions can have names as well. A possible advantage of using an extension that's not a number is that it will be much harder for a user to dial it from her phone and, thus, more secure. We're going to use several named extensions in this example. If you want to be absolutely sure that a malicious user cannot access those named extensions, simply use the trick that the AEL loader uses: *start with a priority other than 1.* You can access the first line of the extension by assigning it a priority label and referencing it via the extension name/priority label combination.

The `login` extension runs some initial checks to verify the pin code entered by the agent. We allow him three tries to enter the correct pin, and if all tries are invalid we send the call to the `login_fail` extension (which we will be writing later):

```
exten => login,1,NoOp() ; set initial counter values
    same => n,Set(PIN_TRIES=1)      ; pin tries counter
    same => n,Set(MAX_PIN_TRIES=3) ; set max number of login attempts
    same => n,Playback(silence/1)  ; play back some silence so first prompt is
                                    ; not cut off
    same => n(get_pin),NoOp()
    same => n,Set(PIN_TRIES=$[${PIN_TRIES} + 1])   ; increase pin try counter
    same => n,Read(PIN_ENTERED,enter-password,${LEN(${${E}_PIN})})
    same => n,GotoIf($["${PIN_ENTERED}" = "${${E}_PIN}"]?valid_login,1)
    same => n,Playback(pin-invalid)
    same => n,GotoIf($[${PIN_TRIES} <= ${MAX_PIN_TRIES}]?get_pin:login_fail,1)
```

If the pin entered matches, we validate the login with the `valid_login` extension. First we utilize the CHANNEL variable to figure out which phone device the agent is calling from. The CHANNEL variable is usually populated with something like SIP/0000FFFF0001-ab4034c, so we make use of the CUT() function to first pull off the SIP/ portion of the string and assign that to LOCATION. We then strip off the -ab4034c part of the string, discard it, and assign the remainder (0000FFFF0001) to the LOCATION variable:

```
exten => valid_login,1,NoOp()
; CUT off the channel technology and assign it to the LOCATION variable
    same => n,Set(LOCATION=${CUT(CHANNEL,/,2)})
; CUT off the unique identifier and save the remainder to the LOCATION variable
    same => n,Set(LOCATION=${CUT(LOCATION,-,1)})
```

We utilize yet another custom function created in the *func_odbc.conf* file, HOT DESK_CHECK_PHONE_LOGINS(), to check if any other users were previously logged into this phone and forgot to log out. If the number of logged-in users is greater than 0 (it should never be more than 1, but we check for higher values anyway and reset those, too), it runs the logic in the `logout_login` extension:

```
; func_odbc.conf
[CHECK_PHONE_LOGINS]
prefix=HOTDESK
dsn=asterisk
```

```
; *** This line should have no line breaks
readsql=SELECT COUNT(status) FROM ast_hotdesk WHERE status = '1' AND
location = '${ARG1}'
```

If there are no other agents logged into the device, we update the login status for this user with the HOTDESK_STATUS() function:

```
; Continuation of the valid_login extension below
    same => n,Set(USERS_LOGGED_IN=${HOTDESK_CHECK_PHONE_
LOGINS(${LOCATION})})
        same => n,GotoIf($[${USERS_LOGGED_IN} > 0]?logout_login,1)
        same => n(set_login_status),NoOp()

    ; Set the status for the phone to '1' and where the agent is logged into
        same => n,Set(HOTDESK_STATUS(${E})=1,${LOCATION})
        same => n,GotoIf($[${ODBCROWS} < 1]?error,1)
        same => n,Playback(agent-loginok)
        same => n,Hangup()
```

We create a write function in *func_odbc.conf* like so:

```
[STATUS]
prefix=HOTDESK
dsn=asterisk

; *** This line should have no line breaks
writesql=UPDATE ast_hotdesk SET status = '${VAL1}',
location = '${VAL2}' WHERE extension = '${ARG1}'
```

The syntax is very similar to the readsql syntax discussed earlier in the chapter, but there are a few new things here, so let's discuss them before moving on.

The first thing you may have noticed is that we now have both ${VALx} and ${ARGx} variables in our SQL statement. These contain the values we pass to the function from the dialplan. In this case, we have two VAL variables and a single ARG variable that were set from the dialplan via this statement:

```
Set(HOTDESK_STATUS(${E})=1,${LOCATION})
```

Notice the syntax is slightly different from that of the read-style function. This signals to Asterisk that you want to perform a write (this is the same syntax as that used for other dialplan functions).

We are passing the value of the ${E} variable to the HOTDESK_STATUS() function, whose value is then accessible in the SQL statement within *func_odbc.conf* with the ${ARG1} variable. We then pass two values: 1 and ${LOCATION}. These are available to the SQL statement in the ${VAL1} and ${VAL2} variables, respectively.

As mentioned previously, if we had to log out one or more agents before logging this one in, we would check this with the logout_login extension. This dialplan logic will utilize the ODBC_FETCH() function to pop information off the information stack returned by the HOTDESK_LOGGED_IN_USER() function. More than likely this will execute only one loop, but it's a good example of how you might update or parse multiple rows in the database.§

The first part of our dialplan returns an ID number that we can use with the ODBC_FETCH() function to iterate through the values returned. We're going to assign this ID to the LOGGED_IN_ID channel variable:

```
same => n,Set(LOGGED_IN_ID=${HOTDESK_LOGGED_IN_USER(${LOCATION})})
```

Here is the logout_login extension, which could potentially loop through multiple rows:

```
exten => logout_login,1,NoOp()
; set all logged-in users on this device to logged-out status
    same => n,Set(LOGGED_IN_ID=${HOTDESK_LOGGED_IN_USER(${LOCATION})})
    same => n(start_loop),NoOp()
    same => n,Set(WHO=${ODBC_FETCH(${LOGGED_IN_ID})})
    same => n,GotoIf($["${ODBC_FETCH_STATUS}" = "FAILURE"]?cleanup)
    same => n,Set(HOTDESK_STATUS(${WHO})=0)          ; log out phone
    same => n,Goto(start_loop)
    same => n(cleanup),ODBCFinish(${LOGGED_IN_ID})
    same => n,Goto(valid_login,set_login_status)     ; return to logging in
```

We assign the first value returned from the database (e.g., the extension 1101) to the WHO channel. Before doing anything, though, we check to see if the ODBC_FETCH() function was successful in returning data. If the ODBC_FETCH_STATUS channel variable contains FAILURE, we have no data to work with, so we move to the cleanup priority label.

If we have data, we then pass the value of ${WHO} as an argument to the HOTDESK_STATUS() function, which contains a value of 0. This is the first value passed to HOTDESK_STATUS() and is shown as ${VAL1} in *func_odbc.conf*, where the function is declared.

If you look at the HOTDESK_STATUS() function in *func_odbc.conf* you will see we could also pass a second value, but we're not doing that here since we want to remove any values from that column in order to log out the user, which setting no value does effectively.

After using HOTDESK_STATUS() to log out the user, we return to the start_loop priority label to loop through all values, which simply executes a NoOp(). After attempting to retrieve a value, we again check ODBC_FETCH_STATUS for FAILURE. If that value is found, we move to the cleanup priority label, where we execute the ODBCFinish() dialplan application to perform cleanup. We then return to the valid_login extension at the set_login_status priority label.

The rest of the context should be fairly straightforward (if some of this doesn't make sense, we suggest you go back and refresh your memory with Chapters 6 and 10). The one trick you may be unfamiliar with could be the usage of the ${ODBCROWS} channel

---

§ Also see "Multirow Functionality with func_odbc" on page 360 for more information and examples of parsing multiple rows returned from the database.

variable, which is set by the HOTDESK_STATUS() function. This tells us how many rows were affected in the SQL UPDATE, which we assume to be 1. If the value of ${ODB CROWS} is less than 1, we assume an error and handle it appropriately:

```
exten => logout,1,NoOp()
    same => n,Set(HOTDESK_STATUS(${E})=0)
    same => n,GotoIf($[${ODBCROWS} < 1]?error,1)
    same => n,Playback(silence/1&agent-loggedoff)
    same => n,Hangup()

exten => login_fail,1,NoOp()
    same => n,Playback(silence/1&login-fail)
    same => n,Hangup()

exten => error,1,NoOp()
    same => n,Playback(silence/1&connection-failed)
    same => n,Hangup()

exten => invalid_user,1,NoOp()
    same => n,Verbose(1,Hot Desk extension ${E} does not exist)
    same => n,Playback(silence/2&invalid)
    same => n,Hangup()
```

We also include the hotdesk_outbound context, which will handle our outgoing calls after we have logged the agent into the system:

```
include => hotdesk_outbound
```

The hotdesk_outbound context utilizes many of the same principles discussed previously, so we won't approach it quite so thoroughly; essentially, this context will catch all numbers dialed from the desk phones. We first set our LOCATION variable using the CHANNEL variable, then determine which extension (agent) is logged into the system and assign that value to the WHO variable. If this variable is NULL, we reject the outgoing call. If it is not NULL, then we get the agent information using the HOTDESK_INFO() function and assign it to several CHANNEL variables, including the context to handle the call with, where we perform a Goto() to the context we have been assigned (which controls our outbound access).

We will make use of the HOTDESK_PHONE_STATUS() dialplan function, which you can define in *func_odbc.conf* like so:

```
[PHONE_STATUS]
prefix=HOTDESK
dsn=asterisk
readsql=SELECT extension FROM ast_hotdesk WHERE status = '1'
readsql+= AND location = '${ARG1}'
```

If we try to dial a number that is not handled by our context (or one of the transitive contexts—i.e., international contains long distance, which also contains local), the built-in extension i is executed, which plays back a message stating that the action cannot be performed and hangs up the call:

```
[hotdesk_outbound]
exten => _X.,1,NoOp()
    same => n,Set(LOCATION=${CUT(CHANNEL,/,2)})
    same => n,Set(LOCATION=${CUT(LOCATION,-,1)})
    same => n,Set(WHO=${HOTDESK_PHONE_STATUS(${LOCATION})})
    same => n,GotoIf($[${ISNULL(${WHO})}]?no_outgoing,1)
    same => n,Set(${WHO}_CID_NAME=${HOTDESK_INFO(cid_name,${WHO})})
    same => n,Set(${WHO}_CID_NUMBER=${HOTDESK_INFO(cid_number,${WHO})})
    same => n,Set(${WHO}_CONTEXT=${HOTDESK_INFO(context,${WHO})})
    same => n,Goto(${${WHO}_CONTEXT},${EXTEN},1)

[international]
exten => _011.,1,NoOp()
    same => n,Set(E=${EXTEN})
    same => n,Goto(outgoing,call,1)

exten => i,1,NoOp()
    same => n,Playback(silence/2&sorry-cant-let-you-do-that2)
    same => n,Hangup()

include => longdistance

[longdistance]
exten => _1NXXNXXXXXX,1,NoOp()
    same => n,Set(E=${EXTEN})
    same => n,Goto(outgoing,call,1)

exten => _NXXNXXXXXX,1,Goto(1${EXTEN},1)

exten => i,1,NoOp()
    same => n,Playback(silence/2&sorry-cant-let-you-do-that2)
    same => n,Hangup()

include => local

[local]
exten => _416NXXXXXX,1,NoOp()
    same => n,Set(E=${EXTEN})
    same => n,Goto(outgoing,call,1)

exten => i,1,NoOp()
    same => n,Playback(silence/2&sorry-cant-let-you-do-that2)
    same => n,Hangup()
```

If the call is allowed to be executed, it is sent to the [outgoing] context for processing and the caller ID name and number are set with the CALLERID() function. The call is then placed via the SIP channel using the service_provider we created in the *sip.conf* file:

```
[outgoing]
exten => call,1,NoOp()
    same => n,Set(CALLERID(name)=${${WHO}_CID_NAME})
    same => n,Set(CALLERID(number)=${${WHO}_CID_NUMBER})
    same => n,Dial(SIP/service_provider/${E})
```

```
    same => n,Playback(silence/2&pls-try-call-later)
    same => n,Hangup()
```

Our `service_provider` might look something like this in *sip.conf*:

```
[service_provider]
type=friend
host=switch1.service_provider.net
defaultuser=my_username
fromuser=my_username
secret=welcome
context=incoming
canreinvite=no
disallow=all
allow=ulaw
```

Now that we've implemented a fairly complex feature in the dialplan with the help of `func_odbc` to retrieve and store data in a remote relational database, hopefully you're starting to get why we think this is so cool. With a handful of self-defined dialplan functions in the *func_odbc.conf* file and a couple of tables in a database, we can create some fairly rich applications!

How many things have you just thought of that you could apply `func_odbc` to?

Using Realtime

The Asterisk Realtime Architecture (ARA) enables you to store the configuration files (that would normally be found in */etc/asterisk*) and their configuration options in a database table. There are two types of realtime: *static* and *dynamic*.

The static version is similar to the traditional method of reading a configuration file, except that the data is read from the database instead.

The dynamic realtime method, which loads and updates the information as it is required, is used for things such as SIP/IAX2 user and peer objects and voicemail.

Making changes to static information requires a reload, just as if you had changed a text file on the system, but dynamic information is polled by Asterisk as needed, so no reload is required when changes are made to this data. Realtime is configured in the *extconfig.conf* file located in the */etc/asterisk* directory. This file tells Asterisk what to load from the database and where to load it from, allowing certain files to be loaded from the database and other files to be loaded from the standard configuration files.

Static Realtime

Static realtime is useful when you want to load from a database the configuration that you would normally place in the configuration files in */etc/asterisk*. The same rules that apply to flat files on your system still apply when using static realtime. For example, after making changes to the configuration you must either run the *reload* command

---

from the Asterisk CLI, or reload the module associated with the configuration file (e.g., using *module reload chan_sip.so*).

When using static realtime, we tell Asterisk which files we want to load from the database using the following syntax in the *extconfig.conf* file:

```
; /etc/asterisk/extconfig.conf
filename.conf => driver,database[,table]
```

 If the table name is not specified, Asterisk will use the name of the file as the table name instead.

The static realtime module uses a specifically formatted table to read the configuration of static files in from the database. Table 16-2 illustrates the columns as they should be defined in your database:

Table 16-2. Table layout and description of ast_config

Column name	Column type	Description
id	Serial, auto-incrementing	An auto-incrementing unique value for each row in the table.
cat_metric	Integer	The weight of the category within the file. A lower metric means it appears higher in the file (see the sidebar on page 370).
var_metric	Integer	The weight of an item within a category. A lower metric means it appears higher in the list (see the sidebar on page 370). This is useful for things like codec order in *sip.conf*, or *iax.conf* where you want disallow=all to appear first (metric of 0), followed by allow=ulaw (metric of 1), then allow=gsm (metric of 2).
filename	Varchar 128	The filename the module would normally read from the hard drive of your system (e.g., *musiconhold.conf*, *sip.conf*, *iax.conf*, etc.).
category	Varchar 128	The section name within the file, such as [general]. Do not include the square brackets around the name when saving to the database.
var_name	Varchar 128	The option on the left side of the equals sign (e.g., disallow is the var_name in disallow=all).
var_val	Varchar 128	The value of an option on the right side of the equals sign (e.g., all is the var_val in disallow=all).
commented	Integer	Any value other than 0 will evaluate as if it were prefixed with a semicolon in the flat file (commented out).

A Word About Metrics

The metrics in static realtime are used to control the order in which objects are read into memory. Think of the cat_metric and var_metric as the original line numbers in the flat file. A higher *cat_metric* is processed first, because Asterisk matches categories from bottom to top. Within a category, through, a lower var_metric is processed first, because Asterisk processes the options top-down (e.g., disallow=all should be set to a value lower than the allow's value within a category to make sure it is processed first).

A simple file we can load from static realtime is the *musiconhold.conf*‖ file. Let's start by moving this file to a temporary location:

```
$ cd /etc/asterisk
$ mv musiconhold.conf musiconhold.conf.old
```

In order for the classes to be removed from memory, we need to restart Asterisk. Then we can verify that our classes are blank by running *moh show classes*:

```
*CLI> core restart now
*CLI> moh show classes
*CLI>
```

Let's put the [default] class back into Asterisk, but now we'll load it from the database. Connect to your database and execute the following INSERT statements:

```
> INSERT INTO ast_config (filename,category,var_name,var_val)
  VALUES ('musiconhold.conf','default','mode','files');

> INSERT INTO ast_config (filename,category,var_name,var_val)
  VALUES ('musiconhold.conf','default','directory','/var/lib/asterisk/moh');
```

You can verify that your values have made it into the database by running a SELECT statement:

```
asterisk=# SELECT filename,category,var_name,var_val FROM ast_config;

    filename       | category       | var_name      | var_val
-------------------+----------------+---------------+-----------------------
 musiconhold.conf  | default        | mode          | files
 musiconhold.conf  | default        | directory     | /var/lib/asterisk/moh
(2 rows)
```

There's one last thing to modify in the *extconfig.conf* file in the */etc/asterisk* directory to tell Asterisk to get the data for *musiconhold.conf* from the database using the ODBC connection. The first column states that we're using the ODBC drivers to connect (*res_odbc.conf*) and that the connection name is asterisk (as defined with [asterisk] in *res_odbc.conf*). Add the following line to the end of the *extconfig.conf* file, and then save it:

‖ The *musiconhold.conf* file can also be loaded via dynamic realtime, but we're using it statically as it's a simple file that makes a good example.

```
musiconhold.conf => odbc,asterisk,ast_config
```

Then connect to the Asterisk console and perform a reload:

```
*CLI> module reload res_musiconhold.so
```

You can now verify that your music on hold classes are loading from the database by running *moh show classes*:

```
*CLI> moh show classes
Class: general
        Mode: files
        Directory: /var/lib/asterisk/moh
```

And there you go: *musiconhold.conf* loaded from the database. If you have issues with the reload of the module loading the data into memory, try restarting Asterisk. You can perform the same steps in order to load other flat files from the database, as needed.

Dynamic Realtime

The dynamic realtime system is used to load objects that may change often, such as SIP/IAX2 users and peers, queues and their members, and voicemail messages. Likewise, when new records are likely to be added on a regular basis, we can utilize the power of the database to let us load this information on an as-needed basis.

All of realtime is configured in the */etc/asterisk/extconfig.conf* file, but dynamic realtime has well-defined configuration names. Defining something like SIP peers is done with the following format:

```
; extconfig.conf
sippeers => driver,database[,table]
```

The table name is optional. If it is omitted, Asterisk will use the predefined name (i.e., sippeers) to identify the table in which to look up the data.

 Remember that we have both SIP peers and SIP users: peers are endpoints we send calls to, and a user is something we receive calls from. A friend is shorthand that defines both.

In our example, we'll be using the ast_sippeers table to store our SIP peer information. So, to configure Asterisk to load all SIP peers from our database using realtime, we would define something like this:

```
; extconfig.conf
sippeers => odbc,asterisk,ast_sipfriends
```

To also load our SIP users from the database, we would define the sipusers object like so:

```
sipusers => odbc,asterisk,ast_sipfriends
```

You may have noticed we used the same table for both the `sippeers` and `sipusers`. This is because there will be a `type` field (just as if we were defining the type in the *sip.conf* file) that will let us define a type of `user`, `peer`, or `friend`. If you unload *chan_sip.so* and then load it back into memory (i.e., using *module unload chan_sip.so* followed by *module load chan_sip.so*) after configuring *extconfig.conf*, you will be greeted with some warnings telling you which columns you're missing for the realtime table. Since we've enabled `sippeers` and `sipusers` in *extconfig.conf*, we will get the following on the console (which has been trimmed due to space requirements):

```
WARNING: Realtime table ast_sipfriends@asterisk requires column
'name', but that column does not exist!

WARNING: Realtime table ast_sipfriends@asterisk requires column
'ipaddr', but that column does not exist!

WARNING: Realtime table ast_sipfriends@asterisk requires column
'port', but that column does not exist!

WARNING: Realtime table ast_sipfriends@asterisk requires column
'regseconds', but that column does not exist!

WARNING: Realtime table ast_sipfriends@asterisk requires column
'defaultuser', but that column does not exist!

WARNING: Realtime table ast_sipfriends@asterisk requires column
'fullcontact', but that column does not exist!

WARNING: Realtime table ast_sipfriends@asterisk requires column
'regserver', but that column does not exist!

WARNING: Realtime table ast_sipfriends@asterisk requires column
'useragent', but that column does not exist!

WARNING: Realtime table ast_sipfriends@asterisk requires column
'lastms', but that column does not exist!
```

As you can see, we are missing several columns from the table `ast_sipfriends`, which we've defined as connecting to the `asterisk` object as defined in *res_odbc.conf*. The next step is to create our `ast_sipfriends` table with all the columns listed by the warning messages, in addition to the following: the `type` column, which is required to define users, peers, and friends; the `secret` column, which is used for setting a password; and the `host` column, which allows us to define whether the peer is dynamically registering to us or has a static IP address. Table 16-3 lists all of the columns that should appear in our table, and their types.

Table 16-3. Minimal sippeers/sipusers realtime table

Column name	Column type
type	Varchar 6
name	Varchar 128

Column name	Column type
secret	Varchar 128
context	Varchar 128
host	Varchar 128
ipaddr	Varchar 128
port	Varchar 5
regseconds	Bigint
defaultuser	Varchar 128
fullcontact	Varchar 128
regserver	Varchar 128
useragent	Varchar 128
lastms	Integer

For each peer you want to register, you need to insert data in the columns type, name, secret, context, host, and defaultuser. The rest of the columns will be populated automatically when the peer registers.

The port, regseconds, and ipaddr fields are required to let Asterisk store the registration information for the peer so it can determine where to send the calls. (Note that if the peer is static, you will have to populate the ipaddr field yourself.) The port field is optional and defaults to the standard port defined in the [general] section, and the regseconds will remain blank. Table 16-4 lists some sample values that we'll use to populate our ast_sipfriends table.

Table 16-4. Example information used to populate the ast_sipfriends table

Column name	Value
type	friend
name	0000FFFF0008
defaultuser	0000FFFF0008
host	dynamic
secret	welcome
context	LocalSets

Prior to registering your peer, though, you need to enable realtime caching in *sip.conf*. Otherwise, the peer will not be loaded into memory, and the registration will not be remembered. If your peers only place calls and don't need to register to your system, you don't need to enable realtime caching because the peers will be checked against the database each time they place a call. However, if you load your peers into memory, the database will only need to be contacted on initial registration, and after the registration expires.

Additional options in *sip.conf* exist for realtime peers. These are defined in the [general] section and described in Table 16-5.

Table 16-5. Realtime options in sip.conf

Configuration option	Description
rtcachefriends	Caches peers in memory on an as-needed basis after they have contacted the server. That is, on Asterisk start, the peers are not loaded into memory automatically; only after a peer has contacted the server (e.g., via a registration or phone call) is it loaded in memory. Values are yes or no.
rtsavesysname	When a peer registers to the system, saves the systemname (as defined in *asterisk.conf*) into the regserver field within the database. (See "Setting the systemname for Globally Unique IDs" on page 375 for more information.) Using regserver is useful when you have multiple servers registering peers to the same table. Values are yes or no.
rtupdate	Sends registration information such as the IP address, the origination port, the registration period, and the username of the user-agent to the database when a peer registers to Asterisk. Values are yes or no, and the default is yes.
rtautoclear	Automatically expires friends on the same schedule as if they had just registered. This causes a peer to be removed from memory when the registration period has expired, until that peer is requested again (e.g., via registration or placing a call). Values are yes, no, or an integer value that causes the peers to be removed from memory after that number of seconds instead of the registration interval.
ignoreregexpire	When enabled, peers are not removed from memory when the registration period expires. Instead, the information is left in memory so that if a call is requested to an endpoint that has an expired registration, the last known information (IP address, port, etc.) will be tried.

After enabling rtcachefriends=yes in *sip.conf* and reloading *chan_sip.so* (using *module reload chan_sip.so*), you can register your peer to Asterisk using realtime, and the peer should then be populated into memory. You will be able to verify this by executing the *sip show peers* command on the Asterisk console:

```
Name/username            Host       Dyn Port Status      Realtime
0000FFFF0008/0000FFFF0008  172.16.0.160  D  5060 Unmonitored Cached RT
```

If you were to look at the table in the database directly, you would see something like this:

```
+--------+---------------+---------+-----------+---------+--------------+
| type   | name          | secret  | context   | host    | ipaddr       |
+--------+---------------+---------+-----------+---------+--------------+
| friend | 0000FFFF0008  | welcome | LocalSets | dynamic | 172.16.0.160 |
+--------+---------------+---------+-----------+---------+--------------+

+------+------------+--------------+------------------------------------+
| port | regseconds | defaultuser  | fullcontact                        |
+------+------------+--------------+------------------------------------+
| 5060 | 1283928895 | 0000FFFF0008 | sip:0000FFFF0008@172.16.0.160:52722 |
+------+------------+--------------+------------------------------------+

+-----------+----------------+--------+
| regserver | useragent      | lastms |
+-----------+----------------+--------+
| NULL      | Zoiper rev.6739 |     0 |
+-----------+----------------+--------+
```

There are many more options for that we can define for SIP friends, such as the caller ID; adding that information is as simple as adding a `callerid` column to the table. See the *sip.conf.sample* file for more options that can be defined for SIP friends.

Storing Call Detail Records (CDRs)

Call detail records (CDRs) contain information about calls that have passed through your Asterisk system. They are discussed further in Chapter 24. Storing CDRs is a popular use of databases in Asterisk, because it makes them easier to manage (for example, you can keep track of many Asterisk systems in a single table). Also, by placing records into a database you open up many possibilities, including building your own web interface for tracking statistics such as call usage and most-called locations, billing, or phone company invoice verification.

Setting the systemname for Globally Unique IDs

A CDR consists of a unique identifier and several fields of information about the call (including the source and destination channel, length of call, last application executed, and so forth). In a clustered set of Asterisk boxes, it is theoretically possible to have duplication among unique identifiers, since each Asterisk system considers only itself. To address this, we can automatically append a system identifier to the front of the unique IDs by adding an option to *etc/asterisk/asterisk.conf*. For each of your boxes, set an identifier by adding something like:

```
[options]
systemname=toronto
```

The best way to store your call detail records is via the `cdr_adaptive_odbc` module. This module allows you to choose which columns of data built into Asterisk are stored in your table, and permits you to add additional columns that you can populate with the `CDR()` dialplan function. You can even store different parts of CDR data to different tables and databases, if that is required.

More information about the standard CDR columns in Asterisk is available in Table 24-2. You can define all or any subset of these records in the database, and Asterisk will work around what is available. You can also add additional columns to store other data relevant to the calls. For example, if you wanted to implement least cost routing (LCR), you could add columns for route, per-minute cost, and per-minute rate. Once you've added those columns, they can be populated via the dialplan by using the `CDR()` function (e.g., `Set(CDR(per_minute_rate)=0.01)`).

After creating your table in the database (which we'll assume you've called cdr), you need to configure the *cdr_adaptive_odbc.conf* file in the */etc/asterisk/* folder. The following example will utilize the asterisk connection we've defined in *res_odbc.conf* and store the data in the cdr table:

```
; cdr_adaptive_odbc.conf
[adaptive_connection]
connection=asterisk
table=cdr
```

Yes, really, that's all you need. After configuring *cdr_adaptive_odbc.conf*, just reload the *cdr_adaptive_odbc.so* module from the Asterisk console by running *module reload cdr_adaptive_odbc.so*. You can verify that the Adaptive ODBC backend has been loaded by running *cdr show status*:

```
*CLI> cdr show status

Call Detail Record (CDR) settings
----------------------------------
  Logging:               Enabled
  Mode:                  Simple
  Log unanswered calls:  No

* Registered Backends
  -------------------
    cdr-syslog
    Adaptive ODBC
    cdr-custom
    csv
    cdr_manager
```

Now place a call that gets answered (e.g., using Playback(), or Dial()ing another channel and answering it). You should get some CDRs stored into your database. You can check by running SELECT * FROM CDR; from your database console.

With the basic CDR information stored into the database, you might want to add some additional information to the cdr table, such as the route rate. You can use the ALTER TABLE directive to add a column called route_rate to the table:

```
sql> ALTER TABLE cdr ADD COLUMN route_rate varchar(10);
```

Now reload the *cdr_adaptive_odbc.so* module from the Asterisk console:

```
*CLI> module reload cdr_adaptive_odbc.so
```

and populate the new column from the Asterisk dialplan using the CDR() function, like so:

```
exten => _NXXNXXXXXX,1,Verbose(1,Example of adaptive ODBC usage)
    same => n,Set(CDR(route_rate)=0.01)
    same => n,Dial(SIP/my_itsp/${EXTEN})
    same => n,Hangup()
```

After the alteration to your database and dialplan, you can place a call and then look at your CDRs. You should see something like the following:

```
+--------------+----------+---------+------------+
| src          | duration | billsec | route_rate |
+--------------+----------+---------+------------+
| 0000FFFF0008 | 37       | 30      | 0.01       |
+--------------+----------+---------+------------+
```

You now have enough information to calculate how much the call should have cost you, which enables you to either bill customers or check your records against what the phone company is sending you, so you can do monthly auditing of your phone bills.

Additional Configuration Options for cdr_adaptive_odbc.conf

Some extra configuration options exist in the *cdr_adaptive_odbc.conf* file that may be useful. The first is that you can define multiple databases or tables to store information into, so if you have multiple databases that need the same information, you can simply define them in *res_odbc.conf*, create tables in the databases, and then refer to them in separate sections of the configuration:

```
[mysql_connection]
connection=asterisk_mysql
table=cdr

[mssql_connection]
connection=production_mssql
table=call_records
```

 If you specify multiple sections using the same connection and table, you will get duplicate records.

Beyond just configuring multiple connections and tables (which of course may or may not contain the same information; the CDR module we're using is adaptive to situations like that), we can define aliases for the built-in variables, such as accountcode, src, dst, billsec, etc.

If we were to add aliases for column names for our MS SQL connection, we might alter our connection definition like so:

```
[mssql_connection]
connection=production_mssql
table=call_records
alias src => Source
alias dst => Destination
alias accountcode => AccountCode
alias billsec => BillableTime
```

In some situations you may specify a connection where you only want to log calls from a specific source, or to a specific destination. We can do this with filters:

```
[logging_for_device_0000FFFF0008]
connection=asterisk_mysql
table=cdr_for_0000FFFF0008
filter src => 0000FFFF0008
```

If you need to populate a certain column with information based on a section name, you can set it statically with the **static** option, which you may utilize with the **filter** option:

```
[mysql_connection]
connection=asterisk_mysql
table=cdr

[filtered_mysql_connection]
connection=asterisk_mysql
table=cdr
filter src => 0000FFFF0008
static "DoNotCharge" => accountcode
```

 In the preceding example you will get duplicate records in the same table, but all the information will be the same except for the populated accountcode column, so you should be able to filter it out using SQL.

ODBC Voicemail

Asterisk enables you to store voicemail inside the database using the ODBC connector. This is useful in a clustered environment where you want to abstract the voicemail data from the local system so that multiple Asterisk boxes have access to the same data. Of course, you have to take into consideration that you are centralizing a part of Asterisk, and you need to take actions to protect that data, such as making regular backups and possibly clustering the database backend using replication.

Asterisk stores each voicemail message inside a Binary Large Object (BLOB). When retrieving the data, it pulls the information out of the BLOB and temporarily stores it on the hard drive while it is being played back to the user. Asterisk then removes the BLOB and the record from the database when the user deletes the voicemail. Many databases, such as MySQL, contain native support for BLOBs, but as you'll see, with PostgreSQL a couple of extra steps are required to utilize this functionality that we'll explore in this section. When you're done, you'll be able to record, play back, and delete voicemail data from the database just as if it were stored on the local hard drive.

This section builds upon previous configuration sections in this chapter. If you have not already done so, be sure to follow the steps in the sections "Installing PostgreSQL for CentOS" on page 342 and "Installing and Configuring ODBC" on page 346 before continuing. In the latter section, be sure you have enabled `ODBC_STORAGE` in the menuselect system under *Voicemail Options*.

Alternate Centralization Method

Storing voicemail in a database is one way to centralize voicemail. Another method is to run a standalone voicemail server, as we discussed in Chapter 8.

Creating the Large Object Type for PostgreSQL

While MySQL has a BLOB (Binary Large OBject) type, we have to tell PostgreSQL how to handle large objects. This includes creating a trigger to clean up the data when we delete from the database a record that references a large object.

Connect to the database as the *asterisk* user from the console:

```
$ psql -h localhost -U asterisk asterisk
Password:
```

 You must be a *superuser* to execute the following code. Also, if you use the *postgres* user to create the table, you will need to use the `ALTER TABLE` SQL directive to change the owner to the *asterisk* user.

At the PostgreSQL console, run the following script to create the large object type:

```
CREATE FUNCTION loin (cstring) RETURNS lo AS 'oidin' LANGUAGE internal
IMMUTABLE STRICT;

CREATE FUNCTION loout (lo) RETURNS cstring AS 'oidout' LANGUAGE internal
IMMUTABLE STRICT;

CREATE FUNCTION lorecv (internal) RETURNS lo AS 'oidrecv' LANGUAGE internal
IMMUTABLE STRICT;

CREATE FUNCTION losend (lo) RETURNS bytea AS 'oidrecv' LANGUAGE internal
IMMUTABLE STRICT;

CREATE TYPE lo ( INPUT = loin, OUTPUT = loout, RECEIVE = lorecv, SEND = losend,
INTERNALLENGTH = 4, PASSEDBYVALUE );

CREATE CAST (lo AS oid) WITHOUT FUNCTION AS IMPLICIT;
CREATE CAST (oid AS lo) WITHOUT FUNCTION AS IMPLICIT;
```

We'll be making use of the PostgreSQL procedural language called pgSQL/PL to create a function. This function will be called from a trigger that gets executed whenever we modify or delete a record in the table used to store voicemail messages. This is so the data is cleaned up and not left as an orphan in the database:

```
CREATE FUNCTION vm_lo_cleanup() RETURNS "trigger"
    AS $$
    declare
      msgcount INTEGER;
    begin
      -- raise notice 'Starting lo_cleanup function for large object with oid
        %',old.recording;
      -- If it is an update action but the BLOB (lo) field was not changed,
        don't do anything
      if (TG_OP = 'UPDATE') then
        if ((old.recording = new.recording) or (old.recording is NULL)) then
          raise notice 'Not cleaning up the large object table,
          as recording has not changed';
          return new;
        end if;
      end if;
      if (old.recording IS NOT NULL) then
        SELECT INTO msgcount COUNT(*) AS COUNT FROM voicemessages WHERE recording
        = old.recording;
        if (msgcount > 0) then
          raise notice 'Not deleting record from the large object table, as object
          is still referenced';
          return new;
        else
          perform lo_unlink(old.recording);
          if found then
            raise notice 'Cleaning up the large object table';
            return new;
          else
            raise exception 'Failed to clean up the large object table';
            return old;
          end if;
        end if;
      else
        raise notice 'No need to clean up the large object table,
        no recording on old row';

        return new;
      end if;
    end$$
    LANGUAGE plpgsql;
```

We're going to create a table called **voicemessages** where the voicemail information will be stored:

```
CREATE TABLE voicemessages
(
  uniqueid serial PRIMARY KEY,
  msgnum int4,
  dir varchar(80),
```

```
    context varchar(80),
    macrocontext varchar(80),
    callerid varchar(40),
    origtime varchar(40),
    duration varchar(20),
    mailboxuser varchar(80),
    mailboxcontext varchar(80),
    recording lo,
    label varchar(30),
    "read" bool DEFAULT false,
    flag varchar(10)
);
```

And now we need to associate a trigger with our newly created table in order to perform cleanup whenever we change or delete a record in the **voicemessages** table:

```
CREATE TRIGGER vm_cleanup AFTER DELETE OR UPDATE ON voicemessages FOR EACH ROW
EXECUTE PROCEDURE vm_lo_cleanup();
```

ODBC Voicemail Storage Table Layout

We'll be utilizing the **voicemessages** table for storing our voicemail information in an ODBC-connected database. Table 16-6 describes the table configuration for ODBC voicemail storage. If you're using a PostgreSQL database, the table definition and large object support were configured in the preceding section.

Table 16-6. ODBC voicemail storage table layout

Column name	Column type
uniqueid	Serial, primary key
dir	Varchar 80
msgnum	Integer
recording	BLOB (Binary Large OBject)
context	Varchar 80
macrocontext	Varchar 80
callerid	Varchar 40
origtime	Varchar 40
duration	Varchar 20
mailboxuser	Varchar 80
mailboxcontext	Varchar 80
label	Varchar 30
read	Boolean, default false[a]
flag	Varchar 10

[a] read is a reserved word in both MySQL and PostgreSQL (and likely other databases), which means you need to escape the column name when you create it. In MySQL this is done with backticks (`) around the word read when you create the table, and in PostgreSQL with double quotes ("). In MS SQL you would use square brackets, e.g., [read].

Configuring voicemail.conf for ODBC Storage

There isn't much to add to the *voicemail.conf* file to enable the ODBC voicemail storage. In fact, it's only three lines! Normally, you probably have multiple format types defined in the [general] section of *voicemail.conf*, but we need to set this to a single format because we can only save one file (format) to the database. The *wav49* format is a compressed WAV file format that should be playable on both Linux and Microsoft Windows desktops.

The odbcstorage option points at the name you defined in the *res_odbc.conf* file (if you've been following along in this chapter, then we called it *asterisk*). The odbctable option refers to the table where voicemail information should be stored. In the examples in this chapter we use the table named voicemessages:

```
[general]
format=wav49
odbcstorage=asterisk
odbctable=voicemessages
```

You may want to create a separate voicemail context, or you can utilize the default voicemail context. Alternatively, you can skip creating a new user and use an existing user, such as 0000FFFF0001. We'll define the mailbox in the default voicemail context like so:

```
[default]
1000 => 1000,J.P. Wiser
```

 You can also use the voicemail definition in *extconfig.conf* to load your users from the database. See "Dynamic Realtime" on page 371 for more information about setting up certain module configuration options in the database, and "Static Realtime" on page 368 for details on loading the rest of the configuration file.

Now connect to your Asterisk console and unload then reload the *app_voicemail.so* module:

```
*CLI> module unload app_voicemail.so
  == Unregistered application 'VoiceMail'
  == Unregistered application 'VoiceMailMain'
  == Unregistered application 'MailboxExists'
  == Unregistered application 'VMAuthenticate'

*CLI> module load app_voicemail.so
  Loaded /usr/lib/asterisk/modules/app_voicemail.so =>
(Comedian Mail (Voicemail System))
  == Registered application 'VoiceMail'
  == Registered application 'VoiceMailMain'
  == Registered application 'MailboxExists'
  == Registered application 'VMAuthenticate'
  == Parsing '/etc/asterisk/voicemail.conf': Found
```

Then verify that your new mailbox loaded successfully:

```
*CLI> voicemail show users for default
Context   Mbox  User                      Zone      NewMsg
default   1000  J.P. Wiser                              0
```

Testing ODBC Voicemail

Let's create some simple dialplan logic to leave and retrieve some voicemail from our test voicemail box. You can use the simple dialplan logic that follows (or, of course, any voicemail delivery and retrieval functionality you defined earlier in this book):

```
[odbc_vm_test]
exten => 100,1,VoiceMail(1000@default)       ; leave a voicemail
exten => 200,1,VoiceMailMain(1000@default)   ; retrieve a voicemail
```

Once you've updated your *extensions.conf* file, be sure to reload the dialplan:

```
*CLI> dialplan reload
```

You can either include the odbc_vm_test context into a context accessible by an existing user, or create a separate user to test with. If you wish to do the latter, you could define a new SIP user in *sip.conf* like so (this will work assuming the phone is on the local LAN):

```
[odbc_test_user]
type=friend
secret=supersecret
context=odbc_vm_test
host=dynamic
qualify=yes
disallow=all
allow=ulaw
allow=gsm
```

 One of the ways that unsavory folks get into systems is via test users that are not immediately removed from the system after testing. Whenever you're utilizing a test extension, you should be doing it on a system that is removed from the Internet, or at the very least, place it into a context that does not have access to outbound dialing and has a strong password.

Don't forget to reload the SIP module:

```
*CLI> module reload chan_sip.so
```

And verify that the SIP user exists:

```
*CLI> sip show users like odbc_test_user
Username        Secret          Accountcode    Def.Context   ACL  NAT
odbc_test_user  supersecret                    odbc_vm_test  No   RFC3581
```

Then configure your phone or client with the username *odbc_test_user* and password *<supersecret>*, and place a call to extension 100 to leave a voicemail. If successful, you should see something like:

```
    -- Executing VoiceMail("SIP/odbc_test_user-10228cac", "1000@default") in new
       stack
    -- Playing 'vm-intro' (language 'en')
    -- Playing 'beep' (language 'en')
    -- Recording the message
    -- x=0, open writing:  /var/spool/asterisk/voicemail/default/1000/tmp/dlZunm
       format: wav49, 0x101f6534
    -- User ended message by pressing #
    -- Playing 'auth-thankyou' (language 'en')
  == Parsing '/var/spool/asterisk/voicemail/default/1000/INBOX/msg0000.txt': Found
```

 At this point you can check the database to verify that your data was successfully written. See the upcoming sections for more information.

Now that you've confirmed everything was stored in the database correctly, you can try listening to it via the VoiceMailMain() application by dialing extension 200:

```
*CLI>
    -- Executing VoiceMailMain("SIP/odbc_test_user-10228cac",
       "1000@default") in new stack
    -- Playing 'vm-password' (language 'en')
    -- Playing 'vm-youhave' (language 'en')
    -- Playing 'digits/1' (language 'en')
    -- Playing 'vm-INBOX' (language 'en')
    -- Playing 'vm-message' (language 'en')
    -- Playing 'vm-onefor' (language 'en')
    -- Playing 'vm-INBOX' (language 'en')
    -- Playing 'vm-messages' (language 'en')
    -- Playing 'vm-opts' (language 'en')
    -- Playing 'vm-first' (language 'en')
    -- Playing 'vm-message' (language 'en')
  == Parsing '/var/spool/asterisk/voicemail/default/1000/INBOX/msg0000.txt': Found
```

Verifying binary data stored in PostgreSQL

To make sure the recording really did make it into the database, use the *psql* application:

```
$ psql -h localhost -U asterisk asterisk
Password:
```

then run a SELECT statement to verify that you have some data in the voicemessages table:

```
localhost=# SELECT uniqueid,dir,callerid,mailboxcontext,recording FROM voicemessages;
uniqueid | dir                                                    | callerid
---------+--------------------------------------------------------+-------------
1        | /var/spool/asterisk/voicemail/default/1000/INBOX | +18005551212


| mailboxcontext | recording |
+----------------+-----------+
| default        | 47395     |
(1 row)
```

If the recording was placed in the database, you should get a row back. You'll notice that the **recording** column contains a number (which will most certainly be different from that listed here), which is really the object ID of the large object stored in a system table. You can verify that the large object exists in this system table with the lo_list command:

```
localhost=# \lo_list
    Large objects
  ID   | Description
-------+-------------
 47395 |
(1 row)
```

What you're verifying is that the object ID in the voicemessages table matches that listed in the large object system table. You can also pull the data out of the database and store it to the hard drive:

```
localhost=# \lo_export 47395 /tmp/voicemail-47395.wav
lo_export
```

Then verify the audio with your favorite audio application, such as *play*:

```
$ play /tmp/voicemail-47395.wav

Input Filename : /tmp/voicemail-47395.wav
Sample Size    : 8-bits
Sample Encoding: wav
Channels       : 1
Sample Rate    : 8000

Time: 00:06.22 [00:00.00] of 00:00.00 (  0.0%) Output Buffer: 298.36K

Done.
```

Verifying binary data stored in MySQL

To verify that your data is being written correctly, you can use the *mysql* application to log into your database and export the voicemail recording to a file:

```
$ mysql -u asterisk -p asterisk
Enter password:
```

Once logged into the database, you can use a SELECT statement to dump the contents of the recording to a file. First, though, make sure you have at least a single recording in your voicemessages table:

```
mysql> SELECT uniqueid, msgnum, callerid, mailboxuser, mailboxcontext, `read`
    -> FROM voicemessages;
```

uniqueid	msgnum	callerid	mailboxuser
1	0	"Leif Madsen" <100>	100
2	1	"Leif Madsen" <100>	100
3	2	"Leif Madsen" <100>	100
5	0	"Julie Bryant" <12565551111>	100

mailboxcontext	read
shifteight.org	0
shifteight.org	0
shifteight.org	0
default	0

 You can also add the recording column to the SELECT statement, but you'll end up with a lot of gibberish on your screen.

Having verified that you have data in your voicemessages table, you can export one of the recordings and play it back from the console.

```
mysql> SELECT recording FROM voicemessages WHERE uniqueid = '5'
    -> DUMPFILE '/tmp/voicemail_recording.wav';
```

 The user you're exporting data with needs to have the FILE permission in MySQL, which means it must have been granted ALL access. If you did not grant ALL privileges to the *asterisk* user, you will need to utilize the *root* user for file export.

Now exit the MySQL console, and use the *play* application from the console (assuming you have speakers and a sound card configured on your Asterisk system, which you might if you are going to use it for overhead paging), or copy the file to another system and listen to it there:

```
$ play /tmp/voicemail_recording.wav

voicemail_recording.wav:

 File Size: 7.28k     Bit Rate: 13.1k
  Encoding: GSM
  Channels: 1 @ 16-bit
Samplerate: 8000Hz
Replaygain: off
  Duration: 00:00:04.44

In:100%  00:00:04.44 [00:00:00.00] Out:35.5k [       |       ] Hd:4.4 Clip:0
Done.
```

Conclusion

In this chapter, we learned about several areas where Asterisk can integrate with a relational database. This is useful for systems where you need to start scaling by clustering multiple Asterisk boxes working with the same centralized information, or when you want to start building external applications to modify information without requiring a reload of the system (i.e., not requiring the modification of flat files).

Interactive Voice Response

*One day Alice came to a fork in the road and saw a
Cheshire cat in a tree. "Which road do I take?"
she asked.*

"Where do you want to go?" was his response.

"I don't know," Alice answered.

"Then," said the cat, "it doesn't matter."

—Lewis Carroll

In this chapter we will talk about IVR. If what you want is an automated attendant, we
have written a chapter for that as well (Chapter 15). The term IVR is often misused to
refer to an automated attendant, but the two are very different things.[*]

What Is IVR?

The purpose of an Interactive Voice Response (IVR) system is to take input from a
caller, perform an action based on that input (commonly, looking up data in an external
system such as a database), and return a result to the caller.[†] Traditionally, IVR systems
have been complex, expensive, and annoying to implement. Asterisk changes all that.

Asterisk blurs the lines between traditional PBXs and IVR systems. The
power and flexibility of the Asterisk dialplan results in a system where
nearly every extension could be considered an IVR in the traditional
sense of the term.

---

[*] We suspect this is because "IVR" is much easier to say than "automated attendant."

[†] In contrast to an auto attendant, the purpose of which is to route calls.

Components of an IVR

The most basic elements of an IVR are quite similar to those of an automated attendant, though the goal is different. We need at least one prompt to tell the caller what the IVR expects from him, a method of receiving input from the caller, logic to verify that the caller's response is valid input, logic to determine what the next step of the IVR should be, and finally, a storage mechanism for the responses, if applicable. We might think of an IVR as a decision tree, although it need not have any branches. For example, a survey may present exactly the same set of prompts to each caller, regardless of what choices the callers make, and the only routing logic involved is whether the responses given are valid for the questions.

From the caller's perspective, every IVR needs to start with a prompt. This initial prompt will tell the caller what the IVR is for and ask the caller to provide the first input. We discussed prompts in the automated attendant in Chapter 15. Later, we'll create a dialplan that will allow you to better manage multiple voice prompts.

The second component of an IVR is a method for receiving input from the caller. Recall that in Chapter 15 we discussed the `Background()` and `WaitExten()` applications for receiving a new extension. While you could create an IVR using `Background()` and `WaitExten()`, it is generally easier and more practical to use the `Read()` application, which handles both the prompt and the capture of the response. The `Read()` application was designed specifically for use with IVR systems. Its syntax is as follows:

```
Read(variable[,filename[&filename2...]][,maxdigits][,option][,attempts][,timeout])
```

The arguments are described in Table 17-1.

Table 17-1. The Read() application

Argument	Purpose	
variable	The variable into which the caller's response is stored. It is best practice to give each variable in your IVR a name that is similar to the prompt associated with that variable. This will help later if, for business reasons or ease of use, you need to reorder the steps of the IVR. Naming your variables var1, var2, etc. may seem easy in the short term, but later in your life cycle it will make fixing bugs more difficult.	
prompt	A file (or list of files, joined together with the & character) to play for the caller, requesting input. Remember to omit the format extension on the end of each filename.	
maxdigits	The maximum number of characters to allow as input. In the case of yes/no and multiple choice questions, it's best practice to limit this value to 1. In the case of larger lengths, the caller may always terminate input by pressing the pound key.	
options	s (skip)	Exit immediately if the channel has not been answered.
	i (indication)	Rather than playing a prompt, play an indication tone of some sort (such as the dialtone).
	n (noanswer)	Read digits from the caller, even if the line is not yet answered.

Argument	Purpose
attempts	The number of times to play the prompt. If the caller fails to enter anything, the Read() application can automatically re-prompt the user. The default is one attempt.
timeout	The number of seconds the caller has to enter his input. The default value in Asterisk is 10 seconds, although it can be altered for a single prompt using this option, or for the entire session by assigning a value using the dialplan function TIMEOUT(response).

Once the input is received, it must be validated. If you do not validate the input, you are most likely going to find your callers complaining of an unstable application. It is not enough to handle the inputs you are expecting; you also need to handle inputs you do not expect. For example, callers may get frustrated and dial 0 when in your IVR; if you've done a good job, you will handle this gracefully and connect them to somebody who can help them, or provide a useful alternative. A well-designed IVR (just like any program) will try to anticipate every possible input and provide mechanisms to gracefully handle that input.

Once the input is validated, you can submit it to an external resource for processing. This could be done via a database query, a submission to a URI, an AGI program, or many other things. This external application should produce a result, which you will want to relay back to the caller. This could be a detailed result, such as "Your account balance is...," or a simple confirmation, such as "Your account has been updated." We can't think of any case where some sort of result returned to the caller is not required.

Sometimes the IVR may have multiple steps, and therefore a result might include a request for more information from the caller in order to move to the next step of the IVR application.

It is possible to design very complex IVR systems, with dozens or even hundreds of possible paths. We've said it before and we'll say it again: people don't like talking to your phone system, regardless of how clever it is. Keep your IVR simple for your callers, and they are much more likely to get some benefit from it.

A Perfectly Tasty IVR

An excellent example of an IVR that people love to use is one that many pizza delivery outfits use: when you call to place your order, an IVR looks up your caller ID and says "If you would like the exact same order as last time, press 1."

That's all it does, and it's perfect.

Obviously, these companies could design massively complex IVRs that would allow you to select each and every detail of your pie ("for seven-grain crust, press 7"), but how many drunken frat boys are going to successfully navigate that?

The best IVRs are the ones that require the least input from the caller. Mash that 1 button and your 'za is on its way! Woo hoo!

IVR Design Considerations

When designing your own IVR, there are some important things to keep in mind. We've put together this list of some things to do, and things not to do in your IVR.

Do

- Keep it simple.
- Have an option to dial 0 to reach a live person.
- Handle errors gracefully.

Don't

- Think that an IVR can completely replace people.
- Use your IVR to show people how clever you are.
- Try to replicate your website with an IVR.
- Bother building an IVR if you can't take numeric input. Nobody wants to have to spell her name on the dialpad of her phone.‡
- Force your callers to listen to advertising. Remember that they can hang up at any moment they wish.

Asterisk Modules for Building IVRs

The "front end" of the IVR (the parts that interact with the callers) can be handled in the dialplan. In theory, it might be possible to build an IVR system using the dialplan alone (perhaps with the *astdb* to store and retrieve data). In practice, your IVR is going to need to communicate with something external to Asterisk.

CURL

The CURL() dialplan function in Asterisk allows you to span entire web applications with a single line of dialplan code. We'll use it in our sample IVR later in this chapter.

While you'll find CURL() itself to be quite simple to use, the creation of the web application will require experience with web development.

func_odbc

Using func_odbc, it is possible to develop extremely complex applications in Asterisk using nothing more than dialplan code and database lookups. If you are not a strong

---

‡ Especially if it's something like Van Meggelen.

programmer but are very adept with Asterisk dialplans and databases, you'll love func_odbc just as much as we do. Check it out in Chapter 16.

AGI

The Asterisk Gateway Interface is such an important part of integrating external applications with Asterisk that we gave it its own chapter. You can find more information in Chapter 21.

AMI

The Asterisk Manager Interface is a socket interface that you can use to get configuration and status information, request actions to be performed, and get notified about things happening to calls. We've written an entire chapter on AMI, as well. You can find more information in Chapter 20.

A Simple IVR Using CURL

The GNU/Linux program cURL is useful for retrieving data from a URI. In Asterisk, CURL() is a dialplan function.

We're going to use CURL() as an example of what an extremely simple IVR can look like. We're going to request our external IP address from the website *http://www.what ismyip.org*.

 In reality, most IVR applications are going to be far more complex. Even most uses of CURL() will tend to be complex, since a URI can return a massive and highly variable amount of data, the vast majority of which will be incomprehensible to Asterisk. The point being that an IVR is not just about the dialplan; it is also very much about the external applications that are triggered by the dialplan, which are doing the real work of the IVR.

Before you can use CURL(), you have to ensure it is installed.

Installing the cURL Module

Installing cURL is easy. If it was not on your system when you last compiled Asterisk, after installing it you'll need to recompile Asterisk so that it can locate the cURL dependencies and compile the *func_curl.so* module.

On CentOS:

```
$ sudo yum -y install libcurl-devel
```

On Ubuntu:

```
$ sudo apt-get install libcurl4-openssl-dev
```

The Dialplan

The dialplan for our example IVR is very simple. The CURL() function will retrieve our IP address from *http://www.whatismyip.org*, and then SayAlpha() will speak the results to the caller:

```
exten => *764,1,Verbose(2, Run CURL to get IP address from whatismyip.org)
    same => n,Answer()
    same => n,Set(MyIPAddressIs=${CURL(http://www.whatismyip.org/)})
    same => n,SayAlpha(${MyIPAddressIs})
    same => n,Hangup()
```

The simplicity of this is impossibly cool. In a traditional IVR system, this sort of thing could take days to program.

A Prompt-Recording Application

In Chapter 15 we created a simple bit of dialplan to record prompts. It was fairly limited in that it only recorded one filename, and thus for each prompt the file needed to be copied before a new prompt could be recorded. Here, we expand upon that to create a complete menu for recording prompts:

```
[prompts]
exten => s,1,Answer
exten => s,n,Set(step1count=0) ; Initialize counters

; If we get no response after 3 times, we stop asking
    same => n(beginning),GotoIf($[${step1count} > 2]?end)
    same => n,Read(which,prompt-instructions,3)
    same => n,Set(step1count=$[${step1count} + 1])

; All prompts must be 3 digits in length
    same => n,GotoIf($[${LEN(${which})} < 3]?beginning)
    same => n,Set(step1count=0) ; We have a successful response, so reset our counters
    same => n,Set(step2count=0)

    same => n(step2),Set(step2count=$[${step2count} + 1])
    same => n,GotoIf($[${step2count} > 2]?beginning) ; No response after 3 tries

; If the file doesn't exist, then don't ask whether to play it
    same => n,GotoIf($[${STAT(f,${which}.wav)} = 0]?recordonly)
    same => n,Background(prompt-tolisten)

    same => n(recordonly),Background(prompt-torecord)
    same => n,WaitExten(10) ; Wait 10 seconds for a response
    same => n,Goto(step2)

exten => 1,1,Set(step2count=0)
```

```
    same => n,Background(${which})
    same => n,Goto(s,step2)

exten => 2,1,Set(step2count=0)
    same => n,Playback(prompt-waitforbeep)
    same => n,Record(${CHANNEL(uniqueid)}.wav)

    same => n(listen),Playback(${CHANNEL(uniqueid)})
    same => n,Set(step3count=0)
    same => n,Read(saveornot,prompt-1tolisten-2tosave-3todiscard,1)
    same => n,GotoIf($["${saveornot}" = "1"]?listen)
    same => n,GotoIf($["${saveornot}" = "2"]?saveit)
    same => n,System(rm -f /var/lib/asterisk/sounds/${CHANNEL(uniqueid)}.wav)
    same => n,Goto(s,beginning)

    same => n(saveit),System(mv -f ${CHANNEL(uniqueid)}.wav ${which}.wav)
    same => n,Playback(prompt-saved)
    same => n,Goto(s,beginning)
```

In this system, the name of the prompt is no longer descriptive, but rather it is a number. This means that you can record a far greater variety of prompts using the same mechanism, but the tradeoff is that your prompts will no longer have descriptive names.

Speech Recognition and Text-to-Speech

Although in most cases an IVR system presents prerecorded prompts to the caller and accepts input by way of the dialpad, it is also possible to: a) generate prompts artificially, popularly known as text-to-speech; and b) accept verbal inputs through a speech recognition engine.

While the concept of being able to have an intelligent conversation with a machine is something sci-fi authors have been promising us for many long years, the actual science of this remains complex and error-prone. Despite their amazing capabilities, computers are ill-suited to the task of appreciating the subtle nuances of human speech.

Having said that, it should be noted that over the last 50 years or so, amazing advances have been made in both text-to-speech and speech recognition. A well-designed system created for a very specific purpose can work very well indeed.

Despite what the marketing people will say, your computer still can't talk to you, and you need to bear this in mind if you are contemplating any sort of system that combines your telephone system with these technologies.

Text-to-Speech

Text-to-speech (also known as speech synthesis) requires that a system be able to artificially construct speech from stored data. While it would be nice if we could simply assign a sound to a letter and have the computer produce each sound as it reads the letters, the written English language is not totally phonetic.

While on the surface, the idea of a speaking computer is very attractive, the reality is that it has limited usefulness. More information about integration of text-to-speech with Asterisk can be found in Chapter 18.

Speech Recognition

As soon as we've convinced computers to talk to us, we will naturally want to be able to talk to them.§ Anyone who has tried to learn a foreign language can begin to recognize the complexity of teaching a computer to understand words; however, speech recognition also has to take into account the fact that before a computer can even attempt the task of understanding the words, it must first convert the audio into a digital format. This challenge is larger than one might at first think. For example, as humans we are naturally able to recognize speech as distinct from, say, the sound of a barking dog or a car horn. For a computer, this is a very complicated thing. Additionally, for a telephone-based speech recognition system, the audio that is received is always going to be of very low fidelity, and thus the computer will have that much less information to work with.‖

Asterisk does not have speech recognition built in, but there are many third-party speech recognition packages that integrate with Asterisk.

Conclusion

Asterisk has become extremely popular as an IVR platform. While the media only really pays attention to Asterisk as a "free PBX," the reality is that Asterisk is quietly taking the IVR industry by storm. Within any respectable-sized organization, it is very likely that the Linux system administrators are using Asterisk to solve telecom problems that previously were either unsolvable or impossibly expensive to solve. This is a stealthy revolution, but no less significant for its relative obscurity.

If you are in the IVR business, you need to get to know Asterisk.

---

§ Actually, most of us talk to our computers, but this is seldom polite.

‖ If the speech recognition has to happen from a cell phone in a noisy conference hall, it becomes near-impossible.

External Services

*Correct me if I'm wrong—the gizmo is connected to the
flingflang connected to the watzis, watzis connected to
the doo-dad connected to the ding dong.*

—Patrick B. Oliphant

Asterisk is pretty nifty all by itself, but one of the most powerful, industry-changing, revolutionary aspects of Asterisk is the sheer number of wonderful ways it may be connected to external applications and services. This is truly unprecedented in the world of telecom. In this chapter we'll explore some popular services and applications that you can integrate with your Asterisk system. Here are some of the external connections we've decided to cover (Asterisk has more, but our editor is waiting for us to finish this edition, which is already the largest Asterisk book yet):

- If you use LDAP in your network (such as with Active Directory), we'll show you how to load your SIP users from your LDAP services.

- For the person on the go with a dynamically changing calendar, we'll sample some ideas on how you can integrate Asterisk with your calendaring server (allowing for automatic call redirection based on your current status).

- If you're a fan of instant messaging, there is a section on how to communicate with Asterisk via the XMPP (Jabber) protocol.

- Skype fans? Asterisk has a channel for that. We'll show you how to get it going.

- If you want to tie your voicemail into your IMAP server, we'll take you through the basics.

- Want to teach your phone system to read? We'll cover the basics of text-to-speech.

There are many more external services that Asterisk can connect to, but these are the ones we feel will give you the best sense of what it takes to integrate an external service with Asterisk.

Calendar Integration

Asterisk can be integrated with several different kinds of calendar formats, such as iCal, CalDAV, MS Exchange (Exchange 2003), and MS Exchange Web Services (Exchange 2007 and later). Integrating Asterisk with your calendar gives you the ability to manipulate call routing based on your current calendar information. For example, if you're not going to be in your office for the afternoon it may make sense for people ringing your desk phone to be routed directly to your voicemail.

Another advantage to calendar integration is the ability to originate calls based on calendar information. For example, if you set up a meeting on your conference server, you can arrange to have a reminder call five minutes before the meeting starts, which then places you into the conference room. We think this type of flexibility and integration is pretty nifty and quite useful.

Compiling Calendaring Support into Asterisk

As there are several modules for calendaring support (allowing us to provide support for different backends, such as MS Exchange, CalDAV, iCal, etc.), you'll need to install the dependencies for the backends you want to support. This modularized setup has the advantage that you only need to install dependencies for the modules you need; also other backends can easily be integrated with the primary calendaring backend in the future.

Because of the different dependencies of each module, we need to check *menuselect* for what needs to be installed for each of the calendaring modules we wish to support. All modules require the neon development library, available from *http://www.webdav.org/ neon/*. res_calendar_ews (Exchange Web Services) requires version 0.29 or later, which means some distributions will require you to compile the neon library from source instead of using the precompiled package available from the distribution.

While the configuration for all the calendaring modules is similar, we'll be discussing CalDAV integration specifically since it is widely supported by a range of calendar software and servers.*

CentOS dependencies

Since all the modules require the neon library, we'll install that first. Be sure to append *.x86_64* to the end of the package name if installing on a 64-bit machine:

```
$ sudo yum install neon-devel
```

---

* And because the authors of this book do not have access to Exchange servers for testing. :)

If you are planning on compiling the `res_calendar_ews` module, you will need to have a version of neon greater than or equal to 0.29. Currently CentOS is shipping with 0.25, so you will have to compile the neon library and link to it from the *configure* script. This can be done via *./ configure --with-neon29=<path to neon>*.

The next step is to install the `libical-devel` dependency. Unfortunately, this module is not shipped with CentOS and requires a third-party repository (see "Third-Party Repositories" on page 46). In this case, we need to install `libical-devel` from the EPEL (Extra Packages for Enterprise Linux) repository:

```
$ sudo yum --enablerepo=epel install libical-devel
```

After installing our dependencies, we can run the *configure* script in our Asterisk source directory and enable both the `res_calendar` and `res_calendar_caldav` modules from within the *Resource Modules* section of *menuselect*.

Ubuntu dependencies

Because all the modules require the neon development library, we're going to install that first. On Ubuntu, you will typically be given several different versions (e.g., on 10.04 we have the option of *libneon* 2.5, 2.6, and 2.7). We're going to install the latest version available to us:

```
$ sudo apt-get install libneon27-dev
```

 If you are planning on compiling the `res_calendar_ews` module, you will need to have neon 0.29 or greater. Currently Ubuntu is shipping with 0.27, so you will have to compile the neon library and link to it from the *configure* script. This can be done via *./configure --with-neon29=<path to neon>*.

With *libneon* installed, we can now install the *libical-dev* package and its dependencies with *apt-get*:

```
$ sudo apt-get install libical-dev
```

After installing our dependencies, we can run the *configure* script in our Asterisk source directory and enable both the `res_calendar` and `res_calendar_caldav` modules from within the *Resource Modules* section of *menuselect*.

Configuring Calendar Support for Asterisk

In this section we're going to discuss how to connect your Asterisk system to a Google calendar. We're using calendars from Google for the simple reason that they don't require any other configuration (such as setting up a calendaring server), which gets us up and running far quicker. Of course, once you're comfortable with configuring calendaring support in Asterisk, you can connect it to any calendaring server you desire.

The first step is to make sure you have a Gmail (*http://www.gmail.com*) account with Google, which will get you access to a calendaring server. Once you've logged into your Gmail account, there should be a link to your calendar in the upper-left corner. Click on the *Calendar* link and insert a couple of items for the next hour or two. When we configure our *calendar.conf* file we'll be instructing Asterisk to check for new events every 15 minutes, and pulling in 60 minutes' worth of data.

Be sure to verify the time on your server. If the time is not in sync with the rest of the world—e.g., if is not updated via the Network Time Protocol (NTP)—your events may not show, or may show at the wrong times. This tip is the result of running into this very issue while testing and documenting. :)

The next step is to configure our *calendar.conf* file for polling our calendar server.

The *calendar.conf.sample* file has several examples for calendaring servers, such as those supplied by Microsoft Exchange–, iCal-, and CalDAV-based calendar servers.

The following configuration will connect to the Google calendaring server and poll for new events every 15 minutes, retrieving 60 minutes' worth of data. Feel free to change these settings as necessary, but be aware that pulling more data (especially if you have multiple calendars for people in your company) will utilize more memory:

```
$ cat >> calendar.conf
[myGoogleCal]
type=caldav
url=https://www.google.com/calendar/dav/<Gmail Email Address>/events/
user=<Gmail Email Address>
secret=<Gmail Password>
refresh=15
timeframe=60
Ctrl+D
```

With your *calendar.conf* file configured, let's load the calendaring modules into Asterisk. First we'll load the *res_calendar.so* module into memory, then we'll follow it up by doing a *module reload*, which will load the sister modules (such as *res_calendar_caldav.so*) correctly.[†]

[†] As of this writing, there is a bug in the process of loading of the calendar modules after Asterisk has been started. It was filed as issue 18067 at *https://issues.asterisk.org* and hopefully will have been resolved by the time you read this. If not, be aware that you may need to restart Asterisk to get the modules loaded into memory correctly.

```
$ asterisk -r
*CLI> module load res_calendar.so
*CLI> module reload
```

After loading the modules we can check to make sure our calendar has connected to the server and been loaded into memory correctly, by executing *calendar show calendars*:

```
*CLI> calendar show calendars
Calendar            Type        Status
--------            ----        ------
myGoogleCal         caldav      busy
```

Our status is currently set to busy (which doesn't have any bearing on our dialplan at the moment, but simply means we have an event that has marked us as busy in the calendar), and we can see the currently loaded events for our time range by running *calendar show calendar <myGoogleCal>* from the Asterisk console:

```
*CLI> calendar show calendar myGoogleCal
Name              : myGoogleCal
Notify channel    :
Notify context    :
Notify extension  :
Notify applicatio :
Notify appdata    :
Refresh time      : 15
Timeframe         : 60
Autoreminder      : 0
Events
------
Summary     : Awesome Call With Russell
Description :
Organizer   :
Location    :
Cartegories :
Priority    : 0
UID         : hlfhcpi0j360j8fteop49cvk68@google.com
Start       : 2010-09-28 08:30:00 AM -0400
End         : 2010-09-28 09:00:00 AM -0400
Alarm       : 2010-09-28 04:20:00 AM -0400
```

The first field in the top section is the *Name* of our calendar. Following that are several *Notify* fields, which are used to dial a destination upon the start of a meeting, that we'll discuss in more detail shortly. The *Refresh time* and *Timeframe* fields are the values we configured for how often to check for new events and how long of a range we should look at for data, respectively. The *Autoreminder* field controls how long prior to an event we should execute the *Notify* options.

If you have not configured any of the *Notify* options but have an alarm set to remind you in the calendar, you may get a WARNING message such as:

```
WARNING[5196]: res_calendar.c:648 do_notify: Channel should be in
form Tech/Dest (was '')
```

The reason for the warning is that an alarm was set for notification about the start of the meeting, but Asterisk was unable to generate a call due to values not being configured to place the call. This warning message can be safely ignored if you don't plan on placing calls for event notifications.

The rest of the screen output is a listing of events available within our *Timeframe*, along with information about the events. The next steps are to look at some dialplan examples of what you can do now that you have your calendaring information in Asterisk, and to configure dialing notifications to remind you about upcoming meetings.

Triggering Calendar Reminders to Your Phone

In this section we'll discuss how to configure the *calendar.conf* file to execute some simple dialplan that will call your phone prior to a calendar event. While the dialplan we'll provide might not be ready for production, it certainly gives you a good taste of the possibilities that exist for triggering calls based on calendar state.

Triggering a wakeup call

In our first example, we're going to call a device and play back a reminder notice for a particular calendar event. It might be useful to get this type of reminder if you're likely to be napping at your desk when your weekly Monday meeting rolls around. To set up a wakeup call reminder, we simply need to add the following lines to our calendar configuration in *calendar.conf*:

```
channel=SIP/0000FFFF0001
app=Playback
appdata=this-is-yr-wakeup-call
```

In your calendar, you need to make sure the event you're adding has an alarm or reminder associated with it. Otherwise, Asterisk won't try to generate a call.

After making this change, reload the *res_calendar.so* module from the Asterisk console:

```
*CLI> module reload res_calendar.so
```

When the event rolls around, Asterisk will generate a call to you and play back the sound file *this-is-yr-wakeup-call*. The output on the console would look like this:

```
 -- Dialing SIP/0000FFFF0001 for notification on calendar myGoogleCal
== Using SIP RTP CoS mark 5
 -- Called 0000FFFF0001
 -- SIP/0000FFFF0001-00000001 is ringing
 -- SIP/0000FFFF0001-00000001 connected line has changed, passing it to
    Calendar/myGoogleCal-5fd3c52
 -- SIP/0000FFFF0001-00000001 answered Calendar/myGoogleCal-5fd3c52
 -- <SIP/0000FFFF0001-00000001> Playing 'this-is-yr-wakeup-call.ulaw'
    (language 'en')
```

> If you modify the calendar event so it's just a couple of minutes in the future, you can trigger the events quickly by unloading and then loading the *res_calendar_caldav.so* module from the Asterisk console. By doing that, you'll trigger Asterisk to generate the call immediately.

Remember that our refresh rate is set to 15 minutes and we're gathering 60 minutes' worth of events. You might have to adjust these numbers if you wish to test this out on your development server.

Scheduling calls between two participants

In this example we're going to show how you can use a combination of some simple dialplan and the `CALENDAR_EVENT()` dialplan function to generate a call between two participants based on the information in the location field. We're going to fill in the location field with 0000FFFF0002, which is the SIP device we wish to call after answering our reminder.

> We haven't specified SIP/0000FFFF0002 directly in the calendar event because we want to be a bit more secure with what we accept. Because we're going to filter out anything but alphanumeric characters, we won't be able to accept a forward slash as the separator between the technology and the location (e.g., SIP/0000FFFF0001). We could certainly allow this, but then we would run the risk of people making expensive outbound calls, especially if a user opens his calendar publicly or is compromised. With the method we're going to employ, we simply limit our risk.

Add the following dialplan to your *extensions.conf* file:

```
[AutomatedMeetingSetup]
exten => start,1,Verbose(2,Triggering meeting setup for two participants)
    same => n,Set(DeviceToDial=${FILTER(0-9A-Za-z,${CALENDAR_EVENT(location)})})
    same => n,Dial(SIP/${DeviceToDial},30)
    same => n,Hangup()
```

When the event time arrives, our device will receive a call, and when that call is answered another call will be placed to the endpoint with which we wish to have our meeting. The console output looks like the following:

```
This is where our calendar triggers a call to our device

    -- Dialing SIP/0000FFFF0001 for notification on calendar myGoogleCal
 == Using SIP RTP CoS mark 5
    -- Called 0000FFFF0001
    -- SIP/0000FFFF0001-00000004 is ringing

And now we have answered the call from Asterisk triggered by an event

    -- SIP/0000FFFF0001-00000004 connected line has changed, passing it to
       Calendar/myGoogleCal-347ec99
    -- SIP/0000FFFF0001-00000004 answered Calendar/myGoogleCal-347ec99

Upon answer, we trigger some dialplan that looks up the endpoint to call

    -- Executing [start@AutomatedMeetingSetup:1] Verbose("SIP/0000FFFF0001-00000004",
       "2,Triggering meeting setup for two participants") in new stack
 == Triggering meeting setup for two participants


This is where we used CALENDAR_EVENT(location) to get the remote device

    -- Executing [start@AutomatedMeetingSetup:2] Set("SIP/0000FFFF0001-00000004",
       "DeviceToDial=0000FFFF0002") in new stack

And now we're dialing that endpoint

    -- Executing [start@AutomatedMeetingSetup:3] Dial("SIP/0000FFFF0001-00000004",
       "SIP/0000FFFF0002,30") in new stack
 == Using SIP RTP CoS mark 5
    -- Called 0000FFFF0002
    -- SIP/0000FFFF0002-00000005 is ringing

The other end answered the call, and Asterisk bridged us together

    -- SIP/0000FFFF0002-00000005 answered SIP/0000FFFF0001-00000004
    -- Locally bridging SIP/0000FFFF0001-00000004 and SIP/0000FFFF0002-00000005
```

Of course, the dialplan could be expanded to prompt the initial caller to acknowledge being ready for the meeting prior to calling the other party. Likewise, we could add some dialplan that plays a prompt to the other caller that lets her know that she has scheduled a meeting and that if she presses 1 she will be connected with the other party immediately. We could even have created a dialplan that would allow the original party to record a message to be played back to the other caller.

Just for fun, we'll show you an example of the functionality we just described. Feel free to modify it to your heart's content:

```
[AutomatedMeetingSetup]
exten => start,1,Verbose(2,Triggering meeting setup for two participants)
```

```
; *** This line should not have any line breaks
    same => n,Read(CheckMeetingAcceptance,to-confirm-wakeup&press-1&otherwise
&press-2,,1)

    same => n,GotoIf($["${CheckMeetingAcceptance}" != "1"]?hangup,1)

    same => n,Playback(silence/1&pls-rcrd-name-at-tone&and-prs-pound-whn-finished)

; We set a random number and assign it to the end of the recording
; so that we have a unique filename in case this is used by multiple
; people at the same time.
;
; We also prefix it with a double underscore because the channel
; variable also needs to be available to the channel we're going to call
;
    same => n,Set(__RandomNumber=${RAND()})
    same => n,Record(/tmp/meeting-invite-${RandomNumber}.ulaw)

    same => n,Set(DeviceToDial=${FILTER(0-9A-Za-z,${CALENDAR_EVENT(location)})})
    same => n,Dial(SIP/${DeviceToDial},30,M(CheckConfirm))
    same => n,Hangup()

exten => hangup,1,Verbose(2,Call was rejected)
    same => n,Playback(vm-goodbye)
    same => n,Hangup()

[macro-CheckConfirm]
exten => s,1,Verbose(2,Allowing called party to accept or reject)
    same => n,Playback(/tmp/meeting-invite-${RandomNumber})

; *** This line should not have any line breaks
    same => n,Read(CheckMeetingAcceptance,to-confirm-wakeup&press-1&otherwise
&press-2,,1)

    same => n,GotoIf($["${CheckMeetingAcceptance}" != "1"]?hangup,1)

exten => hangup,1,Verbose(2,Call was rejected by called party)
    same => n,Playback(vm-goodbye)
    same => n,Hangup()
```

We hope you'll be able to use this simple dialplan example as a jumping-off point. With a little creativity and some dialplan skills, the possibilities are endless!

Calling meeting participants and placing them into a conference

To expand upon the functionality in the previous section, we're going to delve into the logic problem of how you might be able to place multiple participants into a meeting. Our goal is to use our calendar to call us when the meeting is scheduled to start, and then, when we answer, to place calls to all the other members of the conference. As the other participants answer their phones, they will be placed into a virtual conference room, where they will wait for the meeting organizer to join. After all participants have been dialed and answered (or perhaps not answered), the organizer will be placed into the call, at which point the meeting will start.

This type of functionality increases the likelihood that the meeting will start on time, and it means the meeting organizer doesn't have to continually perform roll call as new participants continue to join after the call is supposed to start (which invariably happens, with people's schedules typically being fairly busy).

The dialplan we're going to show you isn't necessarily a polished, production-ready installation (for example, the data returned from the calendar comes from the description field, only deals with device names, and assumes the technology is SIP). However, we've done the hard work for you by developing the Local channel usage, along with the M() flag (macro) usage with Dial(). With some testing and tweaks this code could certainly be developed more fully for your particular installation, but we've kept it general to allow for it to be usable for more people in more situations. The example dialplan looks like this:

```
[AutomatedMeetingSetup]
exten => start,1,Verbose(2,Calling multiple people and placing into a conference)

; Get information from calendar and save that information. Prefix
; CalLocation with an underscore so it is available to the Local
; channel (variable inheritance).
;
    same => n,Set(CalDescription=${CALENDAR_EVENT(description)})
    same => n,Set(_CalLocation=${CALENDAR_EVENT(location)})
    same => n,Set(X=1)

; Our separator is a caret (^), so the description should be in the
; format of:    0000FFFF0001^0000FFFF0002^etc...
;
    same => n,Set(EndPoint=${CUT(CalDescription,^,${X})})

; This loop is used to build the ${ToDial} variable, which contains
; a list of Local channels to be dialed, thereby triggering the multiple
; Originate() actions simultaneously instead of linearly
;
    same => n,While($[${EXISTS(${EndPoint})}])

; This statement must be on a single line
    same => n,Set(ToDial=${IF($[${ISNULL(${ToDial})}]?
                    :${ToDial}&)}Local/${EndPoint}@MeetingOriginator)
    same => n,Set(X=$[${X} + 1])
    same => n,Set(EndPoint=${CUT(CalDescription,^,${X})})
    same => n,EndWhile()

; If no values are passed back, then don't bother dialing
    same => n,GotoIf($[${ISNULL(${ToDial})}]?hangup)
    same => n,Dial(${ToDial})

; After our Dial() statement returns, we should be placed into
; the conference room. We are marked, so the conference can start
; (which is indicated by the 'A' flag to MeetMe).
```

```
;
    same => n,MeetMe(${CalLocation},dA)
    same => n(hangup),Hangup()

[MeetingOriginator]
exten => _[A-Za-z0-9].,1,NoOp()
    same => n,Set(Peer=${FILTER(A-Za-z0-9,${EXTEN})})

; Originate calls to a peer as passed to us from the Local channel. Upon
; answer, the called party should execute the dialplan located at the
; _meetme-XXXX extension, where XXXX is the conference room number.
;
    same => n,Originate(SIP/${Peer},exten,MeetingOriginator,meetme-${CalLocation},1)
    same => n,Hangup()

; Join the meeting; using the 'w' flag, which means 'wait for marked
; user to join before starting'
;
exten => _meetme-XXXX,1,Verbose(2,Joining a meeting)
    same => n,Answer()
    same => n,MeetMe(${EXTEN:7},dw)
    same => n,Hangup()
```

Controlling Calls Based on Calendar Information

Sometimes it is useful to redirect calls automatically—for example, when you're in a meeting, or on vacation. In this section we'll be making use of the CALENDAR_BUSY() dialplan function, which allows us to check the current status of our calendar to determine if we're busy or not. A simple example of this would be to send all calls to voicemail using the busy message whenever an event that marks us as busy has been scheduled.

The following dialplan shows a simple example where we check our calendar for busy status prior to sending a call to a device. Notice that a lot of the information in this example is static, and additional effort would be required to make it dynamic and suitable for production:

```
exten => 3000,1,Verbose(2,Simple calendar busy check example)
    same => n,Set(CurrentExten=${EXTEN})
    same => n,Set(CalendarBusy=${CALENDAR_BUSY(myGoogleCal)})
    same => n,GotoIf($["${CalendarBusy}" = "1"]?voicemail,1)
    same => n,Dial(SIP/0000FFFF0002,30)
    same => n,Goto(voicemail,1)

exten => voicemail,1,Verbose(2,Caller sent to voicemail)

; *** This line should not have any line breaks
    same => n,GotoIf($["${DIALSTATUS}" = "BUSY" |
"${CalendarBusy}" = "1"]?busy:unavail)

    same => n(busy),VoiceMail(${CurrentExten}@shifteight,b)
    same => n,Hangup()
```

```
    same => n(unavail),VoiceMail(${CurrentExten}@shifteight,u)
    same => n,Hangup()
```

And here is a slightly more elaborate section of dialplan that utilizes a few of the tools we've learned throughout the book, including DB_EXISTS(), GotoIf(), and the IF() function:

```
exten => _3XXX,1,Verbose(2,Simple calendar busy check example)
    same => n,Set(CurrentExten=${EXTEN})
    same => n,GotoIf($[${DB_EXISTS(extension/${CurrentExten}/device)}]?:no_device,1)
    same => n,Set(CurrentDevice=${DB_RESULT})
    same => n,GotoIf($[${DB_EXISTS(extension/${CurrentExten}/calendar)}]?:no_calendar)
    same => n,Set(CalendarBusy=${CALENDAR_BUSY(${DB_RESULT})})
    same => n,GotoIf($[${CalendarBusy}]?voicemail,1)
    same => n(no_calendar),Verbose(2,No calendar was found for this user)
    same => n,Dial(SIP/${CurrentDevice},30)
    same => n,Goto(voicemail,1)

exten => voicemail,1,Verbose(2,Sending caller to voicemail)

; *** This line should not have any line breaks
    same => n,GotoIf($[${DB_EXISTS(extension/${CurrentExten}/voicemail_context)}]
?:no_voicemail)

    same => n,Set(VoiceMailContext=${DB_RESULT})
; *** This line should not have any line breaks
    same => n,Set(VoiceMailStatus=${IF($["${DIALSTATUS}" = "BUSY" |
O${CalendarBusy}]?b:u)})
    same => n,VoiceMail(${CurrentExten}@${VoiceMailContext},${VoiceMailStatus})
    same => n,Hangup()

    same => n(no_voicemail),Playback(number-not-answering)
    same => n,Hangup()

exten => no_device,1,Verbose(2,No device found in the DB)
    same => n,Playback(invalid)
    same => n,Hangup()
```

Writing Call Information to a Calendar

Using the CALENDAR_WRITE() function opens some other possibilities in terms of calendar integration. From the Asterisk dialplan, we can insert information into a calendar, which can be consumed by other devices and applications. Our next example is a calendar that tracks call logs. For anyone who may be on the phone a fair amount who needs to track time for clients, writing all calls to a calendar for a visual reference can be useful when verifying things at the end of the day.

We're going to utilize the Google web calendar again for this example, but we're going to create a new, separate calendar just for tracking calls. In order to write to the calendar, we'll need to set up our *calendar.conf* file a little bit differently, by using the CalDAV calendar format. First, though, we need to create our new calendar.

On the left side of the Google calendar interface will be a link labeled *Add*. Clicking on this will open a new window where you can create the calendar. Go ahead and do that now. We've called ours "Phone Calls."

Now we need to enable CalDAV calendar syncing for our calendar. Information about how to do this is located at *http://www.google.com/support/mobile/bin/answer.py?answer=151674*. This page notes that only your primary calendar will be synced to the device, but we want to make sure our calls are logged to a separate calendar so we can easily hide them (and so our smartphone doesn't synchronize the phone's calls either, which may cause confusion). There are two links near the bottom of the page: one for regular Google calendar users, and the other for Google Apps users. Select the appropriate link and open it. You will then be presented with a page that contains your calendars. Select the *Phone Calls* calendar and then select *Save*.

Next up is configuring our *calendar.conf* file for Asterisk. One of the parameters we need is the link to the CalDAV calendar. There is a Calendar ID value that we need that will identify our calendar specifically. To find the calendar ID, click the down arrow beside the calendar name on the lefthand side of the calendar page and select *Calendar Settings*. Near the bottom of the calendar settings will be two rows that contain the icons for sharing the calendar (XML, ICAL, HTML). Beside the first set of icons inside the *Calendar Address* box will be the calendar ID. It will look like this:

```
(Calendar ID: 2hfb6p5974gds924j61cmg4gfd@group.calendar.google.com)
```

If you're setting this up via Google Apps, the calendar ID will be prefixed with your domain name and an underscore (e.g., `shifteight.org_`). Make a note of this string, as we're going to use it next.

Open up the *calendar.conf* file and add a new calendar section. In our case we've called it [`phone_call_calendar`]. You'll recognize the formatting of the calendar from earlier, so we won't go through all the settings here. The key setting to note is the `url` parameter. The format of this parameter is:

```
https://www.google.com/calendar/dav/<calendar_id>/events/
```

We need to replace the `<calendar_id>` with the calendar ID we recently made a note of. The full configuration then ends up looking like so:

```
[phone_call_calendar]
type=caldav

; The URL must be on a single line
url=https://www.google.com/calendar/dav/
    shifteight.org_2hfb6p5974gds924j61cmg4gfd@group.calendar.google.com/events/

user = leif@shifteight.org
secret = my_secret_password
refresh=15
timeframe=120
```

Now that we have our calendar configured, we need to load it into memory, which can be done by reloading the *res_calendar.so* module:

```
*CLI> module reload res_calendar.so
```

Verify that the calendar has been loaded into memory successfully with the *calendar show* command:

```
*CLI> calendar show calendars
Calendar            Type     Status
--------            ----     ------
phone_call_calendar caldav   free
```

With our calendar successfully loaded into memory, we can write some dialplan around our Dial() command to save our call information to the calendar with the CALEN DAR_WRITE() function:

```
[LocalSets]
exten => _NXXNXXXXXX,1,Verbose(2,Outbound calls)
    same => n,Set(CalendarStart=${EPOCH}) ; Used by CALENDAR_WRITE()
    same => n,Set(X=${EXTEN})             ; Used by CALENDAR_WRITE()
    same => n,Dial(SIP/ITSP/${EXTEN},30)
    same => n,NoOp(Handle standard voicemail stuff here)
    same => n,Hangup()

exten => h,1,Verbose(2,Call cleanup)

; Everything that follows must be on a single line
    same => n,Set(CALENDAR_WRITE(phone_call_calendar,summary,description,start,end)=
        OUTBOUND: ${X},Phone call to ${X} lasted for ${CDR(billsec)} seconds.,
        ${CalendarStart},${EPOCH})
```

In our dialplan we've created a simple scenario where we place an outbound call through our Internet Telephony Service Provider (ITSP), but prior to placing the call, we save the epoch[‡] to a channel variable (so we can use it later when we write our calendar entry at the end of the call). After our call, we write our calendar entry to the phone_call_calendar with the CALENDAR_WRITE() dialplan function within the built-in h extension. There are several options we can pass to the calendar, such as the summary, description, and start and end times. All of this information is then saved to the calendar.

We've also used the CDR() dialplan function in our description to show the number of seconds the answered portion of the call lasted for, so we can get a more accurate assessment of whether a call was answered and, if so, how long the answered portion lasted for. We could also be clever and only write to the calendar if ${CDR(billsec)} was greater than 0 by wrapping the Set() application in an ExecIf(); e.g., same => n,ExecIf($[${CDR(billsec)} > 0]?Set(CALENDAR_WRITE...)).

[‡] In Unix, the epoch is the number of seconds that have elapsed since January 1, 1970, not counting leap seconds.

Many possibilities exist for the `CALENDAR_WRITE()` function; this is just one that we've implemented and enjoy.

Conclusion

In this section we've learned how to integrate Asterisk with an external calendar server such as that provided by Google, but the concepts for attaching to other calendaring servers remain the same. We explored how to set up a meeting between two participants, and how to set up a multiparty conference using information obtained from the description field in the calendar. We also looked at how to control calls using the `CALENDAR_BUSY()` function, to redirect calls to voicemail when our current event describes us as busy. By implementing this type of functionality in Asterisk, you can see the power we have to control call flow using external services such as those supplied by a calendar server.

And we didn't even get to dive into every use of the calendar implementation—there exist other calendar functions, such as `CALENDAR_QUERY()`, which allows you to pull back a list of events within a given time period for a particular calendar, and `CALENDAR_QUERY_RESULT()`, which allows you to access the specifics of those calendar events. Additionally, you could create functionality that writes events into your calendar with the `CALENDAR_WRITE()` function: for example, you may wish to develop some dialplan that allows you to set aside blocks of times in your calendar from your phone when you're on the road without access to your laptop. Many possibilities exist, and all it takes is a little creativity.

VoiceMail IMAP Integration

"Unified messaging" has been a buzzword in the telecommunications industry for ages. It's all about integrating services together so users can access the same types of data in multiple locations, using different methods. One of the most touted applications is the integration of email and voicemail. Asterisk has been doing this for years, but many of the larger companies are still trying to get this right. Asterisk has had the ability to send users voicemails via email, using the Mail Transport Agent (MTA) in your Linux distro (this always used to be *sendmail*, but Postfix has become increasingly popular as an MTA). Voicemail to email is one of the oldest features in Asterisk, and it normally works without any configuration at all.[§]

Internet Message Access Protocol (IMAP) integration has existed in Asterisk (and been steadily evolving) since version 1.4. IMAP voicemail integration means your users can access their voicemails via a folder within their email accounts, which gives them the

---

[§] When we say "it works," what we mean is that Asterisk will compose the email and submit it to the MTA, and the email will successfully be passed out of the system. What happens to it after it leaves the system is a bit more complicated, and will often involve spam filters treating the mail as suspect and not actually delivering it. This is not really Asterisk's fault, but it's something you'll have to deal with.

ability to listen to, forward, and mark voicemail messages with the same flexibility that the Asterisk VoiceMail() dialplan application gives. Asterisk will be aware of the statuses of those messages when the users next log in via the phone system.

As the number of administrators integrating Asterisk with their IMAP servers has increased, the number of bugs filed and fixed has first increased and then decreased, to the point where IMAP integration can be considered stable enough for production use. In this section we'll discuss how to compile in IMAP voicemail support and connect your voicemail system to your IMAP server.

Compiling IMAP VoiceMail Support into Asterisk

To get IMAP voicemail support into Asterisk, we need to compile the University of Washington's IMAP library. The UW IMAP toolkit will give us the functionality in Asterisk to connect to our IMAP server. Before compiling the software, though, we need to install some dependencies.

The dependencies for building the IMAP library include the tools required to build Asterisk, but the way we're building it also requires the development libraries for OpenSSL and Pluggable Authentication Modules (PAM). We've included instructions for both CentOS and Ubuntu.

CentOS dependencies

Installing both the OpenSSL and PAM development libraries on CentOS can be done with the following command:

```
$ sudo yum install openssl-devel pam-devel
```

Remember to add *.x86_64* to the name of each package if installing on a 64-bit machine.

Ubuntu dependencies

Installing both the OpenSSL and PAM development libraries on Ubuntu can be done with the following command:

```
$ sudo apt-get install libssl-dev libpam0g-dev
```

If you try to install *libpam-dev* on Ubuntu, it will warn you that *libpam-dev* is a virtual package and that you should explicitly install one of the packages in the list it presents you with (which in our case contained only a single package). If *libpam0g-dev* is not the correct package on your version of Ubuntu, try installing the virtual package. This should give you a list of valid packages for the PAM development library.

Compiling the IMAP library

Now that we have our dependencies satisfied, we can compile the IMAP library that Asterisk will use to connect to our IMAP server.

The first thing to do is change to the *thirdparty* directory located under the *asterisk-complete* directory. If you have not already created this directory, do so now:

```
$ cd ~/src/asterisk-complete
$ mkdir thirdparty
$ cd thirdparty
```

Next up is downloading the IMAP toolkit and compiling it. The next steps will get the latest version of the IMAP toolkit from the University of Washington's server (more information about the toolkit is available at *http://www.washington.edu/imap/*):

```
$ wget ftp://ftp.cac.washington.edu/mail/imap.tar.Z
$ tar zxvf imap.tar.Z
$ cd imap-2007e
```

 The directory name *imap-2007e* may change as new versions of the toolkit become available.

There are a few options we need to pass to the *make* command when building the IMAP library, and the values you should pass will depend on what platform you're building on (32-bit vs. 64-bit), if you need OpenSSL support, and whether you need IPv6 support or just IPv4. Table 18-1 shows some of the various options you could pass on different platforms.

Table 18-1. IMAP library compile time options

Option	Description
EXTRACFLAGS="-fPIC"	Required when building on 64-bit platforms.
EXTRACFLAGS="-I/usr/include/openssl"	Used for building in OpenSSL support.
IP6=4	Many platforms that support IPv6 prefer that method of connection, which may not be desirable for all servers. If you would like to force IPv4 as the preferred connection method, set this option.

If you look in the *Makefile* shipped with the IMAP library, you will find a list of platforms for which the library can be compiled. In our case, we'll be compiling for either CentOS or Ubuntu with PAM support. If you're compiling on other systems, take a look in the *Makefile* for the three-letter code that tells the library how to compile for your platform.

To compile for a 64-bit platform with OpenSSL support and a preference for connecting via IPv4:

```
$ make lnp EXTRACFLAGS="-fPIC -I/usr/include/openssl" IP6=4
```

To compile for a 32-bit platform with OpenSSL support and a preference for connecting via IPv4:

```
$ make lnp EXTRACFLAGS="-I/usr/include/openssl" IP6=4
```

If you don't wish to compile with OpenSSL support, simply remove the `-I/usr/include/openssl` from the `EXTRACFLAGS` option. If you would prefer connecting by IPv6 by default, simply don't specify the `IP6=4` option.

> When installing IMAP support, we have always compiled the c-client library from source. However, it may be available as a package for your distribution. For example, Ubuntu has a *libc-client-dev* package available. It may work and save you some trouble, but we have not tested it.

Compiling Asterisk

After compiling the IMAP library, we need to recompile the *app_voicemail.so* module with IMAP support. The first step is to run the *configure* script and pass it the *--with-imap* option to tell it where the IMAP library exists:

```
$ cd ~/src/asterisk-complete/asterisk/1.8.0
$ ./configure --with-imap=/usr/src/asterisk-complete/thirdparty/imap-2007e/
```

Once the *configure* script has finished executing, we need to enable IMAP voicemail support in *menuselect*:

```
$ make menuselect
```

From the *menuselect* interface, go to *Voicemail Build Options*. Within that menu, you should have the option to select `IMAP_STORAGE`.

> If you don't have the ability to select that option, check to make sure your IMAP library was built successfully (i.e., that you have all the required dependencies installed and that it didn't error out when building) and that you correctly specified the path to the IMAP library when running the *configure* script. You can also verify that the IMAP library was found correctly by looking in the *config.log* file (located in your Asterisk build directory) for IMAP.

After selecting `IMAP_STORAGE`, save and exit from *menuselect* and run *make install*, which will recompile the *app_voicemail.so* module and install it to the appropriate location. The next step is to configure the *voicemail.conf* file located in */etc/asterisk/*.

Configuring Asterisk

Now that we've compiled IMAP support into Asterisk, we need to enable it by connecting to an IMAP-enabled server. There are many IMAP servers that you could use, including those supplied with Microsoft servers, Dovecot (*http://www.dovecot.org*), and Cyrus (*http://www.cyrusimap.org*) on Unix, or a web-based IMAP server such as that supplied by Google's Gmail (*http://www.gmail.com*).‖ Our instructions are going to show how to connect Asterisk to a Gmail account with IMAP enabled, as it requires the least amount of effort to get up and running with IMAP voicemail, but these instructions can easily be adapted for use with any existing IMAP server.

Enabling IMAP on your Gmail account. Enabling IMAP support on your Gmail account is straightforward (see Figure 18-1). Once logged into your account, select *Settings* from the upper-right corner. Then select *Forwarding and POP/IMAP* support from the task bar under the *Settings* header. In the *IMAP Access* section, enable IMAP support by selecting *Enable IMAP*. After enabling it, click the *Save Changes* button at the bottom of the screen.

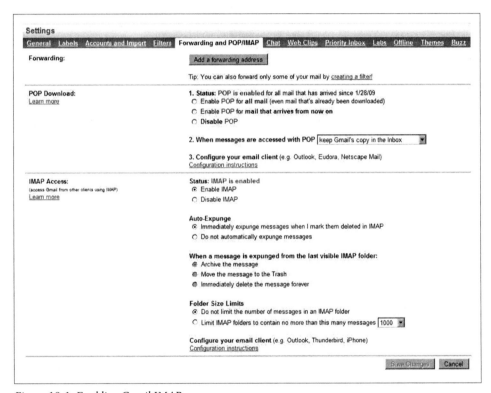

Figure 18-1. Enabling Gmail IMAP

Configuring voicemail.conf for IMAP. To enable our voicemail system to connect to an IMAP system, we need to make sure IMAP support has been built into the *app_voicemail.so* module per the instructions in "Compiling Asterisk" on page 414. With IMAP support

‖ We recently checked out the open source webmail project roundcube (*http://www.roundcube.net*) project as well, and we were quite impressed.

compiled into Asterisk, we just need to instruct the voicemail module how to connect to our IMAP server.

We're going to demonstrate how to connect to an IMAP-enabled Gmail account and use that to store and retrieve our voicemail messages. If you haven't already, read the section "Enabling IMAP on your Gmail account" before proceeding. The final step is configuring *voicemail.conf* to connect to the server.

In *voicemail.conf*, add the following lines to the [general] section. Be sure you only specify a single format (we recommend wav49) for voicemail recordings, and remove any references to ODBC voicemail storage if you've enabled that previously:

```
[general]
format=wav49              ; format to store files
imapserver=imap.gmail.com ; IMAP server location
imapport=993              ; port IMAP server listens to
imapflags=ssl             ; flags required for connecting
expungeonhangup=yes       ; delete messages on hangup
pollmailboxes=yes         ; used for message waiting indication
pollfreq=30               ; how often to check for message changes
```

Before we configure our user for connecting to the Gmail IMAP server, let's discuss the options we've just set in the [general] section. These are the basic options that will get us started; we'll do some more customization shortly, but let's see what we've done so far.

First, the format=wav49 option has declared that we're going to save our files as GSM with a WAV header, which can be played on most desktops (including Microsoft Windows) while retaining a small file size.

Next, we've configured Asterisk to connect to an imapserver located at imap.gmail.com on imapport 993. We've also set imapflags to ssl, as Gmail requires a secure connection. Without the ssl IMAP flag being set, the server will reject our connection attempts (which is why it was important that we compiled our IMAP library with OpenSSL support). Another option that may be required on private IMAP servers such as Dovecot is to specify novalidate-cert for imapflags when an SSL connection is necessary, but the certificate is not generated by a certificate authority.

Next, we've set expungeonhangup=yes, which causes messages marked for deletion to be removed from the server upon hangup from the VoiceMail() application. Without this option, messages are simply marked as read and left on the server until they have been removed via an email application or web interface.

In order to get message waiting indication (MWI) updates correctly, we need to enable pollmailboxes=yes, which causes Asterisk to check with the server for any changes to the status of a message. For example, when someone leaves us a voicemail and we listen to it by opening the message via our email application, the message will be marked as read, but without polling the mailbox Asterisk will have no way of knowing this and will enable the MWI light on the associated device indefinitely. Finally, we've set the related option pollfreq to 30 seconds. This option controls how often Asterisk will ask the server for the status of messages: set it appropriately to control the amount of traffic going to the voicemail server.

Table 18-2 shows some of the other options available to us.

Table 18-2. Additional IMAP voicemail options

Option	Description
imapfolder	Provides the name of the folder in which to store voicemail messages on your IMAP server. By default they are stored in the *INBOX*.[a]
imapgreetings	Defines whether voicemail greetings are stored on the IMAP server or stored locally on the server. Valid values are yes or no.
imapparentfolder	Defines the parent folder on the IMAP server. Usually this configured as *INBOX* on the server, but if it is called something else, you can specify it here.
greetingfolder	Specifies the folder in which to save the voicemail greetings, if you've enabled the imapgreetings option by setting it to yes. By default greetings are saved in the *INBOX*.
authuser	Specifies the master user to use for connecting to your IMAP server, if the server is configured with a single user that has access to all mailboxes.
authpassword	Complement to the authuser directive. See authuser for more information.
opentimeout	Specifies the TCP open timeout (in seconds).
closetimeout	Specifies the TCP close timeout (in seconds).
readtimeout	Specifies the TCP read timeout (in seconds).
writetimeout	Specifies the TCP write timeout (in seconds).

[a] It is important to store your voicemail messages in a folder other than the *INBOX* if the number of messages contained in the *INBOX* could be rather large. Asterisk will try to gather information about all the emails contained in the *INBOX*, and could either time out before retrieving all the information or just take a very long time to store or retrieve voicemail messages, which is not desirable.

With our [general] section configured, let's define a mailbox for connecting to the IMAP server.

In Chapter 8 we defined some users in the [shifteight] voicemail context. Here is the original configuration as defined in that chapter:

```
[shifteight]
100 => 0107,Leif Madsen,leif@shifteight.org
101 => 0523,Jim VanMeggelen,jim@shifteight.org,,attach=no|maxmsg=100
102 => 11042,Tilghman Lesher,,,attach=no|tz=central
```

We're going to modify mailbox 100 in such a way that it connects to the Gmail IMAP server to store and retrieve voicemail messages:

```
[shifteight]
100 => 0107,Leif Madsen,,,|imapuser=leif@shifteight.org|imappassword=secret
```

The *voicemail.conf* file uses both commas and pipes as separators, depending on which field is being used. The first few fields have specific settings in them, and the last field can contain extra information about the mailbox, which is separated by the pipe character (|).

We've removed the email address from the third field because we're not going to use *sendmail* to email us voicemails anymore: they are just going to be stored on the email server directly now. We've configured the mailbox to connect with the IMAP username of `leif@shifteight.org` (because we've enabled Google Apps for the domain that hosts our email) and are connecting using the IMAP password `secret`.

After configuring Asterisk, we need to reload the *app_voicemail.so* module. If you enable console debugging, you should see output similar to the following upon connection to the voicemail server:

```
*CLI> core set debug 10
*CLI> module reload app_voicemail.so
DEBUG[3293]: app_voicemail.c:2734 mm_log: IMAP Info: Trying IP address [74.125.53.109]
DEBUG[3293]: app_voicemail.c:2734 mm_log: IMAP Info: Gimap ready for requests
              from 99.228.XXX.XXX 13if2973206wfc.0
DEBUG[3293]: app_voicemail.c:2757 mm_login: Entering callback mm_login
DEBUG[3293]: app_voicemail.c:2650 mm_exists: Entering EXISTS callback for message 7
DEBUG[3293]: app_voicemail.c:3074 set_update: User leif@shifteight.org mailbox set for
update.
DEBUG[3293]: app_voicemail.c:2510 init_mailstream: Before mail_open, server:
              {imap.gmail.com:993/imap/ssl/user=leif@shifteight.org}INBOX, box:0
DEBUG[3293]: app_voicemail.c:2734 mm_log: IMAP Info: Reusing connection to
              gmail-imap.l.google.com/user="leif@shifteight.org"
```

If you get any ERRORs, check your configuration and verify that the IMAP library is compiled with SSL support. Once *app_voicemail.so* is connected, try leaving yourself a voicemail; then check your voicemail via the Gmail web interface and verify that your message is stored correctly. You should also have an MWI light on your device if it supports it, and if you've configured `mailbox=100@shifteight` for the device in *sip.conf*. If you load the voicemail message envelope and mark it as read, the MWI light should turn off within 30 seconds (or whatever value you set `pollfreq` to in *voicemail.conf*).

Using XMPP (Jabber) with Asterisk

The eXtensible Messaging and Presence Protocol (XMPP) (formerly called Jabber) is used for instant messaging and communicating presence information across networks in near-realtime. Within Asterisk, it is also used for call setup (signaling). There are various cool things we can do with XMPP integration once it's enabled, such as getting a message whenever someone calls us. We can even send messages back to Asterisk, redirecting our calls to voicemail or some other location. Additionally, with `chan_gtalk`, we can accept and place calls over the Google Voice network or accept calls from Google Talk users via the web client.

Compiling Jabber Support into Asterisk

The `res_jabber` module contains various dialplan applications and functions that are useful from the Asterisk dialplan. It is also a dependency of the `chan_gtalk` and `chan_jingle` channel modules. To get started with XMPP integration in Asterisk, we need to compile `res_jabber`.

CentOS dependencies

To install `res_jabber`, we need the `iksemel` development library (*http://code.google.com/ p/iksemel/*). If the OpenSSL development library is installed, `res_jabber` will also utilize that for secure connections (this is recommended). We can install both on CentOS with the following command:

```
$ sudo yum install iksemel-devel openssl-devel
```

> As always, be sure to append *.x86_64* to the module names if installing on a 64-bit machine.

Ubuntu dependencies

To install `res_jabber`, we need the `iksemel` development library. If the OpenSSL development library is installed, `res_jabber` will also utilize that for secure connections (this is recommended). We can install both on Ubuntu with the following command:

```
$ sudo apt-get install libiksemel-dev libssl-dev
```

Jabber Dialplan Commands

Several dialplan applications and functions can be used for communication using the XMPP protocol via Asterisk. We're going to explore how to connect Asterisk to an XMPP server, how to send messages to the client from the dialplan, and how to route calls based on responses to the initially sent messages. By sending a message via XMPP, we're essentially creating a simple screen pop application to let users know when calls are coming to the system.

Connecting to an XMPP server

Before we can start sending messages to our XMPP buddies, we need to connect to an XMPP-enabled server. We're going to utilize the XMPP server at Google, as it is open and easily accessible by anyone. To do so, we need to configure the *jabber.conf* file in our */etc/asterisk/* configuration directory. The following example will connect us to the XMPP server at Google.

 You must already have a Gmail account, which you can get at *http://www.gmail.com*.

Our *jabber.conf* file should look like this:

```
[general]
debug=no
autoprune=no
autoregister=yes
auth_policy=accept

[asterisk]
type=client
serverhost=talk.google.com
username=asterisk@shifteight.org
secret=<super_secret_password>
port=5222
usetls=yes
usesasl=yes
status=available
statusmessage="Ohai from Asterisk"
```

Let's take a quick look at some of the options we just set so you understand what is going on. The options are described in Table 18-3. Note that the first four options are set in the [general] section, and the others are set in the peer section.

Table 18-3. jabber.conf options

Option	Description
debug	Enables/disables XMPP message debugging (which can be quite verbose). Available options are yes or no.
autoprune	Enables/disables autoremoval of users from your buddy list each time *res_jabber.so* connects to your accounts. Do not use this for accounts you might use outside of Asterisk (e.g., your personal account). Available options are yes or no.
autoregister	Specifies whether to automatically register users from your buddy list into memory. Available options are yes or no.
auth_policy	Determines whether or not we should automatically accept subscription requests. Available options are accept or deny.
type	Sets the type of client we will connect as. Available options are client or component. (You will almost always want client.)
serverhost	Indicates which host this connection should connect to (e.g., *talk.google.com*).
username	Provides the username that will be used to connect to the serverhost (e.g., *asterisk@gmail.com*).
secret	Specifies the password that will be used to connect to the serverhost.

Option	Description
port	Indicates which port we will attempt the connection to serverhost on (e.g., *5222*).
usetls	Specifies whether to use TLS or not when connecting to serverhost. Available options are yes or no.
usesasl	Specifies whether to use SASL or not when connecting to serverhost. Available options are yes or no.
status	Defines our default connection status when signed into our account. Available options are: chat, available, away, xaway, and dnd.
statusmessage	Sets a custom status message to use when connected with Asterisk, such as "Connected Via Asterisk". Use double quotes around the message.
buddy	Used to manually add buddies to the list upon connection to the server. You can specify multiple buddies on multiple buddy lines (e.g., buddy=jim@shifteight.org).
timeout	Specifies the timeout (in seconds) that messages are stored on the message stack. Defaults to 5 seconds. This option only applies to incoming messages, which are intended to be processed by the JABBER_RECEIVE() dialplan function.
priority	Defines the priority of this resource in relation to other resources. The lower the number, the higher the priority.

After configuring our *jabber.conf* file, we can load (or reload) the *res_jabber.so* module. We can do this from the console with *jabber reload*:

```
*CLI> jabber reload
Jabber Reloaded.
```

and check the connection with the *jabber show connections* command:

```
*CLI> jabber show connections
Jabber Users and their status:
      User: asterisk@shifteight.org     - Connected
----
    Number of users: 1
```

If you're having problems getting connected, you can try unloading the module and then loading it back into memory. If you're still having problems, you can run the *jabber purge nodes* command to remove any existing or bad connections from memory. Beyond that, check your configuration and verify that you don't have any configuration problems or typos. Once you've gotten connected, you can move on to the next sections, where the fun starts.

Sending messages with JabberSend()

The JabberSend() dialplan application is used for sending messages to buddies from the Asterisk dialplan. You can use this application in any place that you would normally utilize the dialplan, which makes it quite flexible. We're going to use it as a screen pop application for sending a message to a client prior to placing a call to the user's phone. Depending on the client used, you may be able to have the message pop up on the user's screen from the task bar.

Here is a simple example to get us started:

```
[LocalSets]
exten => 104,1,Answer()

; *** This line should not have any line breaks
    same => n,JabberSend(asterisk,jim@shifteight.org,Incoming call from
${CALLERID(all)})

    same => n,Dial(SIP/0000FFFF0002,30)
    same => n,Hangup()
```

This example demonstrates how to use the JabberSend() application to send a message to someone prior to dialing a device. Let's break down the values we've used. The first argument, asterisk, is the section header we defined in the *jabber.conf* file as [asterisk]. In our *jabber.conf* example, we set up a user called *asterisk@shifteight.org* to send messages via the Google XMPP server, and asterisk is the section name we defined. The second argument, jim@shifteight.org, is the buddy we're sending the message to. We can define any buddy here, either as a bare JID (as we've done above) or as a full JID with a resource (e.g., *jim@shifteight.org/laptop*). The third argument to JabberSend() is the message we want to send to the buddy. In this case we're sending Incoming call from ${CALLERID(all)}, with the CALLERID() dialplan function being used to enter the caller ID information in the message.

Obviously, we would have to further build out our dialplan to make this useful: specifically, we'd have to associate the buddy name (e.g., jim@shifteight.org) with the device we're calling (SIP/0000FFFF0002) so that we're sending the message to the correct buddy. You can save these associations in any one of several locations, such as the in AstDB, in a relational database retrieved with func_odbc, or even in a global variable.

Receiving messages with JABBER_RECEIVE()

The JABBER_RECEIVE() dialplan function allows us to receive responses via XMPP messages, capture those responses, and presumably act on them. We would typically use the JABBER_RECEIVE() function in conjunction with the JabberSend() dialplan application, as we are likely to need to send a message to someone and prompt him with the acceptable values he can return. We could use the JABBER_RECEIVE() function either personally, to direct calls to a particular device such as a cell phone or desk phone, or as a text version of an auto attendant to be used when people who are likely to have difficulty hearing the prompts dial in (e.g., users who are deaf or work at noisy job sites). In the latter case, the system would have to be preconfigured to know where to send the messages to, perhaps based on the caller ID of the person calling.

Here is a simple example that sends a message to someone, waits for a response, and then routes the call based on the response:

```
exten => 106,1,Answer()

    ; All text must be on a single line.
```

```
    same => n,JabberSend(asterisk,leif.madsen@gmail.com,Incoming call from
${CALLERID(all)}. Press 1 to route to desk. Press 2 to send to voicemail.)

    same => n,Set(JabberResponse=${JABBER_RECEIVE(asterisk,leif@shifteight.org)})
    same => n,GotoIf($["${JabberResponse}" = "1"]?dial,1)
    same => n,GotoIf($["${JabberResponse}" = "2"]?voicemail,1)
    same => n,Goto(dial,1)

exten => dial,1,Verbose(2,Calling our desk)
    same => n,Dial(SIP/0000FFFF0002,6)
    same => n,Goto(voicemail,1)

exten => voicemail,1,Verbose(2,VoiceMail)

; *** This line should not have any line breaks
    same => n,Set(VoiceMailStatus=${IF($[${ISNULL(${DIALSTATUS})}
| "${DIALSTATUS}" = "BUSY"]?b:u)})

    same => n,Playback(silence/1)
    same => n,VoiceMail(100@lmentinc,${VoiceMailStatus})
    same => n,Hangup()
```

 Unfortunately, the JabberSend() application requires all of the message to be sent on a single line. If you wish to break up the text onto multiple lines, you will need to send it as multiple messages on separate lines using JabberSend().

Our simple dialplan first sends a message to a Jabber account (*leif@shifteight.org*) via our systems' Jabber account (asterisk), as configured in *jabber.conf*. We then use the JABBER_RECEIVE() dialplan function to wait for a response from *leif@shifteight.org*. The default timeout is 5 seconds, but you can specify a different timeout with a third argument to JABBER_RECEIVE(). For example, to wait 10 seconds for a response, we could have used a line like this:

```
Set(JabberResponse=${JABBER_RECEIVE(asterisk,leif@shifteight.org,10)})
```

Once we've either received a response or the timeout has expired, we move on to the next line of the dialplan, which starts checking the response saved to the ${JabberResponse} channel variable. If the value is 1, we continue our dialplan at dial, 1 of the current context. If the response is 2, we continue our dialplan at voicemail,1. If no response (or an unknown response) is received, we continue the dialplan at dial,1.

The dialplan at dial,1 and voicemail,1 should be fairly self-evident. This is a non-production example; some additional dialplan should be implemented to make the values dynamic.

There is a disadvantage to the way we've implemented the JABBER_RECEIVE() function, though. Our function blocks, or waits, for a response from the endpoint. If we set the response value low to minimize delay, we don't give the user we sent the message to much time to respond. However, if we set the response long enough to make it comfortable for the user to send a response, we cause unnecessary delay in calling a device or sending to voicemail.

We can skirt around this issue by using a Local channel. This allows us to execute two sections of dialplan simultaneously, sending a call to the device at the same time we're waiting for a response from JABBER_RECEIVE(). If we get a response from JAB BER_RECEIVE() and we need to do something, we can Answer() the line and cause that section of dialplan to continue. If the device answers the phone, our dialplan with JABBER_RECEIVE() will just be hung up. Let's take a look at a modified dialplan that implements the Local channel:

```
exten => 106,1,Verbose(2,Example using the Local channel)
    same => n,Dial(Local/jabber@${CONTEXT}/n&Local/dial@${CONTEXT}/n)

exten => jabber,1,Verbose(2,Send an XMPP message and expect a response)

; *** This line should not have any line breaks
    same => n,JabberSend(asterisk,leif.madsen@gmail.com,Incoming call from
${CALLERID(all)}. Press 2 to send to voicemail.)

    same => n,Set(JabberResponse=${JABBER_RECEIVE(asterisk,leif@shifteight.org,6)})
    same => n,GotoIf($["${JabberResponse}" = "2"]?voicemail,1)
    same => n,Hangup()

exten => dial,1,Verbose(2,Calling our desk)
    same => n,Dial(SIP/0000FFFF0002,15)
    same => n,Goto(voicemail,1)

exten => voicemail,1,Verbose(2,VoiceMail)
    same => n,Answer()

; *** This line should not have any line breaks
    same => n,Set(VoiceMailStatus=${IF($[${ISNULL(${DIALSTATUS})}
| "${DIALSTATUS}" = "BUSY"]?b:u)})

    same => n,Playback(silence/1)
    same => n,VoiceMail(100@lmentinc,${VoiceMailStatus})
    same => n,Hangup()
```

By adding a Dial() statement at the beginning and shifting our Jabber send and receive functionality into a new extension called jabber, we ensure that we can simultaneously call the dial extension and the jabber extension.

Notice that we removed the Answer() application from the first line of the example. The reason for this is because we want to Answer() the line only after a device has answered (which causes the jabber extension to be hung up); otherwise, we want the voicemail extension to Answer() the line. If the voicemail extension has answered the line, that means either the jabber extension has received a response and was told to Goto() the voicemail extension, or the Dial() to our device timed out, causing the voicemail extension to be executed, thereby causing the line to be Answer()ed.

With the examples provided here serving as a springboard, you should be able to develop rich applications that make use of sending and receiving messages via XMPP servers. Some other dialplan applications and functions exist that may help in the development of your application, such as JABBER_STATUS() (or the JabberStatus() dialplan application), which is used for checking on the status of a buddy; the JabberJoin() and JabberLeave() applications, which are used for joining and leaving XMPP conference rooms; and the JabberSendGroup() application, which allows you to send messages to an XMPP chat room.

chan_gtalk

The chan_gtalk module can be used for connecting to Google Talk (GTalk) clients or for sending and receiving calls via the Google Voice network, which is a PSTN-connected network where you can purchase minutes just like you would from any other ITSP. GTalk is the web-based voice system typically found in GMail web interfaces. Other clients and addons do exist for external applications such as Pidgin, but we'll be testing with the web-based client from Google.

 As of the beginning of 2011, the Google Voice system can only be used in the US.

---

chan_gtalk's Cousin, chan_jingle

Another channel module that is similar to chan_gtalk exists, and that is chan_jingle. Both modules utilize the same type of underlying system: XMPP signaling. However, the chan_jingle module was written long ago and implements a specification that is not widely supported. You may find the implementation of chan_jingle incompatible with many endpoints, and thus at this time you may not find it all that useful. However, work is being done to update chan_jingle to the current specifications, so it may be more useful in a future version of Asterisk.

Before we can get connected to chan_gtalk, we need to make sure we're connected via res_jabber, so if you haven't already done so, review "Connecting to an XMPP server" on page 419 for information about how to connect to the Google XMPP servers.

Configuring gtalk.conf

Once we're connected via res_jabber, we can configure the *gtalk.conf* file, which is used for accepting incoming calls from the Google network. The following configuration enables the guest account, which is required to accept incoming calls. There is currently no support for authenticating incoming calls and then separating and sending them to different contexts, which you may be used to from the configuration of other channel drivers in Asterisk. For now `chan_gtalk` is fairly simple, but future versions of Asterisk may add this feature.

Our *gtalk.conf* file looks like this:

```
[general]
bindaddr=0.0.0.0          ; Address to bind to
allowguests=yes           ; Allow calls from people not in contact list

; Optional arguments
; externip=<external IP of server>
; stunaddr=<stun.yourdomain.tld>

[guest]                   ; special account for options on guest account
disallow=all
allow=ulaw
context=gtalk_incoming
connection=asterisk       ; connection name defined in jabber.conf
```

If your Asterisk system lives behind NAT, you may need to add some additional options to the [general] section in order to place the correct IP address into the headers. If you have a static external IP address, you can use the `externip` option to specify it. Alternatively, you could use the `stunaddr` option to specify the address of your STUN server, which will then look up your address from an external server and place that information into the headers.

> If you configure the `stunaddr` option in *gtalk.conf* and the lookup is successful, it will override any value specified in the `externip` option.

Let's discuss briefly the options we've configured in *gtalk.conf*. In the [general] section, we have set the `bindaddr` option to 0.0.0.0, which means to listen on all interfaces.# You can also specify a single interface to listen on by specifying the IP address of that interface. The next line is `allowguests`, which can be set to either yes or no but is only useful when set to yes. Because the module does not offer the ability to specify different control mechanisms for different users, all users are treated as guests.*

Next we've specified the [guest] account, which will let us accept calls from Google Voice and Google Talk users. This account is only used for incoming calls. When placing outgoing calls, we'll use the account specified in the *jabber.conf* file. Within the [guest] account, we've disabled all codecs with the `disallow=all` option, and then specifically enabled the ulaw codec with `allow=ulaw` on the following line. Incoming calls are then directed to the `gtalk_incoming` context with the `context` option. We

---

#The `chan_gtalk` module only support IPv4 interfaces.

* Future versions of Asterisk may offer more fine-grained control.

specify which account calls will be coming from with the connection option, which we've set to the account created in *jabber.conf*.

The chan_gtalk module does not support reloading the configuration. If you change the configuration, you will have to either restart Asterisk or unload and reload the module, which can only be done when no GTalk calls are up. You can do that using the following commands:

```
*CLI> module unload chan_gtalk.so
*CLI> module load chan_gtalk.so
```

Accepting calls from Google Talk

To allow calls from other Google Talk users, we need to configure our dialplan to accept incoming calls. Inside your *extensions.conf* file, add the [gtalk_incoming] context:

```
[gtalk_incoming]
exten => s,1,Verbose(2,Incoming Gtalk call from ${CALLERID(all)})
    same => n,Answer()
    same => n,Dial(SIP/0000FFFF0001,30)
    same => n,Hangup()
```

We've now configured a simple test dialplan that will send calls to the SIP/0000FFFF0001 device and wait 30 seconds before hanging up the line. The s extension can be used to match any incoming call from Google Talk or Google Voice, but if you have multiple accounts that could be coming into this context, you can match different users by specifying the username portion of the Gmail email address as the extension. So, for example, if we had a user *my_asterisk_user@gmail.com*, the username portion would be my_asterisk_user, and this is what we'd specify in [gtalk_incoming]:

```
[gtalk_incoming]
exten => my_asterisk_user,1,Verbose(2,Gtalk call from ${CALLERID(all)})
    same => n,Answer()
    same => n,Dial(SIP/0000FFFF0001,3)
    same => n,Hangup()
```

The order of rules used for matching incoming calls to chan_gtalk is:

1. Match the username portion of the Gmail account in the context specified for the [guest] account.
2. Match the s extension in the context specified for the [guest] account.
3. Match the s extension in the [default] context.

Accepting calls from Google Voice

The configuration for accepting calls from Google Voice is similar (if not identical) to that for Google Talk, which we set up in the preceding section. A little tip, though, is that sometimes you can't disable the call screening functionality (for some reason we still got it even when we'd disabled it in the Google Voice control panel). If you run into this problem but don't want to have to screen your calls, you can automatically send the DTMF prior to ringing your device by adding the two boldface lines shown here prior to performing the `Dial()`:

```
[gtalk_incoming]
exten => s,1,Verbose(2,Incoming call from ${CALLERID(all)})
    same => n,Answer()
    same => n,Wait(2)
    same => n,SendDTMF(2)
    same => n,Dial(SIP/0000FFFF0001,30)
    same => n,Hangup()
```

Here, we're using the `Wait()` and `SendDTMF()` applications to first wait 2 seconds after answering the call (which is the time when the call screening message will start) and then accept the call automatically (by sending DTMF tones for the number 2). After that, we then send the call off to our device.

Outgoing calls via Google Talk

To place a call to a Google Talk user, configure your dialplan like so:

```
[LocalSets]
exten => 123,1,Verbose(2,Extension 123 calling some_user@gmail.com)
    same => n,Dial(Gtalk/asterisk/some_user@gmail.com,30)
    same => n,Hangup()
```

The `Gtalk/asterisk/some_user@gmail.com` part of the `Dial()` line can be broken into three parts. The first part, `Gtalk`, is the protocol we're using for placing the outgoing call. The second part, `asterisk`, is the account name as defined in the *jabber.conf* file. The last part, `some_user@gmail.com`, is the location we're attempting to place a call to.

Outgoing calls via Google Voice

To place calls using Google Voice to PSTN numbers, create a dialplan like the following:

```
[LocalSets]
exten => _1NXXNXXXXXX,1,Verbose(2,Placing call to ${EXTEN} via Google Voice)
    same => n,Dial(Gtalk/asterisk/+${EXTEN}@voice.google.com)
    same => n,Hangup()
```

Let's discuss the `Dial()` line briefly, so you understand what is going on. We start with `Gtalk`, which is the technology we'll use to place the call. Following that, we have defined the **asterisk** user as the account we'll use to authenticate with when placing our outgoing call (this is configured in *jabber.conf*). Next is the number we're attempting to place a call to, as defined in the `${EXTEN}` channel variable. We've prefixed the `${EXTEN}` channel variable with a plus sign (+), as it's required by the Google network when placing calls. We've also appended `@voice.google.com` to let the Google servers know this is a call that should be placed through Google Voice[†] as opposed to to another Google Talk user.

[†] You may have to purchase credits from Google Voice in the control panel in order to place calls to certain destinations.

Skype Integration

Skype integration now exists with Asterisk through a commercial module from Digium called Skype for Asterisk (SFA).‡ The SFA module loads directly into Asterisk and allows communication with all users on the Skype network directly by using an account created on the Skype Manager. Previous methods were messy, requiring the use of a Windows-based computer running another instance of Skype controlled via an API (application programming interface) and directing media to a sound card and into Asterisk via `chan_oss` or `chan_alsa`. Now, two methods exist: SFA and Skype Connect (formerly known as Skype for SIP).

As an Asterisk module, Skype for Asterisk does have some features that Skype Connect does not, including text chat, presence updates, and the ability to call a Skype user directly rather an via SkypeIn number. Additionally, Skype for Asterisk utilizes Skype's encryption for calls, which provides security benefits without the need to use SRTP (secure RTP) with SIP.

Installation of Skype for Asterisk

Since Skype for Asterisk is a commercial product, documentation for installing and configuring the module is available from Digium directly. For the most up-to-date installation documentation and information about the Skype for Asterisk module, see *http://www.digium.com/en/products/software/skypeforasterisk.php*.

You can download the modules from *http://downloads.digium.com/pub/telephony/sky peforasterisk*, and the registration utility for registering commercial modules from Digium is available at *http://downloads.digium.com/pub/register*.

Using Skype for Asterisk

In this section we'll explore the various ways we can utilize Skype from our dialplan, such as sending calls to and receiving calls from users on the Skype network and exchanging messages with our Skype buddies. We'll also show you how to implement a clever dialplan that will make it easier to call your friends on the Skype network without having to assign everyone an extension number.

Configuring chan_skype.conf

While the *README* file that comes with the Skype module helps to document the *chan_skype.conf* configuration, and the sample *chan_skype.conf* file is well-documented, it is worth showing a simple version of the configuration for the purposes of documenting the usage of Skype from the dialplan.

---

‡ Skype for Asterisk currently retails for $66 per license, and includes a G.729 license. Each license permits one simultaneous call.

Users configured in *chan_skype.conf* must be created with the Skype Manager interface. Personal Skype IDs are not allowed.

Our example Skype user will be *pbx.shifteight.org*. We'll configure this user in *chan_skype.conf*. There are additional options that could be set here, but for our purposes we're keeping it simple:

```
[general]
default_user=pbx.shifteight.org

[pbx.shifteight.org]
secret=my_secret_pass
context=skype_incoming
exten=start
buddy_autoadd=true
```

The default_user option in the [general] section is used to control which account we should use when placing calls via Skype. If we had multiple accounts, the default_user would be used when placing calls unless we specified a different user to place the call as (we'll discuss this further in the next section).

We've also defined the password (secret), the context incoming calls will enter into, and the extension (exten) that will be executed within the context. If we had multiple Skype users and wanted to control all of them from the same context, we could give them each different extension values, such as exten=leifmadsen or exten=russellbryant.

Additionally, we've enabled the ability to automatically add people who contact us to our buddies list.

Placing and receiving calls via Skype

Placing a call to a Skype buddy is relatively straightforward. Like with other channel types in Asterisk, the Skype channel type is used to place calls to endpoints on the Skype network.

Utilizing the Dial() application from the dialplan, we can place calls to other Skype users:

```
[LocalSets]
exten => 100,1,Answer()
    same => n,Dial(Skype/vuc.me,30)
    same => n,Playback(silence/1&user&is-currtly-unavail)
    same => n,Hangup()
```

Our dialplan simply answers the call and attempts to place a call to *vuc.me*,§ wait 30 seconds for that user to answer, and, if there is no answer, play back a message saying that the user is currently unavailable before hanging up. We could, of course, be more elaborate with our dialplan; for example, we could turn this into a `Macro()` or `GoSub()` routine so we just needed to pass in the name of the person we wish to call.

Unfortunately, if you're utilizing a device that only has a number pad for dialing, you'll need to assign extension numbers to all your favorite Skype buddies. However, we've come up with a clever way of reading back your online buddies to you, which we'll describe in "Calling your Skype buddies without assigning extension numbers" on page 433.

If you have a softphone, though, you should have the ability to place calls by dialing names directly. We can use this to our advantage by creating a pattern match in our dialplan with the prefix of `SKYPE`:

```
[LocalSets]
exten => _SKYPE-.,1,Verbose(2,Dialing via Skype)
    same => n,Set(NameToDial=${FILTER(a-zA-Z0-9.,${EXTEN:6})})
    same => n,Playback(silence/1&pls-wait-connect-call)
    same => n,Dial(Skype/${NameToDial},30)
    same => n,Playback(user&is-curntly-unavail)
    same => n,Hangup()
```

By dialing `SKYPE-vuc.me`, we can dial the VoIP Users Conference via Skype from our softphone. The `FILTER()` function is used here to control what we're allowed to pass to the `Dial()` application. If we didn't do any filtering, someone could potentially send a string like `SKYPE-nobody&SIP/my_itsp/4165551212`, replacing the number *4165551212* with a number that is very expensive to call. By using `FILTER()`, we restrict the allowable characters to alphanumeric characters and periods.

After that, we're simply passing the string to the `Dial()` application and waiting for an answer for 30 seconds. If no one answers, an audio message is played back to the caller stating that the user is unavailable and then the call is hung up.

To receive calls, you simply need to configure your user in the *chan_skype.conf* file as described in "Configuring chan_skype.conf" on page 429. Once you've done that, you can configure your dialplan to answer calls like so:

```
[skype_incoming]
exten => start,1,Verbose(2,Incoming Skype Call)
    same => n,Answer()
    same => n,Dial(SIP/0000FFFF0001,30)
    same => n,Playback(user&is-curntly-unavail)
    same => n,Hangup()
```

Obviously, you can change this section of the dialplan to be more elaborate; all we've done is configured the dialplan to call our SIP device at 0000FFFF0001, wait for an answer

---

§ The VUC is the VoIP Users Conference, which runs weekly at 12:00 noon Eastern time (–0500 GMT). More information is available at *http://vuc.me*.

for 30 seconds, and then (if there is no answer or the device is busy or unavailable) play back a prompt that says the user is currently unavailable, followed by a hangup.

We've just shown you how to place and receive calls via Skype. The following sections will show you how to send and receive messages via the Skype network, and how to place calls to your Skype buddies without assigning extension numbers to them.

Sending and receiving messages via Skype

Sending and receiving messages via Skype is similar to doing this via XMPP (Jabber), which we described in "Sending messages with JabberSend()" on page 421 and "Receiving messages with JABBER_RECEIVE()" on page 422, so we won't go into quite the detail in these sections as we did there. Please review the sections about XMPP messaging before continuing, as we'll be using the same basic dialplans to accomplish sending and receiving of messages via Skype, while making adjustments to use the appropriate dialplan applications and functions.

The primary thing to remember is that messages are sent with the dialplan application SkypeChatSend() and received with the dialplan function SKYPE_CHAT_RECEIVE(). Additionally, messages can only be received when the SKYPE_CHAT_RECEIVE() function has been called from the dialplan, and it blocks (does not continue in the dialplan) while waiting for a message.

Sending a message from the dialplan to a Skype buddy is relatively straightforward. Here is a simple dialplan we can use to send a message from Asterisk to someone on the Skype network:

```
[LocalSets]
exten => 104,1,Answer()

; *** This line should not have any line breaks
    same => n,SkypeChatSend(pbx.shifteight.org,tfot.madsen,Incoming call from
${CALLERID(all)})

    same => n,Dial(SIP/0000FFFF0002,30)
    same => n,Hangup()
```

Our dialplan is simple. We created a test extension of 104 that answers the line, then sends a message to Skype user tfot.madsen from the pbx.shifteight.org account (which we configured in the *chan_skype.conf* file). The message sent is "Incoming call from ${CALLERID(all)}", where the caller ID is provided by the CALLERID() function. After sending our message, we then dial the device located at 0000FFFF0002 and hang up if no one answers within 30 seconds.

That's it for sending messages via Skype. Now let's look at some of the ways we can receive messages from Skype. Here is the simple example we explored in "Receiving messages with JABBER_RECEIVE()" on page 422), with a few changes made to reflect the technology. This time, we'll be replacing JabberSend() and JABBER_RECEIVE() with the SkypeChatSend() and SKYPE_CHAT_RECEIVE() dialplan application and function, respectively:

```
exten => 106,1,Answer()

   ; All text must be on a single line.
   same => n,SkypeChatSend(pbx.shifteight.org,tfot.madsen,Incoming call from
${CALLERID(all)}. Press 1 to route to desk. Press 2 to send to voicemail.)

   ; Wait for a response for 6 seconds.
   ; *** This line should not have any line breaks
   same => n,Set(SkypeResponse=
${SKYPE_CHAT_RECEIVE(pbx.shifteight.org,tfot.madsen,6)})

   same => n,GotoIf($["${SkypeResponse}" = "1"]?dial,1)
   same => n,GotoIf($["${SkypeResponse}" = "2"]?voicemail,1)
   same => n,Goto(dial,1)

exten => dial,1,Verbose(2,Calling our desk)
   same => n,Dial(SIP/0000FFFF0002,6)
   same => n,Goto(voicemail,1)

exten => voicemail,1,Verbose(2,VoiceMail)

; *** This line should not have any line breaks
   same => n,Set(VoiceMailStatus=${IF($[${ISNULL(${DIALSTATUS})}]
| "${DIALSTATUS}" = "BUSY"]?b:u)})

   same => n,Playback(silence/1)
   same => n,VoiceMail(100@lmentinc,${VoiceMailStatus})
   same => n,Hangup()
```

There you have it—sending and receiving messages via the Skype network!

 You can also send and receive messages with the Asterisk Manager Interface, the topic of Chapter 20.

We've essentially implemented a screen pop solution for incoming calls, but by allowing messages to be sent back to Asterisk via Skype within a defined period of time, we've also created a solution for redirecting calls prior to ringing any devices. A more functional version of the dynamic routing dialplan we just explored was developed in the section about JABBER_RECEIVE() earlier in this chapter: it used the Local channel to get around the dialplan blocking issue, enabling calls can be routed even after a device has started to be rung.

Calling your Skype buddies without assigning extension numbers

While working on this book, we had some issues with trying to come up with clever ways to use a text-to-speech engine. It seemed that dynamic data would need to be involved for text-to-speech to really make a lot of sense—otherwise, why not use pre-recorded prompts instead? However, an idea finally came to us, based on the fact that having to assign extension numbers to each Skype user we wanted to call was not only cumbersome, but was a mental exercise we weren't willing to take on.

The following dialplan makes use of the SKYPE_BUDDIES() and SKYPE_BUDDY_FETCH() dialplan functions to retrieve all the Skype buddies in memory on the server, and to read those buddies' names back to you along with their statuses. After each buddy name is read, a prompt asking if this is who you wish to call is presented, with the option of asking for another buddy from the list. We've utilized the Festival() application for this example (the configuration and setup of which can be found in "Festival" on page 440) to read back the users' names. Once a buddy has been marked as selected, it is then dialed using the Dial() application.

Our implementation is as follows:

```
[LocalSets]
exten => 75973,1,Verbose(2,Read off list of Skype accounts)
    same => n,Answer()
    same => n,Set(ID=${SKYPE_BUDDIES(pbx.shifteight.org)})
    same => n(new_buddy),Set(ARRAY(buddy,status)=${SKYPE_BUDDY_FETCH(${ID})})
    same => n,GotoIf($[${ISNULL(${buddy})}]?no_more_buddies)
    same => n,Festival(${buddy} is ${status})
    same => n,Read(Answer,if-correct-press&digits/1&otherwise-press&digits/2,1)
    same => n,GotoIf($[${Answer} = 2]?new_buddy)
    same => n,Dial(Skype/${buddy},30)
    same => n,Playback(user&is-curntly-unavail)
    same => n,Hangup()

exten => no_more_buddies,1,Verbose(2,No more buddies to find)
    same => n,Playback(dir-nomore)
    same => n,Hangup()
```

LDAP Integration

Asterisk supports the ability to connect to an existing Lightweight Directory Access Protocol (LDAP) server to load information into your Asterisk server using the Asterisk Realtime Architecture (ARA). The advantage of integrating Asterisk and LDAP will become immediately obvious when you start centralizing your authentication mechanisms to the LDAP server and utilizing it for several applications: you significantly cut down the administrative overhead of managing your users by placing all their information into a central location.

There are both commercial and open source LDAP servers available, the most popular commercial solution likely being that implemented by Microsoft Windows servers. A popular open source LDAP server is OpenLDAP (*http://www.openldap.org*). We will not delve into the configuration of the LDAP server here, but we will show you the schema required to connect Asterisk to your server and to use it to provide SIP connections and voicemail service to your existing user base.

Configuring OpenLDAP

While a discussion of the installation and configuration of an LDAP server is beyond the scope of this chapter, it is certainly applicable to show you how we expanded our initial LDAP schema to include the information required for Asterisk integration. Our initial installation followed instructions from the Ubuntu documentation page located at *https://help.ubuntu.com/10.04/serverguide/C/openldap-server.html*. We only needed to follow the instructions up to and including the *backend.example.com.ldif* import; the next step after importing the backend configuration is installing the Asterisk-related schemas.

If you're following along, with the backend imported, change into your Asterisk source directory. Then copy the *asterisk.ldap-schema* file into the */etc/ldap/schema/* directory:

```
$ cd ~/src/asterisk-complete/asterisk/1.8/contrib/scripts/
$ sudo cp asterisk.ldap-schema /etc/ldap/schema/asterisk.schema
```

With the schema file copied in, restart the OpenLDAP server:

```
$ sudo /etc/init.d/slapd restart
```

Now we're ready to import the contents of *asterisk.ldif* into our OpenLDAP server. The *asterisk.ldif* file is located in the *contrib/scripts/* folder of the Asterisk source directory:

```
$ sudo ldapadd -Y EXTERNAL -H ldapi:/// -f asterisk.ldif
```

We can now continue with the instructions at *https://help.ubuntu.com/10.04/server guide/C/openldap-server.html* and import the *frontend.example.com.ldif* file. Within that file is an initial user, which we can omit for now as we're going to modify the user import portion to include an `objectClass` for Asterisk (i.e., in the example file, the section of text that starts with `uid=john` can be deleted).

We're going to create a user and add the configuration values that will allow the user to register his phone (which will likely be a softphone, since the hardphone on the user's desk will, in most cases, be configured from a central location) via SIP by using his username and password, just as he would normally log in to check email and such.

The configuration file we'll create next will get imported with the *ldapadd* command and will be added into the `people` object unit within the *shifteight.org* space. Be sure to change the values to match those of the user you wish to set up in LDAP and to substitute `dc=shifteight,dc=org` with your own location.

Before we create our file, though, we need to convert the password into an MD5 hash. Asterisk will not authenticate phones using plain-text passwords when connecting via LDAP. We can convert the password using the *md5sum* command:

```
$ echo "my_secret_password" | md5sum
a7be810a28ca1fc0668effb4ea982e58  -
```

We'll insert the returned value (without the hyphen) into the following file within the userPassword field, prefixed with {md5}:

```
$ cat > astuser.ldif

dn: uid=rbryant,ou=people,dc=shifteight,dc=org
objectClass: inetOrgPerson
objectClass: posixAccount
objectClass: shadowAccount
objectClass: AsteriskSIPUser
uid: rbryant
sn: Bryant
givenName: Russell
cn: RussellBryant
displayName: Russell Bryant
uidNumber: 1001
gidNumber: 10001
userPassword: {md5}a7be810a28ca1fc0668effb4ea982e58
gecos: Russell Bryant
loginShell: /bin/bash
homeDirectory: /home/russell
shadowExpire: -1
shadowFlag: 0
shadowWarning: 7
shadowMin: 8
shadowMax: 999999
shadowLastChange: 10877
mail: russell.bryant@shifteight.org
postalCode: 31000
l: Huntsville
o: shifteight
title: Asterisk User
postalAddress:
initials: RB
AstAccountCallerID: Russell Bryant
AstAccountContext: LocalSets
AstAccountDTMFMode: rfc2833
AstAccountMailbox: 101@shifteight
AstAccountNAT: yes
AstAccountQualify: yes
AstAccountType: friend
AstAccountDisallowedCodec: all
AstAccountAllowedCodec: ulaw
AstAccountMusicOnHold: default
```

The one field we should explicitly mention here is the userPassword field. We require that the value in the LDAP server contain the password we're going to authenticate from the phone with to be in the format of an MD5 hash. In versions prior to Asterisk 1.8.0, the prefix of {md5} in front of the hash was required. While it is no longer necessary, it is still recommended.

With the file created, we can add the user to our LDAP server:

```
$ sudo ldapadd -x -D cn=admin,dc=shifteight,dc=org -f astusers.ldif -W
Enter LDAP Password:
adding new entry "uid=rbryant,ou=people,dc=shifteight,dc=org"
```

Our user has now been imported into LDAP. The next step is to configure Asterisk to connect to the LDAP server and allow users to authenticate and register their phones.

Compiling LDAP Support into Asterisk

With our OpenLDAP server configured and the schema imported, we need to install the dependencies for Asterisk and compile the `res_config_ldap` module. This module is the key that will allow us to configure Asterisk realtime for accessing our peers via LDAP.

Once we've installed the dependency, we need to rerun the *./configure* script inside the Asterisk source directory, then verify that the `res_config_ldap` module is selected. Then we can run *make install* to compile and install the new module.

Ubuntu dependencies

On Ubuntu, we need to install the *openldap-dev* package to provide the dependency for the `res_config_ldap` module:

```
$ sudo apt-get install openldap-dev
```

CentOS dependencies

On CentOS, we need to install the *openldap-devel* package to provide the dependency for the `res_config_ldap` module:

```
$ sudo yum install openldap-devel
```

Configuring Asterisk for LDAP Support

Now that we've configured our LDAP server and installed the `res_config_ldap` module, we need to configure Asterisk to support loading of peers from LDAP. To do this, we need to configure the *res_ldap.conf* file to connect to the LDAP server and the *extconfig.conf* file to tell Asterisk what information to get from the LDAP server, and how. Once that is done, we can configure any remaining module configuration files, such as *sip.conf*, *iax.conf*, *voicemail.conf*, and so on, where appropriate. In our example we'll be configuring Asterisk to load our SIP peers from realtime using the LDAP server as our database.

Configuring res_ldap.conf

The *res_ldap.conf.sample* file is a good place to start because it contains a good set of templates. At the top of the file, though, under the [_general] section, we need to configure how Asterisk is going to connect to our LDAP server. Our first option is url, which will determine how to connect to the server. We have defined a connection as ldap://172.16.0.103:389, which will connect to the LDAP server at IP address 172.16.0.103 on port 389. If you have a secure connection to your LDAP server, you can replace ldap:// with ldaps://. Additionally, we have set protocol=3 to state that we're connecting with protocol version 3, which in most (if not all) cases will be correct.

The last three options, basedn, user, and pass, are used for authenticating to our LDAP server. We need to specify:

- The basedn (dc=shifteight,dc=org), which is essentially our domain name
- The user name we're going to authenticate to the LDAP server as (admin)
- The password for the user to authenticate with (canada)

If we put it all together, we end up with something like the following:

```
[_general]
url=ldap://172.16.0.103:389
protocol=3
basedn=dc=shifteight,dc=org
user=cn=admin,dc=shifteight,dc=org
pass=canada
```

Beyond this, in the rest of the sample configuration file we'll see lots of templates we can use for mapping the information in Asterisk onto our LDAP schema. Lets take a look at the first lines of the [sip] template that we'll be using to map the information of our SIP peers into the LDAP database:

```
[sip]
name = cn
amaflags = AstAccountAMAFlags
callgroup = AstAccountCallGroup
callerid = AstAccountCallerID
...
lastms = AstAccountLastQualifyMilliseconds
useragent = AstAccountUserAgent
additionalFilter=(objectClass=AsteriskSIPUser)
```

On the left side we have the field name Asterisk will be looking up, and on the right is the mapping to the LDAP schema for the request. Our first set of fields is mapping the name field to the cn field on the LDAP server. If you look back at the data we imported in "Configuring OpenLDAP" on page 435, you'll see that we have created a user and assigned the value of RussellBryant to the cn field. So, in this case, we're mapping the authentication name (the name field) from the SIP user to the value of the cn field in the LDAP server (RussellBryant).

This goes for the rest of the values all the way down, with some fields (i.e., useragent, lastms, ipaddr, etc.) simply needing to exist so Asterisk can write information (e.g., registration information) to the LDAP server.

Configuring extconfig.conf

Our next step is to tell Asterisk what information to load via realtime and what technology to use. Using the *extconfig.conf* file, we have the option of loading several modules dynamically (and we can also load files statically). For more information about Asterisk realtime, see "Using Realtime" on page 368.

For our example, we're going to configure the `sipusers` and `sippeers` dynamic realtime objects to load our SIP peers from LDAP. In the following example, we have a line like this:

```
ldap,"ou=people,dc=shifteight,dc=org",sip
```

We've specified three arguments. The first is `ldap`, which is the technology we're going to use to connect to our realtime object. There are other technologies available, such as `odbc`, `pgsql`, `curl`, and so on. Our second argument, enclosed in double quotes, specifies which database we're connecting to. In the case of LDAP, we're connecting to the object-unit `people` within the domain `shifteight.org`. Lastly, our third argument, `sip`, defines which template we're using (as defined in *res_ldap.conf*) to map the realtime data to the LDAP database.

> Additionally, you can specify a fourth argument, which is the priority. If you define multiple realtime objects, such as when defining `queues` or `sippeers`, you can utilize the priority argument to control failover if a particular storage engine becomes unavailable. Priorities must start at `1` and increment sequentially.

To define the use of `sipusers` and `sippeers` from the LDAP server, we would enable these lines in *extconfig.conf*:

```
sipusers => ldap,"ou=people,dc=shifteight,dc=org",sip
sippeers => ldap,"ou=people,dc=shifteight,dc=org",sip
```

Configuring sip.conf for realtime

These steps are optional for configuring SIP for realtime, although you will likely expect things to work in the manner we're going to describe. In the *sip.conf* file, we will enable a few realtime options that will cache information into memory as it is loaded from the database. By doing this, we'll allow Asterisk to place calls to devices by simply looking at the information stored in memory. Not only does caching make realtime potentially more efficient, but things like device state updates simply can't work unless the devices are cached in memory.

> A peer is only loaded into memory upon registration of the device or placing a call to the device. If you run the command *sip reload* on the console, the peers will be cleared from memory as well, so you may need to adjust your registration times if that could cause issues in your system.

To enable peer caching in Asterisk, use the `rtcachefriends` option in *sip.conf*:

```
rtcachefriends=yes
```

There are additional realtime options as well, such as `rtsavesysname`, `rtupdate`, `rtauto clear`, and `ignoreregexpire`. These are all explained in the *sip.conf.sample* file located within your Asterisk source.

Text-to-Speech Utilities

Text-to-speech utilities are used to convert strings of words into audio that can be played to your callers. Text-to-speech has been around for many years, and has been continually improving. While we can't recommend text-to-speech utilities to take the place of professionally recorded prompts, they do offer some degree of usefulness in applications where dynamic data needs to be communicated to a caller.

Festival

Festival is one of the oldest running applications for text-to-speech on Linux. While the quality of Festival is not sufficient for us to recommend it for production use, it is certainly a useful way of testing a text-to-speech-based application. If a more polished sound is required for your application, we recommend you look at Cepstral (covered next).

Installing Festival on CentOS

Installing Festival and its dependencies on CentOS is straightforward. Simply use *yum* to install the *festival* package:

```
$ sudo yum install festival
```

Installing Festival on Ubuntu

To install Festival and its dependencies on Ubuntu, simply use *apt-get* to install the *festival* package:

```
$ sudo apt-get install festival
```

Using Festival with Asterisk

With Festival installed, we need to module the *festival.scm* file in order to enable Asterisk to connect to the Festival server. On both CentOS and Ubuntu, the file is located in */usr/share/festival/*. Open the file and place the following text just above the last line, (provide 'festival):

```
(define (tts_textasterisk string mode)
"(tts_textasterisk STRING MODE)
Apply tts to STRING. This function is specifically designed for
use in server mode so a single function call may synthesize the string.
```

```
This function name may be added to the server safe functions."
(let ((wholeutt (utt.synth (eval (list 'Utterance 'Text string)))))
(utt.wave.resample wholeutt 8000)
(utt.wave.rescale wholeutt 5)
(utt.send.wave.client wholeutt)))
```

After adding that, you need to start the Festival server:

```
$ sudo festival_server 2>&1 > /dev/null &
```

Using *menuselect* from your Asterisk source directory, verify that the `app_festival` application has been selected under the *Applications* heading. If it was not already selected, be sure to run *make install* after selecting it to install the `Festival()` dialplan application.

Before you can use the `Festival()` application, you need to tell Asterisk how to connect to the Festival server. The *festival.conf* file is used for controlling how Asterisk connects to and interacts with the Festival server. The sample *festival.conf* file located in the Asterisk source directory is a good place to start, so copy *festival.conf.sample* from the *configs/* subdirectory of your Asterisk source to the */etc/asterisk/* configuration directory now:

```
$ cp ~/asterisk-complete/asterisk/1.8/configs/festival.conf.sample \
/etc/asterisk/festival.conf
```

The default configuration is typically enough to connect to the Festival server running on the local machine, but you can optionally configure parameters such as the `host` where the Festival server is running (if remote), the `port` to connect to, whether to enable caching of files (defaults to `no`), the location of the cache directory (defaults to */tmp*), and the command Asterisk passes to the Festival server.

You can verify that the `Festival()` dialplan application is accessible by running *core show application festival* from the Asterisk console:

```
*CLI> core show application festival
```

If you don't get output, you may need to load the *app_festival.so* module:

```
*CLI> module load app_festival.so
```

Verify that the *app_festival.so* module exists in */usr/lib/asterisk/modules/* if you're still having issues with loading the module.

After loading the `Festival()` application into Asterisk, you need to create a test dialplan extension to verify that `Festival()` is working:

```
[LocalSets]
exten => 203,1,Verbose(2,This is a Festival test)
    same => n,Answer()
    same => n,Playback(silence/1)
    same => n,Festival(Hello World)
    same => n,Hangup()
```

Reload the dialplan with the *dialplan reload* command from the Asterisk console, and test out the connection to Festival by dialing extension 203.

Alternatively, if you're having issues with the Festival server, you could use the following method to generate files with the *text2wave* application supplied with the *festival* package:

```
exten => 202,1,Verbose(2,Trying out Festival)
    same => n,Answer()

; *** This line should not have any line breaks
    same => n,System(echo "This is a test of Festival"
| /usr/bin/text2wave -scale 1.5 -F 8000 -o /tmp/festival.wav)

    same => n,Playback(/tmp/festival)
    same => n,System(rm -f /tmp/festival.wav)
    same => n,Hangup()
```

You should now have enough to get started with generating text-to-speech audio for your Asterisk system. The audio quality is not brilliant, and the speech generated is not clear enough to be easy to understand over a telephone, but for development and testing purposes Festival is an application that can fill the gap until you're ready for a more professional-sounding text-to-speech generator such as Cepstral.

Cepstral

Cepstral is a text-to-speech engine that works in a similar manner as the `Festival()` application in the dialplan, but produces much higher-quality sound. Not only is the quality significantly better, but Cepstral has developed a text-to-speech engine that emulates Allison's voice, so your text-to-speech engine can sound the same as the English sound files that ship with Asterisk by default, to give a consistent experience to the caller.

Cepstral is commercial module, but for around $30 you can have a text-to-speech engine that is clearer, is more consistent with other sound prompts on your system, and provides a more pleasurable experience for your callers. The Cepstral software and installation instructions can be downloaded from the Digium.com webstore at *http://www.digium.com/en/products/software/cepstral.php*.

Conclusion

In this chapter we focused on integrating Asterisk with external services that may not be directly related to generating or handling calls, but do enable tighter coupling with existing services on your network by providing information for call routing, or information about your users from your existing infrastructure.

Fax

Have no fear of perfection. You'll never reach it.

—Salvador Dali

The concept of facsimile transmission has been around for over 100 years, but it was not until the 1980s that the use of fax machines became essential in business. This lasted for perhaps two decades. Then the Internet came along, and very shortly after that, the fax quickly became almost irrelevant.

What Is a Fax?

A fax machine allows a facsimile (copy) of a document to be transmitted across a telephone line. In the Internet age, this sort of functionality seems useless; however, prior to ubiquitous Internet access, this was a very useful thing indeed. Fax machines scan a document into a digital format, transmit the digital information in a manner similar to that used by an analog modem, and then convert and print the received information on the other end.

Ways to Handle Faxes in Asterisk

Asterisk offers the ability to both send and receive faxes, but it should be noted that all Asterisk is doing is the basics of fax transport. This means that providing a complete experience to your users will require external programs and resources beyond what Asterisk delivers.

Asterisk Fax can:

- Recognize an incoming fax connection, and negotiate a session
- Store (receive) the incoming fax as a Tagged Image File Format (TIFF) file
- Accept TIFF files in a fax-compatible format
- Transmit TIFF files to another fax machine

Asterisk Fax cannot:

- Print faxes
- Accept documents for transmission in any format other than TIFF

Receiving is relatively simple, since the format of the document is determined at the sending end, and thus all Asterisk needs to do is store the document.

Transmitting is somewhat more complex, since the transmitting end is responsible for ensuring that the document to be sent is in the correct format for faxing. This typically places a burden on the user to understand how to create a properly formatted document, or requires complex client or server software to handle the formatting (for example, through a print driver installed on the local PC) and placement of the fax job in a location where the server can grab it and transmit it.

spandsp

Initially, the only way to handle faxing in Asterisk was through the spandsp library. spandsp provides a multitude of Digital Signal Processing (DSP) capabilities, but in this context all we are interested in is its fax functionality.

Asterisk has the hooks built in to make use of spandsp, but due to incompatible licenses, the spandsp libraries must be downloaded and compiled separately from Asterisk. Also, since spandsp was not written only for Asterisk, it will not assume that it is being installed for Asterisk. This means that a few extra steps will be required to ensure Asterisk can use spandsp.

Obtaining spandsp

As of this writing, the current version of spandsp is 0.0.6.

Download and extract the spandsp source code as follows:

```
$ mkdir ~/src/asterisk-complete/thirdparty
$ cd ~/src/asterisk-complete/thirdparty
$ wget http://www.soft-switch.org/downloads/spandsp/spandsp-0.0.6pre17.tgz
$ tar zxvf spandsp-0.0.6pre17.tgz
$ cd spandsp-0.0.6
```

Compiling and Installing spandsp

The spandsp software should compile and install with the following commands:

```
$ ./configure
$ make
$ sudo make install
```

This will install the library in the */usr/local/lib/* folder. On many Linux systems this folder is not automatically part of the library path (*libpath*), so it will need to be added manually.

Adding the spandsp Library to Your libpath

In order to make the `spandsp` library visible to all applications on the system, the folder where it is located must be added to the *libpath* for the system. This is typically done by editing files in the */etc/ld.so.conf.d/* directory. You simply need to ensure that one of the files in that directory has `/usr/local/lib` listed. If not, the following command will create a suitable file for you:

```
$ sudo cat >> /etc/ld.so.conf.d/usrlocallib.conf
/usr/local/lib
```

Press Ctrl+D to save the file, then run the *ldconfig* command to refresh the library paths:

```
$ sudo ldconfig
```

You are now ready to recompile Asterisk for spandsp support.

Recompiling Asterisk with spandsp Support

Since the `spandsp` library was probably not installed on the system when Asterisk was first compiled, you will need to do a quick recompile of Asterisk in order to have the spandsp support added:

```
$ cd ~/src/asterisk-complete/asterisk/1.8/
$ ./configure
$ make menuselect
```

Ensure that under the *Resource Modules* heading, the section for spandsp looks like this:

```
[*] res_fax_spandsp
```

If you see this instead:

```
XXX res_fax_spandsp
```

It means that Asterisk was not able to find the `spandsp` library.

Once you have verified that Asterisk can see spandsp, you are ready to recompile. Save and exit from *menuselect*, and run the following:

```
$ make
$ make install
```

You can verify that spandsp is working with Asterisk by issuing the following command from the Asterisk CLI:

```
*CLI> module show like res_fax_spandsp.so
```

At this point the SendFAX() and ReceiveFAX() dialplan applications will be available to you.

Disabling spandsp (Should You Want to Test Digium Fax)

The spandsp library and the Digium fax library, discussed in the next section, are mutually exclusive. If you want to try out the Digium fax product, you will need to ensure that spandsp does not load. To disable spandsp in Asterisk, simply edit your */etc/asterisk/modules.conf* file as follows:

```
noload => res_fax_spandsp.so
```

Save the changes and restart Asterisk.

Digium Fax For Asterisk

Digium Fax For Asterisk (FFA) was developed out of a strong desire from the Asterisk community to have a Digium-supported fax mechanism in Asterisk. Free for single-channel use, this product can also be licensed from Digium to handle more than one simultaneous fax channel.

Obtaining Digium FFA

Digium FFA can be obtained from the Digium website (*http://www.digium.com*). The process of downloading, installing, and registering this library is thoroughly documented in the *Fax For Asterisk Administrator Manual*, which you can also download from Digium. You will need to register at the Digium website in order to obtain your free single-channel fax license key and download the admin manual. There is also a comprehensive *README* file included with the software, which details the steps necessary to get Digium FFA going on your system.

Disabling Digium FFA (Should You Want to Test spandsp)

The Digium FFA library and the spandsp library are mutually exclusive. If you want to try out spandsp, you will need to ensure that Digium FFA does not load. To disable FFA, simply edit your */etc/asterisk/modules.conf* file as follows:

```
noload => res_fax_digium.so
```

Save the changes and restart Asterisk.

Incoming Fax Handling

Received faxes are commonly encoded in Tagged Image File Format (TIFF). This graphics file format, while not as well known as JPEG or GIF, is not as obscure as one might think. In fact, we suspect your computer (whether you're running Windows, Linux, or MacOS) will already have the ability to interpret TIFF files built in. While it has become popular to offer PDF as a delivery format for received faxes, we're not sure this is strictly required, since TIFF is so ubiquitous.

Received faxes will be stored by Asterisk as files. Where those files are stored will depend on several factors, including:

- What software you are using to simulate a fax modem (e.g., IAXmodem, Digium ReceiveFAX, etc.)
- The location in your filesystem that you have configured for storage of received faxes
- Any post-receipt processing you have decided to perform on the files

In the dialplan, you will need to build in enough intelligence to name faxes in such a way that they will be distinct from each other. There are many channel variables and functions that can be used for this purpose, such as the STRFTIME() function. Asterisk can easily handle capturing the fax to a file, but you will need to make sense out of what happens to that file once it is stored on the system.

Fax to TIFF

The Tagged Image File Format is not very well known, but it is actually more common than you might realize, and since it is natively supported on Windows, MacOS, and Linux, TIFF files can be viewed on pretty much any computer with the most basic graphics viewer. A subset of the TIFF file format has for a long time been the de facto file format used for faxes.

Since Asterisk will receive and store faxes in TIFF format, there is no post-processing required. Once the incoming fax call has been completed, the resulting TIFF file can be opened directly from the folder where it was stored (or perhaps emailed to the intended user).

Fax to Email

Once Asterisk has received a fax, the resulting TIFF file needs a way to get to its final destination: a person.

The key consideration is that unless the information that Asterisk knows about the fax is sufficiently detailed, it may not be possible to deduce the intended recipient without having someone actually read the fax (it is common for a fax to have a cover page with the recipient's information written on it, which even the most capable text recognition software would have a difficult time making sense of). In other words, unless you dedicate a DID to each user who might receive a fax, Asterisk isn't going to be able to do much more than send all faxes to a single email address. You could code something in the dialplan to handle this, though, or have an external *cron* job or other daemon handle distributing the received faxes.

A simple dialplan to handle fax to email might look something like this (you will need the mail program *mutt* installed on your system):

```
exten => fax,1,Verbose(3,Incoming fax)
; folder where your incoming faxes will initially be stored
  same => n,Set(FAXDEST=/tmp)

; put a timestamp on this call so the resulting file is unique
  same => n,Set(tempfax=${STRFTIME(,,%C%y%m%d%H%M)})
  same => n,ReceiveFax(${FAXDEST}/${tempfax}.tif)
  same => n,Verbose(3,- Fax receipt completed with status: ${FAXSTATUS})

; *** This line should not have any line breaks
  same => n,System(echo | mutt -a ${FAXDEST}/${tempfax}
-s "received fax" somebody@shifteight.org)
```

Obviously, this sample would not be suitable for production (for example, it does not handle fax failure); however, it would be enough to start prototyping a more fully featured incoming fax handler.

Fax to PDF

Delivering a fax to a user as a PDF is a popular request. Since PCs, Macs, and Linux desktops can natively read TIFF files without any conversion required, this isn't strictly required. Nevertheless, many people insist on this functionality.

You can use a utility such as *ghostscript* to perform the conversion.

Fax Detection

You may have a dedicated phone number for receiving faxes. However, with Asterisk, that is not a requirement. Asterisk has the ability to detect that an incoming call is a fax and can handle it differently in the dialplan. Fax detection is available for both DAHDI and SIP channels. To enable it for DAHDI, set the faxdetect option in */etc/asterisk/chan_dahdi.conf*. In most cases, you should set this option to incoming. Table 19-1 lists the possible values for the faxdetect option in *chan_dahdi.conf*.

Table 19-1. Possible values for the faxdetect option in chan_dahdi.conf

Value	Description
incoming	Enables fax detection on inbound calls. When a fax is detected, applies the faxbuffers option if it has been set and redirects the call to the fax extension in the dialplan. For more information on the faxbuffers option, see "Using Fax Buffers in chan_dahdi.conf" on page 454.
outgoing	Enables fax detection on outbound calls. The dialplan is not executing on an outbound channel. If a fax is detected, the faxbuffers option will be applied and the channel will be redirected and start executing the dialplan at the fax extension.
both	Enables fax detection for both incoming and outgoing calls.
no	Disables fax detection. This is the default.

To enable fax detection for SIP calls, you must set the faxdetect option in */etc/asterisk/ sip.conf*. This option may be set in the [general] section, or for a specific peer. Table 19-2 covers the possible values for the faxdetect option in *sip.conf*.

Table 19-2. Possible values for the faxdetect option in sip.conf

Value	Description
cng	Enables fax detection by watching the audio for a CNG tone. If a CNG tone is detected, redirects the call to the fax extension in the dialplan.
t38	Redirects the call to the fax extension in the dialplan if a T.38 reinvite is received.
yes	Enables both cng and t38 fax detection.
no	Disables fax detection. This is the default.

Outgoing Fax Handling

Transmitting faxes from Asterisk is somewhat more difficult than receiving them. The reason for this is simply due to the fact that the preparation of the fax prior to transmission involves more work. There isn't anything particularly complex about fax transmittal, but you will need to make some design decisions about things like:

- How to get the source fax file formatted for Asterisk
- How to get the source fax file onto the Asterisk system (specifically, into some folder where Asterisk can access it)
- What to do with transmitted faxes (save them? delete them? move them somewhere?)
- How to handle transmission errors

Transmitting a Fax from Asterisk

To transmit a fax from Asterisk, you must have a TIFF file. How you generate this TIFF is important, and may involve many steps. However, from Asterisk's perspective the sending of a fax is fairly straightforward. You simply run the `SendFAX()` dialplan application, passing it the path to a valid TIFF file:

```
exten => faxthis,1,SendFAX(/path/to/fax/file,d)
```

In practice, you will normally want to set some parameters prior to transmission, so a complete extension for sending a fax using Digium's Fax For Asterisk might look something like this:

```
exten => faxthis,1,Verbose(2,Set options and transmit fax)

; some folder where your outgoing faxes will be found
  same => n,Set(faxlocation=/tmp)

; In production you would probably not want to hardcode the filename
  same => n,Set(faxfile=faxfile.tif)
  same => n,Set(FAXOPT(headerinfo)=Fax from ShiftEight.org)
  same => n,Set(FAXOPT(localstationid=4169671111)
  same => n,SendFax(${faxlocation}/${faxfile})
```

File Format for Faxing

The real trick of sending a fax is having a source file that is in a format that the fax engine can handle. At a basic level, these files are known as TIFF files; however, the TIFF spec allows for all sorts of parameters, not all of which are compatible with fax and not many of which are documented in any useful way. Additionally, the types of TIFF formats that spandsp can handle are different from those Digium FFA will handle.

In the absence of one simple, clear specification of what TIFF file format will work for sending faxes from Asterisk, we will instead document what we know to work, and leave it up to the reader to perform any experimentation required to find other ways to generate the TIFF.[*]

Digium's *Fax For Asterisk Administration Manual* documents a process for converting a PDF file into a TIFF using commonly available Linux command-line tools. While kludgy, this method should allow you to build Linux scripts to handle the file conversion, and your users will be able to submit PDFs as fax jobs.

You will need the *ghostscript* PDF interpreter, which can be installed in CentOS by the command:

---

[*] One format we tried was using the Microsoft Office Document Image Writer, which offers "TIFF-monochrome fax" as an output format. This seemed too good to be true, which is exactly what it turned out to be (neither spandsp nor Digium FFA could handle the resulting file). It would have been ideal to have found something common to Windows PCs that could be used by users to "print" an Asterisk-compatible TIFF file.

```
$ sudo yum -y install ghostscript
```

and in Ubuntu with:

```
$ sudo apt-get install ghostscript
```

Once installed, *ghostscript* can convert the PDF into an Asterisk-compatible TIFF file with the following command:

```
$ gs -q -dNOPAUSE -dBATCH -sDEVICE=tiffg4 -sPAPERSIZE=letter -sOutputFile=<dest> <src>
```

Replace *<dest>* with the name of the output file, and specify the location of your source PDF with *<src>*.

The *ghostscript* program should create a TIFF file from your PDF that will be suitable for transmission using Asterisk SendFax().

An Experiment in Email to Fax

Many users would like to be able to send emails as fax documents. The primary challenge with this is ensuring that what the users submit is in a format suitable for faxing. This ultimately requires some form of application development, which is outside the scope of this book.

What we have done is provided a simple example of some methods that at least provide a starting point for delivering email to fax capabilities.

One of the first changes that you would need to make in order to handle this is a change to your */etc/aliases* file, which will redirect incoming faxes to an application that can handle them. We are not actually aware of any app that can do this, so you'll have to write one. The change to your */etc/aliases* file would look something like this:

```
fax:    "| /path/to/program/that/will/handle/incoming/fax/emails"
```

In our case, Russell built a little Python script called *fax.py*, so our */etc/aliases* file would read something like this:

```
fax:    "| /asteriskpbx/fax.py"
```

We have included a copy of the Python script we developed for your reference in Example 19-1. Note that this file is not suitable for production, but merely serves as an example of how a very basic kind of email to fax functionality might be implemented.

Example 19-1. Proof of concept email to fax gateway, fax.py

```
#!/usr/bin/env python
"""Poor Man's Email to Fax Gateway.

This is a proof of concept email to fax gateway.  There are multiple aspects
that would have to be improved for it to be used in a production environment.

Copyright (C) 2010 - Russell Bryant, Leif Madsen, Jim Van Meggelen
Asterisk: The Definitive Guide
"""
```

```
import sys
import os
import email
import base64
import shutil
import socket


AMI_HOST = "localhost"
AMI_PORT = 5038
AMI_USER = "hello"
AMI_PASS = "world"

# This script will pull a TIFF out of an email and save it off to disk to allow
# the SendFax() application in Asterisk to send it.  This is the location on
# disk where the TIFF will be stored.
TIFF_LOCATION = "/tmp/loremipsum.tif"


# Read an email from stdin and parse it.
msg = email.message_from_file(sys.stdin)

# For testing purposes, if you wanted to read an email from a file, you could
# do this, instead.
#try:
#    f = open("email.txt", "r")
#    msg = email.message_from_file(f)
#    f.close()
#except IOError:
#    print "Failed to open email input file."
#    sys.exit(1)

# This next part pulls out a TIFF file attachment from the email and saves it
# off to disk in a format that can be used by the SendFax() application.  This
# part of the script is incredibly non-flexible.  It assumes that the TIFF file
# will be in a specific location in the structure of the message (the second
# part of the payload, after the main body).  Further, it assumes that the
# encoding of the TIFF attachment is base64.  This was the case for the test
# email that we were using that we generated with mutt.  Emails sent by users'
# desktop email clients will vary in _many_ ways.  To be used with user-
# generated emails, this section would have to be much more flexible.
try:
    f2 = open(TIFF_LOCATION, "w")
    f2.write(base64.b64decode(msg.get_payload()[1].get_payload().replace("\n", "")))
    f2.close()
except IOError:
    print "Failed to open file for saving off TIFF attachment."
    sys.exit(1)

# Now that we have a TIFF file to fax, connect to the Asterisk Manager Interface
# to originate a call.
ami_commands = """Action: Login\r
Username: %s\r
Secret: %s\r
```

```
\r
Action: Originate\r
Channel: Local/s@sendfax/n\r
Context: receivefax\r
Extension: s\r
Priority: 1\r
SetVar: SUBJECT=%s\r
\r
Action: Logoff\r
\r
""" % (AMI_USER, AMI_PASS, msg['subject'])

print ami_commands

def my_send(s, data):
    """Ensure that we send out the whole data buffer.
    """
    sent = 0
    while sent < len(data):
        res = s.send(data[sent:])
        if res == 0:
            break
        sent = sent + res

def my_recv(s):
    """Read input until there is nothing else to read.
    """
    while True:
        res = s.recv(4096)
        if len(res) == 0:
            break
        print res

s = socket.socket(socket.AF_INET, socket.SOCK_STREAM)
s.connect((AMI_HOST, AMI_PORT))
my_send(s, ami_commands)
my_recv(s)
s.shutdown(socket.SHUT_RDWR)
s.close()
```

We tested this out at a very rudimentary level, and proved the basic concept. If you want to put email to fax into production, you need to understand that you will have more work to do on the application development side in order to deliver something that will actually be robust enough to turn over to an average group of users.

Fax Pass-Through

In theory, it should be possible to connect a traditional fax machine to an FXS port of some sort and then pass incoming faxes to that device (see Figure 19-1). This concept is attractive for a few reasons:

1. It allows you to integrate existing fax machines with your Asterisk system.
2. It requires far less configuration in the dialplan.

Unfortunately, fax pass-through is not the home run we would like it to be. The analog carrier signal that two fax machines use to communicate is a delicate thing, and any corruption of that signal will often cause a transmission failure. In an Asterisk system performing pass-through, internal timing issues, coupled with signal attenuation, can create an environment that is unstable for fax use, especially for larger (multipage) faxes.

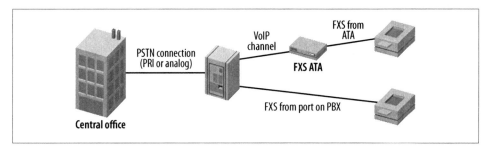

Figure 19-1. Typical fax pass-through

If you are using fax on a casual basis (mostly noncritical, one-page faxes), this sort of setup can work well. If faxing is critical to your business, or you are often expecting multipage faxes, we must reluctantly recommend that you connect your fax machines directly to the PSTN and leave Asterisk out of it.

Using Fax Buffers in chan_dahdi.conf

Many of the problems with fax pass-through are caused by inconsistent timing. Since faxes are more tolerant of latency than voice calls (a fax has to be able to travel halfway around the world, which takes a few dozen milliseconds), the introduction of a buffer in DAHDI (which is strictly used for faxes) has reportedly corrected many of the problems that have plagued fax pass-through.

As of this writing, this is a fairly new configuration option. The currently preferred setting is as follows:

```
faxbuffers => 12,half
```

This would be placed in your */etc/asterisk/chan_dahdi.conf* file and would cause *chan_dahdi* to create a 96 ms buffer for fax calls and delay start of transmission until the buffer was half full.

You would also need to set `faxdetect`, since the fax buffers are part of the `faxdetect` functionality:

```
faxdetect = both
```

We have not extensively tested this capability yet, but anecdotal evidence suggests that this should greatly improve the performance of fax pass-through in Asterisk.

Conclusion

Fax is a technology whose days are behind it. Having said that, it remains popular. Asterisk has some interesting technology built into it that allows you some level of creativity in how you handle faxes. With careful planning and system design, and a patient prototyping and debugging phase, you can use your Asterisk system to handle faxing in creative ways.

Asterisk Manager Interface (AMI)

John Malkovich: I have seen a world that
NO man should see!

Craig Schwartz: Really? Because for most people it's a
rather enjoyable experience.

—*Being John Malkovich*

The Asterisk Manager Interface (AMI) is a system monitoring and management interface provided by Asterisk. It allows live monitoring of events that occur in the system, as well enabling you to request that Asterisk perform some action. The actions that are available are wide-ranging and include things such as returning status information and originating new calls. Many interesting applications have been developed on top of Asterisk that take advantage of the AMI as their primary interface to Asterisk.

Quick Start

This section is for getting your hands dirty with the AMI as quickly as possible. First, put the following configuration in */etc/asterisk/manager.conf*:

```
;
; Turn on the AMI and ask it to only accept connections from localhost.
;
[general]
enabled = yes
webenabled = yes
bindaddr = 127.0.0.1


;
; Create an account called "hello", with a password of "world"
;
[hello]
secret=world
```

This sample configuration is set up to only allow local connections to the AMI. If you intend on making this interface available over a network, it is strongly recommended that you only do so using TLS. The use of TLS is discussed in more detail later in this chapter.

Once the AMI configuration is ready, enable the built-in HTTP server by putting the following contents in */etc/asterisk/http.conf*:

```
;
; Enable the built-in HTTP server, and only listen for connections on localhost.
;
[general]
enabled = yes
bindaddr = 127.0.0.1
```

AMI over TCP

There are multiple ways to connect to the AMI, but a TCP socket is the most common. We will use *telnet* to demonstrate AMI connectivity. This example shows these steps:

1. Connect to the AMI over a TCP socket on port 5038.

2. Log in using the Login action.

3. Execute the Ping action.

4. Log off using the Logoff action.

Here's how the AMI responds to those actions:

```
$ telnet localhost 5038
Trying 127.0.0.1...
Connected to localhost.
Escape character is '^]'.
Asterisk Call Manager/1.1
Action: Login
Username: hello
Secret: world

Response: Success
Message: Authentication accepted

Action: Ping

Response: Success
Ping: Pong
Timestamp: 1282739190.454046

Action: Logoff

Response: Goodbye
Message: Thanks for all the fish.

Connection closed by foreign host.
```

Once you have this working, you have verified that AMI is accepting connections via a TCP connection.

AMI over HTTP

It is also possible to use the AMI over HTTP. In this section we will perform the same actions as before, but over HTTP instead of the native TCP interface to the AMI. The responses will be delivered over HTTP in the same format as the previous example, since the `rawman` encoding type is being used. AMI-over-HTTP responses can be encoded in other formats, such as XML. These response-formatting options are covered in "AMI over HTTP" on page 467.

 Accounts used for connecting to the AMI over HTTP are the same accounts configured in */etc/asterisk/manager.conf*.

This example demonstrates how to access the AMI over HTTP, log in, execute the Ping action, and log off:

```
$ wget "http://localhost:8088/rawman?action=login&username=hello&secret=world" \
> --save-cookies cookies.txt -O -

--2010-08-31 12:34:23--
Resolving localhost... 127.0.0.1
Connecting to localhost|127.0.0.1|:8088... connected.
HTTP request sent, awaiting response... 200 OK
Length: 55 [text/plain]
Saving to: `STDOUT'

Response: Success
Message: Authentication accepted

2010-08-31 12:34:23 (662 KB/s) - written to stdout [55/55]


$ wget "http://localhost:8088/rawman?action=ping" --load-cookies cookies.txt -O -

--2010-08-31 12:34:23--
Resolving localhost... 127.0.0.1
Connecting to localhost|127.0.0.1|:8088... connected.
HTTP request sent, awaiting response... 200 OK
Length: 63 [text/plain]
Saving to: `STDOUT'

Response: Success
Ping: Pong
Timestamp: 1283258063.040293

2010-08-31 12:34:23 (775 KB/s) - written to stdout [63/63]
```

```
$ wget "http://localhost:8088/rawman?action=logoff" --load-cookies cookies.txt -O -

--2010-08-31 12:34:23--
Resolving localhost... 127.0.0.1
Connecting to localhost|127.0.0.1|:8088... connected.
HTTP request sent, awaiting response... 200 OK
Length: 56 [text/plain]
Saving to: `STDOUT'

Response: Goodbye
Message: Thanks for all the fish.

2010-08-31 12:34:23 (696 KB/s) - written to stdout [56/56]
```

The HTTP interface to AMI lets you integrate Asterisk call control into a web service.

Configuration

The section "Quick Start" on page 457 showed a very basic set of configuration files to get you started. However, there are many more options available for the AMI.

manager.conf

The main configuration file for the AMI is */etc/asterisk/manager.conf*. The [general] section contains options (listed in Table 20-1) that control the overall operation of the AMI. Any other sections in the *manager.conf* file will define accounts for logging in and using the AMI.

Table 20-1. Options in the manager.conf [general] section

Option	Value/Example	Description
enabled	yes	Enables the AMI. The default is no.
webenabled	yes	Allows access to the AMI through the built-in HTTP server. The default is no.[a]
port	5038	Sets the port number to listen on for AMI connections. The default is 5038.
bindaddr	127.0.0.1	Sets the address to listen on for AMI connections. The default is to listen on all addresses (0.0.0.0). However, it is highly recommended to set this to 127.0.0.1.
tlsenable	yes	Enables listening for AMI connections using TLS. The default is no. It is highly recommended to only expose connectivity via TLS outside of the local machine.[b]
tlsbindport	5039	Sets the port to listen on for TLS connections to the AMI. The default is 5039.

Option	Value/Example	Description
tlsbindaddr	0.0.0.0	Sets the address to listen on for TLS-based AMI connections. The default is to listen on all addresses (0.0.0.0).
tlscertfile	*/var/lib/asterisk/keys/asterisk.pem*	Sets the path to the server certificate for TLS. This is required if tlsenable is set to yes.
tlsprivatekey	*/var/lib/asterisk/keys/private.pem*	Sets the path to the private key for TLS. If this is not specified, the tlscertfile will be checked to see if it also contains the private key.
tlscipher	<cipher string>	Specifies a list of ciphers for OpenSSL to use. Setting this is optional. To see a list of available ciphers, run *openssl ciphers -v* at the command line.
allowmultiplelogin	no	Allows the same account to make more than one connection at the same time. The default is yes.
displayconnects	yes	Reports connections to the AMI as verbose messages printed to the Asterisk console. This is usually useful, but it can get in the way on a system that uses scripts that make a lot of connections to the AMI. The default is yes.
timestampevents	no	Adds a Unix epoch-based timestamp to every event reported to the AMI. The default is no.
brokeneventsaction	no	Restores previously broken behavior for the Events AMI action, where a response would not be sent in some circumstances. This option is there for the sake of backward-compatibility for applications that worked around a bug and should not be used unless absolutely necessary. The default is no.
channelvars	VAR1,VAR2,VAR3[,VAR4[...]]	Specifies a list of channel variables to include with all manager events that are channel-oriented. The default is to include no channel variables.
debug	no	Enables some additional debugging in the AMI code. This is primarily there for developers of the Asterisk C code. The default is no.
httptimeout	60	Sets the HTTP timeout, in seconds. This timeout affects users of the AMI over HTTP: it sets the Max-Age of the HTTP cookie, sets how long events are cached to allow retrieval of the events over HTTP using the WaitEvents action, and the amount of time that the HTTP server keeps a session alive after completing an AMI action. The default is 60 seconds.

[a] To access the AMI over HTTP, the built-in HTTP server must also be configured in */etc/asterisk/http.conf*.

[b] The OpenSSL development package must be installed for Asterisk to be able to use encryption. On Ubuntu, the package is *libssl-dev*. On CentOS, the package is *openssl-devel*.

The *manager.conf* configuration file also contains the configuration of AMI user accounts. An account is created by adding a section with the username inside square brackets. Within each [*username*] section there are options that can be set that will apply only to that account. Table 20-2 lists the options available in a [*username*] section.

Table 20-2. Options for [username] sections

Option	Value/Example	Description
secret	password	Sets the password used for authentication. This must be set.
deny	0.0.0.0/0.0.0.0	Sets an IP address Access Control List (ACL) for addresses that should be denied the ability to authenticate as this user. By default this option is not set.
permit	192.168.1.0/255.255.255.0	Sets an IP address ACL for addresses that should be allowed to authenticate as this user. As with deny, by default this option is not set. Without these options set, any IP address that can reach the AMI will be allowed to authenticate as this user.
writetimeout	100	Sets the timeout used by Asterisk when writing data to the AMI connection for this user. This option is specified in milliseconds. The default value is 100.
displayconnects	yes	Also available in the [general] section (refer to Table 20-1), but can be controlled on a per-user basis.
read	system,call[,...]	Defines which manager events this user will receive. By default, the user will receive no events. Table 20-3 covers the available permission types for the read and write options.
write	system,call[,...]	Defines which manager actions this user is allowed to execute. By default, the user will not be able to execute any actions. Table 20-3 covers the available permission types for the read and write options.
eventfilter	!Channel: DAHDI*	Used to provide a whitelist- or blacklist-style filtering of manager events before they are delivered to the AMI client application. Filters are specified using a regular expression. A specified filter is a whitelist filter unless preceded by an exclamation point.[a]

[a] If no filters are specified, all events that are allowed based on the read option will be delivered. If only whitelist filters have been specified, only events that match one of the filters will be delivered. If there are only blacklist-style filters, all events that do not match any of the filters will be delivered. Finally, if there is a mix of whitelist- and blacklist-style filters, the whitelist filters will be processed first, and then the blacklist filters.

As discussed in Table 20-2, the read and write options set which manager actions and manager events a particular user has access to. Table 20-3 shows the available permission values that can be specified for these options.

Table 20-3. Available values for AMI user account read/write options

Permission identifier	read	write
all	Shorthand way of specifying that this user should have access to all available privilege options.	Grants user all privilege options.
system	Allows user to receive general system information, such as notifications of configuration reloads.	Allows user to perform system management commands such as *Restart*, *Reload*, or *Shutdown*.
call	Allows user to receive events about channels on the system.	Allows user to set information on channels.
log	Gives user access to logging information.[a]	read-only
verbose	Gives user access to verbose logging information.[b]	read-only
agent	Gives user access to events regarding the status of agents from the app_queue and chan_agent modules.	Enables user to perform actions for managing and retrieving the status of queues and agents.
user	Grants access to user-defined events, as well as events about Jabber/XMPP users.	Lets user perform the UserEvent manager action, which provides the ability to request that Asterisk generate a user-defined event.[c]
config	write-only	Allows user to retrieve, update, and reload configuration files.
command	write-only	Allows user to execute Asterisk CLI commands over the AMI.
dtmf	Allows user to receive events generated as DTMF passes through the Asterisk core.[d]	read-only
reporting	Gives user access to call-quality events, such as jitterbuffer statistics or RTCP reports.	Enables user to execute a range of actions to retrieve statistics and status information from across the system.
cdr	Grants user access to CDR records reported by the cdr_manager module.	read-only
dialplan	Allows user to receive events generated when variables are set or new extensions are created.	read-only
originate	write-only	Allows user to execute the Originate action, which allows an AMI client to request that Asterisk create a new call.
agi	Allows user to receive events generated when AGI commands are processed.	Enables user to perform actions for managing channels that are running AGI in its asynchronous mode. AGI is discussed in more detail in Chapter 21.
cc	Allows user to receive events related to Call Completion Supplementary Services (CCSS).	read-only

Permission identifier	read	write
aoc	Lets user see Advice of Charge events generated as AOC events are received.	Allows user to execute the AOCMessage manager action, for sending out AOC messages.

[a] This level has been defined, but it is not currently used anywhere in Asterisk.

[b] This level has been defined, but it is not currently used anywhere in Asterisk.

[c] The UserEvent action is a useful mechanism for having messages delivered to other AMI clients.

[d] DTMF events will not be generated in a bridged call between two channels unless generic bridging in the Asterisk core is being used. For example, if the DTMF is being transmitted with the media stream and the media stream is flowing directly between the two endpoints, Asterisk will not be able to report the DTMF events.

http.conf

As we've seen, the Asterisk Manager Interface can be accessed over HTTP as well as TCP. To make that work, a very simple HTTP server is embedded in Asterisk. All of the options relevant to the AMI go in the [general] section of */etc/asterisk/http.conf*.

> Enabling access to the AMI over HTTP requires both */etc/asterisk/manager.conf* and */etc/asterisk/http.conf*. The AMI must be enabled in *manager.conf*, with the enabled option set to yes, and the *manager.conf* option webenabled must be set to yes to allow access over HTTP. Finally, the enabled option in *http.conf* must be set to yes to turn on the HTTP server itself.

The available options are listed in Table 20-4:

Table 20-4. Options in the http.conf [general] section

Option	Value/Example	Description
enabled	yes	Enables the built-in HTTP server. The default is no.
bindport	8088	Sets the port number to listen on for HTTP connections. The default is 8088.
bindaddr	127.0.0.1	Sets the address to listen on for HTTP connections. The default is to listen on all addresses (0.0.0.0). However, it is highly recommended to set this to 127.0.0.1.
tlsenable	yes	Enables listening for HTTPS connections. The default is no. It is highly recommended that you only use HTTPS if you wish to expose HTTP connectivity outside of the local machine.[a]
tlsbindport	8089	Sets the port to listen on for HTTPS connections. The default is 8089.
tlsbindaddr	0.0.0.0	Sets the address to listen on for TLS-enabled AMI connections. The default is to listen on all addresses (0.0.0.0).
tlscertfile	*/var/lib/asterisk/ keys/asterisk.pem*	Sets the path to the HTTPS server certificate. This is required if tlsenable is set to yes.
tlsprivate key	*/var/lib/asterisk/ keys/private.pem*	Sets the path to the HTTPS private key. If this is not specified, the tlscert file will be checked to see if it also contains the private key.

Option	Value/Example	Description
tlscipher	`<cipher string>`	Specifies a list of ciphers for OpenSSL to use. Setting this is optional. To see a list of available ciphers, run *openssl ciphers -v* at the command line.

[a] The OpenSSL development package must be installed for Asterisk to be able to use encryption. On Ubuntu, the package is *libssl-dev*. On CentOS, the package is *openssl-devel*.

Protocol Overview

There are two main types of messages on the Asterisk Manager Interface: manager events and manager actions.

Manager events are one-way messages sent from Asterisk to AMI clients to report something that has occurred on the system. See Figure 20-1 for a graphical representation of the transmission of manager events.

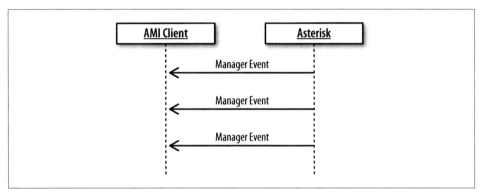

Figure 20-1. Manager events

Manager actions are requests from a client that have associated responses that come back from Asterisk. That is, a manager action may be a request that Asterisk perform some action and return the result. For example, there is an AMI action to originate a new call. See Figure 20-2 for a graphical representation of a client sending manager actions and receiving responses.

Other manager actions are requests for data that Asterisk knows about. For example, there is a manager action to get a list of all active channels on the system: the details about each channel are delivered as a manager event. When the list of results is complete, a final message will be sent to indicate that the end has been reached. See Figure 20-3 for a graphical representation of a client sending this type of manager action and receiving a list of responses.

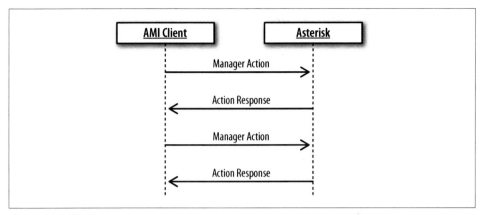

Figure 20-2. Manager actions

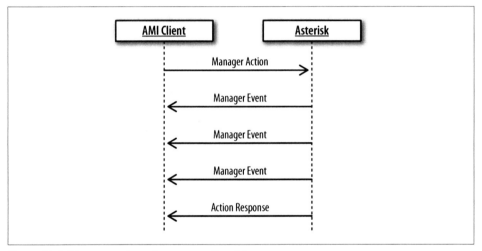

Figure 20-3. Manager actions that return a list of data

Message Encoding

All AMI messages, including manager events, manager actions, and manager action responses, are encoded in the same way. The messages are text-based, with lines terminated by a carriage return and a line-feed character. A message is terminated by a blank line:

```
Header1: This is the first header<CR><LF>
Header2: This is the second header<CR><LF>
Header3: This is the last header of this message<CR><LF>
<CR><LF>
```

Events

Manager events always have an `Event` header and a `Privilege` header. The `Event` header gives the name of the event, while the `Privilege` header lists the privilege levels associated with the event. Any other headers included with the event are specific to the event type. Here's an example:

```
Event: Hangup
Privilege: call,all
Channel: SIP/0004F2060EB4-00000000
Uniqueid: 1283174108.0
CallerIDNum: 2565551212
CallerIDName: Russell Bryant
Cause: 16
Cause-txt: Normal Clearing
```

Actions

When executing a manager action, it *must* include the `Action` header. The `Action` header identifies which manager action is being executed. The rest of the headers are arguments to the manager action. Some headers are required.

 To get a list of the headers associated with a particular manager action, type *manager show command <Action>* at the Asterisk command line. To get a full list of manager actions supported by the version of Asterisk you are running, enter *manager show commands* at the Asterisk CLI.

The final response to a manager action is typically a message that includes the `Response` header. The value of the `Response` header will be `Success` if the manager action was successfully executed. If the manager action was not successfully executed, the value of the `Response` header will be `Error`. For example:

```
Action: Login
Username: russell
Secret: russell

Response: Success
Message: Authentication accepted
```

AMI over HTTP

In addition to the native TCP interface, it is also possible to access the Asterisk Manager Interface over HTTP. Programmers with previous experience writing applications that use web APIs will likely prefer this over the native TCP connectivity.

Authentication and session handling

There are two methods of performing authentication against the AMI over HTTP. The first is to use the `Login` action, similar to authentication with the native TCP interface.

This is the method that was used in the quick-start example, as seen in "AMI over HTTP" on page 459. The second authentication option is HTTP digest authentication.* The next three sections discuss each of the AMI over HTTP encoding options. To indicate that HTTP digest authentication should be used, prefix the encoding type with an **a**.

Once successfully authenticated, Asterisk will provide a cookie that identifies the authenticated session. Here is an example response to the Login action that includes a session cookie from Asterisk:

```
$ curl -v "http://localhost:8088/rawman?action=login&username=hello&secret=world"

* About to connect() to localhost port 8088 (#0)
*   Trying 127.0.0.1... connected
* Connected to localhost (127.0.0.1) port 8088 (#0)
> GET /rawman?action=login&username=hello&secret=worlda HTTP/1.1
> User-Agent: curl/7.19.7 (x86_64-pc-linux-gnu) libcurl/7.19.7
OpenSSL/0.9.8k zlib/1.2.3.3 libidn/1.15
> Host: localhost:8088
> Accept: */*
>
< HTTP/1.1 200 OK
< Server: Asterisk/1.8.0-beta4
< Date: Tue, 07 Sep 2010 11:51:28 GMT
< Connection: close
< Cache-Control: no-cache, no-store
< Content-Length: 55
< Content-type: text/plain
< Cache-Control: no-cache;
< Set-Cookie: mansession_id="0e929e60"; Version=1; Max-Age=60
< Pragma: SuppressEvents
<

Response: Success
Message: Authentication accepted
* Closing connection #0
```

/rawman encoding

The rawman encoding type is what has been used in all the AMI over HTTP examples in this chapter so far. The responses received from requests using rawman are formatted in the exact same way that they would be if the requests were sent over a direct TCP connection to the AMI.

---

* At the time of writing, there is a problem with HTTP digest authentication that prevents it from working properly. Issue 18598 (*https://issues.asterisk.org/view.php?id=18598*) in the Asterisk project issue tracker has been opened for this problem. Hopefully it will be fixed by the time you read this.

/manager encoding

The manager encoding type provides a response in simple HTML form. This interface is primarily useful for experimenting with the AMI. Here is an example Login using this encoding type:

```
$ curl -v "http://localhost:8088/manager?action=login&username=hello&secret=world"

* About to connect() to localhost port 8088 (#0)
*   Trying 127.0.0.1... connected
* Connected to localhost (127.0.0.1) port 8088 (#0)
> GET /manager?action=login&username=hello&secret=world HTTP/1.1
> User-Agent: curl/7.19.7 (x86_64-pc-linux-gnu) libcurl/7.19.7
OpenSSL/0.9.8k zlib/1.2.3.3 libidn/1.15
> Host: localhost:8088
> Accept: */*
>
< HTTP/1.1 200 OK
< Server: Asterisk/1.8.0-beta4
< Date: Tue, 07 Sep 2010 12:19:05 GMT
< Connection: close
< Cache-Control: no-cache, no-store
< Content-Length: 881
< Content-type: text/html
< Cache-Control: no-cache;
< Set-Cookie: mansession_id="139deda7"; Version=1; Max-Age=60
< Pragma: SuppressEvents
<

<title>Asterisk&trade; Manager Interface</title><body bgcolor="#ffffff">
<table align=center bgcolor="#f1f1f1" width="500">
<tr><td colspan="2" bgcolor="#f1f1ff"><h1>Manager Tester</h1></td></tr>
<tr><td colspan="2" bgcolor="#f1f1ff"><form action="manager" method="post">
    Action: <select name="action">
        <option value="">-----&gt;</option>
        <option value="login">login</option>
        <option value="command">Command</option>
        <option value="waitevent">waitevent</option>
        <option value="listcommands">listcommands</option>
    </select>
    or <input name="action"><br/>
    CLI Command <input name="command"><br>
    user <input name="username"> pass <input type="password" name="secret"><br>
    <input type="submit">
</form>
</td></tr>
<tr><td>Response</td><td>Success</td></tr>
<tr><td>Message</td><td>Authentication accepted</td></tr>
<tr><td colspan="2"><hr></td></tr>
* Closing connection #0
</table></body>
```

/mxml encoding

The mxml encoding type provides responses to manager actions encoded in XML. Here is an example Login using the mxml encoding type:

```
$ curl -v "http://localhost:8088/mxml?action=login&username=hello&secret=world"

* About to connect() to localhost port 8088 (#0)
*   Trying 127.0.0.1... connected
* Connected to localhost (127.0.0.1) port 8088 (#0)
> GET /mxml?action=login&username=hello&secret=world HTTP/1.1
> User-Agent: curl/7.19.7 (x86_64-pc-linux-gnu) libcurl/7.19.7
OpenSSL/0.9.8k zlib/1.2.3.3 libidn/1.15
> Host: localhost:8088
> Accept: */*
>
< HTTP/1.1 200 OK
< Server: Asterisk/1.8.0-beta4
< Date: Tue, 07 Sep 2010 12:26:58 GMT
< Connection: close
< Cache-Control: no-cache, no-store
< Content-Length: 146
< Content-type: text/xml
< Cache-Control: no-cache;
< Set-Cookie: mansession_id="536d17a4"; Version=1; Max-Age=60
< Pragma: SuppressEvents
<

<ajax-response>
<response type='object' id='unknown'>
<generic response='Success' message='Authentication accepted' />
</response>
* Closing connection #0
</ajax-response>
```

Manager events

When connected to the native TCP interface for the AMI, manager events are delivered asynchronously. When using the AMI over HTTP, events must be retrieved by polling for them. Events are retrieved over HTTP by executing the WaitEvent manager action. The following example shows how events can be retrieved using the WaitEvent manager action. The steps are:

1. Start an HTTP AMI session using the Login action.
2. Register a SIP phone to Asterisk to generate a manager event.
3. Retrieve the manager event using the WaitEvent action.

The interaction looks like this:

```
$ wget --save-cookies cookies.txt \
> "http://localhost:8088/mxml?action=login&username=hello&secret=world" -O -

<ajax-response>
<response type='object' id='unknown'>
```

```
    <generic response='Success' message='Authentication accepted' />
</response>
</ajax-response>


$ wget --load-cookies cookies.txt "http://localhost:8088/mxml?action=waitevent" -O -

<ajax-response>
<response type='object' id='unknown'>
    <generic response='Success' message='Waiting for Event completed.' />
</response>
<response type='object' id='unknown'>
    <generic event='PeerStatus' privilege='system,all'
            channeltype='SIP' peer='SIP/0000FFFF0004'
            peerstatus='Registered' address='172.16.0.160:5060' />
</response>
<response type='object' id='unknown'>
    <generic event='WaitEventComplete' />
</response>
</ajax-response>
```

Development Frameworks

Many application developers write code that directly interfaces with the AMI. However, there are a number of existing libraries that aim to make writing AMI applications easier. Table 20-5 lists a few that we know are being used successfully. If you search around for Asterisk libraries in any other popular programming language of your choice, you are likely to find one that exists.

Table 20-5. AMI development frameworks

Framework	Language	URL
Adhearsion	Ruby	*http://adhearsion.com/*
StarPy	Python	*http://starpy.sourceforge.net/*
Asterisk-Java	Java	*http://asterisk-java.org/*

CSTA

Computer-Supported Telecommunications Applications (CSTA) is a standard for Computer Telephony Integration (CTI). One of the biggest benefits of CSTA is that it is used by multiple manufacturers. Some of what is provided by CSTA can be mapped to operations available in the AMI. There have been multiple efforts to provide a CSTA interface to Asterisk. One of these efforts is the Open CSTA (*http://opencsta.org*) project. While none of the authors have experience with this CSTA interface to Asterisk, it is certainly worth considering if you have CSTA experience or an existing CSTA application you would like to integrate with Asterisk.

Interesting Applications

Many useful applications have been developed that take advantage of the AMI. Here are a couple of examples.

AsteriskGUI

The AsteriskGUI is an open source PBX administration interface developed by Digium. It is intended for use on small installations. The AsteriskGUI is written entirely in HTML and JavaScript and uses the AMI over HTTP for all interaction with Asterisk. It has been especially popular for use in resource-constrained embedded Asterisk environments, since it does not require additional software to run on the Asterisk server. Figure 20-4 shows a page from the AsteriskGUI.

The AsteriskGUI can be obtained from the Digium subversion server:

```
$ svn co http://svn.digium.com/svn/asterisk-gui/branches/2.0
```

It is also bundled as an option with the AsteriskNOW (*http://www.asterisk.org/asteris know/*) distribution.

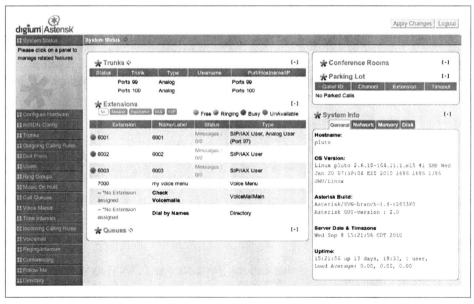

Figure 20-4. AsteriskGUI

Flash Operator Panel

Flash Operator Panel is an application that runs in a web browser using Flash. It is primarily used as an interface to see which extensions are currently ringing or in use. It also includes the ability to monitor conference room and call queue status. Some call actions can be performed as well, such as barging into a call and transferring calls. Figure 20-5 shows a screenshot of the Flash Operator Panel interface.

Downloads and more detailed information on Flash Operator Panel can be found at *http://www.asternic.org*.

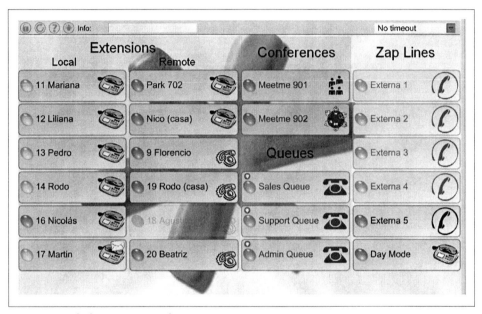

Figure 20-5. Flash Operator Panel

Conclusion

The Asterisk Manager Interface provides an API for monitoring events from an Asterisk system, as well as requesting that Asterisk perform a wide range of actions. An HTTP interface has been provided and a number of frameworks have been developed that make it easier to develop applications. All of this information, as well as the examples we looked at at the end of this chapter, should help get you thinking about what new applications you might be able to build using the Asterisk Manager Interface.

## CHAPTER 21

Asterisk Gateway Interface (AGI)

> *Caffeine. The gateway drug.*
>
> —Eddie Vedder

The Asterisk dialplan has evolved into a simple yet powerful programming interface for call handling. However, many people, especially those with a prior programming background, still prefer implementing their custom call handling in a different programming language. Using another programming language may also allow you to utilize existing code for integration with other systems. The Asterisk Gateway Interface (AGI) allows the development of first-party call control in the programming language of your choice. If you are not interested in implementing call control outside of the native Asterisk dialplan, you may safely skip this chapter.

Quick Start

This section gives a quick example of using the AGI. First, add the following line to */etc/asterisk/extensions.conf*:

```
exten => 500,1,AGI(hello-world.sh)
```

Next, create a *hello-world.sh* script in */var/lib/asterisk/agi-bin/*, as shown in Example 21-1.

Example 21-1. A sample AGI script, hello-world.sh

```
#!/bin/bash

# Consume all variables sent by Asterisk
while read VAR && [ -n ${VAR} ] ; do : ; done

# Answer the call.
echo "ANSWER"
read RESPONSE

# Say the letters of "Hello World"
```

```
echo 'SAY ALPHA "Hello World" ""'
read RESPONSE

exit 0
```

Now, call extension 500 with AGI debugging turned on and listen to Allison spell out *"Hello World"*:

```
*CLI> agi set debug on
AGI Debugging Enabled

    -- Executing [500@phones:1] AGI("SIP/0004F2060EB4-00000009",
       "hello-world.sh") in new stack
    -- Launched AGI Script /var/lib/asterisk/agi-bin/hello-world.sh
<SIP/0004F2060EB4-00000009>AGI Tx >> agi_request: hello-world.sh
<SIP/0004F2060EB4-00000009>AGI Tx >> agi_channel: SIP/0004F2060EB4-00000009
<SIP/0004F2060EB4-00000009>AGI Tx >> agi_language: en
<SIP/0004F2060EB4-00000009>AGI Tx >> agi_type: SIP
<SIP/0004F2060EB4-00000009>AGI Tx >> agi_uniqueid: 1284382003.9
<SIP/0004F2060EB4-00000009>AGI Tx >> agi_version: 1.8.0-beta4
<SIP/0004F2060EB4-00000009>AGI Tx >> agi_callerid: 2563619899
<SIP/0004F2060EB4-00000009>AGI Tx >> agi_calleridname: Russell Bryant
<SIP/0004F2060EB4-00000009>AGI Tx >> agi_callingpres: 0
<SIP/0004F2060EB4-00000009>AGI Tx >> agi_callingani2: 0
<SIP/0004F2060EB4-00000009>AGI Tx >> agi_callington: 0
<SIP/0004F2060EB4-00000009>AGI Tx >> agi_callingtns: 0
<SIP/0004F2060EB4-00000009>AGI Tx >> agi_dnid: 7010
<SIP/0004F2060EB4-00000009>AGI Tx >> agi_rdnis: unknown
<SIP/0004F2060EB4-00000009>AGI Tx >> agi_context: phones
<SIP/0004F2060EB4-00000009>AGI Tx >> agi_extension: 500
<SIP/0004F2060EB4-00000009>AGI Tx >> agi_priority: 1
<SIP/0004F2060EB4-00000009>AGI Tx >> agi_enhanced: 0.0
<SIP/0004F2060EB4-00000009>AGI Tx >> agi_accountcode:
<SIP/0004F2060EB4-00000009>AGI Tx >> agi_threadid: 140071216785168
<SIP/0004F2060EB4-00000009>AGI Tx >>
<SIP/0004F2060EB4-00000009>AGI Rx << ANSWER
<SIP/0004F2060EB4-00000009>AGI Tx >> 200 result=0
<SIP/0004F2060EB4-00000009>AGI Rx << SAY ALPHA "Hello World" ""
    -- <SIP/0004F2060EB4-00000009> Playing 'letters/h.gsm' (language 'en')
    -- <SIP/0004F2060EB4-00000009> Playing 'letters/e.gsm' (language 'en')
    -- <SIP/0004F2060EB4-00000009> Playing 'letters/l.gsm' (language 'en')
    -- <SIP/0004F2060EB4-00000009> Playing 'letters/l.gsm' (language 'en')
    -- <SIP/0004F2060EB4-00000009> Playing 'letters/o.gsm' (language 'en')
    -- <SIP/0004F2060EB4-00000009> Playing 'letters/space.gsm' (language 'en')
    -- <SIP/0004F2060EB4-00000009> Playing 'letters/w.gsm' (language 'en')
    -- <SIP/0004F2060EB4-00000009> Playing 'letters/o.gsm' (language 'en')
    -- <SIP/0004F2060EB4-00000009> Playing 'letters/r.gsm' (language 'en')
    -- <SIP/0004F2060EB4-00000009> Playing 'letters/l.gsm' (language 'en')
    -- <SIP/0004F2060EB4-00000009> Playing 'letters/d.gsm' (language 'en')
<SIP/0004F2060EB4-00000009>AGI Tx >> 200 result=0
    -- <SIP/0004F2060EB4-00000009>AGI Script hello-world.sh completed, returning 0
```

AGI Variants

There are a few variants of AGI that differ primarily in the method used to communicate with Asterisk. It is good to be aware of all of the options so you can make the best choice based on the needs of your application.

Process-Based AGI

Process-based AGI is the simplest variant of AGI. The "quick-start" example at the beginning of this chapter was an example of a process-based AGI script. The AGI script is invoked by using the `AGI()` application from the Asterisk dialplan. The application to run is specified as the first argument to `AGI()`. Unless a full path is specified, the application is expected to exist in the */var/lib/asterisk/agi-bin/* directory. Arguments to be passed to your AGI application can be specified as additional arguments to the `AGI()` application in the Asterisk dialplan. The syntax is:

```
AGI(command[,arg1[,arg2[,...]]])
```

> Ensure that your application has the proper permissions set such that the user the Asterisk process is running as has permissions to execute it. Otherwise, `AGI()` will fail.

Once Asterisk executes your AGI application, communication between Asterisk and your application will take place over *stdin* and *stdout*. More details about this communication will be covered in "AGI Communication Overview" on page 480. For more details about invoking `AGI()` from the dialplan, check the documentation built into Asterisk:

```
*CLI> core show application AGI
```

Pros of process-based AGI
It is the simplest form of AGI to implement.

Cons of process-based AGI
It is the least efficient form of AGI with regard to resource consumption. Systems with high load should consider FastAGI, discussed in "FastAGI—AGI over TCP" on page 478, instead.

EAGI

EAGI (Enhanced AGI) is a slight variant on `AGI()`. It is invoked in the Asterisk dialplan as `EAGI()`. The difference is that in addition to the communication on *stdin* and *stdout*, Asterisk also provides a unidirectional stream of audio coming from the channel on file descriptor 3. For more details on how to invoke `EAGI()` from the Asterisk dialplan, check the documentation built into Asterisk:

```
*CLI> core show application EAGI
```

Pros of Enhanced AGI

It has the simplicity of process-based AGI, with the addition of a simple read-only stream of the channel's audio. This is the only variant that offers this feature.

Cons of Enhanced AGI

Since a new process must be spawned to run your application for every call, it has the same efficiency concerns as regular process-based AGI.

 For an alternative way of getting access to the audio outside of Asterisk, consider using JACK (*http://jackaudio.org/*). Asterisk has a module for JACK integration, called `app_jack`. It provides the `JACK()` dialplan application and the `JACK_HOOK()` dialplan function.

DeadAGI Is Dead

In versions of Asterisk prior to 1.8, there was a dialplan application called `DeadAGI()`. Its purpose was similar to that of `AGI()`, except you used it on a channel that had already been hung up. This would usually be done in the special h extension, when you wanted to use an AGI application to aid in some type of post-call processing. Invoking `Dead AGI()` from the dialplan will still work, but you will get a `WARNING` message in the Asterisk log. It has been deprecated in favor of using `AGI()` in all cases. The code for `AGI()` has been updated so it knows how to correctly adjust its operation after a channel has been hung up.

Pros of DeadAGI

None. It's dead.

Cons of DeadAGI

It's dead. Really, don't use it. If you do, your configuration may break if `Dead AGI()` is completely removed from Asterisk in a future version.

FastAGI—AGI over TCP

FastAGI is the term used for AGI call control over a TCP connection. With process-based AGI, an instance of an AGI application is executed on the system for every call and communication with that application is done over *stdin* and *stdout*. With FastAGI, a TCP connection is made to a FastAGI server. Call control is done using the same AGI protocol, but the communication is over the TCP connection and does not require a new process to be started for every call. The AGI protocol is discussed in more detail in "AGI Communication Overview" on page 480. Using FastAGI is much more scalable than process-based AGI, though it is also more complex to implement.

FastAGI is used by invoking the `AGI()` application in the Asterisk dialplan, but instead of providing the name of the application to execute, you provide an `agi://` URL. For example:

```
exten => 1234,1,AGI(agi://127.0.0.1)
```

The default port number for a FastAGI connection is 4573. A different port number can be appended to the URL after a colon. For example:

```
exten => 1234,1,AGI(agi://127.0.0.1:4574)
```

Just as with process-based AGI, arguments can be passed to a FastAGI application. To do so, add them as additional arguments to the AGI() application, delimited by commas:

```
exten => 1234,1,AGI(agi://192.168.1.199,arg1,arg2,arg3)
```

FastAGI also supports the usage of SRV records if you provide a URL in the form of hagi://. By using SRV records, you can list multiple hosts that Asterisk can attempt to connect to for purposes of high availability and load balancing. In the following example, to find a FastAGI server to connect to, Asterisk will do a DNS lookup for _agi._tcp.shifteight.org:

```
exten => 1234,1,AGI(hagi://shifteight.org)
```

Pros of FastAGI

It's more efficient than process-based AGI. Instead of spawning a process per call, a FastAGI server can handle many calls.

DNS can be used to achieve high availability and load balancing among FastAGI servers to further enhance scalability.

Cons of FastAGI

It is more complex to implement a FastAGI server than to implement a process-based AGI application. However, implementing a TCP server is something that has been done countless times before, so there are many examples available for virtually any programming language.

Async AGI—AMI-Controlled AGI

Async AGI is a newer method of using AGI that was first introduced in Asterisk 1.6.0. The purpose of async AGI is to allow an application that uses the Asterisk Manager Interface (AMI) to asynchronously queue up AGI commands to be executed on a channel. This can be especially useful if you are already making extensive use of the AMI and would like to take advantage of the same application to handle call control, as opposed to writing a detailed Asterisk dialplan or developing a separate FastAGI server.

 More information on the Asterisk Manager Interface can be found in Chapter 20.

Async AGI is invoked by the AGI() application in the Asterisk dialplan. The argument to AGI() should be async:agi, as shown in the following example:

```
exten => 1234,1,AGI(async:agi)
```

Additional information on how to use async AGI over the AMI can be found in the next section.

Pros of async AGI
   An existing AMI application can be used to control calls using AGI commands.

Cons of async AGI
   It is the most complex method of using AGI to implement.

Setting Up /etc/asterisk/manager.conf for Async AGI

"Configuration" on page 460 discusses the configuration options in *manager.conf* in detail. To make use of async AGI, an AMI account must have the **agi** permission for both **read** and **write**. For example, the following user defined in *manager.conf* would be able to both execute AGI manager actions and receive AGI manager events:

```
;
; Define a user called 'hello', with a password of 'world'.
; Give this user read/write permissions for AGI.
;
[hello]
secret = world
read = agi
write = agi
```

AGI Communication Overview

The preceding section discussed the variations of AGI that can be used. This section goes into more detail about how your custom AGI application communicates with Asterisk once `AGI()` has been invoked.

Setting Up an AGI Session

Once `AGI()` or `EAGI()` has been invoked from the Asterisk dialplan, some information is passed to the AGI application to set up the AGI session. This section discusses what steps are taken at the beginning of an AGI session for the different variants of AGI.

Process-based AGI/FastAGI

For a process-based AGI application or a connection to a FastAGI server, the variables listed in Table 21-1 will be the first pieces of information sent from Asterisk to your application. Each variable will be on its own line, in the form:

```
agi_variable: value
```

Table 21-1. AGI environment variables

Variable	Value / Example	Description
agi_request	hello-world.sh	The first argument that was passed to the AGI() or EAGI() application. For process-based AGI, this is the name of the AGI application that has been executed. For FastAGI, this would be the URL that was used to reach the FastAGI server.
agi_channel	SIP/0004F2060EB4-00000009	The name of the channel that has executed the AGI() or EAGI() application.
agi_language	en	The language set on agi_channel.
agi_type	SIP	The channel type for agi_channel.
agi_uniqueid	1284382003.9	The uniqueid of agi_channel.
agi_version	1.8.0-beta4	The Asterisk version in use.
agi_callerid	12565551212	The full caller ID string that is set on agi_channel.
agi_callerid name	Russell Bryant	The caller ID name that is set on agi_channel.
agi_callingpres	0	The caller presentation associated with the caller ID set on agi_channel. For more information, see the output of *core show function CALLERPRES* at the Asterisk CLI.
agi_callingani2	0	The caller ANI2 associated with agi_channel.
agi_callington	0	The caller ID TON (Type of Number) associated with agi_channel.
agi_callingtns	0	The dialed number TNS (Transit Network Select) associated with agi_channel.
agi_dnid	7010	The dialed number associated with agi_channel.
agi_rdnis	unknown	The redirecting number associated with agi_channel.
agi_context	phones	The context of the dialplan that agi_channel was in when it executed the AGI() or EAGI() application.
agi_extension	500	The extension in the dialplan that agi_channel was executing when it ran the AGI() or EAGI() application.
agi_priority	1	The priority of agi_extension in agi_context that executed AGI() or EAGI().
agi_enhanced	0.0	An indication of whether AGI() or EAGI() was used from the dialplan. 0.0 indicates that AGI() was used. 1.0 indicates that EAGI() was used.
agi_accountcode	myaccount	The accountcode associated with agi_channel.
agi_threadid	140071216785168	The threadid of the thread in Asterisk that is running the AGI() or EAGI() application. This may be useful for associating logs generated by the AGI application with logs generated by Asterisk, since the Asterisk logs contain thread IDs.
agi_arg_*<argument number>*	my argument	These variables provide the contents of the additional arguments provided to the AGI() or EAGI() application.

For an example of the variables that might be sent to an AGI application, see the AGI communication debug output in "Quick Start" on page 475. The end of the list of variables will be indicated by a blank line. Example 21-1 handles these variables by reading lines of input in a loop until a blank line is received. At that point, the application continues and begins executing AGI commands.

Async AGI

When you use async AGI, Asterisk will send out a manager event to initiate the async AGI session. Here is an example manager event sent out by Asterisk:

```
Event: AsyncAGI
Privilege: agi,all
SubEvent: Start
Channel: SIP/0000FFFF0001-00000000
Env: agi_request%3A%20async%0Aagi_channel%3A%20SIP%2F0000FFFF0001-00000000%0A \
    agi_language%3A%20en%0Aagi_type%3A%20SIP%0Aagi_uniqueid%3A%201285219743.0%0A \
    agi_version%3A%201.8.0-beta5%0Aagi_callerid%3A%2012565551111%0A \
    agi_calleridname%3A%20Julie%20Bryant%0Aagi_callingpres%3A%200%0A \
    agi_callingani2%3A%200%0Aagi_callington%3A%200%0Aagi_callingtns%3A%200%0A \
    agi_dnid%3A%20111%0Aagi_rdnis%3A%20unknown%0Aagi_context%3A%20LocalSets%0A \
    agi_extension%3A%20111%0Aagi_priority%3A%201%0Aagi_enhanced%3A%200.0%0A \
    agi_accountcode%3A%20%0Aagi_threadid%3A%20-1339524208%0A%0A
```

The value of the Env header in this AsyncAGI manager event is all on one line. The long value of the Env header has been URI encoded.

Commands and Responses

Once an AGI session has been set up, Asterisk begins performing call processing in response to commands sent from the AGI application. As soon as an AGI command has been issued to Asterisk, no further commands will be processed on that channel until the current command has been completed. When it finishes processing a command, Asterisk will respond with the result.

The AGI processes commands in a serial manner. Once a command has been executed, no further commands can be executed until Asterisk has returned a response. Some commands can take a very long time to execute. For example, the EXEC AGI command executes an Asterisk application. If the command is EXEC Dial, AGI communication is blocked until the call is done. If your AGI application needs to interact further with Asterisk at this point, it can do so using the AMI, which is covered in Chapter 20.

A full list of available AGI commands can be retrieved from the Asterisk console by running the command *agi show commands*. These commands are described in Table 21-2. To get more detailed information on a specific AGI command, including syntax information for any arguments that a command expects, use *agi show commands topic <COMMAND>*. For example, to see the built-in documentation for the ANSWER AGI command, you would use *agi show commands topic ANSWER*.

Table 21-2. AGI commands

AGI command	Description
ANSWER	Answer the incoming call.
ASYNCAGI BREAK	End an async AGI session and have the channel return to the Asterisk dialplan.
CHANNEL STATUS	Retrieve the status of the channel. This is used to retrieve the current state of the channel, such as up (answered), down (hung up), or ringing.
DATABASE DEL	Delete a key/value pair from the built-in AstDB.
DATABASE DELTREE	Delete a tree of key/value pairs from the built-in AstDB.
DATABASE GET	Retrieve the value for a key in the AstDB.
DATABASE PUT	Set the value for a key in the AstDB.
EXEC	Execute an Asterisk dialplan application on the channel. This command is very powerful in that between EXEC and GET FULL VARIABLE, you can do anything with the call that you can do from the Asterisk dialplan.
GET DATA	Read digits from the caller.
GET FULL VARIABLE	Evaluate an Asterisk dialplan expression. You can send a string that contains variables and/or dialplan functions, and Asterisk will return the result after making the appropriate substitutions. This command is very powerful in that between EXEC and GET FULL VARIABLE, you can do anything with the call that you can do from the Asterisk dialplan.
GET OPTION	Stream a sound file while waiting for a digit from the caller. This is similar to the Background() dialplan application.
GET VARIABLE	Retrieve the value of a channel variable.
HANGUP	Hang up the channel.[a]
NOOP	Do nothing. You will get a result response from this command, just like any other. It can be used as a simple test of the communication path with Asterisk.
RECEIVE CHAR	Receive a single character. This only works for channel types that support it, such as IAX2 using TEXT frames or SIP using the MESSAGE method.
RECEIVE TEXT	Receive a text message. This only works in the same cases as RECEIVE CHAR.
RECORD FILE	Record the audio from the caller to a file. This is a blocking operation similar to the Record() dialplan application. To record a call in the background while you perform other operations, use EXEC Monitor or EXEC MixMonitor.
SAY ALPHA	Say a string of characters. You can find an example of this in "Quick Start" on page 475. To get localized handling of this and the other SAY commands, set the channel language either in the device configuration file (e.g., *sip.conf*) or in the dialplan, by setting the CHANNEL(language) dialplan function.

AGI command	Description
SAY DIGITS	Say a string of digits. For example, 100 would be said as "one zero zero" if the channel's language is set to English.
SAY NUMBER	Say a number. For example, 100 would be said as "one hundred" if the channel's language is set to English.
SAY PHONETIC	Say a string of characters, but use a common word for each letter (Alpha, Bravo, Charlie...).
SAY DATE	Say a given date.
SAY TIME	Say a given time.
SAY DATETIME	Say a given date and time using a specified format.
SEND IMAGE	Send an image to a channel. IAX2 supports this, but there are no actively developed IAX2 clients that support it that we know of.
SEND TEXT	Send text to a channel that supports it. This can be used with SIP and IAX2 channels, at least.
SET AUTOHANGUP	Schedule the channel to be hung up at a specified point in time in the future.
SET CALLERID	Set the caller ID name and number on the channel.
SET CONTEXT	Set the current dialplan context on the channel.
SET EXTENSION	Set the current dialplan extension on the channel.
SET MUSIC	Start or stop music on hold on the channel.
SET PRIORITY	Set the current dialplan priority on the channel.
SET VARIABLE	Set a channel variable to a given value.
STREAM FILE	Stream the contents of a file to a channel.
CONTROL STREAM FILE	Stream the contents of a file to a channel, but also allow the channel to control the stream. For example, the channel can pause, rewind, or fast forward the stream.
TDD MODE	Toggle the TDD (Telecommunications Device for the Deaf) mode on the channel.
VERBOSE	Send a message to the verbose logger channel. Verbose messages show up on the Asterisk console if the verbose setting is set high enough. Verbose messages will also go to any log file that has been configured for the verbose logger channel in /etc/asterisk/logger.conf.
WAIT FOR DIGIT	Wait for the caller to press a digit.
SPEECH CREATE	Initialize speech recognition. This must be done before using other speech AGI commands.[b]
SPEECH SET	Set a speech engine setting. The settings that are available are specific to the speech recognition engine in use.
SPEECH DESTROY	Destroy resources that were allocated for doing speech recognition. This command should be the last speech command executed.
SPEECH LOAD GRAMMAR	Load a grammar.
SPEECH UNLOAD GRAMMAR	Unload a grammar.
SPEECH ACTIVATE GRAMMAR	Activate a grammar that has been loaded.

AGI command	Description
SPEECH DEACTIVATE GRAMMAR	Deactivate a grammar.
SPEECH RECOGNIZE	Play a prompt and perform speech recognition, as well as wait for digits to be pressed.
GOSUB	Execute a dialplan subroutine. This will perform in the same way as the GoSub() dialplan application.

[a] When the HANGUP AGI command is used, the channel is not immediately hung up. Instead, the channel is marked as needing to be hung up. Your AGI application must exit first before Asterisk will continue and perform the actual hangup process.

[b] While Asterisk includes a core API for handling speech recognition, it does not come with a module that provides a speech recognition engine. Digium currently provides two commercial options for speech recognition: Lumenvox (*http://www.digium.com/en/products/software/lumenvox.php*) and Vestec (*http://www.digium.com/en/products/software/vestec.php*).

Process-based AGI/FastAGI

AGI commands are sent to Asterisk on a single line. The line must end with a single newline character. Once a command has been sent to Asterisk, no further commands will be processed until the last command has finished and a response has been sent back to the AGI application. Here is an example response to an AGI command:

```
200 result=0
```

> The Asterisk console allows debugging the communications with an AGI application. To enable AGI communication debugging, run the *agi set debug on* command. To turn debugging off, use *agi set debug off*. While this debugging mode is on, all communication to and from an AGI application will be printed out to the Asterisk console. An example of this output can be found in "Quick Start" on page 475.

Async AGI

When you're using async AGI, commands are issued by using the AGI manager action. To see the built-in documentation for the AGI manager action, run *manager show command AGI* at the Asterisk CLI. A demonstration will help clarify how AGI commands are executed using the async AGI method. First, an extension is created in the dialplan that runs an async AGI session on a channel:

```
exten => 7011,1,AGI(agi:async)
```

The following shows an example manager action execution and the manager events that are emitted during async AGI processing. After the initial execution of the AGI manager action, there is an immediate response to indicate that the command has been queued up for execution. Later, there is a manager event that indicates that the queued command has been executed. The CommandID header can be used to associate the initial request with the event that indicates that the command has been executed:

```
Action: AGI
Channel: SIP/0004F2060EB4-00000013
ActionID: my-action-id
```

```
CommandID: my-command-id
Command:  VERBOSE "Puppies like cotton candy." 1

Response: Success
ActionID: my-action-id
Message: Added AGI command to queue

Event: AsyncAGI
Privilege: agi,all
SubEvent: Exec
Channel: SIP/0004F2060EB4-00000013
CommandID: my-command-id
Result: 200%20result%3D1%0A
```

The following output is what was seen on the Asterisk console during this async AGI session:

```
    -- Executing [7011@phones:1] AGI("SIP/0004F2060EB4-00000013",
       "agi:async") in new stack
agi:async: Puppies like cotton candy.
    == Spawn extension (phones, 7011, 1) exited non-zero on 'SIP/0004F2060EB4-00000013'
```

Ending an AGI Session

An AGI session ends when your AGI application is ready for it to end. The details about how this happens depend on whether your application is using process-based AGI, FastAGI, or async AGI.

Process-based AGI/FastAGI

Your AGI application may exit or close its connection at any time. As long as the channel has not hung up before your application ends, dialplan execution will continue. If channel hangup occurs while your AGI session is still active, Asterisk will provide notification that this has occurred so that your application can adjust its operation as appropriate.

 This is an area where behavior has changed since Asterisk 1.4. In Asterisk 1.4 and earlier versions, AGI would automatically exit and stop operation as soon as the channel hung up. It now gives your application the opportunity to continue running if needed.

If a channel hangs up while your AGI application is still executing, a couple of things will happen. If an AGI command is in the middle of executing, you may receive a result code of -1. You should not depend on this, though, since not all AGI commands require channel interaction. If the command being executed does not require channel interaction, the result will not reflect the hangup.

The next thing that happens after a channel hangs up is that an explicit notification of the hangup is sent to your application. For process-based AGI, the signal SIGHUP will be sent to the process to notify it of the hangup. For a FastAGI connection, Asterisk will send a line containing the word HANGUP. If you would like to disable having Asterisk send the SIGHUP signal to your process-based AGI application or the HANGUP string to your FastAGI server, you can do so by setting the AGISIGHUP channel variable, as demonstrated in the following short example:

```
;
; Don't send SIGHUP to an AGI process
; or the "HANGUP" string to a FastAGI server.
;
exten => 500,1,Set(AGISIGHUP=no)
    same => n,AGI(my-agi-application)
```

At this point, Asterisk automatically adjusts its operation to be in DeadAGI mode. This just means that an AGI application can run on a channel that has been hung up. The only AGI commands that may be used at this point are those that do not require channel interaction. The documentation for the AGI commands built into Asterisk includes an indication of whether or not each command can be used once the channel has been hung up.

Async AGI

When you're using async AGI, the manager interface provides mechanisms to notify you about channel hangups. When you would like to end an async AGI session for a channel, you must execute the ASYNCAGI BREAK command. When the async AGI session ends, Asterisk will send an AsyncAGI manager event with a SubEvent of End. The following is an example of ending an async AGI session:

```
Action: AGI
Channel: SIP/0004F2060EB4-0000001b
ActionID: my-action-id
CommandID: my-command-id
Command: ASYNCAGI BREAK

Response: Success
ActionID: my-action-id
Message: Added AGI command to queue

Event: AsyncAGI
Privilege: agi,all
SubEvent: End
Channel: SIP/0004F2060EB4-0000001b
```

At this point, the channel returns to the Asterisk dialplan if it has not yet been hung up.

Development Frameworks

There have been a number of efforts to create frameworks or libraries that make AGI programming easier. Table 21-3 lists some of them. If you do not see a library listed here for your preferred programming language, do a quick search for it and you're likely to find one, as others do exist.

Table 21-3. AGI development frameworks

Framework	Language	URL
Adhearsion	Ruby	*http://adhearsion.com/*
StarPy	Python	*http://starpy.sourceforge.net/*
Asterisk-Java	Java	*http://asterisk-java.org/*
Asterisk-perl	Perl	*http://asterisk.gnuinter.net/*
PHPAGI	PHP	*http://phpagi.sourceforge.net/*

Conclusion

AGI provides a powerful interface to Asterisk that allows you to implement first-party call control in the programming language of your choice. There are multiple approaches that you can take to implementing an AGI application. Some approaches can provide better performance, but at the cost of more complexity. AGI provides a programming environment that may make it easier to integrate Asterisk with other systems, or just provide a more comfortable call control programming environment for the experienced programmer.

## CHAPTER 22

Clustering

You cannot eat a cluster of grapes at once, but it is very
easy if you eat them one by one.

—Jacques Roumain

The word "clustering" can mean different things to different people. Some people would say clustering is simply having a replicated system on standby available to be turned on when the primary system fails. To others, clustering is having several systems working in concert with one another, with replicated data, fully redundant, and infinitely expandable. For most people, it's probably somewhere between those two extremes.

In this chapter, we're going to explore the possibilities for clustering that exist with Asterisk at a high level, giving you the knowledge and direction to start planning your system into the future. As examples, we'll discuss some of the tools that we've used in our own large deployments; while there is no single way to go about building an Asterisk cluster, the topologies we'll cover have proven reliable over the years.

Our examples will delve into building a distributed call center, one of the more popular reasons for building a distributed system. In some cases this is necessary simply because a company has satellite offices it wants to tie into the primary system. For others, the goal is to integrate remote employees, or to be able to handle a large number of seats. We'll start by looking at a simple, traditional PBX system, and see how that system can eventually grow into something much larger.

Traditional Call Centers

Most systems deployed before the year 2000 will look quite similar. They will involve a set of phone lines delivered either via a PRI or through an array of analog lines, which connect to a PBX system that delivers calls to handsets that are likely proprietary to the systems deployed. These systems will likely provide a basic set of functions, with extra functions such as voicemail and conferencing being provided through external modules that may cost thousands of dollars. This topology is illustrated in Figure 22-1.

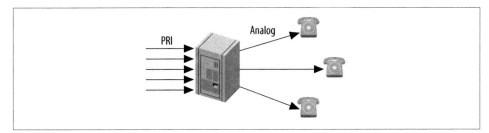

Figure 22-1. Traditional call center

Such systems will utilize a set of rules for delivering calls to agents through the standard automatic call distribution (ACD) rules, and will have little flexibility. It will likely be either impossible or expensive to add remote agents, as the calls would need to be delivered over the PSTN, which utilizes two phone lines: one for the incoming caller to the queue, and another to be delivered to the remote agent (in most cases, the agents just need to reside at the same physical location as the PBX itself).

These traditional phone systems are slowly being phased out, though, as more people start clamoring for the features VoIP brings to the table. And even for systems that aren't going to be using VoIP, solutions like Asterisk bring to the table features that once cost thousands of dollars as an included part of the software.

Of course, with the money invested in expensive hardware in traditional systems, it is natural that organizations with these systems will want to get as much use from them as possible. Plus, simply swapping out an existing system is not only expensive (wiring costs for SIP phones, replacement costs for proprietary handsets, etc.), but may be invasive to the call center, especially if it operates continuously.

Perhaps, though, the time to expand has come, and the existing system is no longer able to keep up with the number of lines required and the number of seats necessary to keep up with demand. In this case, it may be advantageous to look toward a hybrid system, where the existing hardware continues to be used, but new seats and features are added to the system using Asterisk.

Hybrid Systems

A hybrid phone system (Figure 22-2) contains the same functionality and hardware as a traditional phone system, with the exception of another system such as Asterisk being attached to it, thereby providing additional capacity and functionality. Adding Asterisk to a traditional system is typically done via a PRI connection. From the viewpoint of the traditional system, Asterisk will look like another phone company (central office, or CO). Depending on the way the traditional system operates and the services available

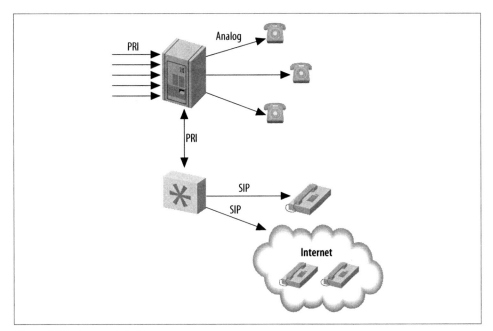

Figure 22-2. Remote hybrid system

to or from the CO, either Asterisk will deliver calls from the PRI through itself and to the existing PBX, or the existing PBX will send calls over the PRI connection to Asterisk, which will then direct the calls to the new endpoints (phones).

With Asterisk in the picture, functionality can be moved piecemeal from the existing PBX system over to Asterisk, which can take on a greater role and command more of the system over time. Eventually the existing PBX system may simply be used as a method for sending calls to the existing handsets on the agents' desks, with those being phased out over time and replaced with SIP-based phones, as the wiring is installed and phones are purchased.

By adding Asterisk to the existing system, we gain a new set of functionality and advantages, such as:

- Support for remote employees: calls are delivered over the existing Internet connection
- Features such as conferencing and voicemail (with the possibility of users being notified via email of new messages)
- Expanded phone lines using VoIP, and a reduction in long-distance costs

Such a system still suffers from a few disadvantages, as all the hardware needs to reside at the call center facility, and we're still restricted to using (relatively) expensive hardware in the Asterisk system for connecting to the traditional PBX. We're moving in the right direction, though, and with the Asterisk system in place we can start the migration over time, limiting interruptions to the business and taking a more gradual approach to training users.

Pure Asterisk, Nondistributed

The next step in our journey is the pure Asterisk system. In this system we've successfully migrated away from the existing PBX system and are now handling all functionality through Asterisk. Our existing PRI has been attached to Asterisk, and we've expanded our capacity by integrating an Internet Telephony Service Provider (ITSP) into our system. All agents are now using SIP phones, and we've even added several remote employees. This topology is illustrated in Figure 22-3.

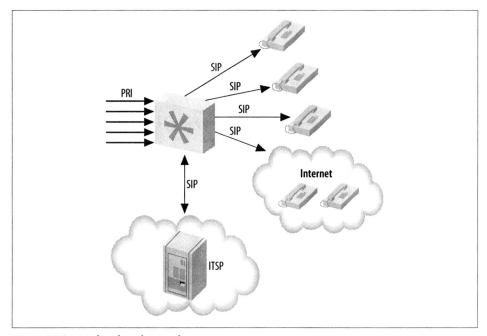

Figure 22-3. Nondistributed Asterisk

Remote employees can be a great advantage for a company. Not only can letting your work from remote locations increase employee morale by alleviating the burden of a potentially long commute, but it allows employees to work in an environment they are comfortable in, which can make them more productive. Furthermore, the call center manager does not have any less control over statistics from employees; their calls can still be monitored for training purposes, and the statistical data gathered look no different to the manager than they do for employees residing at the facility.

A measurable advantage for the company is the reduction in the amount of hardware required to be purchased for each employee. If agents can utilize their existing computer systems, electrical grids, and Internet connections, the company can save a significant amount of money by supporting remote employees. Additionally, those employees can be located across the globe to expand the number of hours your agents are available, thereby allowing you to serve more time zones.

Using this system is simple and efficient, but as the company grows, the system may reach a capacity issue. We'll look at how the system can be expanded later in this chapter.

Asterisk and Database Integration

Integrating Asterisk with a database can add a great deal of functionality to your system. Additionally, it provides a way to build web-based configuration utilities to make the maintenance of an Asterisk system easier. What's more, it allows instant access to information from the dialplan and other parts of the Asterisk system.

Single Database

Adding database integration to Asterisk (Figure 22-4) is a powerful way of gaining access to information that can be manipulated by other means. For example, we can read information about the extensions and devices in the system from a database using the Asterisk Realtime Architecture (discussed in Chapter 16), and we can modify the information stored in the database via an external system, such as a web page.

The integration with the database adds a layer between Asterisk and the web interface that the web designer is familiar with, and allows the manipulation of data in a way that doesn't require any additional skill sets. Knowledge of Asterisk itself is left to the Asterisk administrator, and the web developer can happily work with tools she is familiar with.

Of course, this makes the Asterisk system slightly more complex to build, but integration with a database via ODBC adds all sorts of possibilities (such as hot-desking, discussed in "Getting Funky with func_odbc: Hot-Desking" on page 354). func_odbc is a powerful tool for the Asterisk administrator, providing the ability to build a static dialplan using data that is dynamic in nature. See Chapter 16 for more information about how to integrate Asterisk with a database, and the functionality it provides.

We're also quite fond of the func_curl module, which provides integration with web services over HTTP directly from the dialplan.

With the data abstracted from Asterisk directly, we will now have an easier time moving toward a system that is getting ready to be clustered. We can use something like Linux-HA (*http://www.linux-ha.org/wiki/Main_Page*) to provide automatic failover between systems. While in the event of a failure the calls on the system that failed will be lost, the failover will take only moments (less than a second) to be detected, and the system will appear to its users to be immediately available again. In this configuration, since

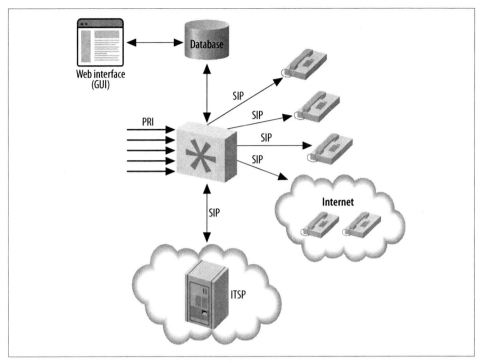

Figure 22-4. Asterisk database integration, single server

our data is abstracted outside of Asterisk, we can use applications such as *unison* (*http://www.cis.upenn.edu/~bcpierce/unison/*) or *rsync* to keep the configuration files synchronized between the primary and the backup system. We could also use *subversion* or *git* to track changes to the configuration files, making it easy to roll back changes that don't work out.

Of course, if our database goes away due to a failure of the hardware or the software, our system will be unavailable unless it is programmed in such a way as to be able to work without the database connection. This could be accomplished either through the use of a local database that simply updates itself periodically from the primary database, or through information programmed directly into the dialplan. In most cases the functionality of the system in this mode will be simpler than when the database was available, but at least the system will not be entirely unusable.

A better solution would be to use a replicated database, which allows data written to one database server to be written to another server at the same time. Asterisk can then fail over to the other database automatically if the primary server becomes unavailable.

Replicated Databases

Using a replicated database provides some redundancy in the backend to help limit the amount of downtime callers and agents experience if a database failure occurs. A master-master database configuration is required so that data can be written to either database and be automatically replicated to the other system, ensuring that we have an exact copy of the data on two physical machines. Another advantage to this approach is that a single system no longer needs to handle all the transactions to the database; the load can be divided among the servers. Figure 22-5 illustrates this distributed design.

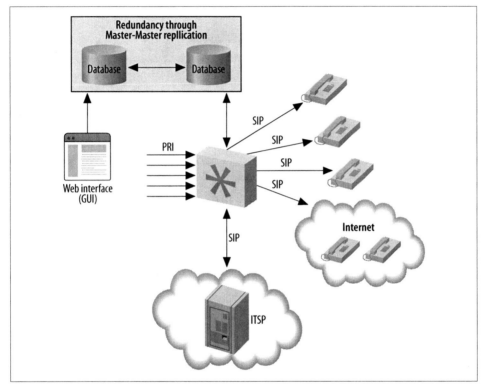

Figure 22-5. Asterisk database integration, distributed database

 We've used MySQL master-master replication before, and it works quite well. It also isn't all that difficult to set up, and several tutorials exist on the Internet. Other database systems will likely contain this functionality as well, especially if you're using a commercial system such as Oracle or MS SQL.

Failover can be done natively in Asterisk, as `res_odbc` and `func_odbc` do contain configuration options that allow you to specify multiple databases. In `res_odbc`, you can specify the preferred order for database connections in case one fails. In `func_odbc`, you can even specify different servers for reading data and writing data through the dialplan functions you create. All of this flexibility allows you to provide a system that works well for your business.

External programs can also be used for controlling failover between systems. The *pen* application (*http://siag.nu/pen/*) is a load balancer for simple TCP applications such as HTTP or SMTP, which allows several servers to appear as one. This means Asterisk only needs to be configured to connect to a single IP address (or hostname); the *pen* application will take care of controlling which server gets used for each request.

Asterisk and Distributed Device States

Device states in Asterisk are important both from a software standpoint (Asterisk might need to know the state of a device or the line on a device in order to know whether a call can be placed to it) and from a user's perspective (for example, a light may be turned on or off to signify whether a particular line is in use, or whether an agent is available for any more calls). From the viewpoint of a queue, it is extremely important to know the status of the device an agent is using in order to determine whether the next caller in the queue can be distributed to that agent. Without knowledge of the device's state, the queue would simply place multiple calls to the same endpoint.

Once you start expanding your single system to multiple boxes (potentially in multiple physical locations, such as remote or satellite offices), you will need to distribute the device state of endpoints between the systems. The kind of implementation that is required will depend on whether you're distributing them between systems on the same LAN (low-latency links) or over a WAN (higher-latency links). We'll discuss two device state distribution methods in this section: OpenAIS for low-latency links, and XMPP for higher-latency links.

Distributing Device States over a LAN

The OpenAIS implementation (*http://www.openais.org/doku.php*) was first added to Asterisk in the 1.6.1 branch, to enable distribution of device state information across servers. The addition of OpenAIS provided great possibilities for distributed systems, as device state awareness is an important aspect of such systems. Previous methods required the use of `GROUP()` and `GROUP_COUNT()` for each channel, with that information queried for over DUNDi. While this approach is useful in some scenarios (we could use this functionality to look up the number of calls our systems are handling and direct calls intelligently to systems handling fewer calls), as a mechanism for determining device state information it is severely lacking.

OpenAIS did give us the first implementation of a system that allows the state of devices and message waiting indications to be distributed among multiple Asterisk systems (see Figure 22-6). The downside of the OpenAIS implementation is that it requires all the

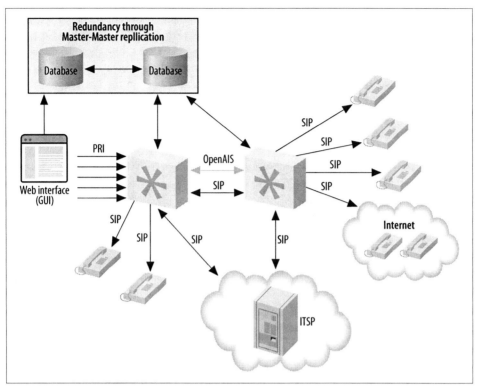

Figure 22-6. Device state distribution with OpenAIS

systems to live on low-latency links, which typically means they all need to reside in the same physical location, attached to the same switch. That said, while the OpenAIS library does not work across physically separate networks, it does allow a Queue() to reside on one system and queue members to reside on another system (or multiple systems). It does this without requiring us to use Local channels and test their availability through other methods, thereby limiting (or eliminating) the number of connection attempts made across the network, and multiple device ringing.

Using OpenAIS has an advantage, in that it is relatively easy to configure and get working. The disadvantage is that it is not distributable over physical locations. As of Asterisk 1.8, though, we can use XMPP for device state distribution over a wide area network, as you'll see in the next section.

More information about configuring distributed device states with OpenAIS is available in "Using OpenAIS" on page 310.

Distributing Device States over a WAN

As of Asterisk 1.8, an implementation that uses XMPP for device state distribution has been added. Because the XMPP protocol is designed for (or at least allows) usage across wide area networks, we can now have Asterisk systems at different physical locations distribute device state information to each other (see Figure 22-7). With the OpenAIS implementation, the library would be used on each system, enabling them to distribute device state information. In the XMPP scenario, a central server (or cluster of servers) is used to distribute the state among all of the Asterisk boxes in the cluster. Currently the best application for doing this is the Tigase XMPP server (*http://www.tigase.org*), because of its support for PubSub events. While other XMPP servers may be supported in the future, only Tigase is known to work at this time.

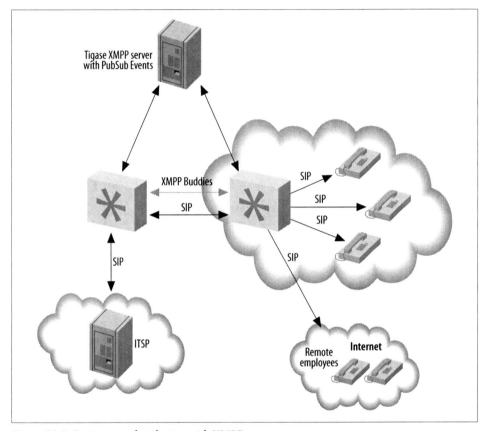

Figure 22-7. Device state distribution with XMPP

With XMPP, the queues can be located in different physical locations, and satellite offices can take calls from the primary office, or vice versa. This provides another layer of redundancy, because if the primary site goes offline and the ITSP is set up in such a way as to fail over to another office, the calls can be distributed among those satellite offices until the primary site goes back online. This is quite exciting for many people, as it adds a layer of functionality that was not previously available, and most of it can be done with relatively minimal configuration.

The advantage to XMPP device state distribution is that it is possible to distribute state to multiple physical locations, which is not possible with OpenAIS. The disadvantage is that it is more complex to set up (since you need an external service running the Tigase XMPP server) than the OpenAIS implementation.

More information about configuring distributed device states with XMPP can be found in "Using XMPP" on page 314.

Multiple Queues, Multiple Sites

Now, let's get creative and use the various tools we've discussed in the previous sections to build a distributed queue infrastructure. Figure 22-8 illustrates a sample setup where we have five Asterisk servers being fronted by another cluster used to distribute/route the calls to the various queues we have set up. Our ITSP sends calls to the routing cluster (which could be something like Kamailio, or even multiple Asterisk servers implementing DUNDi or some other method to route and distribute calls), which then sends the calls as appropriate to one of the three Asterisk systems we have our queues configured on. Each server handles a different queue, such as sales, technical support, and returns. These servers in turn use the agents located at two separate physical locations. The agents' devices are registered to their own local registration servers (which may also perform other functionality).

 We are not showing all aspects of the system, in order to keep the diagram simple, but in this case we would be using the XMPP distributed device state system as we're implying that the agents are distributed across multiple physical sites.

All of the agents at the different locations can be loaded into one or more queues, and because we're distributing device state information, each queue will know the current state of the agents in the queue and will only distribute callers to the agents as appropriate. Beyond that, we can configure penalties for the queues and/or for the agents in order to get the callers to the best agents if they are available, and only use the other agents when all the best agents are in use (for more information on penalties and priorities, refer to "Advanced Queues" on page 283).

We can add more agents to the system by adding more servers to the cluster at either the same location or additional physical locations. We can also expand the number of queues we support by adding more servers, each handling a different queue or queues.

A disadvantage to using this system is the way the Queue() application has been developed. Queue() is one of the older applications in Asterisk, and it has unfortunately not kept up with the pace of development in the realm of device state distribution, so there is no way to distribute the same Queue() across multiple boxes. For example, suppose you have sales queues on two systems. If a caller enters the sales queue on the first

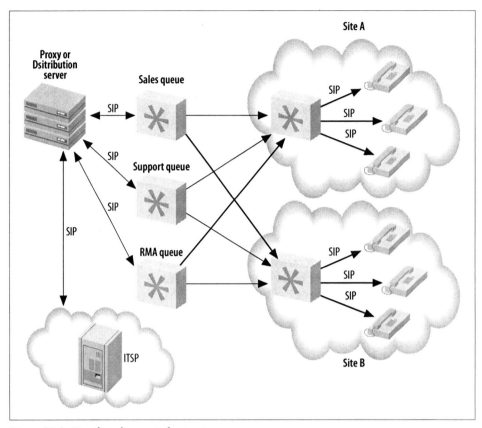

Figure 22-8. Distributed queue infrastructure

Asterisk system, and then another caller enters the sales queue on box two, no information will be distributed between those queues to indicate who is first and who is second in line. The two queues are effectively separate. Perhaps future versions of Asterisk will have the ability to do that, but at this time it is not supported. We mention this so you can plan your system accordingly.

Since queues in some implementations (such as call centers) may be required to handle many calls at once, the processing and load requirements for a single system can be quite steep. Having the ability to tap into the same agent resources across multiple systems means that we can distribute our callers among multiple boxes, significantly lowering the processing requirements placed on any single system. No longer does one system need to do it all—we can break out various components of the system into different servers.

Conclusion

In this chapter we explored how you can transition a traditional (non-Asterisk) telephony system into a distributed call center. Along the way, we've seen how a call center with just a few seats can grow into a system with hundreds of seats in different physical locations.

While the ability to grow your business and plan for the future is crucial, it is also important to not build a system that is more complex than it needs to be. The larger you go, and the more distributed a system you build, the longer it will take to get off the ground and the harder it will be to do all the things that are important when changes occur, such as testing, implementing the changes, and keeping things synchronized. If your system is never going to grow beyond a 40-seat call center, don't build it for 500 seats. All you're doing is adding additional costs and complexity to accommodate a system on a scale that may never be fully realized.

Building a simple system now and planning for the future and how you're going to get there (especially if you can do it in iterations, without having to rip your entire infrastructure apart or start from scratch) will get you up and running that much quicker. As you grow, you can add more pieces, determine if the approach you're taking is correct, and, if not, go back and rework that particular piece. This kind of approach can save you a lot of headaches down the road, when you realize you don't have to redo your entire complex system because of some new requirement that you didn't foresee at the beginning.

We also mentioned some advantages of having a distributed system with remote employees, such as improved employee morale and cost savings. You can use your employees' existing Internet connections, hardware, and electricity, which can save the company money, and your employees will benefit by avoiding the aggravation and costs of commuting to an office every day. While not all situations allow this type of scenario, it is worth exploring whether adding support for remote employees will be useful to your business.

Finally, distributed device state can open up a world of possibilities for your company, allowing it to grow beyond the single Asterisk system that does everything. Breaking out functionality to multiple boxes is now a reality, and can be approached with a measure of confidence not previously seen.

## CHAPTER 23

Distributed Universal Number Discovery (DUNDi)

A community is like a ship; everyone ought to be
prepared to take the helm.

—Henrik Ibsen

Distributed Universal Number Discovery, or DUNDi, is a service discovery protocol that can be used for locating resources at remote locations. The original intention of DUNDi was to permit decentralized routing among many peers using a General Peering Agreement (GPA). The GPA (available at *http://dundi.com/PEERING.pdf*) is intended to take on the role of a centralized control authority with a document to create a trust relationship among the peers in the cloud. While the idea is interesting and sound, the GPA has not taken off. That doesn't mean the DUNDi protocol itself hasn't found a home though: the original intention of DUNDi has been expanded so that now it doesn't just act as a location service, but can be used to request and pass information among peers.

How Does DUNDi Work?

Think of DUNDi as a large phone book that allows you to ask peers if they know of an alternative VoIP route to an extension number or PSTN telephone number.

For example, assume that you are connected to another set of Asterisk boxes listening for and responding to DUNDi requests, and those boxes are in turn connected to other Asterisk boxes listening for and responding to DUNDi requests. Assume also that your system does not have direct access to request anything from the remote servers.

Figure 23-1 illustrates how DUNDi works. You ask your friend Bob if he knows how to reach 4001, an extension to which you have no direct access. Bob replies, "I don't know how to reach that extension, but let me ask my peer, Sally."

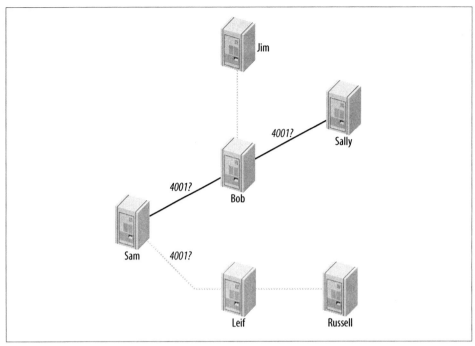

Figure 23-1. DUNDi peer-to-peer request system

Bob asks Sally if she knows how to reach the requested extension, and she responds with, "You can reach that extension at IAX2/dundi:*very_long_password@hostname/ extension*." Bob then stores the address in his database and passes on to you the information about how to reach 4001. With the newfound information, you can then make a separate request to actually place the call to Sally's box in order to reach extension 4001. (DUNDi only helps you find the information you need in order to connect; it does not actually place the call.)

Because Bob has stored the information he found, he'll be able to provide it to any peers who later request the same number from him, so the lookup won't have to go any further. This helps reduce the load on the network and decreases response times for numbers that are looked up often. (However, it should be noted that DUNDi creates a rotating key, and thus stored information is valid for a limited period of time.)

DUNDi performs lookups dynamically, either with a switch => statement in your *extensions.conf* file or with the use of the DUNDILOOKUP() dialplan function.

While DUNDi was originally designed and intended to be used as a peering fabric for the PSTN, it is used most frequently in private networks. If you're the Asterisk administrator of a large enterprise installation (or even an installation with only a pair of Asterisk boxes at different physical locations), you may wish to simplify the administration of extension numbers. DUNDi is a fantastic tool for this, because it allows you to simply share the extensions that have been configured at each location dynamically, by requesting the extension numbers from the remote location when your local box doesn't know how to reach them.

Additionally, if one of the locations had a cheaper route to a PSTN number you wanted to dial, you could request that route in your DUNDi cloud. For example, if one box was located in Vancouver and the other in Toronto, the Vancouver office could send calls destined for the Toronto area across the network using VoIP and out the PRI in Toronto, so they can be placed locally on the PSTN. Likewise, the Toronto office could place calls destined for Vancouver out of the PRI at the Vancouver office.

The dundi.conf File

It is often useful to be aware of the options available to us prior to delving into the configuration file, but feel free to skip this section for now and come back to reference particular options after you've got your initial configuration up and working.

There are three sections in the *dundi.conf* file: the [general] section, the [mappings] section, and the peer definitions, such as [FF:FF:FF:FF:FF:FF]. We'll show the options available for each section in separate tables.

Table 23-1 lists the options available in the [general] section of *dundi.conf*.

Table 23-1. Options available in the [general] section

Option	Description
department	Used when querying a remote system's contact information. An example might be: Communications.
organization	Used when querying a remote system's contact information. An example might be: ShiftEight.org.
locality	Used when querying a remote system's contact information. An example might be: Toronto.
stateprov	Used when querying a remote system's contact information. An example might be: Ontario.
country	Used when querying a remote system's contact information. An example might be: Canada.
email	Used when querying a remote system's contact information. An example might be: support@shifteight.org
phone	Used when querying a remote system's contact information. An example might be: +1-416-555-1212
bindaddr	Used to control which IP address the system will bind to. This can only be an IPv4 address. The default is 0.0.0.0, meaning the system will listen (and respond) on all available interfaces.
port	The port to listen for requests on. The default is 4520.
tos	The Terms of Service or Quality of Service (ToS/QoS) value to be used for requests. See *https://wiki.asterisk.org/wiki/display/AST/IP+Quality+of+Service* for more information about the values available and how to use them.
entityid	The entity ID of the system. Should be an externally (network) facing MAC address. The format is 00:00:00:00:00:00.
cachetime	How long peers should cache our responses for, in seconds. The default is 3600.
ttl	The time-to-live, or, maximum depth to search the network for a response. The maximum wait time for a response is calculated using (2000 + 200 * ttl) ms.

Option	Description
autokill	Used to control how long we wait for an ACK to our DPDISCOVER. Setting this timeout prevents the lookups from stalling due to a latent peer. This can be yes, no, or a numeric value representing the number of milliseconds to wait. You can use the qualify option to enable this per-peer.
secretpath	A rotating key is created and stored within the AstDB. The value is stored in the key 'secret' under the family defined by secretpath. The default secretpath is dundi, resulting in the key being stored in dundi/secret by default.
storehistory	Used to indicate whether or not the history of the last several requests should be stored in memory, along with how long the requests took. Valid values are yes and no (also available using the CLI commands *dundi store history* and *dundi no store history*). This is a debugging tool that is disabled by default due to possible performance impacts.

Table 23-2 lists the options you can configure in the [mappings] section of *dundi.conf*.

Table 23-2. Options available in the [mappings] section

Option	Description
nounsolicited	Used for advertising no unsolicited calls to the returned result. Used in public networks.
nocomunsolicit	Used for advertising no commercial unsolicited calls to the returned result. Used in public networks.
residential	Used to define the route returned as being a residential location. Used in public networks.
commercial	Used to define the route returned as being a commercial location. Used in public networks.
mobile	Used to define the route returned as being a mobile phone. Used in public networks.
nopartial	Used to prevent partial number lookups from being performed against this mapping.
${NUMBER}	Variable that contains the value of the request being looked up.
${IPADDR}	Variable that contains the IP address of the local system. Can be used to dynamically construct mapping responses. Not recommended.
${SECRET}	Variable that contains the value of the rotating secret key as defined in the secretpath location within the AstDB.

Finally, Table 23-3 lists the options available in the peer sections of *dundi.conf*.

Table 23-3. Options available for peer definitions in dundi.conf

Option	Description
inkey	The inbound authentication key.
outkey	The key used for authentication to the remote peer.
host	The hostname or IP address of the remote peer.
port	The port on which to communicate with the remote peer.
order	The search order associated with this peer. Values include primary, secondary, tertiary, and quartiary. Will only search primary peers unless none are available, in which case secondary peers will be searched, and so on.

Option	Description
include	Used to control whether this peer is included in searches for the mapping defined. Can be set to the value of all if used for all mappings.
noinclude	Used to control whether this peer is excluded from searches for the mapping defined. Can be set to all if this peer should be excluded from all lookups.
permit	Used to control whether this peer can perform lookups against a particular mapping. If the value is set to all, this peer can search against all defined mappings.
deny	Used to control which mappings this peer is restricted from searching. The value can be set to all to restrict this peer from being able to perform any lookups against defined mappings.
model	Used to control whether this peer can receive requests (inbound), transmit requests (outbound) or do both (symmetric).
precache	Typically used when we have a node with only a few routes that wants to push those values up to another node that is providing more responses (this is known as precaching, providing an answer when no request has been received). The values include outgoing, incoming, and symmetric. If this is set to outgoing, we push routes to this peer. If set to incoming, we receive routes from this peer. If set to symmetric, we do both.

Configuring Asterisk for Use with DUNDi

There are three files that need to be configured for DUNDi: *dundi.conf*, *extensions.conf*, and *sip.conf*.* The *dundi.conf* file controls the authentication of peers whom we allow to perform lookups through our system. This file also manages the list of peers to whom we might submit our own lookup requests. Since it is possible to run several different networks on the same box, it is necessary to define a different section for each peer, and then configure the networks in which those peers are allowed to perform lookups. Additionally, we need to define which peers we wish to use to perform lookups.

General Configuration

The [general] section of *dundi.conf* contains parameters relating to the overall operation of the DUNDi client and server:

```
; DUNDi configuration file for Toronto
;
[general]
;
department=IT
organization=toronto.example.com
locality=Toronto
stateprov=ON
```

---

* The *dundi.conf* and *extensions.conf* files must be configured. We have chosen to configure *sip.conf* for the purposes of address advertisement on our network, but DUNDi is protocol-agnostic, so *iax.conf*, *h323.conf*, or *mgcp.conf* could be used instead. DUNDi simply performs the lookups; the standard methods of placing calls are still required.

```
country=CA
email=support@toronto.example.com
phone=+14165551212
;
; Specify bind address and port number.  Default is port 4520.
;bindaddr=0.0.0.0
port=4520
entityid=FF:FF:FF:FF:FF:FF
ttl=32
autokill=yes
;secretpath=dundi
```

The entity identifier defined by `entityid` should generally be the Media Access Control
(MAC) address of an interface in the machine. The entity ID defaults to the first Ethernet
address of the server, but you can override this with `entityid`, as long as it is set to the
MAC address of *something* you own. The MAC address of the primary external interface
is recommended. This is the address that other peers will use to identify you.

The time-to-live (`ttl`) field defines how many hops away the peers we receive replies
from can be and is used to break loops. Each time a request is passed on down the line
because the requested number is not known, the value in the TTL field is decreased by
one, much like the TTL field of an ICMP packet. The TTL field also defines the maximum number of seconds we are willing to wait for a reply.

When you request a number lookup, an initial query (called a `DPDISCOVER`) is sent to
your peers requesting that number. If you do not receive an acknowledgment (`ACK`) of
your query (`DPDISCOVER`) within 2000 ms (enough time for a single transmission only)
and `autokill` is set to `yes`, Asterisk will send a `CANCEL` to the peers. (Note that an acknowledgment is not necessarily a reply to the query; it is just an acknowledgment that
the peer has received the request.) The purpose of `autokill` is to keep the lookup from
stalling due to hosts with high latency. In addition to the `yes` and `no` options, you may
also specify the number of milliseconds to wait.

The `pbx_dundi` module creates a rotating key and stores it in the local Asterisk database
(AstDB). The key name `secret` is stored in the `dundi` family. The value of the key can
be viewed with the *database show* command at the Asterisk console. The database
family can be overridden with the `secretpath` option.

We need another peer to interact with, so here's the configuration for the other node:

```
; DUNDi configuration file for Vancouver
;
[general]
;
department=IT
organization=vancouver.example.com
locality=Vancouver
stateprov=BC
country=CA
email=support@vancouver.example.com
phone=+16135551212
;
```

```
; Specify bind address and port number.  Default port is 4520.
;bindaddr=0.0.0.0
port=4520
entityid=00:00:00:00:00:00
ttl=32
autokill=yes
;secretpath=dundi
```

In the next section we'll create our initial DUNDi peers.

Initial DUNDi Peer Definition

A DUNDi peer is identified by the unique layer-two MAC address of an interface on
the remote system. The *dundi.conf* file is where we define what context to search for
peers requesting a lookup and which peers we want to use when doing a lookup for a
particular network. The following configuration is defined in the *dundi.conf* file on our
Toronto system:

```
[00:00:00:00:00:00] ; Vancouver Remote Office
model = symmetric
host = vancouver.example.com
inkey = vancouver
outkey = toronto
qualify = yes
dynamic=yes
```

The remote peer's identifier (MAC address) is enclosed in square brackets ([]). The
inkey and outkey are the public/private key pairs that we use for authentication. Key
pairs are generated with the *astgenkey* script, located in the *~/src/asterisk-complete/
asterisk/1.8/contrib/scripts/* source directory. We use the -*n* flag so that we don't have
to initialize passwords every time we start Asterisk:

```
$ cd /var/lib/asterisk/keys
$ sh ~/src/asterisk-complete/asterisk/1.8/contrib/scripts/astgenkey -n toronto
```

We'll place the resulting keys, *toronto.pub* and *toronto.key*, in our */var/lib/asterisk/
keys/* directory. The *toronto.pub* file is the public key, which we'll post to a web server
so that it is easily accessible for anyone with whom we wish to peer. When we peer, we
can give our peers the HTTP-accessible public key, which they can then place in
their */var/lib/asterisk/keys/* directories (using something like *wget*).

On the Vancouver box, we'll use the following peer configuration in *dundi.conf*:

```
[FF:FF:FF:FF:FF:FF] ; Toronto Remote Office
model = symmetric
host = toronto.example.com
inkey = toronto
outkey = vancouver
qualify = yes
dynamic=yes
```

Then we'll execute the same *astgenkey* script on the Vancouver box to generate the
public and private *vancouver* keys. Finally, we'll place the *toronto.pub* key on the

Vancouver server in */var/lib/asterisk/keys/* and place the *vancouver.pub* file on the Toronto server in the same location.

After downloading the keys, we must reload the *res_crypto.so* and *pbx_dundi.so* modules in Asterisk:

```
toronto*CLI> module reload res_crypto.so
    -- Reloading module 'res_crypto.so' (Cryptographic Digital Signatures)
    -- Loaded PUBLIC key 'vancouver'
    -- Loaded PUBLIC key 'toronto'
    -- Loaded PRIVATE key 'toronto'

vancouver*CLI> module reload res_crypto.so
    -- Reloading module 'res_crypto.so' (Cryptographic Digital Signatures)
    -- Loaded PUBLIC key 'toronto'
    -- Loaded PUBLIC key 'vancouver'
    -- Loaded PRIVATE key 'vancouver'
```

We can verify the keys so we know they're ready to be loaded at any time with the *keys show* CLI command:

```
*CLI> keys show
Key Name             Type     Status           Sum
-------------------- -------- ---------------- --------------------------------
vancouver            PRIVATE  [Loaded]         c02efb448c37f5386a546f03479f7d5e
vancouver            PUBLIC   [Loaded]         0a5e53420ede5c88de95e5d908274fb1
toronto              PUBLIC   [Loaded]         5f806860e0c8219f597f876caa6f2aff

3 known RSA keys.
```

With the keys loaded into memory, we can reload the *pbx_dundi.so* module on both systems in order to peer them together:

```
*CLI> module reload pbx_dundi.so
    -- Reloading module 'pbx_dundi.so' (Distributed Universal Number
       Discovery (DUNDi))
    == Parsing '/etc/asterisk/dundi.conf': Found
```

Finally, we can verify that the systems have peered successfully with *dundi show peers*:

```
toronto*CLI> dundi show peers
EID                   Host               Port   Model      AvgTime  Status
00:00:00:00:00:00     172.16.0.104   (S) 4520   Symmetric  Unavail  OK (3 ms)
1 dundi peers [1 online, 0 offline, 0 unmonitored]
```

Now, with our peers configured and reachable, we need to create the mapping contexts that will control what information will be returned in a lookup.

Creating Mapping Contexts

The *dundi.conf* file defines DUNDi contexts that are mapped to dialplan contexts in your *extensions.conf* file. DUNDi contexts are a way of defining distinct and separate directory service groups. The contexts in the [mapping] section point to contexts in the *extensions.conf* file, which control the numbers that you advertise.

When you create a peer, you need to define which mapping contexts you will allow this peer to search. You do this with the `permit` statement (each peer may contain multiple `permit` statements). Mapping contexts are related to dialplan contexts in the sense that they are a security boundary for your peers. We'll enable our mapping in the next section.

All DUNDi mapping contexts take the form of:

```
dundi_context => local_context,weight,technology,destination[,options]]
```

The following configuration creates a DUNDi mapping context that we'll use to advertise our local extension numbers to the group. We'll add this configuration to the *dundi.conf* file on the Toronto system under the `[mappings]` header. Note that this should all appear on one line:

```
[mappings]

; All on a single line
;
extensions => RegisteredDevices,0,SIP,dundi:very_secret_secret@toronto.example.com/
${NUMBER},nopartial
```

The configuration on the Vancouver system will look like this:

```
[mappings]

; All on a single line
;
extensions => RegisteredDevices,0,SIP,dundi:very_secret_secret@vancouver.example.com/
${NUMBER},nopartial
```

In this example, the mapping context is `extensions`, which points to the `RegisteredDevices` context within *extensions.conf* (providing a listing of extension numbers to reply with: our phone book). Numbers that resolve to the PBX should be advertised with a *weight* of zero (directly connected). Numbers higher than zero indicate an increased number of hops or paths to reach the final destination. This is useful when multiple replies for the same lookup are received at the end that initially requested the number; a path with a lower *weight* will be preferred. We'll look at how to control responses in "Controlling Responses" on page 516.

If we can reply to a lookup, our response will contain the method by which the other end can connect to the system. This includes the technology to use (such as IAX2, SIP, H323, and so on), the username and password with which to authenticate, which host to send the authentication to, and finally the extension number.

Asterisk provides some shortcuts to allow us to create a "template" with which we can build our responses. The following channel variables can be used to construct the template:

${SECRET}
    Replaced with the password stored in the local AstDB. Only used with *iax.conf*.

`${NUMBER}`

The number being requested.

`${IPADDR}`

The IP address to connect to.

 It is generally safest to statically configure the hostname, rather than make use of the `${IPADDR}` variable. The `${IPADDR}` variable will sometimes reply with an address in the private IP space, which is unreachable from the Internet.

With our mapping configured, let's create a simple dialplan context against which we can perform lookups for testing. We'll make this more dynamic in "Controlling Responses" on page 516.

In *extensions.conf*, we can add the following on both systems:

```
[RegisteredDevices]
exten => 1000,1,NoOp()
```

With our dialplan and mappings configured, we need to load them into memory from the CLI:

```
*CLI> dialplan reload
```

```
*CLI> module reload pbx_dundi.so
    -- Reloading module 'pbx_dundi.so' (Distributed Universal Number
       Discovery (DUNDi))
  == Parsing '/etc/asterisk/dundi.conf':   == Found
```

We can verify the mapping was loaded into memory with the *dundi show mappings* command:

```
toronto*CLI> dundi show mappings
DUNDi Cntxt  Weight  Local Cntxt  Options   Tech  Destination
extensions   0       RegisteredDe NONE      SIP   dundi:${SECRET}@172.16.0.
```

With our simple dialplan and mappings configured, we need to define the mappings each of our peers is allowed to use. We'll do this in the next section.

Using Mapping Contexts with Peers

With our mappings defined in the *dundi.conf* file, we need to give our peers permission to use them. Control of the various mappings is done via the `permit`, `deny`, `include`, and `noinclude` options within a peer definition. We use `permit` and `deny` to control whether the remote peer is allowed to search a particular mapping on our local system. We use `include` and `noinclude` to control which peers we will use to perform lookups with in a particular mapping.

Since we only have a single mapping defined (extensions), we're going to permit and include extensions within our peer definitions on both the Toronto and Vancouver systems.

On Toronto, we'll permit Vancouver to search the extensions mapping, and use Vancouver whenever we're performing a lookup within the extensions mapping:

```
[00:00:00:00:00:00] ; Vancouver Remote Office
model = symmetric
host = vancouver.example.com
inkey = vancouver
outkey = toronto
qualify = yes
dynamic=yes
permit=extensions
include=extensions
```

Similarly, we'll permit and include the extensions mapping for the Toronto office on the Vancouver system:

```
[FF:FF:FF:FF:FF:FF] ; Toronto Remote Office
model = symmetric
host = toronto.example.com
inkey = toronto
outkey = vancouver
qualify = yes
dynamic=yes
permit=extensions
include=extensions
```

After modifying the peers, we reload the *pbx_dundi.so* module to have the changes take effect:

```
*CLI> module reload pbx_dundi.so
```

The include and permit configuration can be verified via the *dundi show peer* command on the Asterisk CLI:

```
*CLI> dundi show peer 00:00:00:00:00:00
Peer:    00:00:00:00:00:00
Model:   Symmetric
Host:    172.16.0.104
Port:    4520
Dynamic: no
Reg:     No
In Key:  vancouver
Out Key: toronto
Include logic:
-- include extensions
Query logic:
-- permit extensions
```

Now we can test our lookups. We can do this easily from the Asterisk CLI using the *dundi lookup* command. If we perform a lookup from the Vancouver system, we'll receive a response from the Toronto system with an address we can use to place a call.

We've added the keyword **bypass** to the end of the lookup in order to bypass the cache (in case we wish to perform several tests):

```
vancouver*CLI> dundi lookup 1000@extensions bypass
  1.     0 SIP/dundi:very_secret_secret@172.16.0.161/1000 (EXISTS)
      from ff:ff:ff:ff:ff:ff, expires in 3600 s
DUNDi lookup completed in 12 ms
```

The response of SIP/dundi:*very_secret_secret*@172.16.0.161/1000 gives us an address that we can use to call extension 1000. (Of course, we can't use this address at the moment because we haven't configured any peers on the Toronto (or Vancouver) system to actually receive the call, but at least we have the DUNDi lookup portion working now!) In the next section we'll explore how to receive calls into our system after we've replied to a DUNDi response.

Allowing Remote Connections

Within our *sip.conf* file, we need to enable a peer that we can accept calls from and handle that peer's calls in the dialplan appropriately. The authentication is done using a password as defined in the mapping within *dundi.conf*.

> If you're using *iax.conf*, you can use the **${SECRET}** variable in the mapping in place of the password, which is dynamically replaced with a rotated key and is refreshed every 3600 seconds (1 hour). The value of the secret key is stored in the Asterisk database and is accessed using the **dbsecret** option within the peer definition of *iax.conf*.

Here is the user definition for the *dundi* user as defined in *sip.conf*:

```
[dundi]
type=user
secret=very_secret_secret
context=DUNDi_Incoming
disallow=all
allow=ulaw
allow=alaw
```

The **context** entry, **DUNDi_Incoming**, is where authorized callers are sent in *extensions.conf*. From there, we can control the call just as we would in the dialplan of any other incoming connection.

> We could also use the **permit** and **deny** options for the peer in *sip.conf* to control which IP addresses we'll accept calls from. Controlling the IP addresses will give us an extra layer of security if we're only expecting calls from known endpoints, such as those within our organization.

Be sure to reload *chan_sip.so* to enable the newly created user in *sip.conf*:

```
toronto*CLI> sip reload
```

To accept the incoming calls, define the [DUNDi_Incoming] context in *extensions.conf* and add the following to the Toronto system's dialplan.

```
[DUNDi_Incoming]
exten => 1000,1,Verbose(2,Incoming call from the DUNDi peer)
    same => n,Answer()
    same => n,Playback(silence/1)
    same => n,Playback(tt-weasels)
    same => n,Hangup()
```

Reload the dialplan with *dialplan reload* after saving your changes to *extensions.conf*.

For our first test, we'll create an extension in the LocalSets context and try placing a call to extension 1000 using the information provided via DUNDi:

```
[LocalSets]
exten => 1000,1,Verbose(2,Test extension to place call to remote server)
same => n,Dial(SIP/dundi:very_secret_secret@172.16.0.161/1000,30)
same => n,Hangup()
```

If we reload the dialplan and try testing the extension by dialing 1000, we should be connected to the *tt-weasels* prompt on the remote machine. With our user configured correctly to accept incoming calls, let's make our dialplan and responses more dynamic with some additional tools.

Using dbsecret with iax.conf

If you use the *iax.conf* channel driver, you can authenticate incoming calls using the dbsecret directive in *iax.conf* along with the ${SECRET} variable in your mapping. The use of the ${SECRET} variable in the mapping causes a rotated password to be sent back in the response, which can then be used for authentication via IAX2. Here is an example of an authentication definition in the *iax.conf* file:

```
[dundi]
type=friend
context=DUNDi_Incoming
dbsecret=dundi/secret
disallow=all
allow=ulaw
allow=alaw
```

The password is stored in the AstDB and is rotated every 3600 seconds (1 hour). To use the password in your mappings, change the mappings in *dundi.conf* to use ${SECRET} instead of *very_secret_secret*. This is the mapping we configured on the Vancouver system:

```
[mappings]

; All on a single line
;
extensions => RegisteredDevices,0,SIP,dundi:${SECRET}@vancouver.example.com/
${NUMBER},nopartial
```

Controlling Responses

Responses are controlled with the dialplan. Whenever an incoming request matches the dialplan configured for the mapping (whether the request is for a specific extension or a pattern match), a response will be sent. If the request does not match within the dialplan, no response is sent. In the example we've been building, the extension 1000 is the only extension that can be matched and thus generate a response.

In the next few sections we'll look at some of the methods we can use to control what requests are responded to.

Manually adding responses

The *extensions.conf* file handles what numbers you advertise and what you do with the calls that connect to them.

The simple method to control responses is to simply add them manually to the [Regis teredDevices] context. If we had several extensions at one of our locations, we could add them all to that context:

```
[RegisteredDevices]
exten => 1000,1,NoOp()

exten => 1001,1,NoOp()

exten => 1002,1,NoOp()
```

The NoOp() dialplan application is used here because the matching and responding is done only against the extension number, and no dialplan is executed. While we could overload this context and cause it to also be the destination for our calls, it's not recommended. Other reasons for using the NoOp() application should become clear as we progress.

Using pattern matches

Of course, adding everything we want to respond with manually would be silly, especially if we wanted to advertise a larger set of numbers, such as all numbers for an area code. As mentioned earlier, in our example we might wish to allow our Toronto and Vancouver offices to call out from one another when placing calls that are free or cheap to make from the other location.

We can respond with all of an area code using pattern matches, just as we do in other parts of the dialplan:

```
[RegisteredDevices]
exten => _416NXXXXXX,1,NoOp()
exten => _647NXXXXXX,1,NoOp()
exten => _905NXXXXXX,1,NoOp()
```

We could also advertise a full or partial range of extensions using pattern matches:

```
[RegisteredDevices]
exten => _1[1-3]XX,1,NoOp()  ; extensions 1100->1399

exten => _1[7-9]XX,1,NoOp()  ; extensions 1700->1999
```

Pattern matches are a good way of adding ranges of numbers, but these are still static. In the next section we'll explore how we can add some fluidity to the RegisteredDevices context.

Dynamically adding extension numbers

In some cases, you might want to only advertise extensions at your location that are currently registered to the system. Perhaps we have a salesperson who flies between the Toronto and Vancouver offices, and plugs her laptop into the network and registers at whichever location she is currently at. In that case, we would want to make sure that calls to that person are routed to the appropriate office in order to avoid sending calls across the country unnecessarily.

The regcontext and regexten options in *iax.conf* and *sip.conf* are useful for this. When a peer registers the value associated with regexten for that peer, an extension of that value will be created in the context defined by regcontext. So, for example, if we define regcontext in the [general] section of *sip.conf* to contain RegisteredDevices, and we define the regexten for each peer to contain the extension number of that peer, when the peers register the RegisteredDevices context will be populated automatically for us. We'll modify our *sip.conf* to look like this:

```
[general]
regcontext=RegisteredDevices

[0000FFFF0001](office-phone)
regexten=1001
```

and then reload *chan_sip.so*.

Now, we'll register our device to the system and look at the RegisteredDevices context:

```
*CLI> dialplan show RegisteredDevices
[ Context 'RegisteredDevices' created by 'SIP' ]
  '1001' =>         1. Noop(0000FFFF0001)              [SIP]
  '1002' =>         1. Noop(0000FFFF0002)              [SIP]
```

With our devices registered and the context used for determining when to respond populated, the only task left is to include the LocalSets context within the DUNDi_Incoming context in order to permit routing of calls to the endpoints.

Using dialplan functions in mappings

Sometimes it's useful to utilize a dialplan function within the mappings to control what a peer responds with. Throughout this book we've been touting the advantages of decoupling the user's extension number from the device in order to permit hot-desking.

Because the other end is just going to request an extension number and won't necessarily know the location of the device on our system, we can use the DB() and DB_EXISTS() functions within the mapping to perform a lookup from our AstDB for the device to call.[†]

 Prior to Asterisk version 1.8.3, the maximum length of the *destina tion* field (see "Creating Mapping Contexts" on page 510) was 80 characters, which made the use of nested dialplan functions nearly impossible. As of Asterisk 1.8.3, the maximum length is 512 characters.

First we need to make sure our database is populated with the information we might respond with. While this would normally be done by the dialplan written for the hot-desking implementation, we'll just add the content directly from the Asterisk console for demonstration purposes:

```
*CLI> database put phones 1001/device 0000FFFF0001
Updated database successfully
```

With our database populated, we need to modify our mapping to utilize some dialplan functions that will take the value requested, perform a lookup to our database for that value, and return a value. If no value exists in the database, we'll return the value of None.

Our existing mapping looks like this:

```
[mappings]
; The mapping exists on a single line
extensions => RegisteredDevices,0,SIP,
    dundi:very_secret_secret@toronto.example.com/${NUMBER},nopartial
```

Our current example simply reflects back the same extension number that was requested, along with some authentication information. The number requested is the extension the peer is looking for. However, because we're using hot-desking, the extension number may be located at various phone locations, so we may want to return the device identifier directly.[‡] We can do this by being clever with the use of dialplan functions in our response. While we may not have the full power of the dialplan (multiple lines, complex logic, etc.) at our disposal, we can at least use some of the simpler dialplan functions, such as DB(), DB_EXISTS(), and IF().

We're going to replace ${NUMBER} with the following bit of dialplan logic:

```
${IF($[${DB_EXISTS(phones/${NUMBER}/device)}]?${DB(phones/${NUMBER}/device)}:None)}
```

If we break this down, we end up with an IF() statement that will return either true or false. If false, we return the value of None. If true, we return the value located in the database at phones/${NUMBER}/device (where ${NUMBER} contains the value of 1001 for

---

† ...or func_odbc, or func_curl, or res_ldap (using the REALTIME_FIELD() function).

‡ We've also looked at using the GROUP() and GROUP_COUNT() functions for looking up the current channel usage on a remote system to determine which location to route calls to (the one with the lowest channel usage) as a simple load balancer.

our example) using the DB() function. To determine which value the IF() function will return, we use the DB_EXISTS() function. This function checks whether a value exists at phones/${NUMBER}/device within the AstDB, and returns either 1 or 0 (true or false).

> The DB_EXISTS() function not only returns 1 or 0, but also sets the ${DB_RESULT} channel variable that contains the value inside the database if the return value is 1. However, we can't use that value because the IF() function is evaluated prior to the condition field being evaluated, which means ${DB_RESULT} will be blank. Thus, we need to use the DB() function to look up the value prior to the condition field being evaluated.

After reloading *pbx_dundi.so* from the console (*module reload pbx_dundi.so*), we can perform a lookup from another server and check out the result:

```
vancouver*CLI> dundi lookup 1001@extensions bypass
   1.      0 SIP/dundi:very_awesome_password/0000FFFF0001 (EXISTS)
        from ff:ff:ff:ff:ff:ff, expires in 3600 s
DUNDi lookup completed in 77 ms
```

With dialplan functions, you can make the responses in your dialplans a lot more dynamic. In the next section we'll look at how you can perform these lookups from the dialplan using the DUNDILOOKUP(), DUNDIQUERY(), and DUNDIRESULT() functions.

> When you perform lookups using the example in this chapter, because all the peers in your network will return a result (None, or the value you want), you'll need to use the DUNDIQUERY() and DUNDIRESULT() functions to parse through the list of results returned. The alternative would be to try calling SIP/dundi:very_long_pass@remote_server/None, but this wouldn't be very effective. You might even want to handle the extension None elegantly, in case it gets called.

Performing Lookups from the Dialplan

Performing lookups from the dialplan is really the bread and butter of all of this, because it allows more dynamic routing from within the dialplan. With DUNDi, you can perform lookups and route calls within your cluster using either the DUNDILOOKUP() or DUNDIQUERY() and DUNDIRESULT() functions.

The DUNDILOOKUP() function replaces the old DUNDiLookup() dialplan application, performing nearly the same functionality. With DUNDILOOKUP(), you perform your lookup like you would at the Asterisk console, and the result can then be saved into a channel variable, or used wherever you might use a dialplan function. Here is an example:

```
[TestContext]
exten => 1001,1,Verbose(2,Look up extension 1001)
same => n,Set(DUNDi_Result=${DUNDILOOKUP(1001,extensions,b)})
```

```
same => n,Verbose(2,The result of the lookup was ${DUNDi_Result})
same => n,Hangup()
```

The arguments passed to DUNDILOOKUP() are: *extension,context,options*. Only one option, b, is available for the DUNDILOOKUP() function, and that is used to bypass the local
cache. The advantage to using the DUNDILOOKUP() function is that it is straightforward
and easy to use. The disadvantage is that it will only set the first value returned; if
multiple values are returned, they will be discarded.

 You won't always want to use the bypass option when performing lookups, because the use of the cache is what will lower the number of requests over your network and limit the amount of resources required.
We're using it in our examples simply because it is useful for testing
purposes, so that we know we've returned a result each time rather than
just a cached value from the previous lookup.

To parse through multiple returned values, we need to use the DUNDIQUERY() and DUN
DIRESULT() functions. Each plays an important part in sifting through multiple returned
values from a lookup. The DUNDIQUERY() function performs the initial lookup and saves
the resulting hash into memory. An ID value is then returned, which can be stored in
a channel variable. The ID value returned from the DUNDIQUERY() function can then be
passed to the DUNDIRESULT() function to parse through the returned values from the
query.

Lets take a look at some dialplan that uses these functions:

```
[TestContext]
exten => _1XXX,1,Verbose(2,Looking up results for extension ${EXTEN})

; Perform our lookup and save the resulting ID to DUNDI_ID
    same => n,Set(DUNDI_ID=${DUNDIQUERY(${EXTEN},extensions,b)})
    same => n,Verbose(2,Showing all results returned from the DUNDi Query)

; The DUNDIRESULT() function can return the number of results using 'getnum'
    same => n,Set(NumberOfResults=${DUNDIRESULT(${DUNDI_ID},getnum)})
    same => n,Set(ResultCounter=1)

; If there is less than 1 result, no results were returned
    same => n,GotoIf($[${NumberOfResults} < 1]?NoResults,1)

; The start of our loop showing the returned values
    same => n,While($[${ResultCounter} <= ${NumberOfResults}])

; Save the returned result at position ${ResultCounter} to thisResult
    same => n,Set(thisResult=${DUNDIRESULT(${DUNDI_ID},${ResultCounter})})

; Show the current result on the console
    same => n,Verbose(2,One of the results returned was: ${thisResult})

; Increase the counter by one
    same => n,Set(ResultCounter=${INC(ResultCounter)})
```

```
; End of our loop
   same => n,EndWhile()
   same => n,Playback(silence/1)
   same => n,Playback(vm-goodbye)
   same => n,Hangup()

; If no results were found, execute this dialplan
exten => NoResults,1,Verbose(2,No results were found)
   same => n,Playback(silence/1)
   same => n,Playback(invalid)
   same => n,Hangup()
```

Our example dialplan performs a lookup using the DUNDIQUERY() function and stores the resulting ID value in the DUNDI_ID channel variable. Using the DUNDIRESULT() function and the getnum option, we store the total number of returned results in the Number OfResults channel variable. We then set the ResultCounter channel variable to 1 as our starting position in the loop.

Using GotoIf(), we check if the ${NumberOfResults} returned is less than one and, if so, jump to the NoResults extension, where we Playback() "Invalid extension". If at least one extension is found, we continue on in the dialplan.

Using the While() application, we check if the ${ResultCounter} is less than or equal to the value of ${NumberOfResults}. If that is true, we continue on in the dialplan, and otherwise, we jump to the EndWhile() application.

For each iteration of our loop, the DUNDIRESULT() function is used to save the value at position ${ResultCounter} to the thisResult channel variable. After storing the value, we output it to the Asterisk console using the Verbose() application. Following that, we increase the value of ResultCounter by one using the INC() function. Our loop test is then done again within the While() loop, and the loop will continue while the value of ${ResultCounter} is less than or equal to the value of ${Number OfResults}.

Using the same type of logic, we could check for values other than None and, if such a value is found, ExitWhile() and continue in the dialplan to perform a call to the endpoint. The dialplan logic might look something like this:

```
[subLookupExtension]
exten => _1XXX,1,Verbose(2,Looking up results for extension ${EXTEN})

; Perform our lookup and save the resulting ID to DUNDI_ID
   same => n,Set(DUNDI_ID=${DUNDIQUERY(${EXTEN},extensions,b)})
   same => n,Set(NumberOfResults=${DUNDIRESULT(${DUNDI_ID},getnum)})
   same => n,Set(ResultCounter=1)

; If no results are found, return 'None'
   same => n,GotoIf($[${NumberOfResults} < 1]?NoResults,1)

; Perform our loop
   same => n,While($[${ResultCounter} <= ${NumberOfResults}])

; Get the current value
```

```
    same => n,Set(thisResult=${DUNDIRESULT(${DUNDI_ID},${ResultCounter})})

; If the current value returned is not None, we have a resulting
; location to call and we can exit the loop
    same => n,ExecIf($["${thisResult}" != "None"]?ExitWhile())

; If we made it this far, no value has been returned yet that we want to
; use, so increase the counter and try the next value.
    same => n,Set(ResultCounter=${INC(ResultCounter)})

; End of our loop
    same => n,EndWhile()

; We've made it here because we made it to the end of the loop or we found
; a value we want to return. Check to see which it is. If we just ran out of
; values, return 'None'.
;
    same => n,GotoIf($["${thisResult}" = "None"]?NoResults,1)

; If we make it here, we have a value we want to return.
    same => n,Return(${thisResult})

; If there were no acceptable results, return the value 'None'
    exten => NoResults,1,Verbose(2,No results were found)
same => n,Return(None)
```

With the DUNDIQUERY() and DUNDIRESULT() functions, you have a lot of power to control how to handle the results returned and perform routing logic with those values.

Conclusion

In this chapter we looked at how the DUNDi protocol helps you to perform lookups against the other Asterisk systems in your cluster, to perform dynamic routing. With the use of DUNDi, you can take multiple systems and control when and where calls are placed within them, providing toll-bypass capabilities and even giving your employees the ability to move between physical locations, while also limiting the number of out-of-system hops a call must take to find them.

While the original intention of DUNDi was to help us migrate away from centralized directory services—an intention that has yet to come to fruition—DUNDi is an extremely effective and useful tool that can be put to work in organizations to advertise and route calls dynamically between systems in a cloud environment. DUNDi is a tool that gives great power to Asterisk administrators looking to create a distributed network.

## CHAPTER 24

System Monitoring and Logging

> *Chaos is inherent in all compounded things.*
> *Strive on with diligence.*
>
> —The Buddha

Asterisk comes with several subsystems that allow you to obtain detailed information about the workings of your system. Whether for troubleshooting or for tracking usage for billing or staffing purposes, Asterisk's various monitoring modules can help you keep tabs on the inner workings of your system.

logger.conf

When troubleshooting issues in your Asterisk system, you will find it very helpful to refer to some sort of historical record of what was going on in the system at the time the reported issue occurred. The parameters for the storing of this information are defined in */etc/asterisk/logger.conf*.

Ideally, one might want the system to store a record of each and every thing it does. However, there is a cost to doing this. On a busy system, with full debug logging enabled, it is possible to completely fill the hard drive with logged data within a day or so. It is therefore necessary to achieve a balance between detail and storage requirements.

The */etc/asterisk/logger.conf* file allows you to define all sorts of different levels of logging, to multiple files if desired. This flexibility is excellent, but it can also be confusing.

The format of an entry in the *logger.conf* file is as follows:

```
filename => type[,type[,type[,...]]]
```

There is a sample *logger.conf* file that comes with the Asterisk source, but rather than just copying over the sample file, we recommend that you use the following for your initial *logger.conf* file:

```
[general]

[logfiles]
console => notice,warning,error,dtmf
messages => notice,warning,error
;verbose => notice,warning,error,verbose
```

When you have saved the file, you will need to reload the logger by issuing the following command from the shell:

```
$ asterisk -rx 'logger reload'
```

or from the Asterisk CLI:

```
*CLI> logger reload
```

Verbose Logging: Useful but Dangerous

We struggled with whether to recommend adding the following line to your *logger.conf* file:

```
verbose => notice,warning,error,verbose
```

This is quite possibly one of the most useful debugging tools you have when building and troubleshooting a dialplan, and therefore it is highly recommended. The danger comes from the fact that if you forget to disable this when you are done with your debugging, what you will have done is leave a ticking time bomb in your Asterisk system, which will slowly fill up the hard drive and kill your system one day, several months or years from now, when you are least expecting it.

Use it. It's fantastic. Just remember to turn it off when you're done!

You can specify any filename you want, but the special filename *console* will in fact print the output to the Asterisk CLI, and not to any file on the hard drive. All other filenames will be stored in the filesystem in the directory */var/log/asterisk*. The *logger.conf* types are outlined in Table 24-1.

Table 24-1. logger.conf types

Type	Description
notice	You will see a lot of these during a reload, but they will also happen during normal call flow. A notice is simply any event that Asterisk wishes to inform you of.
warning	A warning represents a problem that could be severe enough to affect a call (including disconnecting a call because call flow cannot continue). Warnings need to be addressed.
error	Errors represent significant problems in the system that must be addressed immediately.

Type	Description
debug	Debugging is only useful if you are troubleshooting a problem with the Asterisk code itself. You would not use debug to troubleshoot your dialplan, but you would use it if the Asterisk developers asked you to provide logs for a problem you were reporting. Do not use debug in production, as the amount of detail stored can fill up a hard drive in a matter of days.[a]
verbose	This is one of the most useful of the logging types, but it is also one of the more risky to leave unattended, due to the possibility of the output filling your hard drive.[b]
dtmf	Logging DTMF can be helpful if you are getting complaints that calls are not routing from the auto attendant correctly.
fax	This type of logging causes fax-related messages from the fax technology backend (res_fax_spandsp or res_fax_digium) to be logged to the fax logger.
*	This will log EVERYTHING (and we mean everything). Do not use this unless you understand the implications of storing this amount of data. It will not end well.

[a] This is not theory. It has happened to us. It was not fun.

[b] It's not as risky as debug, since it'll take months to fill the hard drive, but the danger is that it will happen, say, a year later when you're on summer vacation, and it will not immediately be obvious what the problem is. Not fun.

 There is a peculiarity in Asterisk's logging system that will cause you some consternation if you are unaware of it. The level of logging for the verbose and debug logging types is tied to the verbosity as set in the console. What this means is that if you are logging to a file with the verbose or debug type, and somebody logs into the CLI and issues the command *core set verbose 0*, or *core set debug 0*, the logging of those details to your log file will stop.

Reviewing Asterisk Logs

Searching through log files can be a challenge. The trick is to be able to filter what you are seeing so that you are only presented with information that is relevant to what you are searching for.

To start with, you are going to need to have an approximate idea of the time when the trouble you are looking for occurred. Once you are oriented to the approximate time, you will need to find clues that will help you to identify the call in question. Obviously, the more information you have about the call, the faster you will be able to pin it down.

If, for example, you are doing verbose logging, you should note that each distinct call has a thread identifier, which, when used with *grep*, can often help you to filter out everything that does not relate to the call you are trying to debug. For example, in the following verbose log, we have more than one call in the log, and since the calls are happening at the same time, it can be very confusing to trace one call:

```
$ tail -1000 verbose
[Mar 11 09:38:35] VERBOSE[31362] logger.c:    -- IAX2/shifteight-4 answered Zap/1-1
[Mar 11 09:39:35] VERBOSE[2973] logger.c:    -- Starting simple switch on 'Zap/1-1'
[Mar 11 09:39:35] VERBOSE[31362] logger.c:  == Spawn extension (shifteight, s, 1)
```

```
exited non-zero on 'Zap/1-1'
[Mar 11 09:39:35] VERBOSE[2973] logger.c:      -- Hungup 'Zap/1-1'
[Mar 11 09:39:35] VERBOSE[3680] logger.c:      -- Starting simple switch on 'Zap/1-1'
[Mar 11 09:39:35] VERBOSE[31362] logger.c:     -- Hungup 'Zap/1-1'
```

To filter on one call specifically, we could grep on the thread ID. For example:

> **$ grep *31362* verbose**

which would give us:

```
[Mar 11 09:38:35] VERBOSE[31362] logger.c:     -- IAX2/shifteight-4 answered Zap/1-1
[Mar 11 09:39:35] VERBOSE[31362] logger.c:     == Spawn extension (shifteight, s, 1)
exited non-zero on 'Zap/1-1'
[Mar 11 09:39:35] VERBOSE[31362] logger.c:     -- Hungup 'Zap/1-1'
```

This method does not guarantee that you will see everything relating to one call, since a call could in theory spawn additional threads, but for basic dialplan debugging we find this approach to be very useful.

Logging to the Linux syslog Daemon

Linux contains a very powerful logging engine, which Asterisk is capable of taking advantage of. While a discussion of all the various flavors of *syslog* and all the possible ways to handle Asterisk logging would be beyond the scope of this book, suffice it to say that if you want to have Asterisk send logs to the *syslog* daemon, you simply need to specify the following in your */etc/asterisk/logger.conf* file:

```
syslog.local0 => notice,warning,error ; or whatever type(s) you want to log
```

You will need to have a designation in your *syslog* configuration file[*] named `local0`, which should look something like:

```
local0.*        /var/log/asterisk/syslog
```

 You can use `local0` through `local7` for this, but check your *syslog.conf* file to ensure that nothing else is using one of those *syslog* channels.

syslog[†] will allow much more powerful logging, but it will also require more knowledge than simply allowing Asterisk to log to files.

---

[*] Which will normally be found at */etc/syslog.conf*.

[†] And *rsyslog*, *syslog-ng*, and what-all-else.

Verifying Logging

You can view the status of all your *logger.conf* settings through the Asterisk CLI by issuing the command:

```
*CLI> logger show channels
```

You should see output similar to:

```
Channel                       Type     Status   Configuration
-------                       ----     ------   -------------
syslog.local0                 Syslog   Enabled  - NOTICE WARNING ERROR VERBOSE
/var/log/asterisk/verbose     File     Enabled  - NOTICE WARNING ERROR VERBOSE
/var/log/asterisk/messages    File     Enabled  - NOTICE WARNING ERROR
                              Console  Enabled  - NOTICE WARNING ERROR DTMF
```

Call Detail Records

The CDR system in Asterisk is used to log the history of calls in the system. In some deployments, these records are used for billing purposes. In others, call records are used for analyzing call volumes over time. They can also be used as a debugging tool by Asterisk administrators.

CDR Contents

A CDR has a number of fields that are included by default. Table 24-2 lists them.

Table 24-2. Default CDR fields

Option	Value/Example	Notes
accountcode	12345	An account ID. This field is user-defined and is empty by default.
src	12565551212	The calling party's caller ID number. It is set automatically and is read-only.
dst	102	The destination extension for the call. This field is set automatically and is read-only.
dcontext	PublicExtensions	The destination context for the call. This field is set automatically and is read-only.
clid	"Big Bird" <12565551212>	The full caller ID, including the name, of the calling party. This field is set automatically and is read-only.
channel	SIP/0004F2040808-a1bc23ef	The calling party's channel. This field is set automatically and is read-only.
dstchannel	SIP/0004F2046969-9786b0b0	The called party's channel. This field is set automatically and is read-only.
lastapp	Dial	The last dialplan application that was executed. This field is set automatically and is read-only.

Option	Value/Example	Notes
lastdata	SIP/0004F2046969,30,tT	The arguments passed to the lastapp. This field is set automatically and is read-only.
start	2010-10-26 12:00:00	The start time of the call. This field is set automatically and is read-only.
answer	2010-10-26 12:00:15	The answered time of the call. This field is set automatically and is read-only.
end	2010-10-26 12:03:15	The end time of the call. This field is set automatically and is read-only.
duration	195	The number of seconds between the start and end times for the call. This field is set automatically and is read-only.
billsec	180	The number of seconds between the answer and end times for the call. This field is set automatically and is read-only.
disposition	ANSWERED	An indication of what happened to the call. This may be NO ANSWER, FAILED, BUSY, ANSWERED, or UNKNOWN.
amaflags	DOCUMENTATION	The Automatic Message Accounting (AMA) flag associated with this call. This may be one of the following: OMIT, BILLING, DOCUMENTATION, or Unknown.
userfield	PerMinuteCharge:0.02	A general-purpose user field. This field is empty by default and can be set to a user-defined string.[a]
uniqueid	1288112400.1	The unique ID for the src channel. This field is set automatically and is read-only.

[a] The userfield is not as relevant now as it used to be. Custom CDR variables are a more flexible way to get custom data into CDRs.

All fields of the CDR record can be accessed in the Asterisk dialplan by using the CDR() function. The CDR() function is also used to set the fields of the CDR that are user-defined.

```
exten => 115,1,Verbose(Call start time: ${CDR(start)})
    same => n,Set(CDR(userfield)=zombie pancakes)
```

In addition to the fields that are always included in a CDR, it is possible to add custom fields. This is done in the dialplan by using the Set() application with the CDR() function:

```
exten => 115,1,NoOp()
    same => n,Set(CDR(mycustomfield)=coffee)
    same => n,Verbose(I need some more ${CDR(mycustomfield)})
```

 If you choose to use custom CDR variables, make sure that the CDR backend that you choose is capable of logging them.

To view the built-in documentation for the CDR() function, run the following command at the Asterisk console:

```
*CLI> core show function CDR
```

In addition to the CDR() function, there are some dialplan applications that may be used to influence CDR records. We'll look at these next.

Dialplan Applications

There are a few dialplan applications that can be used to influence CDRs for the current call. To get a list of the CDR applications that are loaded into the current version of Asterisk, we can use the following CLI command:

```
*CLI> core show applications like CDR
     -= Matching Asterisk Applications =-
          ForkCDR: Forks the Call Data Record.
            NoCDR: Tell Asterisk to not maintain a CDR for the current call
         ResetCDR: Resets the Call Data Record.
     -= 3 Applications Matching =-
```

Each application has documentation built into the Asterisk application, which can be viewed using the following command:

```
*CLI> core show application <application name>
```

cdr.conf

The *cdr.conf* file has a [general] section that contains options that apply to the entire CDR system. Additional optional sections may exist in this file that apply to specific CDR logging backend modules. Table 24-3 lists the options available in the [general] section.

Table 24-3. cdr.conf [general] section

Option	Value/ Example	Notes
enable	yes	Enable CDR logging. The default is yes.
unanswered	no	Log unanswered calls. Normally, only answered calls result in a CDR. Logging all call attempts can result in a large number of extra call records that most people do not care about. The default value is no.
endbeforehexten	no	Close out CDRs before running the h extension in the Asterisk dialplan. Normally CDRs are not closed until the dialplan is completely finished running. The default value is no.
initiatedseconds	no	When calculating the billsec field, always round up. For example, if the difference between when the call was answered and when the call ended is 1 second and 1 microsecond, billsec will be set to 2 seconds. This helps ensure that Asterisk's CDRs match the behavior used by telcos. The default value is no.

Option	Value/ Example	Notes
batch	no	Queue up CDRs to be logged in batches instead of logging synchronously at the end of every call. This prevents CDR logging from blocking the completion of the call teardown process within Asterisk. The default value is no, but we recommend turning it on.[a]
size	100	Set the number of CDRs to queue up before they are logged during batch mode. The default value is 100.
time	300	Set the maximum number of seconds that CDRs will wait in the batch queue before being logged. The CDR batch logging process will run at the end of this time period, even if size has not been reached. The default value is 300 seconds.
scheduleronly	no	Set whether CDR batch processing should be done by spawning a new thread, or within the context of the CDR batch scheduler. The default value is no, and we recommend not changing it.
safeshutdown	yes	Block Asterisk shutdown to ensure that all queued CDR records are logged. The default is yes, and we recommend leaving it that way, as this option prevents important data loss.

[a] The disadvantage of enabling this option is that if Asterisk were to crash or die for some reason, the CDR records would be lost, as they are only stored in memory while the Asterisk process exists. See safeshutdown for more information.

Backends

Asterisk CDR backend modules provide a way to log CDRs. Most CDR backends require specific configuration to get them going.

cdr_adaptive_odbc

As the name suggests, the cdr_adaptive_odbc module allows CDRs to be stored in a database through ODBC. The "adaptive" part of the name refers to the fact that it works to adapt to the table structure: there is no static table structure that must be used with this module. When the module is loaded (or reloaded), it reads the table structure. When logging CDRs, it looks for a CDR variable that matches each column name. This applies to both the built-in CDR variables and custom variables. If you want to log the built-in channel CDR variable, just create a column called channel.

Adding custom CDR content is as simple as setting it in the dialplan. For example, if we wanted to log the User-Agent that is provided by a SIP device, we could add that as a custom CDR variable:

```
exten => 105,n,Set(CDR(useragent)=${CHANNEL(useragent)})
```

To have this custom CDR variable inserted into the database by cdr_adaptive_odbc, all we have to do is create a column called useragent.

Multiple tables may be configured in the cdr_adaptive_odbc configuration file. Each goes into its own configuration section. The name of the section can be anything; the module does not use it. Here is an example of a simple table configuration:

```
[mytable]

connection = asterisk
table = asterisk_cdr
```

A more detailed example of setting up a database for logging CDRs can be found in "Storing Call Detail Records (CDRs)" on page 375.

Table 24-4 lists the options that can be specified in a table configuration section in the *cdr_adaptive_odbc.conf* file.

Table 24-4. cdr_adaptive_odbc.conf table configuration options

Option	Value/Example	Notes
connection	pgsql1	The database connection to be used. This is a reference to the configured connection in *res_odbc.conf*. This field is required.
table	asterisk_cdr	The table name. This field is required.
usegmtime	no	Indicates whether to log timestamps using GMT instead of local time. The default value for this option is no.

In addition to the key/value pair fields that are shown in the previous table, *cdr_adaptive_odbc.conf* allows for a few other configuration items. The first is a column alias. Normally, CDR variables are logged to columns of the same name. An alias allows the variable name to be mapped to a column with a different name. The syntax is:

```
alias <CDR variable> => <column name>
```

Here is an example column mapping using the alias option:

```
alias src => source
```

It is also possible to specify a content filter. This allows you to specify criteria that must match for records to be inserted into the table. The syntax is:

```
filter <CDR variable> => <content>
```

Here is an example content filter:

```
filter accountcode => 123
```

Finally, *cdr_adaptive_odbc.conf* allows static content for a column to be defined. This can be useful when used along with a set of filters. This static content can help differentiate records that were inserted into the same table by different configuration sections. The syntax for static content is:

```
static <"Static Content Goes Here"> => <column name>
```

Here is an example of specifying static content to be inserted with CDRs:

```
static "My Content" => my_identifier
```

cdr_csv

The `cdr_csv` module is a very simple CDR backend that logs CDRs into a CSV (comma separated values) file. The file is */var/log/asterisk/cdr-csv/Master.csv*. As long as CDR logging is enabled in *cdr.conf* and this module has been loaded, CDRs will be logged to the *Master.csv* file.

While no options are required to get this module working, there are some options that customize its behavior. These options, listed in Table 24-5, are placed in the [csv] section of *cdr.conf*.

Table 24-5. cdr.conf [csv] section options

Option	Value/Example	Notes
usegmtime	no	Log timestamps using GMT instead of local time. The default is no.
loguniqueid	no	Log the uniqueid CDR variable. The default is no.
loguserfield	no	Log the userfield CDR variable. The default is no.
accountlogs	yes	Create a separate CSV file for each different value of the accountcode CDR variable. The default is yes.

The order of CDR variables in CSV files created by the `cdr_csv` module is:

```
<accountcode>,<src>,<dst>,<dcontext>,<clid>,<channel>,<dstchannel>,<lastapp>, \
    <lastdata>,<start>,<answer>,<end>,<duration>,<billsec>,<disposition>, \
    <amaflags>[,<uniqueid>][,<userfield>]
```

cdr_custom

This CDR backend allows for custom formatting of CDR records in a log file. This module is most commonly used for customized CSV output. The configuration file used for this module is */etc/asterisk/cdr_custom.conf*. A single section called [mappings] should exist in this file. The [mappings] section contains mappings between a filename and the custom template for a CDR. The template is specified using Asterisk dialplan functions.

The following example shows a sample configuration for `cdr_custom` that enables a single CDR log file, *Master.csv*. This file will be created as */var/log/asterisk/cdr-custom/Master.csv*. The template that has been defined uses both the CDR() and CSV_QUOTE() dialplan functions. The CDR() function retrieves values from the CDR being logged. The CSV_QUOTE() function ensures that the values are properly escaped for the CSV file format:

```
[mappings]

Master.csv => ${CSV_QUOTE(${CDR(clid)})},${CSV_QUOTE(${CDR(src)})},
    ${CSV_QUOTE(${CDR(dst)})},${CSV_QUOTE(${CDR(dcontext)})},
    ${CSV_QUOTE(${CDR(channel)})},${CSV_QUOTE(${CDR(dstchannel)})},
    ${CSV_QUOTE(${CDR(lastapp)})},${CSV_QUOTE(${CDR(lastdata)})},
    ${CSV_QUOTE(${CDR(start)})},${CSV_QUOTE(${CDR(answer)})},
```

```
${CSV_QUOTE(${CDR(end)})},${CSV_QUOTE(${CDR(duration)})},
${CSV_QUOTE(${CDR(billsec)})},${CSV_QUOTE(${CDR(disposition)})},
${CSV_QUOTE(${CDR(amaflags)})},${CSV_QUOTE(${CDR(accountcode)})},
${CSV_QUOTE(${CDR(uniqueid)})},${CSV_QUOTE(${CDR(userfield)})}
```

 In the actual configuration file, the value in the *Master.csv* mapping should be on a single line.

cdr_manager

The `cdr_manager` backend emits CDRs as events on the Asterisk Manager Interface (AMI), which we discussed in detail in Chapter 20. This module is configured in the */etc/asterisk/cdr_manager.conf* file. The first section in this file is the [general] section, which contains a single option to enable this module (the default value is no):

```
[general]

enabled = yes
```

The other section in *cdr_manager.conf* is the [mappings] section. This allows for adding custom CDR variables to the manager event. The syntax is:

```
<CDR variable> => <Header name>
```

Here is an example of adding two custom CDR variables:

```
[mappings]

rate => Rate
carrier => Carrier
```

With this configuration in place, CDR records will appear as events on the manager interface. To generate an example manager event, we will use the following dialplan example:

```
exten => 110,1,Answer()
    same => n,Set(CDR(rate)=0.02)
    same => n,Set(CDR(carrier)=BS&S)
    same => n,Hangup()
```

This is the command used to execute this extension and generate a sample manager event:

```
*CLI> console dial 110@testing
```

Finally, this is an example manager event produced as a result of this test call:

```
Event: Cdr
Privilege: cdr,all
AccountCode:
Source:
Destination: 110
DestinationContext: testing
```

```
CallerID:
Channel: Console/dsp
DestinationChannel:
LastApplication: Hangup
LastData:
StartTime: 2010-08-23 08:27:21
AnswerTime: 2010-08-23 08:27:21
EndTime: 2010-08-23 08:27:21
Duration: 0
BillableSeconds: 0
Disposition: ANSWERED
AMAFlags: DOCUMENTATION
UniqueID: 1282570041.3
UserField:
Rate: 0.02
Carrier: BS&S
```

cdr_mysql

This module allows posting of CDRs to a MySQL database. We recommend that new installations use `cdr_adaptive_odbc` instead.

cdr_odbc

This module enables the legacy ODBC interface for CDR logging. New installations should use `cdr_adaptive_odbc` instead.

cdr_pgsql

This module allows posting of CDRs to a PostgreSQL database. We recommend that new installations use `cdr_adaptive_odbc` instead.

cdr_radius

The `cdr_radius` backend allows posting of CDRs to a RADIUS server. When using this module, each CDR is reported to the RADIUS server as a single stop event. This module is configured in the *etc/asterisk/cdr.conf* file. Options for this module are placed in a section called `[radius]`. The available options are listed in Table 24-6.

Table 24-6. cdr.conf [radius] section options

Option	Value/Example	Notes
usegmtime	no	Enables logging of timestamps using GMT instead of local time. The default is yes.
loguniqueid	no	Enables logging of the `uniqueid` CDR variable. The default is yes.
loguserfield	no	Enables logging of the `userfield` CDR variable. The default is yes.
radiuscfg	/etc/radiusclient-ng/ radiusclient.conf	Sets the location of the *radiusclient-ng* configuration file. The default is */etc/radiusclient-ng/radiusclient.conf*.

cdr_sqlite

This module allows posting of CDRs to a SQLite database using SQLite version 2. Unless you have a specific need for SQLite version 2 as opposed to version 3, we recommend that all new installations use `cdr_sqlite3_custom`.

This module requires no configuration to work. If the module has been compiled and loaded into Asterisk, it will insert CDRs into a table called *cdr* in a database located at */var/log/asterisk/cdr.db*.

cdr_sqlite3_custom

This CDR backend inserts CDRs into a SQLite database using SQLite version 3. The database created by this module lives at */var/log/asterisk/master.db*. This module requires a configuration file, */etc/asterisk/cdr_sqlite3_custom.conf*. The configuration file identifies the table name, as well as customizes which CDR variables will be inserted into the database:

```
[master]

table = cdr

;
; List the column names to use when inserting CDRs.
;
columns => calldate, clid, dcontext, channel, dstchannel, lastapp, lastdata,
    duration, billsec, disposition, amaflags, accountcode, uniqueid, userfield,
    test


;
; Map CDR contents to the previously specified columns.
;
values => '${CDR(start)}','${CDR(clid)}','${CDR(dcontext)}','${CDR(channel)}',
    '${CDR(dstchannel)}','${CDR(lastapp)}','${CDR(lastdata)}','${CDR(duration)}',
    '${CDR(billsec)}','${CDR(disposition)}','${CDR(amaflags)}',
    '${CDR(accountcode)}','${CDR(uniqueid)}','${CDR(userfield)}','${CDR(test)}'
```

 In the *cdr_sqlite3_custom.conf* file, the contents of the columns and val ues options must each be on a single line.

cdr_syslog

This module allows logging of CDRs using *syslog*. To enable this, first add an entry to the system's *syslog* configuration file, */etc/syslog.conf*. For example:

```
local4.*        /var/log/asterisk/asterisk-cdr.log
```

The Asterisk module has a configuration file, as well. Add the following section to */etc/asterisk/cdr_syslog.conf*:

```
[cdr]

facility = local4
priority = info
template = "We received a call from ${CDR(src)}"
```

Here is an example *syslog* entry using this configuration:

```
$ cat /var/log/asterisk/asterisk-cdr.log

Aug 12 19:17:36 pbx cdr: "We received a call from 2565551212"
```

cdr_tds

The cdr_tds module uses the FreeTDS library to post CDRs to a Microsoft SQL Server or Sybase database. It is possible to use FreeTDS with *unixODBC*, so we recommend using cdr_adaptive_odbc instead of this module.

Example Call Detail Records

We will use the cdr_custom module to illustrate some example CDR records for different call scenarios. The configuration used for */etc/asterisk/cdr_custom.conf* is shown in "cdr_custom" on page 532.

Single-party call

In this example, we'll show what a CDR looks like for a simple one-party call. Specifically, we will use the example of a user calling in to check her voicemail. Here is the extension from */etc/asterisk/extensions.conf*:

```
exten => *98,1,VoiceMailMain(@${GLOBAL(VOICEMAIL_CONTEXT)})
```

This is the CDR from */var/log/asterisk/cdr-custom/Master.csv* that was created as a result of calling this extension:

```
"""Console"" <2565551212>","2565551212","*98","UserServices","Console/dsp","",
    "VoiceMailMain","@shifteight.org","2010-08-16 01:08:44","2010-08-16 01:08:44",
    "2010-08-16 01:08:53","9","9","ANSWERED","DOCUMENTATION","","1281935324.0","",0
```

Two-party call

For this next example, we show what a CDR looks like for a simple two-party call. We'll have one SIP phone place a call to another SIP phone. The call is answered and then hung up after a short period of time. Here is the extension that was dialed:

```
exten => 101,1,Dial(SIP/0000FFFF0002)
```

Here is the CDR that was logged to *Master.csv* as a result of this call:

```
"""Console"" <2565551212>","2565551212","101","LocalSets","Console/dsp",
    "SIP/0000FFFF0002-00000000","Dial","SIP/0000FFFF0002","2010-08-16 01:16:10",
    "2010-08-16 01:16:16","2010-08-16 01:16:29","19","13","ANSWERED",
    "DOCUMENTATION","","1281935770.2","",2
```

Caveats

The CDR system in Asterisk works very well for fairly simple call scenarios. However, as call scenarios get more complicated, involving calls to multiple parties, transfers, parking, and other such features, the CDR system starts to fall short. Many users report that the records do not show all of the information that they expect. Many bug fixes have been made to address some of the issues, but the cost of regressions or changes in behavior when making changes in this area is very high since these records are used for billing.

As a result, the Asterisk development team has become increasingly resistant to making additional changes to the CDR system. Instead, a new system, channel event logging (CEL), has been developed that is intended to help address logging of more complex call scenarios. Bear in mind that call detail records are simpler and easier to consume, though, so we still recommend using CDRs if they suit your needs.

CEL (Channel Event Logging)

Channel event logging (CEL) is a new system that was created to provide a more flexible means of logging the details of complex call scenarios. Instead of collapsing a call down to a single log entry, a series of events are logged for the call. This provides a more accurate picture of what has happened to the call, at the expense of a more complex log.

Channel Event Types

Each CEL record represents an event that occurred for a channel in the Asterisk system. Table 24-7 lists the events that are generated by Asterisk as calls are processed.

Table 24-7. CEL event types

CEL event type	Description
CHAN_START	A channel has been created.
CHAN_END	A channel has been destroyed.
LINKEDID_END	The last channel with a given `linkedid` has been destroyed.
ANSWER	A channel has been answered. On a channel created for an outbound call, this event will be generated when the remote end answers.
HANGUP	A channel has hung up. Generally, this event will be followed very shortly by a CHAN_END event. The difference is that this event occurs as soon as a hangup request is received, whereas

CEL event type	Description
	CHAN_END occurs after Asterisk has completed post-call cleanup and all resources associated with that channel have been released.
APP_START	A tracked application has started executing on a channel. Tracked applications are set in the main CEL configuration file, which is covered in "cel.conf" on page 540.
APP_END	A tracked application has stopped executing on a channel.
PARK_START	A channel has been parked.
PARK_END	A channel has left the parking lot.
BRIDGE_START	A channel bridge has started. This event occurs when two channels are bridged together by an application such as Dial() or Queue().
BRIDGE_END	A channel bridge has ended.
BRIDGE_UPDATE	An update to a bridge has occurred. This event will reflect if a channel's name or other information has changed during a bridge.
BLINDTRANSFER	A channel has executed a blind transfer.
ATTENDEDTRANSFER	A channel has executed an attended transfer.
USER_DEFINED	A user-defined channel event has occurred. These events are generated by using the CELGenUser Event() application.

There are some more events that have been defined, but are not yet used anywhere in the Asterisk code. Presumably, some future version will generate these events in the right place. They are listed in Table 24-8.‡

Table 24-8. Defined but unused CEL event types

CEL event type	Description
CONF_ENTER	A channel has connected to a conference room.
CONF_EXIT	A channel has left a conference room.
CONF_START	A conference has started. This event occurs at the time the first channel enters a conference room.
CONF_END	A conference has ended. This event occurs at the time the last channel leaves a conference room.
3WAY_START	A three-way call has started.
3WAY_END	A three-way call has ended.
TRANSFER	A generic transfer has been executed.
HOOKFLASH	A channel has reported a hookflash event.

‡ If you submit a patch to add any of these events to the code and reference this footnote, Russell will send you a free Asterisk t-shirt. Footnote bribery!

Channel Event Contents

Each CEL event contains the fields listed in Table 24-9:

Table 24-9. CEL event fields

Field name	Value/Example	Notes
eventtype	CHAN_START	The name of the event. The list of events that may occur can be found in Table 24-7.
eventtime	2010-08-19 07:27:19	The time that the event occurred.
cidname	Julie Bryant	The caller ID name set on the channel associated with this event.
cidnum	18435551212	The caller ID number set on the channel associated with this event.
cidani	18435551212	The Automatic Number Identification (ANI) number set on the channel associated with this event.
cidrdnis	18435551234	The redirecting number set on the channel associated with this event.
ciddnid	18435550987	The dialed number set on the channel associated with this event.
exten	101	The extension in the dialplan that is currently being executed.
context	LocalSets	The context for the extension in the dialplan that is currently being executed.
channame	SIP/0004F2060EB4-00000010	The name of the channel associated with this event.
appname	Dial	The name of the dialplan application currently being executed.
appdata	SIP/0004F2060E55	The arguments that were passed to the dialplan application that is currently being executed.
amaflags	DOCUMENTATION	The Automatic Message Accounting (AMA) flag associated with this call. This may be one of the following: OMIT, BILLING, DOCUMENTATION, or Unknown.
accountcode	1234	An account ID. This field is user-defined and is empty by default.
uniqueid	1282218999.18	The unique ID for the channel that is associated with this event.
userfield	I like waffles!	User-defined event content.
linkedid	1282218999.18	The per-call ID. This ID helps tie together multiple events from multiple channels that are all a part of the same logical call. The ID comes from the uniqueid of the first channel in the call.
peer	SIP/0004F2060E55-00000020	The name of the channel bridged to the channel identified by channame.

Some of the contents of a CEL event are user-defined. For example, the userfield is user-defined and will be empty by default. To set it to something, use the CHANNEL() dialplan function. Here is an example of setting the userfield for a channel:

```
exten => 101,1,Set(CHANNEL(userfield)=I like waffles!)
```

Dialplan Applications

The CEL system includes a single dialplan application that lives in the *app_celgenu-serevent.so* module. This application is used to generate custom user-defined events of the type EV_USER_EVENT. A practical example of using this would be for logging a caller's choices in a menu:

```
exten => 7,1,CELGenUserEvent(MENU_CHOICE,Caller chose option 7)
```

For full current details on the syntax of the CELGenUserEvent() application, use the built-in documentation from the Asterisk CLI:

```
*CLI> core show application CELGenUserEvent
```

cel.conf

The CEL system has a single configuration file, */etc/asterisk/cel.conf*. All options set here affect CEL processing, regardless of which logging backend modules are in use. Table 24-10 shows the options that exist in this file. All options should be set in the [general] section of the configuration file.

Table 24-10. cel.conf [general] section options

Option	Value/Example	Notes
enable	yes	Enables/disables CEL. The default is no.
apps	dial,queue	Sets which dialplan applications to track. The default is to track no applications. EV_APP_START and EV_APP_END events will be generated when channels start and stop executing any tracked application.
events	CHAN_START,CHAN_END, ANSWER,HANGUP	Lists which events to generate. This is useful if you are only interested in a subset of the events generated by CEL. If you would like to see all events, set this option to ALL. The default value is to generate no events.
dateformat	%F %T	Specifies the format for the date when a CEL event includes a timestamp. For syntax information, see the manpage for strftime by running *man strftime* at the command line. The default format for the CEL timestamp is seconds.microseconds since the epoch.

 At a minimum, to start using CEL, you must set the enable and events options in */etc/asterisk/cel.conf*.

Backends

As with the CDR system, there are a number of backend modules available for logging CEL events. In fact, all of the CEL backend modules were derived from CDR modules, so their configuration is very similar. In addition to the configuration options for *cel.conf*, which were described in the previous section, these modules require configuration to make them operate.

cel_odbc

The *cel_odbc.so* module provides the ability to log CEL events to a database using ODBC. This module is not quite as adaptive as the CDR adaptive ODBC backend. For CEL events, there are no custom variables. However, this module will still adapt to the structure of the database, in that it will log the fields of CEL events for which there are corresponding columns and will not produce an error if there is not a column for every field. The configuration for this module goes in */etc/asterisk/cel_odbc.conf*.

Multiple tables may be configured in the `cel_odbc` configuration file. Each goes into its own configuration section. The name of the section can be anything; the module does not use it. Here is an example of a simple table configuration:

```
[mytable]

connection = asterisk
table = asterisk_cel
```

The `cel_odbc` module will use the following columns, if they exist (see the table following this list for a set of mappings between event types and their integer value that will be inserted into the database):

- `eventtype`
- `eventtime`
- `userdeftype`
- `cid_name`
- `cid_num`
- `cid_ani`
- `cid_rdnis`
- `cid_dnid`
- `exten`
- `context`
- `channame`
- `appname`
- `appdata`
- `accountcode`
- `peeraccount`
- `uniqueid`
- `linkedid`
- `amaflags`

- userfield
- peer

Table 24-11 shows the mapping between event types and their integer values that will be inserted into the **eventtype** column of the database.

Table 24-11. Event type to integer value mappings for the eventtype column

Event type	Integer value
CHANNEL_START	1
CHANNEL_END	2
HANGUP	3
ANSWER	4
APP_START	5
APP_END	6
BRIDGE_START	7
BRIDGE_END	8
CONF_START	9
CONF_END	10
PARK_START	11
PARK_END	12
BLINDTRANSFER	13
ATTENDEDTRANSFER	14
TRANSFER	15
HOOKFLASH	16
3WAY_START	17
3WAY_END	18
CONF_ENTER	19
CONF_EXIT	20
USER_DEFINED	21
LINKEDID_END	22
BRIDGE_UPDATE	23
PICKUP	24
FORWARD	25

Table 24-12 shows the options that can be specified in a table configuration section in the *cel_odbc.conf* file.

Table 24-12. cel_odbc.conf table configuration

Option	Value/Example	Notes
connection	pgsql1	Specifies the database connection to be used. This is a reference to the configured connection in *res_odbc.conf*. This field is required.
table	asterisk_cdr	Specifies the table name. This field is required.
usegmtime	no	Enables/disables logging of timestamps using GMT instead of local time. The default value for this option is no.

In addition to the key/value pair fields that are shown in the previous table, *cel_odbc.conf* allows for a few other configuration items. The first is a column alias. Normally, CEL fields are logged to columns of the same name. An alias allows the variable name to be mapped to a column with a different name. The syntax is:

```
alias <CEL field> => <column name>
```

Here is an example column mapping using the alias option:

```
alias exten => extension
```

It is also possible to specify a content filter. This allows you to specify criteria that must match for records to be inserted into the table. The syntax is:

```
filter <CEL field> => <content>
```

Here is an example content filter:

```
filter appname => Dial
```

Finally, *cel_odbc.conf* allows static content to be specified for a column. This can be useful when used along with a set of filters. This static content can help differentiate records that were inserted into the same table by different configuration sections. The syntax for static content is:

```
static <"Static Content Goes Here"> => <column name>
```

Here is an example of specifying static content to be inserted with a CEL event:

```
static "My Content" => my_identifier
```

cel_custom

This CEL backend allows for custom formatting of CEL events in a log file. It is most commonly used for customized CSV output. The configuration file used for this module is */etc/asterisk/cel_custom.conf*. A single section called [mappings] should exist in this file. This section contains mappings between filenames and the custom templates for CEL events. The templates are specified using Asterisk dialplan functions and a few special CEL variables.

The following example shows a sample configuration for `cel_custom` that enables a single CEL log file, *Master.csv*. This file will be created as */var/log/asterisk/cel-custom/Master.csv*. The template that has been defined uses the `CHANNEL()`, `CALLERID()`, and `CSV_QUOTE()` dialplan functions. The `CSV_QUOTE()` function ensures that the values are properly escaped for the CSV file format. This example also references some special CEL variables, which are listed in Table 24-13.

Table 24-13. CEL variables available for use in [mappings]

CEL variable	Value/Example	Description
${eventtype}	CHAN_START	The name of the CEL event.
${eventtime}	1281980238.660403	The timestamp of the CEL event. The timestamp is given in the default format in this example.
${eventextra}	Whiskey Tango Foxtrot	Custom data included with a CEL event. Extra data is usually included when CELGenUserEvent() is used.

Here is the example */etc/asterisk/cel_custom.conf* file:

```
[mappings]

Master.csv => ${CSV_QUOTE(${eventtype})},${CSV_QUOTE(${eventtime})},
    ${CSV_QUOTE(${CALLERID(name)})},${CSV_QUOTE(${CALLERID(num)})},
    ${CSV_QUOTE(${CALLERID(ANI)})},${CSV_QUOTE(${CALLERID(RDNIS)})},
    ${CSV_QUOTE(${CALLERID(DNID)})},${CSV_QUOTE(${CHANNEL(exten)})},
    ${CSV_QUOTE(${CHANNEL(context)})},${CSV_QUOTE(${CHANNEL(channame)})},
    ${CSV_QUOTE(${CHANNEL(appname)})},${CSV_QUOTE(${CHANNEL(appdata)})},
    ${CSV_QUOTE(${CHANNEL(amaflags)})},${CSV_QUOTE(${CHANNEL(accountcode)})},
    ${CSV_QUOTE(${CHANNEL(uniqueid)})},${CSV_QUOTE(${CHANNEL(linkedid)})},
    ${CSV_QUOTE(${CHANNEL(peer)})},${CSV_QUOTE(${CHANNEL(userfield)})},
    ${CSV_QUOTE(${eventextra})}
```

 In the actual configuration file, the value in the *Master.csv* mapping should be on a single line.

cel_manager

The `cel_manager` backend emits CEL events on the Asterisk Manager Interface (we discussed the AMI in detail in Chapter 20). This module is configured in the */etc/asterisk/cel.conf* file. This file should contain a single section called [`manager`], which contains a single option to enable this module. The default value is `no`, but you can enable it as follows:

```
[manager]

enabled = yes
```

With this configuration in place, CEL events will appear as events on the manager interface. To generate example manager events, we will use the following dialplan example:

```
exten => 111,1,Answer()
    same => n,CELGenUserEvent(Custom Event,Whiskey Tango Foxtrot)
    same => n,Hangup()
```

This is the command used to execute this extension and generate sample CEL events:

```
*CLI> console dial 111@testing
```

Finally, this is one of the example manager events produced as a result of this test call:

```
Event: CEL
Privilege: call,all
EventName: CHAN_START
AccountCode:
CallerIDnum:
CallerIDname:
CallerIDani:
CallerIDrdnis:
CallerIDdnid:
Exten: 111
Context: testing
Channel: Console/dsp
Application:
AppData:
EventTime: 2010-08-23 08:14:51
AMAFlags: NONE
UniqueID: 1282569291.1
LinkedID: 1282569291.1
Userfield:
Peer:
```

cel_pgsql

This module allows posting of CEL events to a PostgreSQL database. We recommend that new installations use cel_odbc instead.

cel_radius

The cel_radius backend allows posting of CEL events to a RADIUS server. When using this module, each CEL event is reported to the RADIUS server as a single stop event. This module is configured in the */etc/asterisk/cel.conf* file. The options for this module, listed in Table 24-14, are placed in a section called [radius].

Table 24-14. Available options in the cel.conf [radius] section

Option	Value/Example	Notes
usegmtime	no	Logs timestamps using GMT instead of local time. The default is yes.
radiuscfg	/etc/radiusclient-ng/ radiusclient.conf	Sets the location of the radiusclient-ng configuration file. The default is */etc/radiusclient-ng/radiusclient.conf*.

cel_sqlite3_custom

This CEL backend inserts CEL events into a SQLite database using SQLite version 3. The database created by this module lives at */var/log/asterisk/master.db*. The configuration file for this module, */etc/asterisk/cel_sqlite3_custom.conf*, identifies the table name, as well as customizes which CEL variables will be inserted into the database. It looks like this:

```
[master]

table = cel


;
; List the column names to use when inserting CEL events.
;
columns => eventtype, eventtime, cidname, cidnum, cidani, cidrdnis, ciddnid,
    context, exten, channame, appname, appdata, amaflags, accountcode, uniqueid,
    userfield, peer


;
; Map CEL event contents to the previously specified columns.
;
values => '${eventtype}','${eventtime}','${CALLERID(name)}','${CALLERID(num)}',
    '${CALLERID(ANI)}','${CALLERID(RDNIS)}','${CALLERID(DNID)}',
    '${CHANNEL(context)}','${CHANNEL(exten)}','${CHANNEL(channame)}',
    '${CHANNEL(appname)}','${CHANNEL(appdata)}','${CHANNEL(amaflags)}',
    '${CHANNEL(accountcode)}','${CHANNEL(uniqueid)}','${CHANNEL(userfield)}',
    '${CHANNEL(peer)}'
```

 In the *cel_sqlite3_custom.conf* file, the contents of the columns and values options must appear on a single line.

cel_tds

The cel_tds module uses the FreeTDS library to post CEL events to a Microsoft SQL Server or Sybase database. It is possible to use FreeTDS with *unixODBC*, so we recommend using cel_odbc instead of this module.

Example Channel Events

Now we will show you some example sets of call events from the CEL system. The cel_custom module will be used for its simplicity. The configuration used for */etc/asterisk/cel_custom.conf* is the same as shown in "cel_custom" on page 543. Additionally, the following configuration was used for */etc/asterisk/cel.conf*:

```
[general]

enable = yes
```

```
apps = Dial,Playback
events = ALL
```

Single-party call

In this example, a single phone calls into an extension that plays back a prompt that says "Hello World." This is the dialplan:

```
exten => 200,1,Answer()
    same => n,Playback(hello-world)
    same => n,Hangup()
```

Here are the CEL events that are logged as a result of making this call:

```
"CHAN_START","1282062437.436130","Julie Bryant","12565553333","","","","200",
    "LocalSets","SIP/0000FFFF0003-00000010","","","3","","1282062437.17",
    "1282062437.17","",""

"ANSWER","1282062437.436513","Julie Bryant","12565553333","12565553333","",
    "200","200","LocalSets","SIP/0000FFFF0003-00000010","Answer","","3","",
    "1282062437.17","1282062437.17","",""

"APP_START","1282062437.501868","Julie Bryant","12565553333","12565553333",
    "","200","200","LocalSets","SIP/0000FFFF0003-00000010","Playback",
    "hello-world","3","","1282062437.17","1282062437.17","",""

"APP_END","1282062439.008997","Julie Bryant","12565553333","12565553333","",
    "200","200","LocalSets","SIP/0000FFFF0003-00000010","Playback",
    "hello-world","3","","1282062437.17","1282062437.17","",""

"HANGUP","1282062439.009127","Julie Bryant","12565553333","12565553333","",
    "200","200","LocalSets","SIP/0000FFFF0003-00000010","","","3","",
    "1282062437.17","1282062437.17","",""

"CHAN_END","1282062439.009666","Julie Bryant","12565553333","12565553333",
    "","200","200","LocalSets","SIP/0000FFFF0003-00000010","","","3","",
    "1282062437.17","1282062437.17","",""

"LINKEDID_END","1282062439.009707","Julie Bryant","12565553333",
    "12565553333","","200","200","LocalSets","SIP/0000FFFF0003-00000010","",
    "","3","","1282062437.17","1282062437.17","",""
```

Two-party call

For the second example, one phone will call another via extension 101. This results in a call that has two channels that are bridged together. Here is the extension that was called in the dialplan:

```
exten => 101,1,Dial(SIP/0000FFFF0001)
```

Here are the CEL events that are generated as a result of making this call:

```
"CHAN_START","1282062455.574611","Julie Bryant","12565553333","","","","101",
    "LocalSets","SIP/0000FFFF0003-00000011","","","3","","1282062455.18",
    "1282062455.18","",""
```

```
"APP_START","1282062455.574872","Julie Bryant","12565553333","12565553333","",
   "101","101","LocalSets","SIP/0000FFFF0003-00000011","Dial",
   "SIP/0000FFFF0001","3","","1282062455.18","1282062455.18","",""

"CHAN_START","1282062455.575044","Candice Yant","12565551111","","","","s",
   "LocalSets","SIP/0000FFFF0001-00000012","","","3","","1282062455.19",
   "1282062455.18","",""

"ANSWER","1282062458.068134","","101","12565551111","","","101","LocalSets",
   "SIP/0000FFFF0001-00000012","AppDial","(Outgoing Line)","3","",
   "1282062455.19","1282062455.18","",""

"ANSWER","1282062458.068361","Julie Bryant","12565553333","12565553333","",
   "101","101","LocalSets","SIP/0000FFFF0003-00000011","Dial",
   "SIP/0000FFFF0001","3","","1282062455.18","1282062455.18","",""

"BRIDGE_START","1282062458.068388","Julie Bryant","12565553333",
   "12565553333","","101","101","LocalSets","SIP/0000FFFF0003-00000011",
   "Dial","SIP/0000FFFF0001","3","","1282062455.18","1282062455.18","",""

"BRIDGE_END","1282062462.965704","Julie Bryant","12565553333","12565553333",
   "","101","101","LocalSets","SIP/0000FFFF0003-00000011","Dial",
   "SIP/0000FFFF0001","3","","1282062455.18","1282062455.18","",""

"HANGUP","1282062462.966097","","101","12565551111","","","","LocalSets",
   "SIP/0000FFFF0001-00000012","AppDial","(Outgoing Line)","3","",
   "1282062455.19","1282062455.18","",""

"CHAN_END","1282062462.966119","","101","12565551111","","","","LocalSets",
   "SIP/0000FFFF0001-00000012","AppDial","(Outgoing Line)","3","",
   "1282062455.19","1282062455.18","",""

"APP_END","1282062462.966156","Julie Bryant","12565553333","12565553333","",
   "101","101","LocalSets","SIP/0000FFFF0003-00000011","Dial",
   "SIP/0000FFFF0001","3","","1282062455.18","1282062455.18","",""

"HANGUP","1282062462.966215","Julie Bryant","12565553333","12565553333",
   "","101","101","LocalSets","SIP/0000FFFF0003-00000011","","","3","",
   "1282062455.18","1282062455.18","",""

"CHAN_END","1282062462.966418","Julie Bryant","12565553333","12565553333",
   "","101","101","LocalSets","SIP/0000FFFF0003-00000011","","","3","",
   "1282062455.18","1282062455.18","",""

"LINKEDID_END","1282062462.966441","Julie Bryant","12565553333",
   "12565553333","","101","101","LocalSets","SIP/0000FFFF0003-00000011",
   "","","3","","1282062455.18","1282062455.18","",""
```

Blind transfer

In this final example, a transfer will be executed. The call is started by calling a phone via extension 102. That call is then transferred to another phone at extension 101. Here is the relevant dialplan:

```
exten => 101,1,Dial(SIP/0000FFFF0001)
exten => 102,1,Dial(SIP/0000FFFF0002)
```

Here are the CEL events logged as a result of this call scenario:

```
"CHAN_START","1282062488.028200","Julie Bryant","12565553333","","","",
    "102","LocalSets","SIP/0000FFFF0003-00000013","","","3","",
    "1282062488.20","1282062488.20","",""

"APP_START","1282062488.028464","Julie Bryant","12565553333","12565553333",
    "","102","102","LocalSets","SIP/0000FFFF0003-00000013","Dial",
    "SIP/0000FFFF0002","3","","1282062488.20","1282062488.20","",""

"CHAN_START","1282062488.028762","Brooke Brown","12565552222","","","",
    "s","LocalSets","SIP/0000FFFF0002-00000014","","","3","","1282062488.21",
    "1282062488.20","",""

"ANSWER","1282062492.565759","","102","12565552222","","","102","LocalSets",
    "SIP/0000FFFF0002-00000014","AppDial","(Outgoing Line)","3","",
    "1282062488.21","1282062488.20","",""

"ANSWER","1282062492.565973","Julie Bryant","12565553333","12565553333","",
    "102","102","LocalSets","SIP/0000FFFF0003-00000013","Dial",
    "SIP/0000FFFF0002","3","","1282062488.20","1282062488.20","",""

"BRIDGE_START","1282062492.566001","Julie Bryant","12565553333",
    "12565553333","","102","102","LocalSets","SIP/0000FFFF0003-00000013",
    "Dial","SIP/0000FFFF0002","3","","1282062488.20","1282062488.20","",""

"CHAN_START","1282062497.940687","","","","","","s","LocalSets",
    "AsyncGoto/SIP/0000FFFF0002-00000014","","","3","","1282062497.22",
    "1282062488.20","",""

"BLINDTRANSFER","1282062497.940925","Julie Bryant","12565553333","12565553333","",
    "102","102","LocalSets","SIP/0000FFFF0003-00000013","Dial","SIP/0000FFFF0002",
    "3","","1282062488.20","1282062488.20",
    "AsyncGoto/SIP/0000FFFF0002-00000014<ZOMBIE>",""

"BRIDGE_END","1282062497.940961","Julie Bryant","12565553333","12565553333","",
    "102","102","LocalSets","SIP/0000FFFF0003-00000013","Dial",
    "SIP/0000FFFF0002","3","","1282062488.20","1282062488.20","",""

"APP_START","1282062497.941021","","102","12565552222","","","101","LocalSets",
    "SIP/0000FFFF0002-00000014","Dial","SIP/0000FFFF0001","3","",
    "1282062497.22","1282062488.20","",""

"CHAN_START","1282062497.941207","Candice Yant","12565551111","","","","s",
    "LocalSets","SIP/0000FFFF0001-00000015","","","3","","1282062497.23",
    "1282062488.20","",""

"HANGUP","1282062497.941361","","","","","","","LocalSets",
    "AsyncGoto/SIP/0000FFFF0002-00000014<ZOMBIE>","AppDial",
    "(Outgoing Line)","3","","1282062488.21","1282062488.20","",""

"CHAN_END","1282062497.941380","","","","","","","LocalSets",
    "AsyncGoto/SIP/0000FFFF0002-00000014<ZOMBIE>","AppDial","(Outgoing Line)",
```

```
        "3","","1282062488.21","1282062488.20","",""


"APP_END","1282062497.941415","Julie Bryant","12565553333","12565553333","",
    "102","102","LocalSets","SIP/0000FFFF0003-00000013","Dial",
    "SIP/0000FFFF0002","3","","1282062488.20","1282062488.20","",""


"HANGUP","1282062497.941453","Julie Bryant","12565553333","12565553333",
    "","102","102","LocalSets","SIP/0000FFFF0003-00000013","","","3","",
    "1282062488.20","1282062488.20","",""


"CHAN_END","1282062497.941474","Julie Bryant","12565553333","12565553333",
    "","102","102","LocalSets","SIP/0000FFFF0003-00000013","","","3","",
    "1282062488.20","1282062488.20","",""


"ANSWER","1282062500.559578","","101","12565551111","","","101","LocalSets",
    "SIP/0000FFFF0001-00000015","AppDial","(Outgoing Line)","3","",
    "1282062497.23","1282062488.20","",""


"BRIDGE_START","1282062500.559720","","102","12565552222","","","101","LocalSets",
    "SIP/0000FFFF0002-00000014","Dial","SIP/0000FFFF0001","3","","1282062497.22",
    "1282062488.20","",""


"BRIDGE_END","1282062512.742600","","102","12565552222","","","101","LocalSets",
    "SIP/0000FFFF0002-00000014","Dial","SIP/0000FFFF0001","3","","1282062497.22",
    "1282062488.20","",""


"HANGUP","1282062512.743006","","101","12565551111","","","","LocalSets",
    "SIP/0000FFFF0001-00000015","AppDial","(Outgoing Line)","3","","1282062497.23",
    "1282062488.20","",""


"CHAN_END","1282062512.743211","","101","12565551111","","","","LocalSets",
    "SIP/0000FFFF0001-00000015","AppDial","(Outgoing Line)","3","","1282062497.23",
    "1282062488.20","",""


"APP_END","1282062512.743286","","102","12565552222","","","101","LocalSets",
    "SIP/0000FFFF0002-00000014","Dial","SIP/0000FFFF0001","3","","1282062497.22",
    "1282062488.20","",""


"HANGUP","1282062512.743346","","102","12565552222","","","101","LocalSets",
    "SIP/0000FFFF0002-00000014","","","3","","1282062497.22","1282062488.20",
    "",""


"CHAN_END","1282062512.743371","","102","12565552222","","","101","LocalSets",
    "SIP/0000FFFF0002-00000014","","","3","","1282062497.22","1282062488.20",
    "",""


"LINKEDID_END","1282062512.743391","","102","12565552222","","","101",
    "LocalSets","SIP/0000FFFF0002-00000014","","","3","","1282062497.22",
    "1282062488.20","",""
```

SNMP

The Simple Network Management Protocol (SNMP) is a standardized protocol for network management. It is very commonly used and implemented across many applications and network devices. Platforms such as OpenNMS (*http://www.opennms.org/*),[§] an open source network management platform, use SNMP (among other things). Asterisk supports SNMP through the `res_snmp` module. This section discusses the installation and configuration of `res_snmp`, as well as how it can be utilized by a platform like OpenNMS.

Installing the SNMP Module for Asterisk

By default, Asterisk will not compile the SNMP development module, since a dependency needs to be satisfied first.

CentOS dependency

In CentOS, you simply need to install the *net-snmp-devel* package:

```
$ sudo yum install net-snmp-devel
```

See the upcoming section "Recompiling Asterisk with the res_snmp module" on page 552 for a description of how to recompile Asterisk with SNMP support.

Ubuntu dependency

Under Ubuntu, the following package needs to be installed:

```
$ sudo apt-get install snmp libsnmp-dev snmpd
```

> Both the *snmp* and *snmpd* packages need to be installed explicitly on Ubuntu, as they are not dependencies of the SNMP development libraries, like they are on CentOS. The *snmp* package installs SNMP tools like *snmpwalk* that we'll need, and the *snmpd* package installs the SNMP daemon.

See the next section for a description of how to recompile Asterisk with SNMP support.

[§] OpenNMS is certainly not the only platform that could be used with the `res_snmp` module. However, we chose to discuss it here for a number of reasons. First, OpenNMS is a very good network management platform that has Asterisk-specific integration. Second, it's open source and 100% free. Lastly, Jeff Gehlbach of OpenNMS has contributed to the development of Asterisk, most notably making significant improvements to the SNMP support. He was also nice enough to help us get all of this stuff working so that we could document it.

Recompiling Asterisk with the res_snmp module

Once you've satisfied the dependencies for SNMP, you can recompile Asterisk with SNMP support:

```
$ cd /usr/src/asterisk-complete/asterisk/1.8/
$ ./configure
$ make menuselect  # verify that res_snmp is selected under Resource Modules
$ sudo make install
```

You then need to copy the sample config file over to the */etc/asterisk* folder:

```
$ sudo cp /usr/src/asterisk-complete/asterisk/1.8/configs/res_snmp.conf.sample \
/etc/asterisk/res_snmp.conf
```

We'll talk about configuring this file for use with OpenNMS (*http://www.opennms .org*) in the next section.

Configuring SNMP for Asterisk Using OpenNMS

The OpenNMS (*http://www.opennms.org*) project provides an open-source network management platform that has Asterisk support built right in. There are a few steps that must be taken to enable this support, though. In this section we'll take you through what you need to do to get your Asterisk server talking to OpenNMS.

Installing OpenNMS

The OpenNMS wiki has detailed instructions for installing OpenNMS, which you can find at *http://opennms.org/wiki/Installation:Yum*.

> OpenNMS should normally not be installed on your Asterisk server. You will want to designate a separate machine as your OpenNMS server.

Since the OpenNMS wiki provides all the required instructions, we'll leave it to the experts to lead you through the first part of the installation. Once you've installed OpenNMS, come back here and we'll take you through how to configure it to work with Asterisk.

The instructions for installing OpenNMS on the wiki use SNMPv2c, which is not a secure method of abstracting data from the SNMP protocol. Since we want to build a secure system, our instructions will show you how to enable SNMPv3 support.[‖]

However, because SNMPv3 can be a bit of an unwieldy beast, and because you may not wish to enable SNMPv3 for some reason (e.g., if your version of SNMP was not

---

‖ Additionally, you can find a blog post on enabling SNMPv3 for OpenNMS at *http://www.opennms.org/wiki/ SNMPv3_protocol_configuration*.

compiled with OpenSSL support), we will provide instructions for configuring the SNMP daemon for both SNMPv2c and SNMPv3.

It tends to be easier to configure the system for SNMPv2c first and then update it to SNMPv3, as the steps to get SNMPv3 set up properly are more complex.

Editing /etc/asterisk/res_snmp.conf to work with your OpenNMS server

In the *etc/asterisk/res_snmp.conf* file that you've copied over from your source directory, there are two lines you must uncomment:

```
[general]
;subagent=yes
;enabled=yes
```

Modify the *res_snmp.conf* file so both the SNMP client and the subagent are enabled:

```
[general]
subagent=yes
enabled=yes
```

After modifying this file, you will need to reload the *res_snmp.so* module in order for the changes to take effect:

```
*CLI> module unload res_snmp.so
```

```
Unloaded res_snmp.so
 Unloading [Sub]Agent Module
   == Terminating SubAgent
```

```
*CLI> module load res_snmp.so
```

```
Loaded res_snmp.so
   == Parsing '/etc/asterisk/res_snmp.conf':   == Found
 Loading [Sub]Agent Module
 Loaded res_snmp.so => (SNMP [Sub]Agent for Asterisk)
   == Starting SubAgent
```

Editing /etc/snmp/snmpd.conf to work with your OpenNMS server

Now you can modify the *etc/snmp/snmpd.conf* file for SNMP on the host machine. Rename the current example configuration file and create a new *snmpd.conf* file:

```
$ cd /etc/snmp
$ sudo mv snmpd.conf snmpd.sample
```

The first thing to do is to add the permissions control to the file. We suggest you read the *etc/snmpd/snmp.sample* file that you just renamed to get a better idea of how the permissions are being set up. Then, add the following to your *snmpd.conf* file:

```
$ sudo cat > snmpd.conf

com2sec notConfigUser  default       public

group   notConfigGroup v1           notConfigUser
group   notConfigGroup v2c          notConfigUser
```

```
view all     included  .1
view system included  .iso.org.dod.internet.mgmt.mib-2.system

access notConfigGroup ""        any        noauth      exact  all      none none

syslocation Caledon, ON
syscontact Leif Madsen lmadsen@shifteight.org
```

Ctrl+D

 The syslocation and syscontact lines are not necessary, but they can make it easier to identify a particular server if you're monitoring several nodes.

Now we need to enable the AgentX subagent support so information about our Asterisk system can be found:

```
$ sudo cat >> snmpd.conf

master agentx
agentXSocket /var/agentx/master
agentXPerms 0660 0775 nobody root

sysObjectID .1.3.6.1.4.1.22736.1
```

Ctrl+D

By adding the master agentx line and the agentX options, we've enabled Asterisk to communicate with the SNMP daemon. The agentXPerms option is stating that Asterisk is running as *root*. If your Asterisk system is running in a different group, change *root* to the group that Asterisk is running as.

Just below the AgentX configuration, we added the sysObjectID option. The purpose of adding the sysObjectID string is so OpenNMS will know that this host system is running Asterisk, allowing it to dynamically grab additional graphing information.

Once you've performed these configuration steps, you need to restart the SNMP daemon:

```
$ sudo /etc/init.d/snmpd restart
```

To verify that the information can be polled correctly, utilize the *snmpwalk* application:

```
$ snmpwalk -On -v2c -c public 127.0.0.1 .1.3.6.1.4.1.22736
```

You should get several lines of information flowing across your screen if your configuration is correct, much like the following:

```
.1.3.6.1.4.1.22736.1.5.4.1.4.3 = INTEGER: 2
.1.3.6.1.4.1.22736.1.5.4.1.4.4 = INTEGER: 2
.1.3.6.1.4.1.22736.1.5.4.1.4.5 = INTEGER: 1
.1.3.6.1.4.1.22736.1.5.4.1.4.6 = INTEGER: 1
```

```
.1.3.6.1.4.1.22736.1.5.4.1.5.1 = INTEGER: 1
...etc
```

At this point your host system should be ready for OpenNMS to connect and gather the information it needs. Proceed by adding a node to the system and filling in the appropriate information. After a period of time, OpenNMS will poll the host system and have access to the Asterisk statistics. You should be able to click on *Resource Graphs* after selecting the node you created and see a selection of graphs available, such as SIP, DAHDI, Local, etc.

Enabling SNMPv3

Enabling SNMPv3 allows you to securely connect and transmit data from the SNMP daemon to the SNMP client. This may not be necessary in a local environment, especially if the client and the daemon are running on the same machine. However, when traversing public networks it is important to secure this data.

First, you need to stop the SNMP daemon, using the *init* script or the *service* command:

```
$ sudo /etc/init.d/snmpd stop
```

After stopping the daemon, you need to add the initial user to the */var/net-snmp/ snmpd.conf* file. This file is dynamic in nature and should only be modified when the SNMP daemon has been stopped.

You need to create a bootstrap user that you will be able to use to create the administration user in the next step. Add the following line to the */var/net-snmp/snmpd.conf* file:

```
createUser initial MD5 setup_passphrase DES
```

 We're modifying the *snmpd.conf* file we created in "Editing /etc/snmp/ snmpd.conf to work with your OpenNMS server" on page 553.

After adding it, save the file and exit. Next, you need to add permissions to the */etc/ snmp/snmpd.conf* file. Add the following line to give the *initial* user read/write permissions:

```
rwuser initial
```

After making these modifications, you can restart the SNMP daemon:

```
$ sudo /etc/init.d/snmpd start
```

```
Starting snmpd:                                          [  OK  ]
```

 If you don't see the [OK] part after Starting snmpd:, you have likely made a mistake somewhere. Stop the daemon and try again.

Using your *initial* user, you now need to create a user for OpenNMS to connect to. You'll do this with the *snmpusm#* application. Execute the following command, which will clone the *opennmsUser* user from the *initial* user. We configured the password setup_passphrase for the authentication and privacy settings when we added the *initial* user to the */var/net-snmp/snmpd.conf* file:

```
$ sudo snmpusm -v3 -u initial -n "" -l authPriv \
-a MD5 -A setup_passphrase \
-x DES -X setup_passphrase \
localhost create opennmsUser initial

User successfully created.
```

Now change the passphrase for the opennmsUser with the following command:

```
$ sudo snmpusm -v 3 -u initial -n "" -l authPriv \
-a MD5 -A setup_passphrase \
-x DES -X setup_passphrase \
-Ca -Cx localhost passwd setup_passphrase \
Op3nNMSv3 opennmsUser

SNMPv3 Key(s) successfully changed.
```

 The password we've assigned to the opennmsUser, Op3nNMSv3, is intended solely as an example, and should definitely not be used. Change it to something else that is secure.

You can now test to make sure you're getting results from your user by utilizing the *snmpwalk* application:

```
$ sudo snmpwalk -v 3 -u opennmsUser -n "" -l authPriv \
-a MD5 -A Op3nNMSv3 \
-x DES -X Op3nNMSv3 \
localhost ifTable

IF-MIB::ifIndex.1 = INTEGER: 1
IF-MIB::ifIndex.2 = INTEGER: 2
IF-MIB::ifIndex.3 = INTEGER: 3
IF-MIB::ifDescr.1 = STRING: lo
IF-MIB::ifDescr.2 = STRING: eth0
IF-MIB::ifDescr.3 = STRING: sit0
```

Now that you have data being returned, lock down the */etc/snmp/snmpd.conf* file to make sure only the opennmsUser can read data from the SNMP daemon.

The file will look quite similar to the one in the previous section, for configuring SNMPv2c. We've commented out the lines you no longer need with the hash symbol (#) and added new group and access lines to control access to the SNMP daemon:

#USM means User-based Security Model.

```
com2sec notConfigUser  default       public

#group   notConfigGroup v1            notConfigUser
#group   notConfigGroup v2c           notConfigUser
group    notConfigGroup usm           opennmsUser

#view    systemview    included   .1.3.6.1.2.1.1
#view    systemview    included   .1.3.6.1.2.1.25.1.1
view     all           included   .1

#access  notConfigGroup ""     any     noauth   exact  all   none none
access   notConfigGroup ""     usm     priv     exact  all   none none


syslocation Caledon, ON
syscontact Leif Madsen lmadsen@shifteight.org


#rwuser initial

master agentx
agentXSocket /var/agentx/master
agentXPerms 0660 0775 nobody root

sysObjectID .1.3.6.1.4.1.22736.1
```

You'll also notice we've commented out the rwuser initial line as we no longer need to permit full read/write access to the SNMP daemon. Permitting read/write access to the *initial* user is only necessary when making changes using the *snmpusm* application.

On the group line, we've configured the system to use usm (the User-based Security Model) and permitted the opennmsUser to connect. We control how it can connect with the access line, where we've enabled access via the notConfigGroup using usm and the priv model, which makes sure we connect using both authentication and privacy settings. These in turn make sure that we authenticate securely and transmit data encrypted.

After modifying the */etc/snmp/snmpd.conf* file, restart the SNMP daemon one last time:

```
$ sudo /etc/init.d/snmpd restart
```

Then verify that you can still access data via *snmpwalk*:

```
$ sudo snmpwalk -v 3 -u opennmsUser -n "" -l authPriv \
-a MD5 -A Op3nNMSv3 \
-x DES -X Op3nNMSv3 \
localhost ifTable
```

and that Asterisk is still able to connect via AgentX with *snmpwalk*:

```
$ sudo snmpwalk -v 3 -u opennmsUser -n "" -l authPriv \
-a MD5 -A Op3nNMSv3 \
```

```
-x DES -X Op3nNMSv3 \
localhost .1.3.6.1.4.1.22736
```

If all goes well, you should get lots of lines back, including:

```
SNMPv2-SMI::enterprises.22736.1.5.4.1.2.1 = STRING: "SIP"
SNMPv2-SMI::enterprises.22736.1.5.4.1.2.2 = STRING: "IAX2"
SNMPv2-SMI::enterprises.22736.1.5.4.1.2.3 = STRING: "Bridge"
SNMPv2-SMI::enterprises.22736.1.5.4.1.2.4 = STRING: "MulticastRTP"
SNMPv2-SMI::enterprises.22736.1.5.4.1.2.5 = STRING: "DAHDI"
SNMPv2-SMI::enterprises.22736.1.5.4.1.2.6 = STRING: "Local"
SNMPv2-SMI::enterprises.22736.1.5.4.1.3.1 = STRING: "Session Initiation Protocol(SIP)"
SNMPv2-SMI::enterprises.22736.1.5.4.1.3.2 = STRING: "Inter Asterisk eXchange Driver"
SNMPv2-SMI::enterprises.22736.1.5.4.1.3.3 = STRING: "Bridge Interaction Channel"
SNMPv2-SMI::enterprises.22736.1.5.4.1.3.4 = STRING: "Multicast RTP Paging Channel"
SNMPv2-SMI::enterprises.22736.1.5.4.1.3.5 = STRING: "DAHDI Telephony Driver"
SNMPv2-SMI::enterprises.22736.1.5.4.1.3.6 = STRING: "Local Proxy Channel Driver"
```

If you don't get data returned right away, it could be because the Asterisk *res_snmp.so* module has not reconnected to the SNMP daemon. You can force this either by restarting Asterisk, or by unloading and reloading the *res_snmp.so* module from the Asterisk CLI.

Monitoring Asterisk with OpenNMS

Once you've installed OpenNMS and configured Asterisk with the `res_snmp` module, you can use OpenNMS to monitor your Asterisk server. You can configure what statistics are monitored, as well as what notifications you would like to receive based on those statistics. Exploring the capabilities of OpenNMS is left as an exercise for the reader. However, we have included a few graphs to demonstrate some of the basic information you can collect from an Asterisk server. These graphs come from an Asterisk server that is not very heavily loaded, but they still give a good indication of what you might see.

Figure 24-1 contains a graph of how many channels were active in Asterisk at different times.

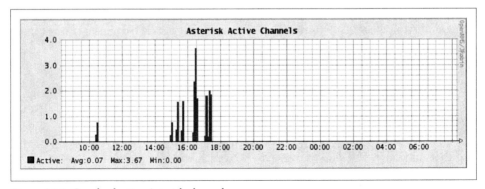

Figure 24-1. Graph of active Asterisk channels

Figure 24-2 shows a graph of active channels of a specific type. In this case, we're looking at how many DAHDI channels are active on the system. Monitoring DAHDI channels is particularly interesting, since DAHDI channels are generally mapped to physical resources, and a predefined number of channels are available. It would be very useful to monitor DAHDI channel utilization and get notified when usage passes a particular threshold, as this might be a signal that additional capacity needs to be added.

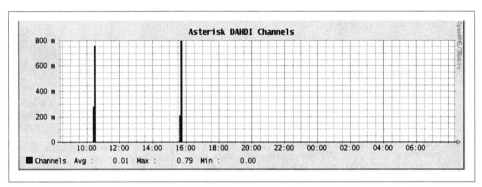

Figure 24-2. Graph of active DAHDI channels

Finally, Figure 24-3 shows network interface utilization. As you can see, there were spikes in the traffic flowing into and out of the system when SIP calls were in progress.

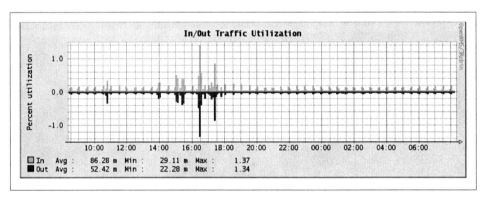

Figure 24-3. Graph of traffic on a network interface

Conclusion

Asterisk is very good at allowing you to keep track of many different facets of its operation, from simple call detail records to full debugging of the running code. These various mechanisms will help you in your efforts to manage your Asterisk PBX, and they represent one of the ways in which Asterisk is vastly superior to most (if not all) traditional PBXs.

## CHAPTER 25

Web Interfaces

A point of view can be a dangerous luxury when
substituted for insight and understanding.

—Marshall McLuhan

Before you get too excited, this chapter is not going to talk about dialplan configuration GUIs such as FreePBX or Digium's Asterisk-GUI. We recognize that much of the success of Asterisk is due to the success of FreePBX-based projects such as AsteriskNOW, Trixbox, and PBX in a Flash, but in this book our focus is on Asterisk. As such, we will not be discussing any GUIs that essentially remove your relationship with the dialplan. It's not that we're against these things, but simply that we have only so much space in this book, and our goal is to look at Asterisk from the bottom up. Most Asterisk GUI projects hide the inner workings of Asterisk behind an interface, and for this reason they are not compatible with the goals of this book.

Our discussion of Asterisk web interfaces, therefore, will focus on interfaces to components other than the dialplan.

The FreePBX Dialplan GUI

Now that we've promised not to talk about dialplan interfaces, we feel it would be wrong to say nothing at all about FreePBX, the juggernaut of the Asterisk community. This interface (which is at the heart of many of the most popular Asterisk distributions, such as AsteriskNOW, PBX in a Flash, and Trixbox), is unarguably a very large part of why Asterisk has been as successful as it has. With the FreePBX interface, you can configure and manage many aspects of an Asterisk system without touching a single configuration file. While we purists may like everyone to work only with the config files, we recognize that for many, learning Linux and editing these files by hand is simply not going to happen. For those folks, there is FreePBX, and it has our respect for the important contributions it has made to the success of Asterisk.

What we will do in this chapter is introduce a few projects that provide web interfaces into other parts of the system, and a selection of web-driven applications that are significant, useful, or recommended. In general, we have tended to focus on free and open source applications, but we will mention some commercial products where we feel it's warranted.

There are many third-party applications that have been developed for Asterisk. The ones described here are among the best, at the time of this writing.

Flash Operator Panel

The Flash Operator Panel (or FOP, as it's more commonly known) is an interface primarily for the use of switchboard operators. FOP uses Adobe Flash to present an interface through a web browser, and connects to Asterisk through the Asterisk Manager Interface (see Chapter 20 for a discussion of the AMI).

There are two versions of the Flash Operator Panel: the original release (which so far is at version 0.30, and is now a maintenance release only), and FOP2 (shown in Figure 25-1), which is a vast improvement over the original FOP, but requires the purchase of a license for any system with more than 15 extensions.

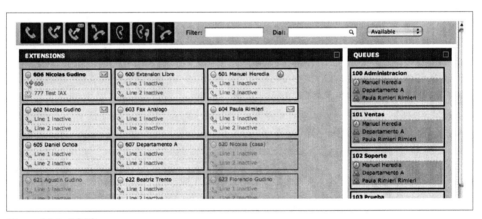

Figure 25-1. FOP2

You can find FOP at *http://www.asternic.org*, and FOP2 at *http://www.fop2.com*.

Queue Status and Reporting

In most call centers, it is not enough simply to be able to route calls correctly. Of equal importance to most queues is the ability for supervisory and management staff to determine how the queue and the agents are performing. For this, two things will be of benefit: live queue status information, and some manner of reporting package.

Queue Status Display

Queue status will often be displayed on a large, wall-mounted panel or a reader board. Here are some of the kinds of information that might be included:

* Number of agents logged in
* Number of callers holding
* Number of calls in progress
* Current longest hold time
* Average hold time
* Abandon rate
* Service level

Other information might be desired as well; the goal of a queue status display is to present to both supervisory staff and queue agents a quick visual indication of the state of the queue at this particular moment in time.

Additionally, individual group or agent performance metrics may be displayed, as an informational tool.

The Asternic Call Center Stats software is available in an open source "lite" version that provides a basic status display. There are also several commercial products that offer this functionality.

Queue Reporting

Queue reporting consists of reports and graphs that supervisory personnel can use to look at queue and agent performance from a historical perspective. Many of the metrics will be similar to those of the status display; however, the goal of reporting is to allow management to monitor staffing levels, identify problems, and analyze trends.

We discussed a few queue reporting interfaces in Chapter 13.

Call Detail Records

While Asterisk does a good-enough job of generating and storing CDRs, the records are in a very raw format, which makes it difficult to perform any sort of analysis on them.

Enter the CDR reporting package. In the 1990s, when long-distance rates were complex and expensive, an entire subindustry was spawned by companies looking to help other companies make sense out of complex long-distance rates. Nowadays, with long-distance being far less expensive, as well as generally simpler in terms of pricing model, there is less need for detailed analysis of call records. Nevertheless, many of these highly experienced companies have added support for Asterisk CDR analysis, and thus if you want excellent reporting capabilities, you will find a huge industry with many experienced participants.

For a simple interface to the call records, a popular program is CDR-Stats (*http://www.cdr-stats.org*), which is the successor to the hugely popular Asterisk-Stat package. This open source reporting interface provides a simple way to examine call detail records, and some basic metrics on calling patterns.

A2Billing

The A2Billing project is not simply a billing interface for Asterisk: it is, in fact, a complete VoIP carrier-in-a-box. This complex and comprehensive product delivers much of the technology you would need to allow you to provide a VoIP reseller service.*

The A2Billing platform has been generously released under the AGPL as open source. The sponsor of the A2Billing project, Star2Billing, offers consultancy services to get you up to speed faster.

Conclusion

In this brief chapter we have provided some pointers to popular graphical applications that can be used in conjunction with Asterisk. While we didn't cover them in detail, we do acknowledge the importance of FreePBX and the AsteriskGUI, which are both GUI open source projects that provide a PBX configuration interface on top of Asterisk. If a full GUI solution for simple PBX configuration interests you, we encourage you to take a look at them. To give them a try, we recommend using the AsteriskNOW distribution, which provides both FreePBX and AsteriskGUI as GUI options.

---

* It cannot provide you with business savvy, experience in running a phone company, or automatic security, though, so please don't think that all you have to do is download A2Billing and you can take on AT&T!

## CHAPTER 26

Security

*We spend our time searching for security and
hate it when we get it.*

—John Steinbeck

Security for your Asterisk system is critical, especially if the system is exposed to the Internet. There is a lot of money to be made by attackers in exploiting systems to make free phone calls. This chapter provides advice on how to provide stronger security for your VoIP deployment.

Scanning for Valid Accounts

If you expose your Asterisk system to the public Internet, one of the things you will almost certainly see is a scan for valid accounts. Example 26-1 contains log entries from one of the authors' production Asterisk systems.* This scan began with checking various common usernames, then later went on to scan for numbered accounts. It is common for people to name SIP accounts the same as extensions on the PBX. This scan takes advantage of that fact. This leads to our first tip for Asterisk security:

> *Tip #1:* Use non-numeric usernames for your VoIP accounts to make them harder to guess. For example, in parts of this book we use the MAC address of a SIP phone as its account name in Asterisk.

Example 26-1. Log excerpts from account scanning

```
[Aug 22 15:17:15] NOTICE[25690] chan_sip.c: Registration from
'"123"<sip:123@127.0.0.1>' failed for '203.86.167.220:5061' - No matching peer
found
[Aug 22 15:17:15] NOTICE[25690] chan_sip.c: Registration from
'"1234"<sip:1234@127.0.0.1>' failed for '203.86.167.220:5061' - No matching peer
found
[Aug 22 15:17:15] NOTICE[25690] chan_sip.c: Registration from
```

---

* The real IP address has been replaced with 127.0.0.1 in the log entries.

'"12345"<sip:12345@127.0.0.1>' failed for '203.86.167.220:5061' - No matching peer found
[Aug 22 15:17:15] NOTICE[25690] chan_sip.c: Registration from
'"123456"<sip:123456@127.0.0.1>' failed for '203.86.167.220:5061' - No matching peer found
[Aug 22 15:17:15] NOTICE[25690] chan_sip.c: Registration from
'"test"<sip:test@127.0.0.1>' failed for '203.86.167.220:5061' - No matching peer found
[Aug 22 15:17:15] NOTICE[25690] chan_sip.c: Registration from
'"sip"<sip:sip@127.0.0.1>' failed for '203.86.167.220:5061' - No matching peer found
[Aug 22 15:17:15] NOTICE[25690] chan_sip.c: Registration from
'"user"<sip:user@127.0.0.1>' failed for '203.86.167.220:5061' - No matching peer found
[Aug 22 15:17:16] NOTICE[25690] chan_sip.c: Registration from
'"admin"<sip:admin@127.0.0.1>' failed for '203.86.167.220:5061' - No matching peer found
[Aug 22 15:17:16] NOTICE[25690] chan_sip.c: Registration from
'"pass"<sip:pass@127.0.0.1>' failed for '203.86.167.220:5061' - No matching peer found
[Aug 22 15:17:16] NOTICE[25690] chan_sip.c: Registration from
'"password"<sip:password@127.0.0.1>' failed for '203.86.167.220:5061' - No matching peer found
[Aug 22 15:17:16] NOTICE[25690] chan_sip.c: Registration from
'"testing"<sip:testing@127.0.0.1>' failed for '203.86.167.220:5061' - No matching peer found
[Aug 22 15:17:16] NOTICE[25690] chan_sip.c: Registration from
'"guest"<sip:guest@127.0.0.1>' failed for '203.86.167.220:5061' - No matching peer found
[Aug 22 15:17:16] NOTICE[25690] chan_sip.c: Registration from
'"voip"<sip:voip@127.0.0.1>' failed for '203.86.167.220:5061' - No matching peer found
[Aug 22 15:17:16] NOTICE[25690] chan_sip.c: Registration from
'"account"<sip:account@127.0.0.1>' failed for '203.86.167.220:5061' - No matching peer found

...

[Aug 22 15:17:17] NOTICE[25690] chan_sip.c: Registration from
'"100"<sip:100@127.0.0.1>' failed for '203.86.167.220:5061' - No matching peer found
[Aug 22 15:17:17] NOTICE[25690] chan_sip.c: Registration from
'"101"<sip:101@127.0.0.1>' failed for '203.86.167.220:5061' - No matching peer found
[Aug 22 15:17:17] NOTICE[25690] chan_sip.c: Registration from
'"102"<sip:102@127.0.0.1>' failed for '203.86.167.220:5061' - No matching peer found
[Aug 22 15:17:17] NOTICE[25690] chan_sip.c: Registration from
'"103"<sip:103@127.0.0.1>' failed for '203.86.167.220:5061' - No matching peer found
[Aug 22 15:17:17] NOTICE[25690] chan_sip.c: Registration from
'"104"<sip:104@127.0.0.1>' failed for '203.86.167.220:5061' - No matching peer found
[Aug 22 15:17:17] NOTICE[25690] chan_sip.c: Registration from
'"105"<sip:105@127.0.0.1>' failed for '203.86.167.220:5061' - No matching peer found

These account scans take advantage of the fact that the response that comes back from the server for a registration attempt will differ depending on whether or not the account exists. If the account exists, the server will request authentication. If the account does not exist, the server will immediately deny the registration attempt. This behavior is just how the protocol is defined. This leads us to our second tip for Asterisk security:

> *Tip #2:* Set `alwaysauthreject` to yes in the `[general]` section of */etc/asterisk/sip.conf*. This option tells Asterisk to respond as if every account is valid, which makes scanning for valid usernames useless.

Authentication Weaknesses

The first section of this chapter discussed scanning for usernames. Even if you have usernames that are difficult to guess, it is critical that you have strong passwords as well. If an attacker is able to obtain a valid username, he will attempt to brute-force the password. Strong passwords make this much more difficult to do.

The default authentication scheme for both the SIP and IAX2 protocols is weak. Authentication is done using an MD5 challenge and response mechanism. If an attacker is able to capture any call traffic, such as a SIP call made from a laptop on an open wireless network, it will be much easier to work on brute-forcing the password since it will not require authentication requests to the server.

> *Tip #3:* Use strong passwords. There are countless resources available on the Internet that help define what constitutes a strong password. There are also many strong password generators available. Use them!

IAX2 provides the option of using key-based authentication, as well as full encryption of a call. The SIP support in Asterisk includes TLS support, which provides encryption for the SIP signaling.

> *Tip #4:* If you are using IAX2, use key-based authentication. This is a much stronger authentication method than the default MD5-based challenge-response method. For further enhanced security with IAX2, use the option to encrypt the entire call. If you are using SIP, use TLS to encrypt the SIP signaling. This will prevent an attacker from capturing a successful authentication exchange with the server.

For more information about setting up IAX2 or SIP encryption, see Chapter 7.

Fail2ban

The last two sections discussed attacks involving scanning for valid usernames and brute-forcing passwords. Fail2ban (*http://www.fail2ban.org*) is an application that can watch your Asterisk logs and update firewall rules to block the source of an attack in response to too many failed authentication attempts.

> *Tip #5:* Use Fail2ban when exposing Voice over IP services on untrusted networks to automatically update the firewall rules to block the sources of attacks.

Installation

Fail2ban is available as a package in many distributions. Alternatively, you can install it from source by downloading it from the Fail2ban website. To install it on Ubuntu, use the following command:

```
$ sudo apt-get install fail2ban
```

To install Fail2ban on CentOS, you must have the EPEL repository enabled. For more information on the EPEL repository, see "Third-Party Repositories" on page 46. Once the repository is enabled, Fail2ban can be installed by running the following command:

```
$ sudo yum install fail2ban
```

 The installation of Fail2ban from a package will include an *init* script to ensure that it runs when the machine boots up. If you install from source, make sure that you take the necessary steps to ensure that Fail2ban is always running.

iptables

For Fail2ban to be able to do anything useful after it detects an attack, you must also have *iptables* installed. To install it on Ubuntu, use the following command:

```
$ sudo apt-get install iptables
```

To install *iptables* on CentOS, use this command:

```
$ sudo yum install iptables
```

You can verify that *iptables* has been installed by running the *iptables* command. The *-L* option requests that the current firewall rules be displayed. In this case, there are no rules configured:

```
$ sudo iptables -L

Chain INPUT (policy ACCEPT)
target     prot opt source               destination

Chain FORWARD (policy ACCEPT)
target     prot opt source               destination

Chain OUTPUT (policy ACCEPT)
target     prot opt source               destination
```

Sending email

It is interesting and useful to allow Fail2ban to email the system administrator when it bans an IP address. For this to work, an MTA must be installed. If you are not sure which one to use, the one used during testing for writing this chapter was Postfix. To install Postfix on Ubuntu, use the following command. You may be asked to answer a couple of questions by the installer:

```
$ sudo apt-get install postfix
```

To install Postfix on CentOS, use this command:

```
$ sudo yum install postfix
```

To test the installation of your MTA, you can send a quick email using *mutt*. To install it, use the same installation commands as given for installing Postfix, but substitute *mutt* for the package name. Then run the following commands to test the MTA:

```
$ echo "Just testing." > email.txt
$ mutt -s "Testing" youraddress@shifteight.org < email.txt
```

Configuration

The first file that must be set up is the Asterisk logging configuration file. Here are the contents of */etc/asterisk/logger.conf* on a working system. Ensure that you at least have dateformat and messages set, as those are required for Fail2ban:

```
[general]

dateformat = %F %T

[logfiles]

console => notice,warning,error,debug
messages => notice,warning,error
```

The next configuration file that must be created is the one that teaches Fail2ban what to watch out for in Asterisk log files. Place the following contents in a new file called */etc/fail2ban/filter.d/asterisk.conf*:

```
[INCLUDES]

# Read common prefixes. If any customizations available -- read them from
# common.local
#before = common.conf


[Definition]

#_daemon = asterisk

# Option:  failregex
# Notes.:  regex to match the password failures messages in the logfile. The
#          host must be matched by a group named "host". The tag "<HOST>" can
#          be used for standard IP/hostname matching and is only an alias for
#          (?:::f{4,6}:)?(?P<host>\S+)
# Values:  TEXT
#

# *** All lines below should start with NOTICE
#
failregex = NOTICE.* .*: Registration from '.*' failed for '<HOST>'
# - Wrong password
```

```
                    NOTICE.* .*: Registration from '.*' failed for '<HOST>'
          # - No matching peer found

                    NOTICE.* .*: Registration from '.*' failed for '<HOST>'
          # - Username/auth name mismatch

                    NOTICE.* .*: Registration from '.*' failed for '<HOST>'
          # - Device does not match ACL

                    NOTICE.* <HOST> failed to authenticate as '.*'$
                    NOTICE.* .*: No registration for peer '.*' \(from <HOST>\)
                    NOTICE.* .*: Host <HOST> failed MD5 authentication for '.*' (.*)
                    NOTICE.* .*: Failed to authenticate user .*@<HOST>.*

          # Option:  ignoreregex
          # Notes.:  regex to ignore. If this regex matches, the line is ignored.
          # Values:  TEXT
          #
          ignoreregex =
```

Next, you must enable the new Asterisk filter that you just created. To do so, append
the following contents to */etc/fail2ban/jail.conf*. You will need to modify the **dest** and
sender options to specify the appropriate email addresses for the To and From headers:

```
[asterisk-iptables]

enabled  = true
filter   = asterisk
action   = iptables-allports[name=ASTERISK, protocol=all]
           sendmail-whois[name=ASTERISK, dest=me@shifteight.org,
           sender=fail2ban@shifteight.org]
logpath  = /var/log/asterisk/messages
maxretry = 5
bantime = 259200
```

Finally, there are a couple of options in the [DEFAULT] section of */etc/fail2ban/jail.conf*
that should be updated. The ignoreip option specifies a list of IP addresses that should
never be blocked. It is a good idea to list your IP address(es) here so that you never
accidentally block yourself if you make a mistake while trying to set up a phone, for
example.[†] You should consider adding other IP addresses as well, such as that of your
SIP provider. The whitelisting of good IP addresses protects you against abuse of your
Fail2ban configuration. A clever attacker could cause a denial of service by crafting a
series of packets that will result in Fail2ban blocking the IP address of their choice.

The destemail option should be set, as well. This address will be used for emails not
specific to the Asterisk filter such as the email Fail2ban sends out when it first starts
up. Here's how you configure these options:

---

† Leif learned this one the hard way. He thought his PBX was down, while Russell and Jim had no problems
  connecting to the conference bridge. It turned out that Fail2ban had banned him from his own PBX.

```
[DEFAULT]

# Multiple addresses can be specified, separated by a space.
ignoreip = 127.0.0.1 10.1.1.1

destemail = youraddress@shifteight.org
```

Encrypted Media

Be aware that the audio for a Voice over IP call is typically transmitted in an unencrypted format. Anyone that can capture the traffic can listen to the audio of the phone call. Luckily, Asterisk supports encrypting the media of VoIP calls. If you are using SIP, you can encrypt the media using SRTP. IAX2 supports fully encrypting calls, as well. Detailed information on encrypting media can be found in Chapter 7.

> Tip #6: Encrypt the media for calls on untrusted networks using SRTP or IAX2 encryption.

Dialplan Vulnerabilities

The Asterisk dialplan is another area where taking security into consideration is critical. The dialplan can be broken down into multiple contexts to provide access control to extensions. For example, you may want to allow your office phones to make calls out through your service provider. However, you do not want to allow anonymous callers that come into your main company menu to be able to then dial out through your service provider. Use contexts to ensure that only the callers you intend have access to services that cost you money.

> Tip #7: Build dialplan contexts with great care. Also, avoid putting any extensions that could cost you money in the [default] context.

One of the more recent Asterisk dialplan vulnerabilities to have been discovered and published is the idea of dialplan injection. A dialplan injection vulnerability begins with an extension that has a pattern that ends with the match-all character, a period. Take this extension as an example:

```
exten => _X.,1,Dial(IAX2/otherserver/${EXTEN},30)
```

The pattern for this extension matches all extensions (of any length) that begin with a digit. Patterns like this are pretty common and convenient. The extension then sends this call over to another server using the IAX2 protocol, with a dial timeout of 30 seconds. Note the usage of the ${EXTEN} variable here. That's where the vulnerability exists.

In the world of Voice over IP, there is no reason that a dialed extension must be numeric. In fact, it is quite common using SIP to be able to dial someone by name. Since it is possible for non-numeric characters to be a part of a dialed extension, what would happen if someone sent a call to this extension?

```
1234&DAHDI/g1/12565551212
```

A call like this is an attempt at exploiting a dialplan injection vulnerability. In the previous extension definition, once `${EXTEN}` has been evaluated, the actual `Dial()` statement that will be executed is:

```
exten => _X.,1,Dial(IAX2/otherserver/1234&DAHDI/g1/12565551212,30)
```

If the system has a PRI configured, this call will cause a call to go out on the PRI to a number chosen by the attacker, even though you did not explicitly grant access to the PRI to that caller. This problem can quickly cost you a whole lot of money.

There are (at least) two approaches for avoiding this problem. The first and easiest approach is to always use strict pattern matching. If you know the length of extensions you are expecting and only expect only numeric extensions, use a strict numeric pattern match. For example, this would work if you are expecting four-digit numeric extensions only:

```
exten => _XXXX,1,Dial(IAX2/otherserver/${EXTEN},30)
```

The other approach to mitigating dialplan injection vulnerabilities is by using the `FILTER()` dialplan function. Perhaps you would like to allow numeric extensions of any length. `FILTER()` makes that easy to achieve safely.

```
exten => _X.,1,Set(SAFE_EXTEN=${FILTER(0-9,${EXTEN})})
    same => n,Dial(IAX2/otherserver/${SAFE_EXTEN},30)
```

For more information about the syntax for the `FILTER()` dialplan function, see the output of the *core show function FILTER* command at the Asterisk CLI.

> Tip #8: Be wary of dialplan injection vulnerabilities. Use strict pattern matching or use the `FILTER()` dialplan function to avoid these problems.

Securing Asterisk Network APIs

FastAGI and the AMI are two network-based APIs commonly used in Asterisk deployments. For more details on AGI, see Chapter 21. For more information on the AMI, see Chapter 20.

In the case of FastAGI, there is no encryption or authentication available. It is up to you as the administrator to ensure that the only communication allowed to the FastAGI server is from Asterisk.

The AMI protocol includes authentication, but it is very weak. Further, the data exchanged via the AMI is often sensitive, from a privacy standpoint. It is critical to secure AMI connectivity. It is best to only expose the AMI on trusted networks. If it must be exposed to an untrusted networks, we recommend only allowing connections using SSL.

It is critical to understand what power the AMI provides. If an AMI user is granted all permissions that are available, that user will be able to run arbitrary commands on your system. If the account has the ability to update configuration files, it will be able to add an extension to the dialplan that runs the `System()` application, enabling it to run any command it wants. If it also has access to originate calls, it can originate a call to that extension, resulting in the execution of that command. Be careful when opening up AMI access on your system and restrict what permissions are granted to each account in */etc/asterisk/manager.conf*.

> *Tip #9:* Secure Asterisk network APIs. Use firewall rules to restrict access to your FastAGI server. Use encryption on the AMI. Restrict access provided to AMI accounts as much as possible.

IAX2 Denial of Service

While SIP is a text-based protocol, IAX2 is a binary encoded protocol. The IAX2 standard is RFC 5456 (*http://www.rfc-editor.org/rfc/rfc5456.txt*). Every IAX2 packet contains a call number that is used to associate the packet with an active call. This is analogous to the `Call-ID` header in SIP. An IAX2 call number, is a 15-bit field. It is large enough to deal with the number of calls that will be practical on one system. Unfortunately, it is also small enough that it is pretty easy for an attacker to send enough small packets to consume all available call numbers on a system for a short period of time, resulting in a denial of service attack.

The IAX2 support in Asterisk has been modified to automatically protect against this type of attack. This protection is referred to as *call token support* and requires a three-way handshake to occur before a call number is allocated. However, older versions of Asterisk and some non-Asterisk IAX2 implementations may not support this, so there are a number of options that let you tweak the behavior.

By default, the security mechanisms are enabled and no configuration changes are required. If for some reason you would like to disable call token support completely, you can do so by using the following configuration in */etc/asterisk/iax.conf*:

```
[general]

calltokenoptional = 0.0.0.0/0.0.0.0
maxcallnumbers = 16382
```

With the default configuration, a host that can pass the call token exchange can still consume the call number table. The call token exchange ensures that call numbers are only allocated once we know we have not received a request with a spoofed source IP address. Once we know a request is legitimate, enforcing resource limits per host is achievable. Consider the following options in *iax.conf*:

```
[general]

; Set the default call number limit per host
maxcallnumbers = 16

[callnumberlimits]
```

```
; Set a different call number limit for all hosts in a
; specified range.
192.168.1.0/255.255.255.0 = 1024

[some_peer]

; A dynamic peer's address is not known until that peer
; registers. A call number limit can be specified in the
; peer's section instead of the callnumberlimits section.

type = peer
host = dynamic
maxcallnumbers = 512
```

If a peer does not yet support call token validation, but you would like to turn it on as soon as you detect that the peer has been upgraded to support it, there is an option that allows for this behavior:

```
[some_other_peer]

requirecalltoken = auto
```

If you would like to allow guest access over IAX2, you will most likely want to disable call token validation for unauthenticated calls. This will ensure that the largest number of people can call your system over IAX2. However, if you do so, you should also set the option that provides a global limit to how many call numbers can be consumed by hosts that did not pass call token validation:

```
[general]

maxcallnumbers_nonvalidated = 2048

[guest]

type = user
requirecalltoken = no
```

If at any time you would like to see some statistics on call number usage on your system, execute the *iax2 show callnumber usage* command at the Asterisk CLI.

> *Tip #10:* Be happy knowing that IAX2 has been updated to secure itself from denial of service attacks due to call number exhaustion. If you must turn off these security features in some cases, use the options provided to limit your exposure to an attack.

Other Risk Mitigation

There are a couple more useful features in Asterisk that can be used to mitigate the risk of attacks. The first is to make use of the **permit** and **deny** options to build access control lists (ACLs) for privileged accounts. Consider a PBX that has SIP phones on a local network, but also accepts SIP calls from the public Internet. Calls coming in over the Internet are only granted access to the main company menu, while local SIP phones have the ability to make outbound calls that cost you money. In this case, it is a very good idea to set ACLs to ensure that only devices on your local network can use the accounts for the phones. Here is an example of doing that in */etc/asterisk/sip.conf*:

```
[phoneA] ; Use a better account name than this.

type = friend

; Start by denying everyone.
deny = 0.0.0.0/0.0.0.0

; Allow connections that originate from 192.168.X.X to attempt
; to authenticate against this account.
permit = 192.168.0.0/255.255.0.0
```

The permit and deny options are accepted almost everywhere that connections to IP services are configured. Another useful place for ACLs is in */etc/asterisk/manager.conf*, to restrict AMI accounts to the single host that is supposed to be using the manager interface.

Tip #11: Use ACLs when possible on all privileged accounts for network services.

Another way you can mitigate security risk is by configuring call limits. The recommended method for implementing call limits is to use the GROUP() and GROUP_COUNT() dialplan functions. Here is an example that limits the number of calls from each SIP peer to no more than two at a time:

```
exten => _X.,1,Set(GROUP(users)=${CHANNEL(peername)})

; *** This line should have no line breaks
    same => n,NoOp(There are ${GROUP_COUNT(${CHANNEL(peername)})}
calls for account ${CHANNEL(peername)}.)

    same => n,GotoIf($[${GROUP_COUNT(${CHANNEL(peername)})} > 2]?denied:continue)
    same => n(denied),NoOp(There are too many calls up already.  Hang up.)
    same => n,HangUp()
    same => n(continue),NoOp(continue processing call as normal here ...)
```

Tip #12: Use call limits to ensure that if an account is compromised, it cannot be used to make hundreds of phone calls at a time.

Resources

Sometimes there are security vulnerabilities that require modifications to the Asterisk source code to resolve. When those issues are discovered, the Asterisk development team puts out new releases that contain only fixes for the security issues, to allow for quick and easy upgrades. When this occurs, the Asterisk development team also publishes a security advisory document that discusses the details of the vulnerability. We recommend that you subscribe to the *asterisk-announce http://lists.digium.com/mailman/listinfo/asterisk-announce* mailing list to make sure that you know about these issues when they come up.

One of the most popular tools for SIP account scanning and password cracking is SIPVicious (*http://sipvicious.org*). We strongly encourage that you take a look at it and use it to audit your own systems. If your system is exposed to the Internet, others will likely run it against your system, so make sure that you do it first.

Another resource for all things VoIP security–related is the VOIPSEC mailing list on VOIPSA.org (*http://voipsa.org/*). The website contains some additional resources, as well.

Finally, *http://www.infiltrated.net/voipabuse/* has some useful information. The author provides a list of addresses known to be the source of VoIP attacks, as well as instructions on how to block all addresses on this list. The author also provides a sample script called AntiToll (*http://www.infiltrated.net/antitoll*), which blocks all addresses outside of the United States.

Conclusion—A Better Idiot

There is a maxim in the technology industry that states, "As soon as something is made idiot-proof, nature will invent a better idiot." The point of this statement is that there is no development effort that can be considered complete. There is always room for improvement.

When it comes to security, you must always bear in mind that the people who are looking to take advantage of your system are highly motivated. No matter how secure your system is, somebody will always be looking to crack it.

We're not advocating paranoia, but we are suggesting that what we have written here is by no means the final word on VoIP security. While we have tried to be as comprehensive as we can be in this book, you must accept responsibility for the security of your system.

As free Internet calling becomes more common, the criminals will be working hard to find weaknesses, and exploit them.

## CHAPTER 27

Asterisk: A Future for Telephony

Now this is not the end. It is not even the beginning of the end. But it is, perhaps, the end of the beginning.

—Winston Churchill

We have arrived at the final chapter of this book. We've covered a lot, but we hope that we have made it clear that this book has merely scratched the surface of this phenomenon called Asterisk. To wrap things up, we want to spend some time exploring what we might see from Asterisk and open source telephony in the near future.

When we wrote the first edition of *Asterisk: The Future of Telephony*, we confidently asserted that open source communications engines such as Asterisk would cause a shift in thinking that would transform the telecommunications industry. In many ways, our belief has been proven correct. While the telecom industry still has much evolving to do, Asterisk has played a key role in fomenting a shift in thinking that has affected the entire industry.

The Problems with Traditional Telephony

Although Alexander Graham Bell is most famously remembered as the father of the telephone,[*] the reality is that during the latter half of the 1800s, dozens of minds were working toward the goal of carrying voice over telegraph lines. These people were mostly business-minded folks, looking to create a product through which they might make their fortunes.

We have come to think of traditional telephone companies as monopolies, but this was not true in their early days. The early history of telephone service took place in a very competitive environment, with new companies springing up all over the world, often with little or no respect for the patents they might be violating. Many famous monopolies got their start through the waging (and winning) of patent wars.

---

[*] Ever heard of Elisha Gray or Antonio Meucci?

It's interesting to contrast the history of the telephone with the history of Linux and the Internet. While the telephone was created as a commercial exercise, and the telecom industry was forged through lawsuits and corporate takeovers, Linux and the Internet arose out of the academic community, which has always valued the sharing of knowledge over profit.

The cultural differences are obvious. Telecommunications technologies tend to be closed, confusing, and expensive, while networking technologies are comparatively open, well documented, and competitive.

Closed Thinking

If one compares the culture of the telecommunications industry to that of the Internet, it is sometimes difficult to believe the two are related. The technology of the Internet was designed in large part by academics and enthusiasts, whereas contributing to the development of the PSTN is impossible for any individual to contemplate. This is an exclusive club; membership is not open to just anyone.†

Although the ITU is the United Nations's sanctioned body responsible for international telecommunications, many of the VoIP protocols (SIP, MGCP, RTP, STUN) come not from the ITU, but rather from the IETF (which publishes all of its standards free to all, and allows anyone to submit an Internet Draft for consideration).

Open protocols such as SIP may have a tactical advantage over ITU protocols such as H.323 due to the ease with which one can obtain them.‡ Although H.323 is widely deployed by carriers as a VoIP protocol in the backbone, it is much more difficult to find H.323-based endpoints; newer products are far more likely to support SIP.

The success of the IETF's open approach has not gone unnoticed by the ITU. Since the first edition of this book, the ITU has made all of the ITU-T and ITU-R recommendations available as free downloads in PDF form from its website (*http://www.itu.int*).

As for Asterisk, it embraces both the past and the future—H.323 support is available, although the community has for the most part shunned H.323 in favor of the IETF protocol SIP and the darling of the Asterisk community, IAX.

Limited Standards Compliancy

One of the oddest things about all the standards that exist in the world of legacy telecommunications is the various manufacturers' seeming inability to implement them

---

† Contrast this with the IETF's membership page, which states: "The IETF is not a membership organization (no cards, no dues, no secret handshakes :-)...It is open to any interested individual...Welcome to the IETF." Talk about community!

‡ Many people who are familiar with both protocols suggest that H.323 is in fact technically superior. Betamax, anyone?

consistently. Each manufacturer desires a total monopoly, so the concept of interoperability tends to take a back seat to being first to market with a creative new idea.

The ISDN protocols are a classic example of this. Deployment of ISDN was (and in many ways still is) a painful and expensive proposition, as each manufacturer decided to implement it in a slightly different way. ISDN could very well have helped to usher in a massive public data network, 10 years before the Internet. Unfortunately, due to its cost, complexity, and compatibility issues, ISDN never delivered much more than voice, with the occasional video or data connection for those willing to pay. ISDN is quite common (especially in Europe, and in North America in larger PBX implementations), but it is not delivering anywhere near the capabilities that were envisioned for it.

As VoIP becomes more and more ubiquitous, the need for ISDN will disappear.

Slow Release Cycles

It can take months, or sometimes years, for the big guys to admit to a trend, let alone release a product that is compatible with it. It seems that before a new technology can be embraced, it must be analyzed to death, and then it must pass successfully through various layers of bureaucracy before it is even scheduled into the development cycle. Months or even years must pass before any useful product can be expected. When those products are finally released, they are often based on hardware that is obsolete; they also tend to be expensive and to offer no more than a minimal feature set.

These slow release cycles simply don't work in today's world of business communications. On the Internet, new ideas can take root in a matter of weeks and become viable in extremely short periods of time. Since every other technology must adapt to these changes, so too must telecommunications.

Open source development is inherently better able to adapt to rapid technological change, which gives it an enormous competitive advantage.

The spectacular crash of the telecom industry may have been caused in large part by an inability to change. Perhaps that continued inability is why recovery has been so slow. Now, there is no choice: change, or cease to be. Community-driven technologies such as Asterisk are seeing to that.

Refusing to Let Go of the Past and Embrace the Future

Traditional telecommunications companies have lost touch with their customers. While the concept of adding functionality beyond the basic telephone is well understood, the idea that the user should be the one defining this functionality is not.

Nowadays, people have nearly limitless flexibility in every other form of communication. They simply cannot understand why telecommunications cannot be delivered as flexibly as the industry has been promising for so many years. The concept of flexibility is not familiar to the telecom industry, and very well might not be until open source products such as Asterisk begin to transform the fundamental nature of the industry. This is a revolution similar to the one Linux and the Internet willingly started over 10 years ago (and IBM unwittingly started with the PC, 15 years before that). What is this revolution? The commoditization of telephony hardware and software, enabling a proliferation of tailor-made telecommunications systems.

Paradigm Shift

In his article "Paradigm Shift" (*http://tim.oreilly.com/articles/paradigmshift_0504 .html*), Tim O'Reilly talks about a paradigm shift that is occurring in the way technology (both hardware and software) is delivered.[§] O'Reilly identifies three trends: *the commoditization of software*, *network-enabled collaboration*, and *software customizability (software as a service)*. These three concepts provide evidence to suggest that open source telephony is an idea whose time has come.

The Promise of Open Source Telephony

> Every good work of software starts by scratching a developer's personal itch.
>
> —Eric S. Raymond, *The Cathedral and the Bazaar*

In his book *The Cathedral and the Bazaar* (O'Reilly), Eric S. Raymond explains that "Given enough eyeballs, all bugs are shallow." The reason open source software development produces such consistent quality is simple: crap can't hide.

The Itch That Asterisk Scratches

In this era of custom database and website development, people are not only tired of hearing that their telephone system "can't do that," but quite frankly just don't believe it. The creative needs of the customers, coupled with the limitations of the technology, have spawned a type of creativity born of necessity: telecom engineers are like contestants in an episode of *Junkyard Wars*, trying to create functional devices out of a pile of mismatched components.

The development methodology of a proprietary telephone system dictates that it will have a huge number of features, and that the number of features will in large part determine the price. Manufacturers will tell you that their products give you hundreds of features, but if you only need five of them, who cares? Worse, if there's one missing feature you really can't do without, the value of that system will be diluted by the fact that it can't completely address your needs.

---

§ Much of the following section is merely our interpretation of O'Reilly's article. To get the full gist of these ideas, the full read is highly recommended.

The fact that a customer might only need five out of five hundred features is ignored, and that customer's desire to have five unavailable features that address the needs of his business is dismissed as unreasonable.|| Until flexibility becomes standard, telecom will remain stuck in the last century—all the VoIP in the world notwithstanding.

Asterisk addresses that problem directly, and solves it in a way that few other telecom systems can. This is extremely disruptive technology, in large part because it is based on concepts that have been proven time and time again: "the closed-source world cannot win an evolutionary arms race with open-source communities that can put orders of magnitude more skilled time into a problem."#

Open Architecture

One of the stumbling blocks of the traditional telecommunications industry has been its apparent refusal to cooperate with itself. The big telecommunications giants have all been around for over a hundred years. The concept of closed, proprietary systems is so ingrained in their culture that even their attempts at standards compliancy are tainted by their desire to get the jump on the competition, by adding that one feature that no one else supports. For an example of this thinking, one simply has to look at the VoIP products being offered by the telecom industry today. While they claim standards compliance, the thought that you would actually expect to be able to connect a Cisco phone to a Nortel switch, or that an Avaya voicemail system could be integrated via IP to a Siemens PBX, is not one that bears discussing.

In the computer industry, things are different. Twenty years ago, if you bought an IBM server, you needed an IBM network and IBM terminals to talk to it. Now, that IBM server is likely to interconnect to Dell terminals though a Cisco network (and run Linux, of all things). Anyone can easily think of thousands of variations on this theme. If any one of these companies were to suggest that we could only use their products with whatever they told us, they would be laughed out of business.

The telecommunications industry is facing the same changes, but it's in no hurry to accept them. Asterisk, on the other hand, is in a big hurry to not only accept change, but embrace it.

Cisco, Nortel, Avaya, and Polycom IP phones (to name just a few) have all been successfully connected to Asterisk systems. There is no other PBX in the world today that can make this claim. None. *Openness is the power of Asterisk.*

|| From the perspective of the closed-source industry, their attitude is understandable. In his book *The Mythical Man-Month: Essays on Software Engineering* (Addison-Wesley), Fred Brooks opined that "the complexity and communication costs of a project rise with the square of the number of developers, while work done only rises linearly." Without a community-based development methodology, it is very difficult to deliver products that at best are little more than incremental improvements over their predecessors, and at worst are merely collections of patches.

#Eric S. Raymond, *The Cathedral and the Bazaar.*

Standards Compliance

In the past few years, it has become clear that standards evolve at such a rapid pace that to keep up with them requires an ability to quickly respond to emerging technology trends. Asterisk, by virtue of being an open source, community-driven development effort, is uniquely suited to the kind of rapid development that standards compliance demands.

Asterisk does not focus on cost-benefit analysis or market research. It evolves in response to whatever the community finds exciting—or necessary.

Lightning-Fast Response to New Technologies

After Mark Spencer attended his first SIP Interoperability Test (SIPIT) event, he had a rudimentary but working SIP stack for Asterisk coded within a few days. This was before SIP had emerged as the protocol of choice in the VoIP world, but he saw its value and momentum and ensured that Asterisk would be ready.

This kind of foresight and flexibility is typical in an open-source development community (and very unusual in a large corporation).

Passionate Community

The *Asterisk-Users* list receives over three hundred email messages per day. Over ten thousand people are subscribed to it. This kind of community support is unheard of in the world of proprietary telecommunications, while in the open source world it is commonplace.

The very first AstriCon event was expected to attract one hundred participants. Nearly five hundred showed up (far more wanted to but couldn't attend). This kind of community support virtually guarantees the success of an open source effort.

Some Things That Are Now Possible

So what sorts of things can be built using Asterisk? Let's look at some of the things we've come up with.

Legacy PBX migration gateway

Asterisk can be used as a fantastic bridge between an old PBX and the future. You can place it in front of the PBX as a gateway (and migrate users off the PBX as needs dictate), or you can put it behind the PBX as a peripheral application server. You can even do both at the same time, as shown in Figure 27-1.

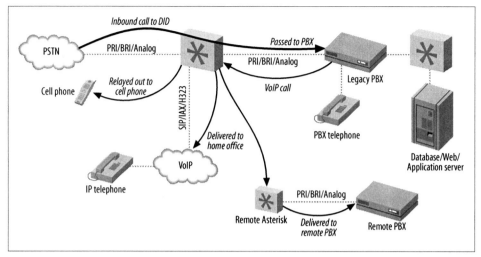

Figure 27-1. Asterisk as a PBX gateway

Here are some of the options you can implement:

Keep your old PBX, but evolve to IP

Companies that have spent vast sums of money in the past few years buying pro-prietary PBX equipment want a way out of proprietary jail, but they can't stomach the thought of throwing away all of their otherwise functioning equipment. No problem—Asterisk can solve all kinds of problems, from replacing a voicemail system to providing a way to add IP-based users beyond the nominal capacity of the system.

Find-me-follow-me

Provide the PBX a list of numbers where you can be reached, and it will ring them all whenever a call to your DID (Direct Inward Dialing, a.k.a. phone) number arrives. Figure 27-2 illustrates this technology.

VoIP calling

If a legacy telephony connection from an Asterisk PBX to an old PBX can be es-tablished, Asterisk can provide access to VoIP services, while the old PBX continues to connect to the outside world as it always has. As a gateway, Asterisk simply needs to emulate the functions of the PSTN, and the old PBX won't know that anything has changed. Figure 27-3 shows how you can use Asterisk to VoIP-enable a legacy PBX.

Low-barrier IVR

Many people confuse Interactive Voice Response (IVR) systems with automated at-tendants (AAs). Since the automated attendant was the very first thing IVR was used for, this is understandable. Nevertheless, to the telecom industry, the term IVR repre-sents far more than an AA. An AA generally does little more than present a way for

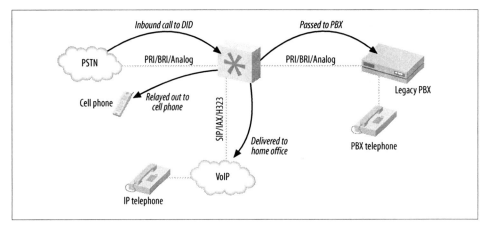

Figure 27-2. Find-me-follow-me

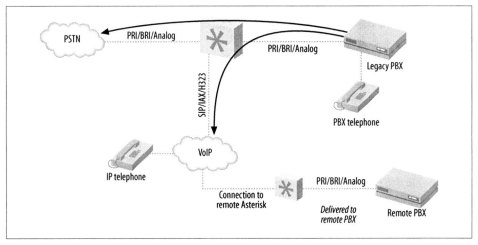

Figure 27-3. VoIP-enabling a legacy PBX

callers to be transferred to extensions, and it is built into most proprietary voicemail systems—but IVR can be so much more.

IVR systems are generally very expensive, not only to purchase, but also to configure. A custom IVR system will usually require connectivity to an external database or application. Asterisk is arguably the perfect IVR, as it embraces the concepts of connectivity to databases and applications at its deepest level.

Here are a few examples of relatively simple IVRs an Asterisk system could be used to create:

Weather reporting

Using the Internet, you can obtain text-based weather reports from around the world in a myriad of ways. Capturing these reports and running them through a purpose-built parser (Perl would probably eat this up) would allow the information to be available to the dialplan. Asterisk's sound library already contains all the required prompts, so it would not be an onerous task to produce an interactive menu to play current forecasts for anywhere in the world.

Math programs

Ed Guy (the architect of Pulver's FWD network) did a presentation at AstriCon 2004 in which he talked about a little math program he'd cooked up for his daughter to use. The program took him no more than an hour to write. What it did was present her with a number of math questions, the answers to which she keyed into the telephone. When all the questions were tabulated, the system presented her with her score. This extremely simple Asterisk application would cost tens of thousands of dollars to implement on any closed PBX platform, assuming it could be done at all. As is so often the case, things that are simple for Asterisk would be either impossible or massively expensive with any other IVR system.

Distributed IVR

The cost of a proprietary IVR system is such that when a company with many small retail locations wants to provide IVR, it is forced to transfer callers to a central server to process the transactions. With Asterisk, it becomes possible to distribute the application to each node, and thus handle the requests locally. Literally thousands of little Asterisk systems deployed at retail locations across the world could serve up IVR functionality in a way that would be impossible to achieve with any other system. No more long-distance transfers to a central IVR server, no more huge trunking facility dedicated to the task—more power with less expense.

These are three rather simple examples of the potential of Asterisk.

Conference rooms

This little gem is going to end up being one of the killer functions of Asterisk. In the Asterisk community, people find themselves using conference rooms more and more, for purposes such as these:

- Small companies need an easy way for business partners to get together for a chat.
- Sales teams want to have weekly meetings where reps can dial in from wherever they are.
- Development teams need to designate a common place and time to update each other on progress.

Home automation

Asterisk is still too much of an über-geek's tool to be able to serve in the average home, but with no more than average Linux and Asterisk skills, the following things become plausible:

Monitoring the kids
> Parents who want to check up on the babysitter (or the kids home alone) could dial an extension context protected by a password. Once authenticated, a two-way audio connection would be created to all the IP phones in the house, allowing Mom and Dad to listen for trouble. Creepy? Yes. But an interesting concept nonetheless.

Locking down your phones
> Going out for the night? Don't want the babysitter tying up the phone? No problem! A simple tweak to the dialplan, and the only calls that can be made are to 911, your cell phone, and the pizza parlor. Any other call attempt will get the recording "We are paying you to babysit our kids, not make personal calls."
>
> Pretty evil, huh?

Controlling the alarm system
> You get a call while on vacation from your mom who wants to borrow some cooking utensils. She forgot her key, and is standing in front of the house shivering. Piece of cake: a call to your Asterisk system, a quick digit string into the context you created for the purpose, and your alarm system is instructed to disable the alarm for 15 minutes. Mom better get her stuff and get out quick, though, or the cops'll be showing up!

Managing teenagers' calls
> How about allocating a specific phone-time limit to your teenagers? To use the phone, they have to enter their access codes. They can earn extra minutes by doing chores, scoring all As, dumping that annoying bum with the bad haircut—you get the idea. Once they've used up their minutes...click...you get your phone back.
>
> Incoming calls can be managed as well, via caller ID. "Donny, this is Suzy's father. She is no longer interested in seeing you, as she has decided to raise her standards a bit. Also, you should consider getting a haircut."

The Future of Asterisk

We've come to love the Internet, both because it is so rich in content and inexpensive and, perhaps more importantly, because it allows us to define how we communicate. As its ability to carry richer forms of media advances, we'll find ourselves using it more and more. Once Internet voice delivers quality that rivals (or betters) the capabilities of the PSTN, the phone company had better look for another line of business. The PSTN will cease to exist; all its complexity will be absorbed into the Internet, as just one more technology. As with most of the rest of the Internet, open source technologies will lead this transformation.

Speech Processing

The dream of having our technical inventions talk to us is older than the telephone itself. Each new advance in technology spurs a new wave of eager experimentation. Generally, results never quite meet expectations, possibly because as soon as a machine says something that *sounds* intelligent, most people assume that it *is* intelligent.

People who program and maintain computers realize their limitations, and thus tend to allow for their weaknesses. Everybody else just expects their computers and software to work. The amount of thinking a user must do to interact with a computer is often inversely proportional to the amount of thinking the design team did. Simple interfaces belie complex design decisions.

The challenge, therefore, is to design a system that has anticipated the most common desires of its users, and can also adroitly handle unexpected challenges.

Festival

The Festival text-to-speech server can transform text into spoken words. While this is a whole lot of fun to play with, there are many challenges to overcome (for more on integrating Festival with Asterisk, refer back to "Text-to-Speech Utilities" on page 440).

For Asterisk, an obvious value of text-to-speech might be the ability to have your telephone system read your emails back to you. If you've noticed the somewhat poor grammar, punctuation, and spelling typically found in email messages these days, you can perhaps appreciate the challenges this poses.

One cannot help but wonder if the emergence of text-to-speech will inspire a new generation of people dedicated to proper writing. Seeing spelling and punctuation errors on the screen is frustrating enough—having to hear a computer speak such things will require a level of Zazen that few possess.

Speech recognition

If text-to-speech is rocket science, speech recognition is science fiction.

Speech recognition can actually work very well, but unfortunately this is generally true only if you provide it with the right conditions—and the right conditions are not those found on a telephone network. Even a perfect PSTN connection is considered to be at the lowest acceptable limit for accurate speech recognition. Add in compressed and lossy VoIP connections, or a cell phone, and you will discover far more limitations than uses.

Asterisk now has an entire speech API, so that outside companies (or even open source projects) can tie their speech recognition engines into Asterisk. One company that has done this is LumenVox. By using LumenVox's speech recognition engine along with

Asterisk, you can make voice-driven menus and IVR systems in record time! For more information, see *http://www.lumenvox.com*.

High-Fidelity Voice

As we gain access to more and more bandwidth, it becomes less and less easy to understand why we still use low-fidelity codecs. Many people do not realize that Skype provides higher fidelity than a telephone; it's a large part of the reason why Skype has a reputation for sounding so good.

If you were ever to phone CNN, wouldn't you love to hear James Earl Jones's mellifluous voice saying "This is CNN," instead of some tinny electronic recording? And if you think Allison Smith* sounds good through the phone, you should hear her in person!

In the future, we will expect, and get, high-fidelity voice through our communications equipment.

As more and more hardware vendors start building support for high-fidelity voice into their VoIP hardware, you'll see more support in Asterisk for making better-than-PSTN-quality calls.

Video

While most of this book focuses on audio, video is also supported in many ways within Asterisk. Video support is not complete, however. The problem is not so much one of functionality as one of bandwidth and processing power. Asterisk 1.10 is expected to contain better support for handling media, including video.

The challenge of videoconferencing

The concept of videoconferencing has been around since the invention of the cathode ray tube. The telecom industry has been promising a videoconferencing device in every home for decades.

As with so many other communications technologies, if you have videoconferencing in your house, you are probably running it over the Internet, with a simple, inexpensive webcam. Still, it seems that people see videoconferencing as a bit gimmicky. Yes, you can see the person you're talking to, but there's something missing.

---

* Allison Smith is The Voice of Asterisk—it is her voice in all of the system prompts. To have Allison produce your own prompt, simply visit *http://www.theivrvoice.com*.

Why we love videoconferencing

Videoconferencing promises a richer communications experience than the telephone. Rather than simply hearing a disembodied voice, you have access to all the nuances of speech that come from face-to-face communication.

Why videoconferencing may never totally replace voice

There are some challenges to overcome, though, and not all of them are technical.

Consider this: using a plain telephone, people working from their home offices can have business conversations, unshowered, in their underwear, feet on the desk, coffee in hand—if they use a telephone. A similar video conversation would require half an hour of grooming to prepare for, and couldn't happen in the kitchen, on the patio, or... well, you get the idea.

Also, the promise of eye-to-eye communication over video will never happen as long as the focal points of the participants are not in line with the cameras. If you look at the camera, your audience will see you looking at them, but you won't see them. If you look at your screen to see whom you are talking to, the camera will show you looking down at something—not at your audience. That looks impersonal. Perhaps if a video-phone could be designed like a Tele-Prompt-R, where the camera was behind the screen, it wouldn't feel so unnatural. As it stands, there's something psychological that's missing. Video ends up being a gimmick.

Wireless

Since Asterisk is fully VoIP-enabled, wireless is all part of the package.

WiFi

WiFi is going to be the office mobility solution for VoIP phones. This technology is already quite mature. The biggest hurdle is the cost of handsets, which can be expected to improve as competitive pressure from around the world drives down prices.

WiMAX

Since we are so bravely predicting so many things, it's not hard to predict that WiMAX spells the beginning of the end for traditional cellular telephone networks.

With wireless Internet access within the reach of most communities, what value will there be in expensive cellular service?

Unified Messaging

This is a term that has been hyped by the telecom industry for years, but adoption has been far slower than predicted.

Unified messaging refers to the concept of tying voice and text-messaging systems into one. With Asterisk, the two don't need to be artificially combined, as Asterisk already treats them the same way.

Just by examining the terms, *unified* and *messaging*, we can see that the integration of email and voicemail must be merely the beginning—unified messaging needs to do a lot more than just that if it is to deserve its name.

Perhaps we need to define "messaging" as communication that does not occur in real time. In other words, when you send a message, you expect that the reply may take moments, minutes, hours, or even days to arrive. You compose what you wish to say, and your audience is expected to compose a reply.

Contrast this with conversing, which happens in real time. When you talk to someone on a telephone connection, you expect no more than a few seconds' delay before the response arrives.

Several years ago, Tim O'Reilly delivered a speech entitled "Watching the Alpha Geeks: OS X and the Next Big Thing" (*http://www.macdevcenter.com/pub/a/mac/2002/05/14/oreilly_wwdc_keynote.html*), in which he talked about someone piping IRC through a text-to-speech engine. One could imagine doing the reverse as well, allowing us to join an IRC or instant messaging chat over a WiFi phone, with our Asterisk PBX providing the speech-to-text-to-speech translations.

Peering

As monopoly networks such as the PSTN give way to community-based networks like the Internet, there will be a period of time where it is necessary to interconnect the two. While the traditional providers would prefer that the existing model be carried into the new paradigm, it is increasingly likely that telephone calls will become little more than another application the Internet happily carries.

But a challenge remains: how to manage the telephone numbering plan with which we are all familiar and comfortable?

E.164

The ITU defined a numbering plan in its E.164 specification. If you've used a telephone to make a call across the PSTN, you can confidently state that you are familiar with the concept of E.164 numbering. Prior to the advent of publicly available VoIP, nobody cared about E.164 except the telephone companies—nobody needed to.

Now that calls are hopping from PSTN to Internet to who-knows-what, some consideration must be given to E.164.

ENUM

In response to this challenge, the IETF has sponsored the Electronic NUmber Mapping (ENUM) working group, the purpose of which is to map E.164 numbers into the Domain Name System (DNS).

While the concept of ENUM is sound, it requires cooperation from the telecom industry to achieve success. However, cooperation is not what the telecom industry is famous for, and thus far ENUM has foundered.

e164.org

The folks at *http://e164.org* are trying to contribute to the success of ENUM. You can log onto this site, register your phone number, and inform the system of alternative methods of communicating with you. This means that someone who knows your phone number can connect a VoIP call to you, as the *http://e164.org* DNS zone will provide the IP addressing and protocol information needed to connect to your location.

As more and more people publish VoIP connectivity information, fewer and fewer calls will be connected through the PSTN.

Challenges

As is true with any worthwhile thing, Asterisk will face challenges. Let's take a glance at what some of them may be.

Too much change, too few standards

These days, the Internet is changing so fast, and offers so much diverse content, that it is impossible for even the most attentive geek to keep on top of it all. While this is as it should be, it also means that an enormous amount of technology churn is an inevitable part of keeping any communications system current.

Toll fraud

As long as long-distance calls cost money, there will be criminals who will wish to steal. Toll fraud is nothing new, but with many unsecured Asterisk systems now on the Internet, the popularity of scripts to find these systems and compromise them has exploded. Administrators of Internet-connected telephone systems will need to carefully design their security to ensure that any calls made from their systems are made only by authorized users.

VoIP spam

Yes, it's coming. There will always be people who believe they have the right to inconvenience and harass others in their pursuit of money. Efforts are under way to try to address this, but only time will tell how efficacious they will be.

Fear, uncertainty, and doubt

The industry is making the transition from ignorance to laughter. If Gandhi is correct, we can expect the fight to begin soon.

As their revenue streams become increasingly threatened by open source telephony, the traditional industry players are certain to mount a fear campaign, in hopes of undermining the revolution.

Bottleneck engineering

There is a rumor that the major network providers will artificially cripple VoIP traffic by tagging and prioritizing the traffic of their premium VoIP services and, worse, detecting and bumping any VoIP traffic generated by services not approved by them.

Some of this is already taking place, with service providers blocking traffic of certain types through their networks, ostensibly as some public service (such as blocking popular file-sharing services to protect us from piracy). In the United States, the FCC has taken a clear stand on the matter and fined companies that engage in such practices. In the rest of the world, regulatory bodies are not always as accepting of VoIP.

What seems clear is that the community and the network will find ways around blockages, just as they always have.

Regulatory wars

A former chairman of the United States Federal Communications Commission, Michael Powell delivered a gift that may well have altered the path of the VoIP revolution. Rather than attempting to regulate VoIP as a telecom service, he championed the concept that VoIP represents an entirely new way of communicating and requires its own regulatory space in which to evolve.

VoIP will become regulated, but not everywhere as a telephony service. Some of the regulations that may be created include:

Presence information for emergency services
    One of the characteristics of a traditional PSTN circuit is that it is always in the same location. This is very helpful to emergency services, as they can pinpoint the location of a caller by identifying the address of the circuit from which the call was placed. The proliferation of cell phones has made this much more difficult to achieve, since a cell phone does not have a known address. A cell phone can be plugged into any network and can register to any server. If the phone does not

identify its physical location, an emergency call from it will provide no clue as to where the caller is. VoIP creates similar challenges.

Call monitoring for law enforcement agencies

Law enforcement agencies have always been able to obtain wiretaps on traditional circuit-switched telephone lines. While regulations are being enacted that are designed to achieve the same end on the network, the technical challenges of delivering this functionality will probably never be completely solved. People value their privacy, and the more governments want to stifle it, the more effort will be put toward maintaining it.

Anti-monopolistic practices

These practices are already being seen in the US, with fines being levied against network providers who attempt to filter traffic based on content.

When it comes to regulation, Asterisk is both a saint and a devil: a saint because it feeds the poor, and a devil because it empowers the phrackers and spammers like nothing ever has. The regulation of open source telephony may in part be determined by how well the community regulates itself. Concepts such as DUNDi, which incorporate anti-spam processes, are an excellent start. On the other hand, concepts such as caller ID–spoofing are ripe with opportunities for abuse.

Quality of service

Due to the best-effort reality of the TCP/IP-based Internet, it is not yet known how increasing real-time VoIP traffic will affect overall network performance. Currently, there is so much excess bandwidth in the backbone that best-effort delivery is generally quite good indeed. Still, it has been proven time and time again that whenever we are provided with more bandwidth, we figure out a way to use it up. The 1-MB DSL connection undreamt of five years ago is now barely adequate.

Perhaps a corollary of Moore's Law[†] will apply to network bandwidth. QoS may become moot, due to the network's ability to deliver adequate performance without any special processing. Organizations that require higher levels of reliability may elect to pay a premium for a higher grade of service. Perhaps the era of paying by the minute for long-distance connections will give way to paying by the millisecond for guaranteed low latency, or by the percentage point for reduced packet loss. Premium services will offer the five-nines[‡] reliability the traditional telecom companies have always touted as their advantage over VoIP.

[†] Gordon Moore wrote a paper in 1965 that predicted the doubling of transistors on a processor every few years.

[‡] This term refers to 99.999%, which is touted as the reliability of traditional telecom networks. Achieving five nines requires that service interruptions for an entire year total no more than 5 minutes and 15 seconds. Many people believe that VoIP will need to achieve this level of reliability before it can fully replace the PSTN. Many other people believe that the PSTN doesn't even come close to five-nines reliability. This could have been an excellent term to describe high reliability, but marketing departments abuse it far too frequently.

Complexity

Open systems require new approaches to solution design. Just because the hardware and software are cheap doesn't mean the solution will be. Asterisk does not come out of the box ready to run; an Asterisk system has to be designed and built, and then maintained. While the base software is free, and the hardware costs will be based on commodity pricing, it is fair to say that the configuration costs for a highly customized system will be a sizable part of the overall solution cost. In fact, in many cases, because of Asterisk's high degree of complexity and configurability, the cost will be more than would be expected with a traditional PBX.

The rule of thumb is generally considered to be something like this: if it can be done in the dialplan, the system design will be roughly the same as for any similarly featured traditional PBX. Beyond that, only experience will allow one to accurately estimate the time required to build a system.

There is much to learn.

Opportunities

Open source telephony creates limitless opportunities. Here are some of the more compelling ones.

Tailor-made private telecommunications networks

Some people will tell you that price is the key, but we believe that the real reason Asterisk will succeed is because it is now possible to build a telephone system as one would a website: with complete, total customization of each and every facet of the system. Customers have wanted this for years. Only Asterisk can deliver.

Low barrier to entry

Anyone can contribute to the future of communicating. It is now possible for someone with an old $200 PC to develop a communications system that has the intelligence to rival the most expensive proprietary systems. Granted, the hardware would not be production-ready, but there is no reason the software couldn't be. This is one of the reasons why closed systems will have a hard time competing. The sheer number of people who have access to the required equipment is impossible to equal in a closed shop.

Hosted solutions of similar complexity to corporate websites

The design of a PBX was always a kind of art form, but before Asterisk, the art lay in finding creative ways to overcome the limitations of the technology. With limitless technology, those same creative skills can now be properly applied to the task of completely answering the needs of the customer. Open source telephony engines such as Asterisk will enable this. Telecom designers will dance for joy, as their considerable creative skills will now actually serve the needs of their customers, rather than being focused on managing kludge.

Proper integration of communications technologies

Ultimately, the promise of open source comes to nothing if it cannot fulfill the need people have to solve problems. The closed industries lost sight of the customer, and tried to fit the customer to the product.

Open source telephony brings voice communications in line with other information technologies. It is finally possible to properly begin the task of integrating email, voice, video, and anything else we might conceive of over flexible transport networks (whether wired or wireless), in response to the needs of the user, not the whims of monopolies.

Welcome to the future of telecom!

Understanding Telephony

Utility is when you have one telephone, luxury is when
you have two, opulence is when you have three—and
paradise is when you have none.

—Doug Larson

In this appendix, we are going to talk about some of the technologies of the traditional telephone network—especially those that people most commonly want to connect to Asterisk. (We'll discuss Voice over IP in Appendix B.)

While tomes could be written about the technologies in use in telecom networks, the material included here was chosen based on our experiences in the community, which helped us to define the specific items that might be most useful. Although this knowledge may not be strictly required in order to configure your Asterisk system, it will be of great benefit when interconnecting to systems (and talking with people) from the world of traditional telecommunications.

Analog Telephony

The purpose of the Public Switched Telephone Network (PSTN) is to establish and maintain audio connections between two endpoints in order to carry speech.

Although humans can perceive sound vibrations in the range of 20–20,000 Hz,* most of the sounds we make when speaking tend to be in the range of 250–3,000 Hz. Since the purpose of the telephone network is to transmit the sounds of people speaking, it was designed with a bandwidth of somewhere in the range of 300–3,500 Hz. This

---

* If you want to play around with what different frequencies look like on an oscilloscope, grab a copy of Sound Frequency Analyzer, from Reliable Software. It's a really simple and fun way to visualize what sounds "look" like. The spectrograph gives a good picture of the complex harmonics our voices can generate, as well as an appreciation for the background sounds that always surround us. You should also try the delightfully annoying NCH Tone Generator, from NCH Swift Sound.

limited bandwidth means that some sound quality will be lost (as anyone who's had to listen to music on hold can attest to), especially in the higher frequencies.

Parts of an Analog Telephone

An analog phone is composed of five parts: the ringer, the dialpad, the hybrid (or network), and the hook switch and handset (both of which are considered parts of the hybrid). The ringer, the dialpad, and the hybrid can operate completely independently of one another.

Ringer

When the central office (CO) wants to signal an incoming call, it will connect an alternating current (AC) signal of roughly 90 volts to your circuit. This will cause the bell in your telephone to produce a ringing sound. (In electronic telephones, this ringer may be a small electronic warbler rather than a bell. Ultimately, a ringer can be anything that is capable of reacting to the ringing voltage; for example, strobe lights are often employed in noisy environments such as factories.)

 Ringing voltage can be hazardous. Be very careful to take precautions when working with an in-service telephone line.

Many people confuse the AC voltage that triggers the ringer with the direct current (DC) voltage that powers the phone. Remember that a ringer needs an alternating current in order to oscillate (just as a church bell won't ring if you don't supply the movement), and you've got it.

In North America, the number of ringers you can connect to your line is dependent on the Ringer Equivalence Number (REN) of your various devices. (The REN must be listed on each device.) The total REN for all devices connected to your line cannot exceed 5.0. An REN of 1.0 is equivalent to an old-fashioned analog set with an electromechanical ringer. Some electronic phones have RENs of 0.3 or even less. If you connect too many devices that require too much current, you will find that none of them will be able to ring.

Dialpad

When you place a telephone call, you need some way of letting the network know the address of the party you wish to reach. The dialpad is the portion of the phone that provides this functionality. In the early days of the PSTN, dialpads were in fact rotary devices that used pulses to indicate digits. This was a rather slow process, so the telephone companies eventually introduced touch-tone dialing. With touch-tone—also known as Dual-Tone Multi Frequency (DTMF)—dialing, the dialpad consists of 12 buttons. Each button has two frequencies assigned to it (see Table A-1).

Table A-1. DTMF digits

	1209 Hz	1336 Hz	1477 Hz	1633 Hz [a]
697 Hz	1	2	3	A
770 Hz	4	5	6	B
852 Hz	7	8	9	C
941 Hz	*	0	#	D

[a] Notice that this column contains letters that are not typically present as keys on a telephone dialpad. They are part of the DTMF standard nonetheless, and any proper telephone contains the electronics required to create them, even if it doesn't contain the buttons themselves. (These buttons actually do exist on some telephones, which are mostly used in military and government applications.)

When you press a button on your dialpad, the two corresponding frequencies are transmitted down the line. The far end can interpret these frequencies and note which digit was pressed.

Hybrid (or network)

The hybrid is a type of transformer that handles the need to combine the signals transmitted and received across a single pair of wires in the PSTN and two pairs of wires in the handset. One of the functions the hybrid performs is regulating *sidetone*, which is the amount of your transmitted signal that is returned to your earpiece; its purpose is to provide a more natural-sounding conversation. Too much sidetone, and your voice will sound too loud; too little, and you'll think the line has gone dead.

Hook switch (or switch hook). This device signals the state of the telephone circuit to the CO. When you pick up your telephone, the hook switch closes the loop between you and the CO, which is seen as a request for a dialtone. When you hang up, the hook switch opens the circuit, which indicates that the call has ended.†

The hook switch can also be used for signaling purposes. Some electronic analog phones have a button labeled *Link* that causes an event called a *flash*. You can perform a flash manually by depressing the hook switch for a duration of between 200 and 1,200 milliseconds. If you leave it down for longer than that, the carrier may assume you've hung up. The purpose of the *Link* button is to handle this timing for you. If you've ever used call waiting or three-way calling on an analog line, you have performed a hook-switch flash for the purpose of signaling the network.

Handset. The handset is composed of the transmitter and receiver. It performs the conversion between the sound energy humans use and the electrical energy the telephone network uses.

---

† When referring to the state of an analog circuit, people often speak in terms of "off-hook" and "on-hook." When your line is "off-hook," your telephone is "on" a call. If your phone is "on-hook," the telephone is essentially "off," or idle.

Tip and Ring

In an analog telephone circuit, there are two wires. In North America, these wires are referred to as Tip and Ring.‡ This terminology comes from the days when telephone calls were connected by live operators sitting at cord boards. The plugs that they used had two contacts—one located at the tip of the plug and the other connected to the ring around the middle (Figure A-1).

Figure A-1. Tip and Ring

The Tip lead is the positive polarity wire. In North America, this wire is typically green and provides the return path. The Ring wire is the negative polarity wire. In North America, this wire is normally red. For modern Cat 5 and 6 cables, the Tip is usually the white wire, and Ring is the colored wire. When your telephone is on-hook, this wire will have a potential of –48V DC with respect to Tip. Off-hook, this voltage drops to roughly –7V DC.

Digital Telephony

Analog telephony is almost dead.

In the PSTN, the famous Last Mile is the final remaining piece of the telephone network still using technology pioneered well over a hundred years ago.§

One of the primary challenges when transmitting analog signals is that all sorts of things can interfere with those signals, causing low volume, static, and all manner of other undesired effects. Instead of trying to preserve an analog waveform over distances that may span thousands of miles, why not simply measure the characteristics of the original sound and send that information to the far end? The original waveform wouldn't get there, but all the information needed to reconstruct it would.

This is the principle of all digital audio (including telephony): sample the characteristics of the source waveform, store the measured information, and send that data to the far

---

‡ They may have other names elsewhere in the world (such as "A" and "B").

§ "The Last Mile" is a term that was originally used to describe the only portion of the PSTN that had not been converted to fiber optics: the connection between the central office and the customer. The Last Mile is more than that, however, as it also has significance as a valuable asset of the traditional phone companies; they own a connection into your home. The Last Mile is becoming more and more difficult to describe in technical terms, as there are now so many ways to connect the network to the customer. As a thing of strategic value to telecom, cable, and other utilities, its importance is obvious.

---

end. Then, at the far end, use the transmitted information to generate a completely new audio signal that has the same characteristics as the original. The reproduction is so good that the human ear can't tell the difference.

The principal advantage of digital audio is that the sampled data can be mathematically checked for errors all along the route to its destination, ensuring that a perfect duplicate of the original arrives at the far end. Distance no longer affects quality, and interference can be detected and eliminated.

Pulse-Code Modulation

There are several ways to digitally encode audio, but the most common method (and the one used in telephony systems) is known as Pulse-Code Modulation (PCM). To illustrate how this works, let's go through a few examples.

Digitally encoding an analog waveform

The principle of PCM is that the amplitude[||] of the analog waveform is sampled at specific intervals so that it can later be re-created. The amount of detail that is captured is dependent both on the bit resolution of each sample and on how frequently the samples are taken. A higher bit resolution and a higher sampling rate will provide greater accuracy, but more bandwidth will be required to transmit this more detailed information.

To get a better idea of how PCM works, consider the waveform displayed in Figure A-2.

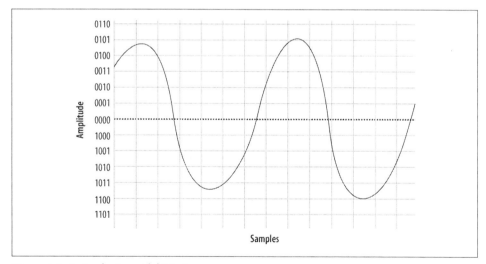

Figure A-2. A simple sinusoidal (sine) wave

[||] Amplitude is essentially the power or strength of the signal. If you have ever held a skipping rope or garden hose and given it a whip, you have seen the resultant wave. The taller the wave, the greater the amplitude.

To digitally encode the wave, it must be sampled on a regular basis, and the amplitude of the wave at each moment in time must be measured. The process of slicing up a waveform into moments in time and measuring the energy at each moment is called *quantization*, or *sampling*.

The samples will need to be taken frequently enough and will need to capture enough information to ensure that the far end can re-create a sufficiently similar waveform. To achieve a more accurate sample, more bits will be required. To explain this concept, we will start with a very low resolution, using 4 bits to represent our amplitude. This will make it easier to visualize both the quantization process itself and the effect that resolution has on quality.

Figure A-3 shows the information that will be captured when we sample our sine wave at 4-bit resolution.

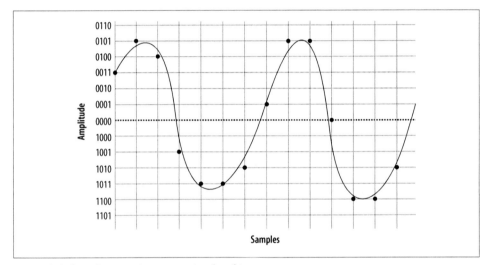

Figure A-3. Sampling our sine wave using four bits

At each time interval, we measure the amplitude of the wave and record the corresponding intensity—in other words, we sample it. You will notice that the 4-bit resolution limits our accuracy. The first sample has to be rounded to 0011, and the next quantization yields a sample of 0101. Then comes 0100, followed by 1001, 1011, and so forth. In total, we have 14 samples (in reality, several thousand samples must be taken per second).

If we string together all the values, we can send them to the other side as:

    0011 0101 0100 1001 1011 1011 1010 0001 0101 0101 0000 1100 1100 1010

On the wire, this code might look something like Figure A-4.

When the far end's digital-to-analog (D/A) converter receives this signal, it can use the information to plot the samples, as shown in Figure A-5.

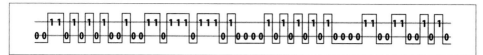

Figure A-4. PCM encoded waveform

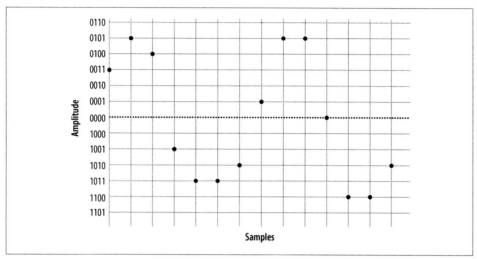

Figure A-5. Plotted PCM signal

From this information, the waveform can be reconstructed (see Figure A-6).

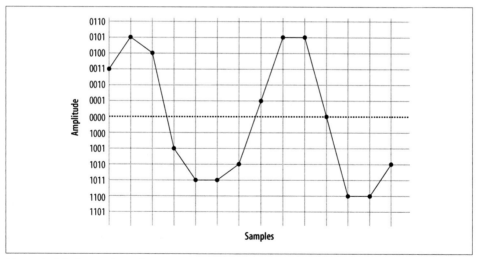

Figure A-6. Delineated signal

As you can see if you compare Figure A-2 with Figure A-6, this reconstruction of the waveform is not very accurate. This was done intentionally, to demonstrate an important point: the quality of the digitally encoded waveform is affected by the resolution and rate at which it is sampled. At too low a sampling rate, and with too low a sample resolution, the audio quality will not be acceptable.

Increasing the sampling resolution and rate

Let's take another look at our original waveform, this time using 5 bits to define our quantization intervals (Figure A-7).

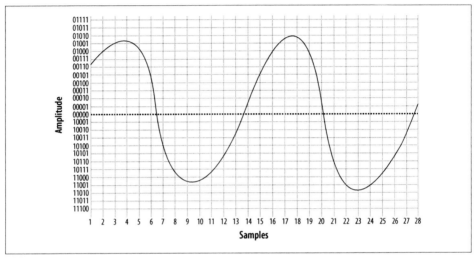

Figure A-7. The same waveform, on a higher-resolution overlay

In reality, there is no such thing as 5-bit PCM. In the telephone network, PCM samples are encoded using 8 bits. Other digital audio methods may employ 16 bits or more.

We'll also double our sampling frequency. The points plotted this time are shown in Figure A-8.

We now have twice the number of samples, at twice the resolution. Here they are:

```
00111 01000 01001 01001 01000 00101 10110 11000 11001 11001 11000 10111
10100 10001 00010 00111 01001 01010 01001 00111 00000 11000 11010 11010
11001 11000 10110 10001
```

When received at the other end, that information can now be plotted as shown in Figure A-9.

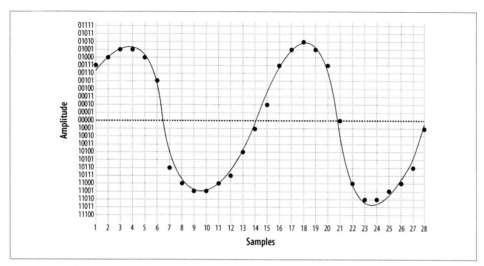

Figure A-8. The same waveform at double the resolution

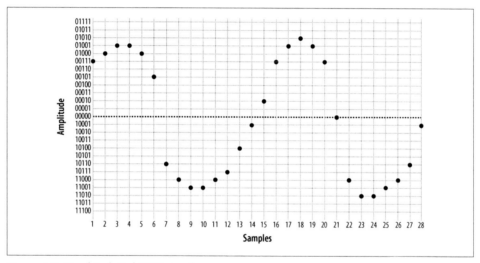

Figure A-9. Five-bit plotted PCM signal

From this information, the waveform shown in Figure A-10 can then be generated.

As you can see, the resultant waveform is a far more accurate representation of the original. However, you can also see that there is still room for improvement.

 Note that 40 bits were required to encode the waveform at 4-bit resolution, while 156 bits were needed to send the same waveform using 5-bit resolution (and also doubling the sampling rate). The point is, there is a tradeoff: the higher the quality of audio you wish to encode, the more bits are required to do it, and the more bits you wish to send (in real time, naturally), the more bandwidth you will need to consume.

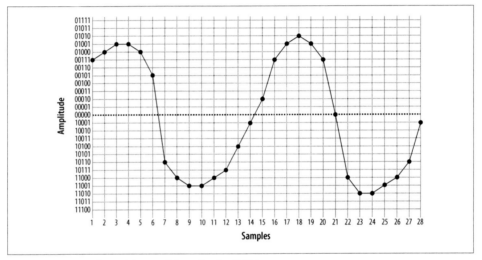

Figure A-10. Waveform delineated from five-bit PCM

Nyquist's Theorem

So how much sampling is enough? That very same question was considered in the 1920s by an electrical engineer (and AT&T/Bell employee) named Harry Nyquist. Nyquist's Theorem states: "When sampling a signal, the *sampling frequency* must be greater than twice the bandwidth of the input signal in order to be able to reconstruct the original perfectly from the sampled version."#

In essence, what this means is that to accurately encode an analog signal you have to sample it twice as often as the total bandwidth you wish to reproduce. Since the

---

#Nyquist published two papers, "Certain Factors Affecting Telegraph Speed" (1924) and "Certain Topics in Telegraph Transmission Theory" (1928), in which he postulated what became known as Nyquist's Theorem. Proven in 1949 by Claude Shannon ("Communication in the Presence of Noise"), it is also referred to as the Nyquist-Shannon sampling theorem.

telephone network will not carry frequencies below 300 Hz and above 4,000 Hz, a sampling frequency of 8,000 samples per second will be sufficient to reproduce any frequency within the bandwidth of an analog telephone. Keep that 8,000 samples per second in mind; we're going to talk about it more later.

Logarithmic companding

So, we've gone over the basics of quantization, and we've discussed the fact that more quantization intervals (i.e., a higher sampling rate) give better quality but also require more bandwidth. Lastly, we've discussed the minimum sampling rate needed to accurately measure the range of frequencies we wish to be able to transmit (in the case of the telephone, it's 8,000 Hz). This is all starting to add up to a fair bit of data being sent on the wire, so we're going to want to talk about companding.

Companding is a method of improving the dynamic range of a sampling method without losing important accuracy. It works by quantizing higher amplitudes in a much coarser fashion than lower amplitudes. In other words, if you yell into your phone, you will not be sampled as cleanly as you will be when speaking normally. Yelling is also not good for your blood pressure, so it's best to avoid it.

Two companding methods are commonly employed: μlaw* in North America, and alaw in the rest of the world. They operate on the same principles but are otherwise not compatible with each other.

Companding divides the waveform into *cords*, each of which has several *steps*. Quantization involves matching the measured amplitude to an appropriate step within a cord. The value of the band and cord numbers (as well as the sign—positive or negative) becomes the signal. The following diagrams will give you a visual idea of what companding does. They are not based on any standard, but rather were made up for the purpose of illustration (again, in the telephone network, companding will be done at an 8-bit, not 5-bit, resolution).

Figure A-11 illustrates 5-bit companding. As you can see, amplitudes near the zero-crossing point will be sampled far more accurately than higher amplitudes (either positive or negative). However, since the human ear, the transmitter, and the receiver will also tend to distort loud signals, this isn't really a problem.

A quantized sample might look like Figure A-12. It yields the following bit stream:

```
00000 10011 10100 10101 01101 00001 00011 11010 00010 00001 01000 10011
10100 10100 00101 00100 00101 10101 10011 10001 00011 00001 00000 10100
10010 10101 01101 10100 00101 11010 00100 00000 01000
```

* μlaw is often referred to as "ulaw" because, let's face it, how many of us have μ keys on our keyboards? μ is in fact the Greek letter Mu; thus, you will also see μlaw written (more correctly) as "Mu-law." When spoken, it is correct to confidently say "Mew-law," but if folks look at you strangely, and you're feeling generous, you can help them out and tell them it's "ulaw." Many people just don't appreciate trivia.

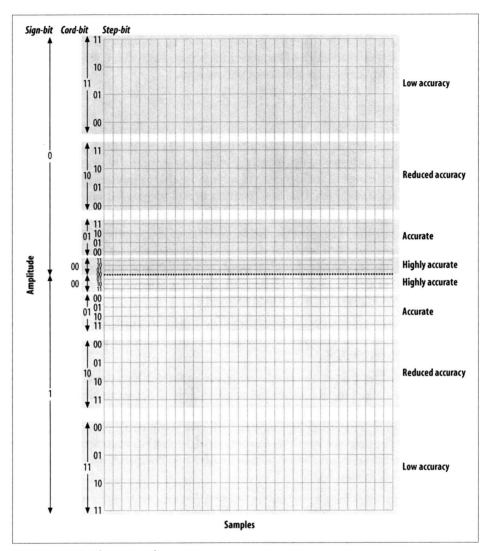

Figure A-11. Five-bit companding

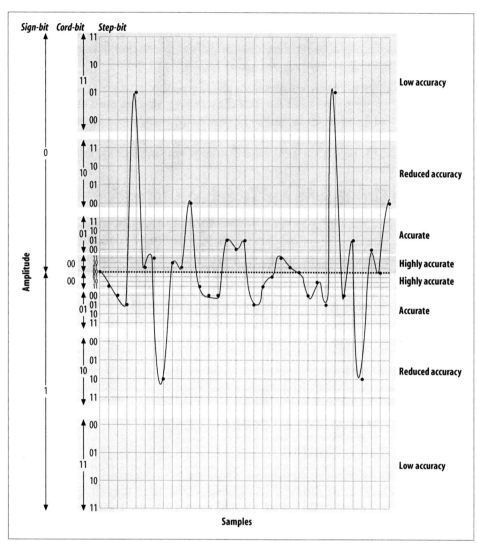

Figure A-12. Quantized and companded at 5-bit resolution

Aliasing

If you've ever watched the wheels on a wagon turn backward in an old Western movie, you've seen the effects of aliasing. The frame rate of the movie cannot keep up with the rotational frequency of the spokes, and a false rotation is perceived.

In a digital audio system (which the modern PSTN arguably is), aliasing always occurs if frequencies that are greater than one-half the sampling rate are presented to the analog-to-digital (A/D) converter. In the PSTN, that includes any audio frequencies above 4,000 Hz (half the sampling rate of 8,000 Hz). This problem is easily corrected by passing the audio through a low-pass filter[†] before presenting it to the A/D converter.[‡]

The Digital Circuit-Switched Telephone Network

For over a hundred years, telephone networks were exclusively circuit-switched. What this meant was that for every telephone call made, a dedicated connection was established between the two endpoints, with a fixed amount of bandwidth allocated to that circuit. Creating such a network was costly, and where distance was concerned, using that network was costly as well. Although we are all predicting the end of the circuit-switched network, many people still use it every day, and it really does work rather well.

Circuit Types

In the PSTN, there are many different sizes of circuits serving the various needs of the network. Between the central office and a subscriber, one or more analog circuits, or a few dozen channels delivered over a digital circuit, generally suffice. Between PSTN offices (and with larger customers), fiber-optic circuits are generally used.

The humble DS-0—The foundation of it all

Since the standard method of digitizing a telephone call is to record an 8-bit sample 8,000 times per second, we can see that a PCM-encoded telephone circuit will need a bandwidth of eight times 8,000 bits per second, or 64,000 bps. This 64-Kbps channel is referred to as a *DS-0* (that's "Dee-Ess-Zero"). The DS-0 is the fundamental building block of all digital telecommunications circuits.

[†] A low-pass filter, as its name implies, allows through only frequencies that are lower than its cut off frequency. Other types of filters are high-pass filters (which remove low frequencies) and band-pass filters (which filter out both high and low frequencies).

[‡] If you ever have to do audio recordings for a system, you might want to take advantage of the band-pass filter that is built into most telephone sets. Doing a recording using even high-end recording equipment can pick up all kinds of background noise that you don't even hear until you downsample, at which point the background noise produces aliasing (which can sound like all kinds of weird things). Conversely, the phone records in the correct format already, so the noise never enters the audio stream. Having said all that, no matter what you use to do recordings, avoid environments that have a lot of background noise. Typical offices can be a lot noisier than you'd think, as HVAC equipment can produce noise that we don't even realize is there.

Even the ubiquitous analog circuit is sampled into a DS-0 as soon as possible. Sometimes this happens where your circuit terminates at the central office, and sometimes well before.[§]

T-carrier circuits

The venerable T1 is one of the more recognized digital telephony terms. A T1 is a digital circuit consisting of 24 DS-0s multiplexed together into a 1.544-Mbps bit stream.[ǁ] This bit stream is properly defined as a *DS-1*. Voice is encoded on a T1 using the µlaw companding algorithm.

The European version of the T1 was developed by the European Conference of Postal and Telecommunications Administrations[#] (CEPT), and was first referred to as a *CEPT-1*. It is now called an *E1*.

The E1 is composed of 32 DS-0s, but the method of PCM encoding is different: E1s use alaw companding. This means that connecting between an E1-based network and a T1-based network will always require a transcoding step. Note that an E1, although it has 32 channels, is also considered a *DS-1*. It is likely that E1 is far more widely deployed, as it is used everywhere in the world except North America and Japan.

The various other T-carriers (T2, T3, and T4) are multiples of the T1, each based on the humble DS-0. Table A-2 illustrates the relationships between the different T-carrier circuits.

Table A-2. T-carrier circuits

Carrier	Equivalent data bitrate	Number of DS-0s	Data bitrate
T1	24 DS-0s	24	1.544 Mbps
T2	4 T1s	96	6.312 Mbps
T3	7 T2s	672	44.736 Mbps
T4	6 T3s	4,032	274.176 Mbps

At densities above T3, it is very uncommon to see a T-carrier circuit. For these speeds, optical carrier (OC) circuits may be used.

---

§ Digital telephone sets (including IP sets) do the analog-to-digital conversion right at the point where the handset plugs into the phone, so the DS-0 is created right at the phone set.

ǁ The 24 DS-0s use 1.536 Mbps, and the remaining .008 Mbps is used by framing bits.

#Conférence Européenne des Administrations des Postes et des Télécommunications.

SONET and OC circuits

The Synchronous Optical Network (SONET) was developed out of a desire to take the T-carrier system to the next technological level: fiber optics. SONET is based on the bandwidth of a T3 (44.736 Mbps), with a slight overhead making it 51.84 Mbps. This is referred to as an *OC-1* or *STS-1*. As Table A-3 shows, all higher-speed OC circuits are multiples of this base rate.

Table A-3. OC circuits

Carrier	Equivalent data bitrate	Number of DS-0s	Data bitrate
OC-1	1 DS-3 (plus overhead)	672	51.840 Mbps
OC-3	3 DS-3s	2,016	155.520 Mbps
OC-12	12 DS-3s	8,064	622.080 Mbps
OC-48	48 DS-3s	32,256	2488.320 Mbps
OC-192	192 DS-3s	129,024	9953.280 Mbps

SONET was created in an effort to standardize optical circuits, but due to its high cost, coupled with the value offered by many newer schemes, such as Dense Wave Division Multiplexing (DWDM), there is some controversy surrounding its future.

Digital Signaling Protocols

As with any circuit, it is not enough for the circuits used in the PSTN to just carry (voice) data between endpoints. Mechanisms must also be provided to pass information about the state of the channel between the endpoints. (Disconnect and answer supervision are two examples of basic signaling that might need to take place; caller ID is an example of a more complex form of signaling.)

Channel Associated Signaling (CAS)

Also known as robbed-bit signaling, CAS is what you will use to transmit voice on a T1 when ISDN is not available. Rather than taking advantage of the power of the digital circuit, CAS simulates analog channels. CAS works by stealing bits from the audio stream for signaling purposes. Although the effect on audio quality is not really noticeable, the lack of a powerful signaling channel limits your flexibility.

When configuring a CAS T1, the signaling options at each end must match. E&M (Ear & Mouth or recEive & transMit) signaling is generally preferred, as it offers the best supervision. Having said that, in an Asterisk environment the most likely reason for you to use CAS would be for a channel bank, which means you are most likely going to have to use FXS signaling.

CAS is very rarely used on PSTN circuits anymore, due to the superiority of ISDN-PRI. One of the limitations of CAS is that it does not allow the dynamic assignment of channels to different functions. Also, caller ID information (which may not even be supported) has to be sent as part of the audio stream. CAS is commonly used on the T1 link in channel banks.

ISDN

The Integrated Services Digital Network (ISDN) has been around for more than 20 years. Because it separates the channels that carry the traffic (the bearer channels, or B-channels) from the channel that carries the signaling information (the D-channel), ISDN allows for the delivery of a much richer set of features than CAS.

In the beginning, ISDN promised to deliver much the same sort of functionality that the Internet has given us, including advanced capabilities for voice, video, and data transfer. Unfortunately, rather than ratifying a standard and sticking to it, the respective telecommunications manufacturers all decided to add their own tweaks to the protocol, in the belief that their versions were superior and would eventually come to dominate the market. As a result, getting two ISDN-compliant systems to connect to each other was often a painful and expensive task. The carriers who had to implement and support this expensive technology, in turn, priced it so that it was not rapidly adopted. Currently, ISDN is rarely used for much more than basic trunking—in fact, the acronym ISDN has become a joke in the industry: "It Still Does Nothing."

Having said that, ISDN has become quite popular for trunking, and it is now (mostly) standards-compliant. If you have a PBX with more than a dozen lines connected to the PSTN, there's a very good chance that you'll be running an ISDN-*PRI* (Primary Rate Interface) circuit. Also, in places where DSL and cable access to the Internet are not available (or are too expensive), an ISDN-*BRI* (Basic Rate Interface) circuit might provide you with an affordable 128-Kbps connection. In much of North America, the use of BRI for Internet connectivity has been deprecated in favor of DSL and cable modems (and it is never used for voice), but in many European countries it has almost totally replaced analog circuits.

ISDN-BRI/BRA. The Basic Rate Interface (or Basic Rate Access) flavor of ISDN is designed to service small endpoints such as workstations.

This flavor is often referred to simply as "ISDN," but this can be a source of confusion, as ISDN is a protocol, not a type of circuit (not to mention that PRI circuits are also correctly referred to as ISDN!).

A Basic Rate ISDN circuit consists of two 64-Kbps B-channels controlled by a 16-Kbps D-channel, for a total of 144 Kbps.

Basic Rate ISDN has been a source of much confusion during its life, due to problems with standards compliance, technical complexity, and poor documentation. Still, many European telecos have widely implemented ISDN-BRI, and thus it is more popular in Europe than in North America.

ISDN-PRI/PRA. The Primary Rate Interface (or Primary Rate Access) flavor of ISDN is used to provide ISDN service over larger network connections. A Primary Rate ISDN circuit uses a single DS-0 channel as a signaling link (the D-channel); the remaining channels serve as B-channels.

In North America, Primary Rate ISDN is commonly carried on one or more T1 circuits. Since a T1 has 24 channels, a North American PRI circuit typically consists of 23 B-channels and 1 D-channel. For this reason, PRI is often referred to as 23B+D.[*]

 In Europe, a 32-channel E1 circuit is used, so a Primary Rate ISDN circuit is referred to as 30B+D (the final channel is used for synchronization).

Primary Rate ISDN is very popular, due to its technical benefits and generally competitive pricing at higher densities. If you believe you will require more than a dozen or so PSTN lines, you should look into Primary Rate ISDN pricing.

From a technical perspective, ISDN-PRI is always preferable to CAS.

Signaling System 7

Signaling System 7 (SS7) is the signaling system used by carriers. It is conceptually similar to ISDN, and it is instrumental in providing a mechanism for the carriers to transmit the additional information ISDN endpoints typically need to pass. However, the technology of SS7 is different from that of ISDN; one big difference is that SS7 runs on a completely separate network than the actual trunks that carry the calls.

SS7 support in Asterisk is on the horizon, as there is much interest in making Asterisk compatible with the carrier networks. An open source version of SS7 (*http://www .openss7.org*) exists, but work is still needed for full SS7 compliance, and as of this writing it is not known whether this version will be integrated with Asterisk. Another promising source of SS7 support comes from Sangoma Technologies, which offers SS7 functionality in many of its products.

It should be noted that adding support for SS7 in Asterisk is not going to be as simple as writing a proper driver. Connecting equipment to an SS7 network will not be possible without that equipment having passed extremely rigorous certification processes. Even then, it seems doubtful that any traditional carrier is going to be in a hurry to allow such a thing to happen, mostly for strategic and political reasons.

---

[*] PRI is actually quite a bit more flexible than that, as it is possible to span a single PRI circuit across multiple T1 spans. This can give rise, for example, to a 47B+D circuit (where a single D-channel serves two T1s) or a 46B+2D circuit (where primary and backup D-channels serve a pair of T1s). You will sometimes see PRI described as nB+nD, because the number of B- and D-channels is, in fact, quite variable. For this reason, you should never refer to a T1 carrying PRI as "a PRI." For all you know, the PRI circuit spans multiple T1s, as is common in larger PBX deployments.

Packet-Switched Networks

In the mid-1990s, network performance improved to the point where it became possible to send a stream of media information in real time across a network connection. Because the media stream is chopped up into segments, which are then wrapped in an addressing envelope, such connections are referred to as *packet-based*. The challenge, of course, is to send a flood of these packets between two endpoints, ensuring that the packets arrive in the same order in which they were sent, in less than 150 milliseconds, with none lost. This is the essence of Voice over IP.

Conclusion

This appendix has explored the technologies currently in use in the PSTN. In Appendix B, we will discuss protocols for VoIP: the carrying of telephone connections across IP-based networks. These protocols define different mechanisms for carrying telephone conversations, but their significance is far greater than just that. Bringing the telephone network into the data network will finally erase the line between telephones and computers, which holds the promise of a revolutionary evolution in the way we communicate.

Protocols for VoIP

The Internet is a telephone system that's gotten uppity.

—Clifford Stoll

The telecommunications industry spans over 100 years, and Asterisk integrates most—if not all—of the major technologies that it has made use of over the last century. To make the most out of Asterisk, you need not be a professional in all areas, but understanding the differences between the various codecs and protocols will give you a greater appreciation and understanding of the system as a whole.

This appendix explains Voice over IP and what makes VoIP networks different from the traditional circuit-switched voice networks that were the topic of Appendix A. We will explore the need for VoIP protocols, outlining the history and potential future of each. We'll also look at security considerations and these protocols' abilities to work within topologies such as Network Address Translation (NAT). The following VoIP protocols will be discussed (some more briefly than others):

- IAX
- SIP
- H.323
- MGCP
- Skinny/SCCP
- UNISTIM

Codecs are the means by which analog voice can be converted to a digital signal and carried across the Internet. Bandwidth at any location is finite, and the number of simultaneous conversations any connection can carry is directly related to the type of codec implemented. We'll also explore the differences between the following codecs in regard to bandwidth requirements (compression level) and quality:

- G.711
- G.726

- G.729A
- GSM
- iLBC
- Speex
- MP3

We will then conclude the appendix with a discussion of how voice traffic can be routed reliably, what causes echo and how to deal with it, and how Asterisk controls the authentication of inbound and outbound calls.

The Need for VoIP Protocols

The basic premise of VoIP is the packetization[*] of audio streams for transport over Internet Protocol-based networks. The challenges to accomplishing this relate to the manner in which humans communicate. Not only must the signal arrive in essentially the same form that it was transmitted in, but it needs to do so in less than 150 milliseconds. If packets are lost or delayed, there will be degradation to the quality of the communications experience, meaning that two people will have difficulty in carrying on a conversation.

The transport protocols that collectively are called "the Internet" were not originally designed with real-time streaming of media in mind. Endpoints were expected to resolve missing packets by waiting longer for them to arrive, requesting retransmission, or, in some cases, considering the information to be gone for good and simply carrying on without it. In a typical voice conversation, these mechanisms will not serve. Our conversations do not adapt well to the loss of letters or words, nor to any appreciable delay between transmittal and receipt.

The traditional PSTN was designed specifically for the purpose of voice transmission, and it is perfectly suited to the task from a technical standpoint. From a flexibility standpoint, however, its flaws are obvious to even people with a very limited understanding of the technology. VoIP holds the promise of incorporating voice communications into all of the other protocols we carry on our networks, but due to the special demands of a voice conversation, special skills are needed to design, build, and maintain these networks.

The problem with packet-based voice transmission stems from the fact that the way in which we speak is totally incompatible with the way in which IP transports data. Speaking and listening consist of the relaying of a stream of audio, whereas the Internet protocols are designed to chop everything up, encapsulate the bits of information into

---

[*] This word hasn't quite made it into the dictionary, but it is a term that is becoming more and more common. It refers to the process of chopping a steady stream of information into discrete chunks (or *packets*), suitable for delivery independently of one another.

thousands of packages, and then deliver each package in whatever way possible to the far end. Clearly, some way of dealing with this is required.

VoIP Protocols

The mechanism for carrying a VoIP connection generally involves a series of signaling transactions between the endpoints (and gateways in between), culminating in two persistent media streams (one for each direction) that carry the actual conversation. There are several protocols in existence to handle this. In this section, we will discuss some that are important to VoIP in general and to Asterisk specifically.

IAX (The "Inter-Asterisk eXchange" Protocol)

If you claim to be one of the folks in the know when it comes to Asterisk, your test will come when you have to pronounce the name of this protocol. It would seem that you should say "eye-ay-ex," but this hardly rolls off the tongue very well.[†] Fortunately, the proper pronunciation is in fact "eeks."[‡] IAX is an open protocol, meaning that anyone can download and develop for it.[§]

In Asterisk, IAX is supported by the *chan_iax2.so* module.

History

The IAX protocol was developed by Digium for the purpose of communicating with other Asterisk servers (hence the Inter-Asterisk eXchange protocol). It is very important to note that IAX is not at all limited to Asterisk. The standard is open for anyone to use, and it is supported by many other open source telecom projects, as well as by several hardware vendors. IAX is a transport protocol (much like SIP) that uses a single UDP port (4569) for both the channel signaling and media streams. As discussed later in this appendix, this makes it easier to manage when behind NATed firewalls.

IAX also has the unique ability to trunk multiple sessions into one dataflow, which can result in a tremendous bandwidth advantage when sending a lot of simultaneous channels to a remote box. Trunking allows multiple media streams to be represented with a single datagram header, which lowers the overhead associated with individual channels. This helps to lower latency and reduce the processing power and bandwidth required, allowing the protocol to scale much more easily with a large number of active channels between endpoints. If you have a large quantity of IP calls to pass between two endpoints, you should take a close look at IAX trunking.

[†] It sounds like the name of a Dutch football team.

[‡] Go ahead. Say it. That sounds much better, doesn't it?

[§] Officially, the current version is IAX2 (officially standardized by the IETF in RFC 5456), but all support for IAX1 has been dropped, so whether you say "IAX" or "IAX2," it is expected that you are talking about version 2.

Future

Since IAX was optimized for voice, it has received some criticism for not better supporting video—but in fact, IAX holds the potential to carry pretty much any media stream desired. Because it is an open protocol, future media types are certain to be incorporated as the community desires them.

Security considerations

IAX includes the ability to authenticate in three ways: plain text, MD5 hashing, and RSA key exchange. This, of course, does nothing to encrypt the media path or headers between endpoints. Many solutions involve using a Virtual Private Network (VPN) appliance or software to encrypt the stream in another layer of technology, which requires the endpoints to pre-establish a method of configuring and opening these tunnels. However, IAX is now also able to encrypt the streams between endpoints with dynamic key exchange at call setup (using the configuration option `encryp tion=aes128`), allowing the use of automatic key rollover.

IAX and NAT

The IAX2 protocol was deliberately designed to work from behind devices performing NAT. The use of a single UDP port for both signaling and transmission of media also keeps the number of holes required in your firewall to a minimum. These considerations have helped make IAX one of the easiest protocols (if not the easiest) to implement in secure networks.

SIP

The Session Initiation Protocol (SIP) has taken the telecommunications industry by storm. SIP has pretty much dethroned the once-mighty H.323 as the VoIP protocol of choice—certainly at the endpoints of the network. The premise of SIP is that each end of a connection is a peer; the protocol negotiates capabilities between them. What makes SIP compelling is that it is a relatively simple protocol, with a syntax similar to that of other familiar protocols such as HTTP and SMTP. SIP is supported in Asterisk with the *chan_sip.so* module.‖

History

SIP was originally submitted to the Internet Engineering Task Force (IETF) in February 1996 as "draft-ietf-mmusic-sip-00." The initial draft looked nothing like the SIP we

---

‖ Having just called SIP simple, it should be noted that it is by no means lightweight. It has been said that if one were to read all of the IETF RFCs that are relevant to SIP, one would have more than 3,000 pages of reading to do. SIP is quickly earning a reputation for being far too bloated, but that does nothing to lessen its popularity.

---

know today and contained only a single request type: a call setup request. In March 1999, after 11 revisions, SIP RFC 2543 was born.

At first, SIP was all but ignored, as H.323 was considered the protocol of choice for VoIP transport negotiation. However, as the buzz grew, SIP began to gain in popularity, and while there may be a lot of different factors that accelerated its growth, we'd like to think that a large part of its success is due to its freely available specification.

SIP is an application-layer signaling protocol that uses the well-known port 5060 for communications. SIP can be transported with either the UDP or TCP transport-layer protocols. Asterisk does not currently have a TCP implementation for transporting SIP messages, but it is possible that future versions may support it (and patches to the code base are gladly accepted). SIP is used to "establish, modify, and terminate multimedia sessions such as Internet telephony calls."#

SIP does not transport media (i.e., voice) between endpoints. Instead, the Real-time Transport Protocol (RTP) is used for this purpose. RTP uses high-numbered, unprivileged ports in Asterisk (10,000 through 20,000, by default).

A common topology to illustrate SIP and RTP, commonly referred to as the "SIP trapezoid," is shown in Figure B-1. When Alice wants to call Bob, Alice's phone contacts her proxy server, and the proxy tries to find Bob (often connecting through his proxy). Once the phones have started the call, they communicate directly with each other (if possible), so that the data doesn't have to tie up the resources of the proxy.

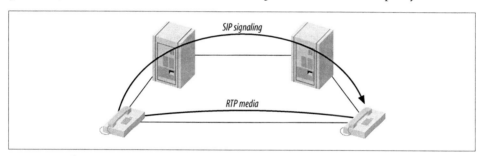

Figure B-1. The SIP trapezoid

SIP was not the first, and is not the only, VoIP protocol in use today (others include H.323, MGCP, IAX, and so on), but currently it seems to have the most momentum with hardware vendors. The advantages of the SIP protocol lie in its wide acceptance and architectural flexibility (and, we used to say, simplicity!).

Future

SIP has earned its place as the protocol that justified VoIP. All new user and enterprise products are expected to support SIP, and any existing products will now be a tough

#RFC 3261, "SIP: Session Initiation Protocol," p. 9, Section 2.

sell unless a migration path to SIP is offered. SIP is widely expected to deliver far more than VoIP capabilities, including the ability to transmit video, music, and any type of real-time multimedia. While its use as a ubiquitous general-purpose media transport mechanism seems doubtful, SIP is unarguably poised to deliver the majority of new voice applications for the next few years.

Security considerations

SIP uses a challenge/response system to authenticate users. An initial INVITE is sent to the proxy with which the end device wishes to communicate. The proxy then sends back a 407 Proxy Authorization Request message, which contains a random set of characters referred to as a *nonce*. This nonce is used along with the password to generate an MD5 hash, which is then sent back in the subsequent INVITE. Assuming the MD5 hash matches the one that the proxy generated, the client is then authenticated.

Denial of service (DoS) attacks are probably the most common type of attack on VoIP communications. A DoS attack can occur when a large number of invalid INVITE requests are sent to a proxy server in an attempt to overwhelm the system. These attacks are relatively simple to implement, and their effects on the users of the system are immediate. SIP has several methods of minimizing the effects of DoS attacks, but ultimately they are impossible to prevent.

SIP implements a scheme to guarantee that a secure, encrypted transport mechanism (namely Transport Layer Security, or TLS) is used to establish communication between the caller and the domain of the callee. Beyond that, the request is sent securely to the end device, based upon the local security policies of the network. Note that the encryption of the media (that is, the RTP stream) is beyond the scope of SIP itself and must be dealt with separately.

More information regarding SIP security considerations, including registration hijacking, server impersonation, and session teardown, can be found in Section 26 of SIP RFC 3261.

SIP and NAT

Probably the biggest technical hurdle SIP has to conquer is the challenge of carrying out transactions across a NAT layer. Because SIP encapsulates addressing information in its data frames, and NAT happens at a lower network layer, the addressing information is not automatically modified, and thus the media streams will not have the correct addressing information needed to complete the connection when NAT is in place. In addition to this, the firewalls normally integrated with NAT will not consider the incoming media stream to be part of the SIP transaction, and will block the connection. Newer firewalls and session border controllers (SBCs) are SIP-aware, but this is still considered a shortcoming in this protocol, and it causes no end of trouble to network professionals needing to connect SIP endpoints using existing network infrastructure.

H.323

This International Telecommunication Union (ITU) protocol was originally designed to provide an IP transport mechanism for videoconferencing. It has become the standard in IP-based video-conferencing equipment, and it briefly enjoyed fame as a VoIP protocol as well. While there is much heated debate over whether SIP or H.323 (or IAX) will come to dominate the VoIP protocol world, in Asterisk, H.323 has largely been deprecated in favor of IAX and SIP. H.323 has not enjoyed much success among users and enterprises, although it might still be the most widely used VoIP protocol among carriers.

The three versions of H.323 supported in Asterisk are handled by the modules *chan_h323.so* (supplied with Asterisk), *chan_oh323.so* (available as a free addon), and *chan_ooh323.so* (supplied in *asterisk-addons*).

 You have probably used H.323 without even knowing it—Microsoft's NetMeeting client is arguably the most widely deployed H.323 client.

History

H.323 was developed by the ITU in May 1996 as a means to transmit voice, video, data, and fax communications across an IP-based network while maintaining connectivity with the PSTN. Since that time, H.323 has gone through several versions and annexes (which add functionality to the protocol), allowing it to operate in pure VoIP networks and more widely distributed networks.

Future

The future of H.323 is a subject of debate. If the media is any measure, it doesn't look good for H.323; it hardly ever gets mentioned (certainly not with the regularity of SIP). H.323 is often regarded as technically superior to SIP, but, that sort of thing is seldom the deciding factor in whether a technology enjoys success. One of the factors that makes H.323 unpopular is its complexity (although many argue that the once-simple SIP is starting to suffer from the same problem).

H.323 still carries by far the majority of worldwide carrier VoIP traffic, but as people become less and less dependent on traditional carriers for their telecom needs, the future of H.323 becomes more difficult to predict with any certainty. While H.323 may not be the protocol of choice for new implementations, we can certainly expect to have to deal with H.323 interoperability issues for some time to come.

Security considerations

H.323 is a relatively secure protocol and does not require many security considerations beyond those that are common to any network communicating with the Internet. Since H.323 uses the RTP protocol for media communications, it does not natively support encrypted media paths. The use of a VPN or other encrypted tunnel between endpoints is the most common way of securely encapsulating communications. Of course, this has the disadvantage of requiring the establishment of these secure tunnels between endpoints, which may not always be convenient (or even possible). As VoIP becomes used more often to communicate with financial institutions such as banks, we're likely to require extensions to the most commonly used VoIP protocols to natively support strong encryption methods.

H.323 and NAT

The H.323 standard uses the Internet Engineering Task Force (IETF) RTP protocol to transport media between endpoints. Because of this, H.323 has the same issues as SIP when dealing with network topologies involving NAT. The easiest method is to simply forward the appropriate ports through your NAT device to the internal client.

To receive calls, you will always need to forward TCP port 1720 to the client. In addition, you will need to forward the UDP ports for the RTP media and RTCP control streams (see the manual for your device for the port range it requires). Older clients, such as Microsoft NetMeeting, will also require TCP ports forwarded for H.245 tunneling (again, see your client's manual for the port number range).

If you have a number of clients behind the NAT device, you will need to use a *gatekeeper* running in proxy mode. The gatekeeper will require an interface attached to the private IP subnet and the public Internet. Your H.323 client on the private IP subnet will then register to the gatekeeper, which will proxy calls on the clients' behalf. Note that any external clients that wish to call you will also be required to register with the proxy server.

At this time, Asterisk can't act as an H.323 gatekeeper. You'll have to use a separate application, such as the open source OpenH323 Gatekeeper (*http://www.gnugk.org*), for this purpose.

MGCP

The Media Gateway Control Protocol (MGCP) also comes to us from the IETF. While MGCP deployment is more widespread than one might think, it is quickly losing ground to protocols such as SIP and IAX. Still, Asterisk loves protocols, so naturally it has rudimentary support for it.

MGCP is defined in RFC 3435.* It was designed to make the end devices (such as phones) as simple as possible, and have all the call logic and processing handled by media gateways and call agents. Unlike SIP, MGCP uses a centralized model. MGCP phones cannot directly call other MGCP phones; they must always go through some type of controller.

Asterisk supports MGCP through the *chan_mgcp.so* module, and the endpoints are defined in the configuration file *mgcp.conf*. Since Asterisk provides only basic call agent services, it cannot emulate an MGCP phone (to register to another MGCP controller as a user agent, for example).

If you have some MGCP phones lying around, you will be able to use them with Asterisk. If you are planning to put MGCP phones into production on an Asterisk system, keep in mind that the community has moved on to more popular protocols, and you will therefore need to budget your software support needs accordingly. If possible (for example, with Cisco phones), you should upgrade MGCP phones to SIP.

Proprietary Protocols

Finally, let's take a look at two proprietary protocols that are supported in Asterisk.

Skinny/SCCP

The Skinny Client Control Protocol (SCCP) is proprietary to Cisco VoIP equipment. It is the default protocol for endpoints on a Cisco Call Manager PBX.† Skinny is supported in Asterisk, but if you are connecting Cisco phones to Asterisk, it is generally recommended that you obtain SIP images for any phones that support this and connect via SIP instead.

UNISTIM

Asterisk's Support for Nortel's proprietary VoIP protocol, UNISTIM, makes it the first PBX in history to natively support proprietary IP terminals from the two biggest players in VoIP: Nortel and Cisco. UNISTIM support is totally experimental and does not yet work well enough to put it into production, but the fact that somebody has taken the trouble to implement it demonstrates the power of the Asterisk platform.

Codecs

Codecs are generally understood to be various mathematical models used to digitally encode (and compress) analog audio information. Many of these models take into account the human brain's ability to form an impression from incomplete information.

---

* RFC 3435 obsoletes RFC 2705.

† Cisco has recently announced that it will be migrating toward SIP in its future products.

We've all seen optical illusions; likewise, voice-compression algorithms take advantage of our tendency to interpret what we *believe* we should hear, rather than what we *actually* hear.‡ The purpose of the various encoding algorithms is to strike a balance between efficiency and quality.§

Originally, the term *codec* referred to a COder/DECoder: a device that converts between analog and digital. Now, the term seems to relate more to COmpression/DECompression.

Before we dig into the individual codecs, take a look at Table B-1—it's a quick reference that you may want to refer back to.

Table B-1. Codec quick reference

Codec	Data bitrate (Kbps)	License required?
G.711	64 Kbps	No
G.726	16, 24, 32, or 40 Kbps	No
G.729A	8 Kbps	Yes (no for pass-through)
GSM	13 Kbps	No
iLBC	13.3 Kbps (30-ms frames) or 15.2 Kbps (20-ms frames)	No
Speex	Variable (between 2.15 and 22.4 Kbps)	No
G.722	64 Kbps	No

G.711

G.711 is the fundamental codec of the PSTN. In fact, if someone refers to PCM (discussed in Appendix A) with respect to a telephone network, you are allowed to think of G.711. Two companding methods are used: μlaw in North America and alaw in the rest of the world. Either one delivers an 8-bit word transmitted 8,000 times per second. If you do the math, you will see that this requires 64,000 bits to be transmitted per second.

Many people will tell you that G.711 is an uncompressed codec. This is not exactly true, as companding is considered a form of compression. What is true is that G.711 is the base codec from which all of the others are derived.

---

‡ Read the following: "Aoccdrnig to rsereach at an Elingsh uinervtisy, it deosn't mttaer in waht oredr the ltteers in a wrod are, the olny iprmoetnt tihng is taht frist and lsat ltteres are in the rghit pclae. The rset can be a toatl mses and you can sitll raed it wouthit a porbelm. Tihs is bcuseae we do not raed ervey lteter by istlef, but the wrod as a wlohe." (The source of this quote is unknown.) We do the same thing with sound: if there is enough information, our brains can fill in the gaps.

§ On an audio CD, quality is far more important than saving bandwidth, so the audio is quantized at 16 bits (times 2, as it's stereo), with a sampling rate of 44,100 Hz. Considering that the CD was invented in the late 1970s, this was quite impressive stuff back then. The telephone network does not require this level of quality (and needs to optimize bandwidth), so telephone signals are encoded using 8 bits, at a sampling frequency of 8,000 Hz.

G.711 imposes minimal (almost zero) load on the CPU.

G.726

This codec has been around for some time (it used to be G.721, which is now obsolete), and it is one of the original compressed codecs. It is also known as Adaptive Differential Pulse-Code Modulation (ADPCM), and it can run at several bitrates. The most common rates are 16 Kbps, 24 Kbps, and 32 Kbps. As of this writing, Asterisk supports only the ADPCM-32 rate, which is far and away the most popular rate for this codec.

G.726 offers nearly identical quality to G.711, but it uses only half the bandwidth. This is possible because rather than sending the result of the quantization measurement, it sends only enough information to describe the difference between the current sample and the previous one. G.726 fell from favor in the 1990s due to its inability to carry modem and fax signals, but because of its bandwidth/CPU performance ratio it is now making a comeback. G.726 is especially attractive because it does not require a lot of computational work from the system.

G.729A

Considering how little bandwidth it uses, G.729A delivers impressive sound quality. It does this through the use of Conjugate-Structure Algebraic-Code-Excited Linear Prediction (CS-ACELP).[||] Because of patents, you can't use G.729A without paying a licensing fee; however, it is extremely popular and is well supported on many different phones and systems.

To achieve its impressive compression ratio, this codec requires an equally impressive amount of effort from the CPU. In an Asterisk system, the use of heavily compressed codecs will quickly bog down the CPU.

G.729A uses 8 Kbps of bandwidth.

GSM

The Global System for Mobile Communications (GSM) codec is the darling of Asterisk. This codec does not come encumbered with a licensing requirement the way that G.729A does, and it offers outstanding performance with respect to the demand it places on the CPU. The sound quality is generally considered to be of a lesser grade

[||] CELP is a popular method of compressing speech. By mathematically modeling the various ways humans make sounds, a codebook of sounds can be built. Rather than sending an actual sampled sound, a code corresponding to the sound is determined. CELP codecs take this information (which by itself would produce a very robot-like sound) and attempt to add the personality back in. (Of course, there is much more to it than that.) Jason Woodward's *Speech Coding* page (*http://www-mobile.ecs.soton.ac.uk/speech_codecs/*) is a source of helpful information for the non-mathematically inclined. This is fairly heavy stuff, though, so wear your thinking cap.

than that produced by G.729A, but much of this comes down to personal opinion; be sure to try it out. GSM operates at 13 Kbps.

iLBC

The Internet Low Bitrate Codec (iLBC) provides an attractive mix of low bandwidth usage and quality, and it is especially well suited to sustaining reasonable quality on lossy network links.

Naturally, Asterisk supports it (and support elsewhere is growing), but it is not as popular as the ITU codecs, and thus may not be compatible with common IP telephones and commercial VoIP systems. IETF RFCs 3951 and 3952 have been published in support of iLBC, and iLBC is on the IETF standards track.

Because iLBC uses complex algorithms to achieve its high levels of compression, it has a fairly high CPU cost in Asterisk.

While you are allowed to use iLBC without paying royalty fees, the holder of the iLBC patent, Global IP Sound (GIPS), wants to know whenever you use it in a commercial application. The way you do that is by downloading and printing a copy of the iLBC license, signing it, and returning it to GIPS. If you want to read about iLBC and its license, you can do so at *http://www.ilbcfreeware.org*.

iLBC operates at 13.3 Kbps (30-ms frames) and 15.2 Kbps (20-ms frames).

Speex

Speex is a variable bitrate (VBR) codec, which means that it is able to dynamically modify its bitrate to respond to changing network conditions. It is offered in both narrowband and wideband versions, depending on whether you want telephone quality or better.

Speex is a totally free codec, licensed under the *Xiph.org* variant of the BSD license.

An Internet draft for Speex is available, and more information about Speex can be found at its home page (*http://www.speex.org*).

Speex can operate at anywhere from 2.15 to 22.4 Kbps, due to its variable bitrate.

G.722

G.722 is an ITU-T standard codec that was approved in 1988. The G.722 codec produces a much higher-quality voice in the same space as G.711 (64 Kbps) and is starting to become popular among VoIP device manufacturers. The patents for G.722 have expired, so it is freely available. If you have access to devices that support G.722, you'll be impressed by the quality improvement.

MP3

Sure thing, MP3 is a codec. Specifically, it's the Moving Picture Experts Group Audio Layer 3 Encoding Standard.# With a name like that, it's no wonder we call it MP3! In Asterisk, the MP3 codec is typically used for music on hold (MoH). MP3 is not a telephony codec, as it is optimized for music, not voice; nevertheless, it's very popular with VoIP telephony systems as a method of delivering MoH.

 Be aware that music cannot usually be broadcast without a license. Many people assume that there is no legal problem with connecting a radio station or CD as a music on hold source, but this is very rarely true.

Quality of Service

Quality of Service, or *QoS* as it's more popularly termed, refers to the challenge of delivering a time-sensitive stream of data across a network that was designed to deliver data in an ad hoc, best-effort sort of way. Although there is no hard rule, it is generally accepted that if you can deliver the sound produced by the speaker to the listener's ear within 150 milliseconds, a normal flow of conversation is possible. When delay exceeds 300 milliseconds, it becomes difficult to avoid interrupting each other. Beyond 500 milliseconds, normal conversation becomes increasingly awkward and frustrating.

In addition to getting it there on time, it is also essential to ensure that the transmitted information arrives intact. Too many lost packets will prevent the far end from completely reproducing the sampled audio, and gaps in the data will be heard as static or, in severe cases, entire missed words or sentences. Even packet loss of 5 percent can severely impede a VoIP network.

TCP, UDP, and SCTP

If you're going to send data on an IP-based network, it will be transported using one of the three transport protocols discussed here.

Transmission Control Protocol

The Transmission Control Protocol (TCP) is almost never used for VoIP, for while it does have mechanisms in place to ensure delivery, it is not inherently in any hurry to do so. Unless there is an extremely low-latency interconnection between the two endpoints, TCP will tend to cause more problems than it solves.

\#If you want to learn all about MPEG audio, do a web search for Davis Pan's paper titled "A Tutorial on MPEG/Audio Compression."

The purpose of TCP is to guarantee the delivery of packets. In order to do this, several mechanisms are implemented, such as packet numbering (for reconstructing blocks of data), delivery acknowledgment, and re-requesting of lost packets. In the world of VoIP, getting the packets to the endpoint quickly is paramount—but 20 years of cellular telephony has trained us to tolerate a few lost packets.[*]

TCP's high processing overhead, state management, and acknowledgment of arrival work well for transmitting large amounts of data, but they simply aren't efficient enough for real-time media communications.

User Datagram Protocol

Unlike TCP, the User Datagram Protocol (UDP) does not offer any sort of delivery guarantee. Packets are placed on the wire as quickly as possible and released into the world to find their way to their final destinations, with no word back as to whether they got there or not. Since UDP itself does not offer any kind of guarantee that the data will arrive,[†] it achieves its efficiency by spending very little effort on what it is transporting.

 TCP is a more "socially responsible" protocol because the bandwidth is more evenly distributed to clients connecting to a server. As the percentage of UDP traffic increases, it is possible that a network could become overwhelmed.

Stream Control Transmission Protocol

Approved by the IETF as a proposed standard in RFC 2960, SCTP is a relatively new transport protocol. From the ground up, it was designed to address the shortcomings of both TCP and UDP, especially as related to the types of services that used to be delivered over circuit-switched telephony networks.

Some of the goals of SCTP were:

• Better congestion-avoidance techniques (specifically, avoiding denial of service attacks)

• Strict sequencing of data delivery

• Lower latency for improved real-time transmissions

---

[*] The order of arrival is important in voice communication, because the audio will be processed and sent to the caller ASAP. However, with a jitter buffer the order of arrival isn't as important, as it provides a small window of time in which the packets can be reordered before being passed on to the caller.

[†] Keep in mind that the upper-layer protocols or applications can implement their own packet-acknowledgment systems.

By addressing the major shortcomings of TCP and UDP, SCTP's developers hoped to create a robust protocol for the transmission of SS7 and other types of PSTN signaling over an IP-based network.

Differentiated Service

Differentiated service, or DiffServ, is not so much a QoS mechanism as a method by which traffic can be flagged and given specific treatment. Obviously, DiffServ can help to provide QoS by allowing certain types of packets to take precedence over others. While this will certainly increase the chance of a VoIP packet passing quickly through each link, it does not guarantee anything.

Guaranteed Service

The ultimate guarantee of QoS is provided by the PSTN. For each conversation, a 64-Kbps channel is completely dedicated to the call; the bandwidth is guaranteed. Similarly, protocols that offer guaranteed service can ensure that a required amount of bandwidth is dedicated to the connection being served. As with any packetized networking technology, these mechanisms generally operate best when traffic is below maximum levels. When a connection approaches its limits, it is next to impossible to eliminate degradation.

MPLS

Multiprotocol Label Switching (MPLS) is a method for engineering network traffic patterns independent of layer-3 routing tables. The protocol works by assigning short labels (MPLS frames) to network packets, which routers then use to forward the packets to the MPLS egress router, and ultimately to their final destinations. Traditionally, routers make an independent forwarding decision based on an IP table lookup at each hop in the network. In an MPLS network, this lookup is performed only once, when the packet enters the MPLS cloud at the ingress router. The packet is then assigned to a stream, referred to as a Label Switched Path (LSP), and identified by a label. The label is used as a lookup index in the MPLS forwarding table, and the packet traverses the LSP independent of layer-3 routing decisions. This allows the administrators of large networks to fine-tune routing decisions and make the best use of network resources. Additionally, information can be associated with a label to prioritize packet forwarding.

RSVP

MPLS contains no method to dynamically establish LSPs, but you can use the Reservation Protocol (RSVP) with MPLS. RSVP is a signaling protocol used to simplify the establishment of LSPs and to report problems to the MPLS ingress router. The advantage of using RSVP in conjunction with MPLS is the reduction in administrative overhead. If you don't use RSVP with MPLS, you'll have to go to every single router and configure the labels and each path manually. Using RSVP makes the network more dynamic by distributing control of labels to the routers. This enables the network to become more responsive to changing conditions, because it can be set up to change the paths based on certain conditions, such as a certain path going down (perhaps due to a faulty router). The configuration within the router will then be able to use RSVP to distribute new labels to the routers in the MPLS network, with no (or minimal) human intervention.

Best Effort

The simplest, least expensive approach to QoS is not to provide it at all—the "best effort" method. While this might sound like a bad idea, it can in fact work very well. Any VoIP call that traverses the public Internet is almost certain to be best-effort, as QoS mechanisms are not yet common in this environment.

Echo

You may not realize it, but echo has been a problem in the PSTN for as long as there have been telephones. You probably haven't often experienced it, because the telecom industry has spent large sums of money designing expensive echo-cancellation devices. Also, when the endpoints are physically close—e.g., when you phone your neighbor down the street—the delay is so minimal that anything you transmit will be returned back so quickly that it will be indistinguishable from the sidetone‡ normally occurring in your telephone. So, the fact of the matter is that there is echo on your local calls much of the time, but you cannot perceive it with a regular telephone because it happens almost instantaneously. It may help you to understand this if you consider that when you stand in a room and speak, everything you say echos back to you off of the walls and ceiling (and possibly the floor, if it's not carpeted), but this does not cause any problems because it happens so fast you do not perceive a delay.

The reason that VoIP telephone systems such as Asterisk can experience echo is that the addition of a VoIP telephone introduces a slight delay. It takes a few milliseconds for the packets to travel from your phone to the server (and vice versa). Suddenly there is an appreciable delay, which allows you to perceive the echo that was always there, but never really noticeable.

Why Echo Occurs

Before we discuss measures to deal with echo, let's first take a look at why echo occurs in the analog world.

---

‡ As discussed in Appendix A, sidetone is a function in your telephone that returns part of what you say back to your own ear, to provide a more natural-sounding conversation.

If you hear echo, it's not your phone that's causing the problem; it's the far end of the circuit. Conversely, echo heard on the far end is being generated at your end. Echo can be caused by the fact that an analog local loop circuit has to transmit and receive on the same pair of wires. If this circuit is not electrically balanced, or if a low-quality telephone is connected to the end of the circuit, signals it receives can be reflected back, becoming part of the return transmission. When this reflected circuit gets back to you, you will hear the words you spoke just moments before. Humans will perceive an echo beyond a certain amount of delay (possibly as low as 20 milliseconds for some people). This echo will become annoying as the delay increases.

In a cheap telephone, it is possible for echo to be generated in the body of the handset. This is why some cheap IP phones can cause echo even when the entire end-to-end connection does not contain an analog circuit.[§] In the VoIP world, echo is usually introduced either by an analog circuit somewhere in the connection, or by a cheap endpoint reflecting back some of the signal (e.g., feedback through a hands-free or poorly designed handset or headset). The greater the latency on the network, the more annoying this echo can be.

Managing Echo on DAHDI Channels

You can enable and disable echo cancellation for DAHDI interfaces in the *chan_dahdi.conf* file. The default configuration enables echo cancellation with echocan cel=yes. echocancelwhenbridged=yes will enable echo cancellation for TDM bridged calls. While bridged calls should not require echo cancellation, this may improve call quality.

When echo cancellation is enabled, the echo canceller learns of echo on the line by listening for it throughout the duration of the call. Consequently, echo may be heard at the beginning of a call and lessen after a period of time. To avoid this situation, you can employ a method called *echo training*, which will mute the line briefly at the beginning of a call, and send a tone from which the amount of echo on the line can be determined. This allows Asterisk to deal with the echo more quickly. Echo training can be enabled with echotraining=yes.

Hardware Echo Cancellation

The most effective way to handle echo cancellation is not in software. If you are planning on deploying a good-quality system, spend the extra money and purchase cards for the system that have onboard hardware echo cancellation. These cards are a bit more expensive, but they quickly pay for themselves in terms of reduced load on the CPU, as well as reduced load on you due to fewer user complaints.

[§] Actually, the handset in any phone, be it traditional or VoIP, is an analog connection.

Asterisk and VoIP

It should come as no surprise that Asterisk loves to talk VoIP. But in order to do so, Asterisk needs to know which function it is to perform: that of client, server, or both. One of the most complex and often confusing concepts in Asterisk is the configuration of inbound and outbound authentication.

Users and Peers and Friends—Oh My!

Connections that authenticate to us, or that we authenticate, are defined in the *iax.conf* and *sip.conf* files as *users* and *peers*. Connections that do both may be defined as *friends*. When determining which way the authentication is occurring, it is always important to view the direction of the channels from Asterisk's viewpoint, as connections are accepted and created by the Asterisk server.

Users

A connection defined as a user is any system/user/endpoint that we allow to connect to us. Keep in mind that a user definition does not provide a method with which to call that user; the user type is used simply to create a channel for incoming calls.∥ A user definition will require a context name to be defined to indicate where the incoming authenticated call will enter the dialplan (in *extensions.conf*).

Peers

A connection defined as a peer type is an outgoing connection. Think of it this way: *users* place calls to us, while we place calls to our *peers*. Since peers do not place calls to us, a peer definition does not typically require the configuration of a context name. However, there is one exception: if calls that originate from your system are returned to your system in a loopback, the incoming calls (which originate from a SIP proxy, not a user agent) will be matched on the peer definition. The default context should handle these incoming calls appropriately, although it's preferable for contexts to be defined for them on a per-peer basis.

In order to know where to send a call to a host, we must know its location in relation to the Internet (that is, its IP address). The location of a peer may be defined either statically or dynamically. A dynamic peer is configured with host=dynamic under the peer definition heading. Because the IP address of a dynamic peer may change constantly, it must register with the Asterisk box so calls can successfully be routed to

---

∥ In SIP, this is not *always* the case. If the endpoint is a SIP proxy service (as opposed to a user agent), Asterisk will authenticate based on the peer definition, matching the IP address and port in the Contact field of the SIP header against the hostname (and port, if specified) defined for the peer (if the port is not specified, the one defined in the [general] section will be used).

it. If the remote end is another Asterisk box, the use of a `register` statement is required, as discussed in the next section.

Friends

Defining a type as a `friend` is a shortcut for defining it as both a `user` and a `peer`. However, connections that are both `users` and `peers` aren't always defined this way, because defining each direction of call creation individually (using both a `user` and a `peer` definition) allows more granularity and control over the individual connections.

Figure B-2 shows the flow of authentication control in relation to Asterisk.

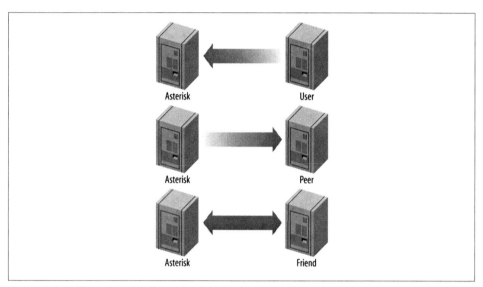

Figure B-2. Call origination relationships of users, peers, and friends to Asterisk

register Statements

A `register` statement is a way of telling a remote peer where your Asterisk box is in relation to the Internet. Asterisk uses `register` statements to authenticate to remote providers when you are employing a dynamic IP address, or when the provider does not have your IP address on record. There are situations when a `register` statement is not required, but to demonstrate when a `register` statement *is* required, let's look at an example.

Say you have a remote peer that is providing DID services to you. When someone calls the number +1-800-555-1212, the call goes over the physical PSTN network to your service provider and into its Asterisk server, possibly over its T1 connection. This call is then routed to your Asterisk server via the Internet.

Your service provider will have a definition in either its *sip.conf* or *iax.conf* configuration file (depending on whether you are connecting with the SIP or IAX protocol, respectively) for your Asterisk server. If you only receive calls from this provider, you will define it as a `user` (if it is another Asterisk system, you might be defined in its system as a `peer`).

Now let's say that your box is on your home Internet connection, with a dynamic IP address. Your service provider has a static IP address (or perhaps a fully qualified domain name), which you place in your configuration file. Since you have a dynamic address, your service provider specifies `host=dynamic` in its configuration file. In order to know where to route your +1-800-555-1212 call, your service provider needs to know where you are located in relation to the Internet. This is where the `register` statement comes into use.

The `register` statement is a way of authenticating and telling your `peer` where you are. In the [`general`] section of your configuration file, you place a statement similar to this:

```
register => username:secret@my_remote_peer
```

You can verify a successful registration with the use of the *iax2 show registry* and *sip show registry* commands at the Asterisk console.

VoIP Security

We can barely scratch the surface of the complex matter of VoIP security in this appendix; therefore, before we dig in, we want to steer you in the direction of the VoIP Security Alliance (*http://www.voipsa.org*). This fantastic resource contains an excellent mailing list, white papers, howtos, and a general compendium of all matters relating to VoIP security. Just as email has been abused by the selfish and criminal, so too will voice. The fine folks at VoIPSA are doing what they can to ensure that we address these challenges now, before they become an epidemic. In the realm of books on the subject, we recommend the most excellent *Hacking Exposed VoIP* by David Endler and Mark Collier (McGraw-Hill Osborne Media). If you are responsible for deploying any VoIP system, you need to be aware of this stuff.

Spam over Internet Telephony (SPIT)

We don't want to think about this, but we know it's coming. The simple fact is that there are people in this world who lack certain social skills, and that coupled with a kind of mindless greed, means that these folks think nothing of flooding the Internet with massive volumes of email. These same types of characters will think little of doing the same with voice. We already know what it's like to get inundated with telemarketing calls; try to imagine what might happen when those telemarketers realize they can send voice spam at almost no cost. Regulation has not stopped email spam, and it will probably not stop voice spam, so it will be up to us to prevent it.

Encrypting Audio with Secure RTP

If you can sniff the packets coming out of an Asterisk system, you can extract the audio from the RTP streams. This data can be fed offline to a speech processing system, which can listen for keywords such as "credit card number" or "PIN" and present the data it gathers to someone who has an interest in it. The stream can also be evaluated to see if there are DTMF tones embedded in it, which is dangerous because many services ask for passwords and credit card information to be input via the dialpad. In business, strategic information could also be gleaned from captured audio.

Using Secure RTP can combat this problem by encrypting the RTP streams. More information about SRTP is available in "Encrypting SIP calls" on page 150.

Spoofing

In the traditional telephone network, it is very difficult to successfully adopt someone else's identity. Your activities can (and will) be traced back to you, and the authorities will quickly put an end to the fun. In the world of IP, it is much easier to remain anonymous. As such, it is no stretch to imagine that there are hordes of enterprising criminals out there who will be only too happy to make calls to your credit card company or bank, pretending to be you. If a trusted mechanism is not discovered to combat spoofing, we will quickly learn that we cannot trust VoIP calls.

What Can Be Done?

The first thing to keep in mind when considering security on a VoIP system is that VoIP is based on network protocols, and needs be evaluated from that perspective. This is not to say that traditional telecom security should be ignored, but we need to pay attention to the underlying network.

Basic network security

One of the most effective things that can be done is to secure access to the voice network. The use of firewalls and VLANs are examples of how this can be achieved. By default, the voice network should be accessible only to those things that have a need. For example, if you do not have any softphones in use, do not allow client PCs access to the voice network.

Segregating voice and data traffic. Unless there is a need to have voice and data on the same network, there may be some value in keeping them separate (this can have other benefits as well, such as simplifying QoS configurations). It is not unheard of to build the internal voice network on a totally separate LAN, using existing CAT3 cabling and terminating on inexpensive network switches. This configuration can even be less expensive.

DMZ. Placing your VoIP system in a demilitarized zone (DMZ) can provide an additional layer of protection for your LAN, while still allowing connectivity for relevant applications. Should your VoIP system be compromised, it will be much more difficult to use it to launch an attack on the rest of your network, since it is not trusted. Regardless of whether you deploy within a DMZ, any abnormal traffic coming out of the system should be considered suspect.

Server hardening. Hardening your Asterisk server is critical. Not only are there performance benefits to doing this (running nonessential processes can eat up valuable CPU and RAM resources), but the elimination of anything not required will reduce the chance that an exploited vulnerability in the operating system can be used to gain access and launch an attack on other parts of your network.

Running Asterisk as non-*root* is an essential part of system hardening. See Chapter 3 for more information.

Encryption

Asterisk 1.8 includes the ability to use both SIP TLS for the encryption of signaling and SRTP for the encryption of the media between endpoints. More information about encrypting SIP calls can be found in "Encrypting SIP calls" on page 150. Asterisk has also supported encryption between endpoints using IAX2 since version 1.4). Information about enabling encryption across IAX2 trunks can be found in "IAX encryption" on page 154.

Physical security

Physical security should not be ignored. All terminating equipment (such as switches, routers, and the PBX itself) should be secured in an environment that can only be accessed by authorized persons. At the user end (such as under desks), it can be more difficult to deliver physical security, but if the network responds only to devices that it is familiar with (e.g., restricting DHCP to devices whose MAC addresses are known), the risk of unauthorized intrusions can be mitigated somewhat.

Conclusion

Over the last couple of years the telecom industry has embraced VoIP, which sets Asterisk up to do quite well. While Asterisk has been doing VoIP for years (well over a decade now), the integration of VoIP and traditional telephony into a single, powerful platform has made Asterisk a major player in the telecommunications industry.

Preparing a System for Asterisk

Very early on, I knew that someday in some "perfect" future out there over the horizon, it would be common- place for computers to handle all of the necessary pro- cessing functionality internally, making the necessary external hardware to connect up to telecom interfaces very inexpensive and, in some cases, trivial.

—Jim Dixon, "The History of Zapata Telephony and How It Relates to the Asterisk PBX"

By this point, you must be anxious to get your Asterisk system up and running. For a mission-critical deployment, however, some thought must be given to the environment in which the Asterisk system will run. Make no mistake: Asterisk, being a very flexible piece of software, will happily and successfully install on nearly any Linux platform you can conceive of, and several non-Linux platforms as well.* However, to arm you with an understanding of the type of operating environment Asterisk will really thrive in, this appendix will discuss issues you need to be aware of in order to deliver a reliable, well-designed system.

In terms of its resource requirements, Asterisk's needs are similar to those of an em- bedded, real-time application. This is due in large part to its need to have priority access to the processor and system buses. It is, therefore, imperative that any functions on the system not directly related to the call-processing tasks of Asterisk be run at a low pri- ority, if at all. On smaller systems and hobby systems, this might not be as much of an issue. However, on high-capacity systems, performance shortcomings will manifest as audio quality problems for users, often experienced as echo, static, and the like. The symptoms will resemble those experienced on a cell phone when going out of range,

---

* People have successfully compiled and run Asterisk on WRAP boards, Linksys WRT54G routers, Soekris systems, Pentium 100s, PDAs, Apple Macs, Sun SPARCs, laptops, and more. Of course, whether you would *want* to put such a system into production is another matter entirely. (Actually, the AstLinux distribution, by Kristian Kielhofner, runs very well indeed on the Soekris 4801 board. Once you've grasped the basics of Asterisk, this is something worth looking into further. Check out *http://www.astlinux.org*.)

although the underlying causes will be different. As loads increase, the system will have increasing difficulty maintaining connections. For a PBX, such a situation is nothing short of disastrous, so careful attention to performance requirements is a critical consideration during the platform selection process.

Table C-1 lists some very basic guidelines that you'll want to keep in mind when planning your system. The next section takes a close look at the various design and implementation issues that will affect its performance. Keep in mind that no guide can tell you exactly how many calls a server can handle. There are an incredibly high number of variables that can affect the answer to the question of how many calls Asterisk can handle. The only way to figure out how many calls a server can handle is to test it yourself in your own environment.

 The size of an Asterisk system is actually not dictated by the number of users or sets, but rather by the number of simultaneous calls it will be expected to support. These numbers are very conservative, so feel free to experiment and see what works for you.

Table C-1. System requirement guidelines

Purpose	Number of channels	Minimum recommended
Hobby system	No more than 5	400-MHz x86, 256 MB RAM
SOHO system (small office/home office—less than three lines and five sets)	5 to 10	1-GHz x86, 512 MB RAM
Small business system	Up to 25	3-GHz x86, 1 GB RAM
Medium to large system	More than 25	Dual CPUs, possibly also multiple servers in a distributed architecture

With large Asterisk installations, it is common to deploy functionality across several servers. One or more central units will be dedicated to call processing; these will be complemented by one or more ancillary servers handling peripherals (such as a database system, a voicemail system, a conferencing system, a management system, a web interface, a firewall, and so on). As is true in most Linux environments, Asterisk is well suited to growing with your needs: a small system that used to be able to handle all your call-processing and peripheral tasks can be distributed among several servers when increased demands exceed its abilities. Flexibility is a key reason why Asterisk is extremely cost-effective for rapidly growing businesses; there is no effective maximum or minimum size to consider when budgeting the initial purchase. While some scalability is possible with most telephone systems, we have yet to hear of one that can scale as flexibly as Asterisk. Having said that, distributed Asterisk systems are not simple to design—this is not a task for someone new to Asterisk.

If you are sure that you need to set up a distributed Asterisk system, you will want to study the DUNDi protocol, the Asterisk Realtime Architecture (ARA), func_odbc, and the various other database tools at your disposal. This will help you to abstract the data your system requires from the dialplan logic your Asterisk systems will utilize, creating a generic set of dialplan logic that can be used across multiple boxes. This in turn will enable you to scale more simply by adding additional boxes to the system. However, this is far beyond the scope of this book and will be left as an exercise for the reader. If you want a teaser of some tools you can use for scaling, see Chapter 22.

Server Hardware Selection

The selection of a server is both simple and complicated: simple because, really, any x86-based platform will suffice, but complicated because the reliable performance of your system will depend on the care that is put into the platform design. When selecting your hardware, you must carefully consider the overall design of your system and what functionality you need to support. This will help you determine your requirements for the CPU, motherboard, and power supply. If you are simply setting up your first Asterisk system for the purpose of learning, you can safely ignore the information in this section. If, however, you are building a mission-critical system suitable for deployment, these are issues that require some thought.

Performance Issues

Among other considerations, when selecting the hardware for an Asterisk installation you must bear in mind this critical question: how powerful must the system be? This is not an easy question to answer, because the manner in which the system is to be used will play a big role in the resources it will consume. There is no such thing as an Asterisk performance-engineering matrix, so you will need to understand how Asterisk uses the system in order to make intelligent decisions about what kinds of resources will be required. You will need to consider several factors, including:

The maximum number of concurrent connections the system will be expected to support
Each connection will increase the workload on the system.

The percentage of traffic that will require processor-intensive DSP of compressed codecs (such as G.729 and GSM)
The digital signal processing (DSP) work that Asterisk performs in software can have a staggering impact on the number of concurrent calls it will support. A system that might happily handle 50 concurrent G.711 calls could be brought to its knees by a request to conference together 10 G.729 compressed channels. We talk more about G.729, GSM, G.711, and many other codecs in Appendix B.

Whether conferencing will be provided, and what level of conferencing activity is expected
Will the system be used heavily? Conferencing requires the system to transcode and mix each individual incoming audio stream into multiple outgoing streams. Mixing multiple audio streams in near-real time can place a significant load on the CPU.

Echo cancellation
Echo cancellation may be required on any call where a Public Switched Telephone Network (PSTN) interface is involved. Since echo cancellation is a mathematical function, the more of it the system has to perform, the higher the load on the CPU will be.† Some telephony hardware vendors offer hardware-based echo cancellation to remove the burden of this task from the host CPU. Echo cancellation is discussed briefly later in this appendix and in more depth in Appendix B.

Dialplan scripting logic
Whenever Asterisk has to pass call control to an external program, there is a performance penalty. As much logic as possible should be built into the dialplan. If external scripts are used, they should be designed with performance and efficiency as critical considerations.

As for the exact performance impact of these factors, it's difficult to know for sure. The effect of each is known in general terms, but an accurate performance calculator has not yet been successfully defined. This is partly because the effect of each component of the system is dependent on numerous variables, such as CPU power, motherboard chipset and overall quality, total traffic load on the system, Linux kernel optimizations, network traffic, number and type of PSTN interfaces, and PSTN traffic—not to mention any non-Asterisk services the system is performing concurrently. Let's take a look at the effects of several key factors:

Codecs and transcoding
Simply put, a *codec* (short for coder/decoder, or compression/decompression) is a set of mathematical rules that define how an analog waveform will be digitized. The differences between the various codecs are due in large part to the levels of compression and quality that they offer. Generally speaking, the more compression that's required, the more work the DSP must do to code or decode the signal. Uncompressed codecs, therefore, put far less strain on the CPU (but require more network bandwidth). Codec selection must strike a balance between bandwidth and processor usage. For more on codecs, see Appendix B.

Central processing unit (and floating point unit)
A CPU is composed of several components, one of which is the floating point unit (FPU). The speed of the CPU, coupled with the efficiency of its FPU, will play a significant role in the number of concurrent connections a system can effectively

---

† Roughly 30 MHz of CPU power per channel.

support. The next section ("Choosing a Processor" on page 644) offers some general guidelines for choosing a CPU that will meet the needs of your system.

Other processes running concurrently on the system

Being Unix-like, Linux is designed to be able to multitask several different processes. A problem arises when one of those processes (such as Asterisk) demands a very high level of responsiveness from the system. By default, Linux will distribute resources fairly among every application that requests them. If you install a system with many different server applications, those applications will each be allowed their fair use of the CPU. Since Asterisk requires frequent high-priority access to the CPU, it does not get along well with other applications, and if Asterisk must coexist with other apps, the system may require special optimization. This primarily involves the assignment of priorities to various applications in the system and, during installation, careful attention to which applications are installed as services.

Kernel optimizations

A kernel optimized for the performance of one specific application is something that very few Linux distributions offer by default, and thus it requires some thought. At the very minimum—whichever distribution you choose—you should download and compile on your platform a fresh copy of the Linux kernel (available from *http://www.kernel.org*). You may also be able to acquire patches that will yield performance improvements, but these are considered hacks to the officially supported kernels.

IRQ latency

Interrupt request (IRQ) latency is basically the delay between the moment a peripheral card (such as a telephone interface card) requests the CPU to stop what it's doing and the moment when the CPU actually responds and is ready to handle the task. Asterisk's peripherals (especially the DAHDI cards) have historically been intolerant of IRQ latency, though there have been extensive improvements in DAHDI to help improve these issues. This is not due to any problem with the cards, but rather is part of the nature of how a software-based TDM engine has to work. If we buffer the TDM data and send it on the bus as a larger packet, that may be more efficient from a system perspective, but it will create a delay between the time the audio is received on the card, and when it is delivered to the CPU. This makes real-time processing of TDM data next to impossible. In the design of DAHDI, it was decided that sending the data every 1 ms would create the best tradeoff, but a side effect of this is that any card in the system that uses the DAHDI interface is going to ask the system to process an interrupt every millisecond. This used to be a factor on older motherboards, but it has largely ceased to be a cause for concern.

Linux has historically had problems with its ability to service IRQs quickly; this problem has caused enough trouble for audio developers that several patches have been created to address this shortcoming. So far, there has been some mild controversy over how to incorporate these patches into the Linux kernel.

Kernel version
Asterisk is officially supported on Linux version 2.6. Almost all of Asterisk itself does not really care about the kernel version, but DAHDI requires 2.6.

Linux distribution
Linux distributions are many and varied. Asterisk should work on all of them. Choose the one that you are most comfortable with.

Choosing a Processor

Since the performance demands of Asterisk will generally involve a large number of math calculations, it is essential that you select a processor with a powerful FPU. The signal processing that Asterisk performs can quickly demand a staggering quantity of complex mathematical computations from the CPU. The efficiency with which these tasks are carried out will be determined by the power of the FPU within the processor.

Actually naming a best processor for Asterisk in this book would fly in the face of Moore's Law. Even in the time between the authoring and publishing of this book, processor speeds will undergo rapid improvements, as will Asterisk's support for various architectures. Obviously, this is a good thing, but it also makes the giving of advice on the topic a thankless task. Naturally, the more powerful the FPU is, the more concurrent DSP tasks Asterisk will be able to handle, so that is the ultimate consideration. When you are selecting a processor, the raw clock speed is only part of the equation. How well it handles floating-point operations will be a key differentiator, as DSP operations in Asterisk will place a large demand on that process.

Both Intel and AMD CPUs have powerful FPUs. Current-generation chips from either of those manufacturers can be expected to perform well.[‡]

The obvious conclusion is that you should get the most powerful CPU your budget will allow. However, don't be too quick to buy the most expensive CPU out there. You'll need to keep the requirements of your system in mind; after all, a Formula 1 Ferrari is ill-suited to the rigors of rush-hour traffic. Slower CPUs will often run cooler, so you might be able to build a lower-powered, fanless Asterisk system for a small office, which could work well in a dusty environment, for example.

[‡] If you want to be completely up-to-the-minute on which CPUs are leading the performance race, surf on over to Tom's Hardware (*http://www.tomshardware.com*) or AnandTech (*http://www.anandtech.com*), where you will find a wealth of information about both current and out-of-date CPUs, motherboards, and chipsets.

To attempt to provide you with a frame of reference from which you can contemplate your platform decision, we have chosen to define three sizes of Asterisk systems: small, medium, and large.

Small systems

Small systems (up to 10 phones) are not immune to the performance requirements of Asterisk, but the typical load that will be placed on a smaller system will generally fall within the capabilities of a modern processor.

If you are building a small system from older components you have lying around, be aware that the resulting system cannot be expected to perform at the same level as a more powerful machine, and performance will begin to degrade under a much lighter load. Hobby systems can be run successfully on very low-powered hardware, although this is by no means recommended for anyone who is not a whiz at Linux performance tuning.§

If you are setting up an Asterisk system for the purposes of learning, you will be able to build a fully featured platform using a relatively low-powered CPU. The authors of this book run several Asterisk lab systems with 433-MHz to 700-MHz Celeron processors, but the workload of these systems is minimal (never more than two concurrent calls).

AstLinux and Asterisk on OpenWRT

If you are really comfortable working with Linux on embedded platforms, you will want to join the *AstLinux* mailing list and run Kristian Kielhofner's creation, AstLinux, or get yourself a Linksys WRT54GL and install Brian Capouch's version of Asterisk for that platform.

These projects strip Asterisk down to its essentials, and allow incredibly powerful PBX applications to be deployed on very inexpensive hardware.

While both projects require a fair amount of knowledge and effort on your part, they also share a huge coolness factor, are extremely popular, and are of excellent quality.

Medium systems

Medium-sized systems (from 10 to 50 phones) are where performance considerations will be the most challenging to resolve. Generally, these systems will be deployed on one or two servers only, and thus each machine will be required to handle more than one specific task. As loads increase, the limits of the platform will become increasingly

---

§ Greg Boehnlein once compiled and ran Asterisk on a 133-MHz Pentium system, but that was mostly as an experiment. Performance problems are far more likely in such conditions, and properly configuring such a system requires an expert knowledge of Linux. We do not recommend running Asterisk on anything less than a 500-MHz system (for a production system, 2 GHz might be a sensible minimum). Still, we think the fact that Asterisk is so flexible is remarkable.

stressed. Users may begin to perceive quality problems without realizing that the system is not faulty in any way, but simply exceeding its capacity. These problems will get progressively worse as more and more load is placed on the system, with the user experience degrading accordingly. It is critical that performance problems be identified and addressed before users notice them.

Monitoring performance on these systems and quickly acting on any developing trends is key to ensuring that a quality telephony platform is provided.

Large systems

Large systems (more than 120 channels) can be distributed across multiple systems and sites, and performance concerns can be managed through the addition of machines. Very large Asterisk systems have been created in this way.

Building a large system requires an advanced level of knowledge in many different disciplines. We will not discuss it in detail in this book, other than to say that the issues you'll encounter will be similar to those encountered during any deployment of multiple servers handling a single, distributed task.

Choosing a Motherboard

Just to get any anticipation out of the way, we also cannot recommend specific motherboards in this book. With new motherboards coming out on a weekly basis, any recommendations we could make would be rendered moot by obsolescence before the published copy hit the shelves. Not only that, but motherboards are like automobiles: while they are all very similar in principle, the difference is in the details. And as Asterisk is a performance application, the details matter.

What we will do, therefore, is give you some idea of the kinds of motherboards that can be expected to work well with Asterisk, and the features that will make for a good motherboard. The key is to have both stability and high performance. Here are some guidelines to follow:

- The various system buses must provide the minimum possible latency. If you are planning a PSTN connection using analog or PRI interfaces (discussed later in this appendix), having DAHDI cards in the system will generate 1,000 interrupt requests per second. Having devices on the bus that interfere with this process will result in degradation of call quality. Chipsets from Intel (for Intel CPUs) and nVidia nForce (for AMD CPUs) seem to score the best marks in this area. Review the specific chipset of any motherboard you are evaluating to ensure that it does not have known problems with IRQ latency.

- If you are running DAHDI cards in your system, you will want to ensure that your BIOS allows you maximum control over IRQ assignment. As a rule, high-end motherboards will offer far greater flexibility with respect to BIOS tweaking; value-priced boards will generally offer very little control. This may be a moot point, however, as APIC-enabled motherboards turn IRQ control over to the operating system.

- Server-class motherboards generally implement a different PCI standard than workstation-class motherboards. While there are many differences, the most obvious and well known is that the two versions have different voltages. Depending on which cards you purchase, you will need to know if you require 3.3V or 5V PCI slots.‖ Figure C-1 shows the visual differences between 3.3V and 5V slots. Most server motherboards will have both types, but workstations will typically have only the 5V version.

 There is some evidence that suggests connecting together two completely separate, single-CPU systems may provide far more benefits than simply using two processors in the same machine. You not only double your CPU power, but you also achieve a much better level of redundancy at a similar cost to a single-chassis, dual-CPU machine. Keep in mind, though, that a dual-server Asterisk solution will be more complex to design than a single-machine solution.

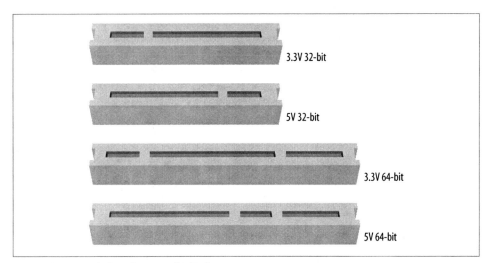

Figure C-1. Visual identification of PCI slots

- Consider using multiple processors, or processors with multiple cores. This will provide an improvement in the system's ability to handle multiple tasks. For Asterisk, this will be of special benefit in the area of floating-point operations.

‖ With the advent of PCI-X and PCI-Express, it is becoming harder and harder to select a motherboard with the correct type of slots. Be very certain that the motherboard you select has the correct type and quantity of card slots for your hardware. Keep in mind that most companies that produce hardware cards for Asterisk offer PCI and PCI-Express versions, but it's still up to you to make sure they make sense in whatever motherboard and chassis combination you choose.

- If you need a modem, install an external unit that connects to a serial port. If you must have an internal modem, you will need to ensure that it is not a so-called "Win-modem"—it must be a completely self-sufficient unit (note that these are very difficult, if not impossible, to find).

- Consider that with built-in networking, if you have a network component failure, the entire motherboard will need to be replaced. On the other hand, if you install a peripheral Network Interface Card (NIC), there may be an increased chance of failure due to the extra mechanical connections involved. It can also be useful to have separate network cards serving sets and users (the internal network) and VoIP providers and external sites (the external network). NICs are cheap; we suggest always having at least two.

- The stability and quality of your Asterisk system will be dependent on the components you select for its architecture. Asterisk is a beast, and it expects to be fed the best. As with just about anything, high cost is not always synonymous with quality, but you will want to become a connoisseur of computer components.

Having said all that, we need to get back to the original point: Asterisk can and will happily install on pretty much any system that will run Linux. The lab systems used to write this book, for example, included everything from a Linksys WRT to a dual-Xeon locomotive.# We have not experienced any performance or stability problems running less than five concurrent telephone connections. For the purposes of learning, do not be afraid to install Asterisk on whatever system you can scrounge up. When you are ready to put your system into production, however, you will need to understand the ramifications of the choices you make with respect to your hardware.

Power Supply Requirements

One often-overlooked component in a PC is the power supply (and the supply of power). For a telecommunications system,* these components can play a significant role in the quality of the user experience.

Computer power supplies

The power supply you select for your system will play a vital role in the stability of the entire platform. Asterisk is not a particularly power-hungry application, but anything relating to multimedia (whether it be telephony, professional audio, video, or the like) is generally sensitive to power *quality*.

#OK, it wasn't *actually* a locomotive, but it sure sounded like one. Does anyone know where to get quiet CPU fans for Xeon processors? It's getting too loud in the lab here.

* Or any system that is expected to process audio.

This oft-neglected component can turn an otherwise top-quality system into a poor performer. By the same token, a top-notch power supply might enable an otherwise cheap PC to perform like a champ.

The power supplied to a system must provide not only the energy a system needs to perform its tasks but also stable, clean signal lines for all of the voltages the system expects from it.

Spend the money and get a top-notch power supply (gamers are pretty passionate about this sort of thing, so there are lots of choices out there).

Redundant power supplies

In a carrier-grade or high-availability environment, it is common to deploy servers that use a redundant power supply. Essentially, this involves two completely independent power supplies, either one of which is capable of meeting the power requirements of the system.

If this is important to you, keep in mind that best practices suggest that to be properly redundant, these power supplies should be connected to completely independent uninterruptible power supplies (UPSs) that are in turn fed by totally separate electrical circuits. In truly mission-critical environments (such as hospitals), even the main electrical feeds into the building are redundant, and diesel-powered generators are on-site to generate electricity during extended power failures (such as the one that hit Northeastern North America on August 15, 2003).

Environment

Your system's environment consists of all of those factors that are not actually part of the server itself but nevertheless play a crucial role in the reliability and quality that can be expected from the system. Electrical supplies, room temperature and humidity, sources of interference, and security are all factors that should be contemplated.

Power Conditioning and Uninterruptible Power Supplies

When selecting the power sources for your system, consideration should be given not only to the amount of power the system will use, but also to the manner in which this power is delivered.

Power is not as simple as voltage coming from the outlet in the wall, and you should never just plug a production system into whatever electrical source is near at hand.[†] Giving some consideration to the supply of power to your system can ensure that you provide a far more stable power environment, leading to a far more stable system.

---

† Okay, look, you *can* plug it in wherever you'd like, and it'll probably work, but if your system has strange stability problems, please give this section another read. Deal?

One of the benefits of clean power is a reduction in heat, which means less stress on components, leading to a longer life expectancy.

Properly grounded, conditioned power feeding a premium-quality power supply will ensure a clean *logic ground* (a.k.a. 0-volt) reference[‡] for the system and keep electrical noise on the motherboard to a minimum. These are industry-standard best practices for this type of equipment, which should not be neglected. A relatively simple way to achieve this is through the use of a *power-conditioned* UPS.[§]

Power-conditioned UPSs

The UPS is well known for its role as a battery backup, but the power-conditioning benefits that high-end UPS units also provide are less well understood.

Power conditioning can provide a valuable level of protection from the electrical environment by regenerating clean power through an isolation transformer. A quality power conditioner in your UPS will eliminate most electrical noise from the power feed and help to ensure a rock-steady supply of power to your system.

Unfortunately, not all UPS units are created equal; many of the less expensive units do not provide clean power. What's worse, manufacturers of these devices will often promise all kinds of protection from surges, spikes, overvoltages, and transients. While such devices may protect your system from getting fried in an electrical storm, they will not clean up the power being fed to your system, and thus will do nothing to contribute to stability.

Make sure your UPS is *power conditioned*. If it doesn't say exactly that, it isn't.

Grounding

Voltage is defined as the difference in electrical potential between two points. When considering a *ground* (which is basically nothing more than an electrical path to earth), the common assumption is that it represents 0 volts. But if we do not define that 0V in *relation* to something, we are in danger of assuming things that may not be so. If you measure the voltage between two grounding references, you'll often find that there is a voltage potential between them. This voltage potential between grounding points can be significant enough to cause logic errors—or even damage—in a system where more than one path to ground is present.

[‡] In electronic devices, a binary zero (0) is generally related to a 0-volt signal, while a binary one (1) can be represented by many different voltages (commonly between 2.5 and 5 volts). The grounding reference that the system will consider 0 volts is often referred to as the *logic ground*. A poorly grounded system might have electrical potential on the logic ground to such a degree that the electronics mistake a binary zero for a binary one. This can wreak havoc with the system's ability to process instructions.

[§] It is a common misconception that all UPSs provide clean power. This is not at all true.

One of the authors recalls once frying a sound card he was trying to connect to a friend's stereo system. Even though both the computer and the stereo were in the same room, more than 6 volts of difference was measured between the ground conductors of the two electrical outlets they were plugged into! The wire between the stereo and the PC (by way of the sound card) provided a path that the voltage eagerly followed, thus frying a sound card that was not designed to handle that much current on its signal leads. Connecting both the PC and the stereo to the same outlet fixed the problem.

When considering electrical regulations, the purpose of a ground is primarily human safety. In a computer, the ground is used as a 0V logic reference. An electrical system that provides proper safety will not always provide a proper logic reference—in fact, the goals of safety and power quality are sometimes in disagreement. Naturally, when a choice must be made, safety has to take precedence.

 Since the difference between a binary zero and a binary one is represented in computers by voltage differences of sometimes less than 3V, it is entirely possible for unstable power conditions caused by poor grounding or electrical noise to cause all kinds of intermittent system problems. Some power and grounding advocates estimate that more than 80 percent of unexplained computer glitches can be traced to power quality. Most of us blame Microsoft.

Modern switching power supplies are somewhat isolated from power quality issues, but any high-performance system will always benefit from a well-designed power environment. In mainframes, proprietary PBXs, and other expensive computing platforms, the grounding of the system is never left to chance. The electronics and frames of these systems are always provided with a dedicated ground that does not depend on the safety grounds supplied with the electrical feed.

Regardless of how much you are willing to invest in grounding, when you specify the electrical supply to any PBX, ensure that the electrical circuit is completely dedicated to your system (as discussed in the next section) and that an insulated, isolated grounding conductor is provided. This can be expensive to provision, but it will contribute greatly to a quality power environment for your system.‖

‖ On a hobby system, this is probably too much to ask, but if you are planning on using Asterisk for anything important, at least be sure to give it a fighting chance; don't put anything like air conditioners, photocopiers, laser printers, or motors on the same circuit. The strain such items place on your power supply will shorten its life expectancy.

It is also vital that each and every peripheral you connect to your system be connected to the same electrical receptacle (or, more specifically, the same ground reference). This will cut down on the occurrence of ground loops, which can cause anything from buzzing and humming noises to damaged or destroyed equipment.

Electrical Circuits

If you've ever seen the lights dim when an electrical appliance kicks in, you've seen the effect that a high-energy device can have on an electrical circuit. If you were to look at the effects of a multitude of such devices, each drawing power in its own way, you would see that the harmonically perfect 50- or 60-Hz sine wave you may think you're getting with your power is anything but. Harmonic noise is extremely common on electrical circuits , and it can wreak havoc on sensitive electronic equipment. For a PBX, these problems can manifest as audio problems, logic errors, and system instability.

Ideally, you should never install a server on an electrical circuit that is shared with other devices. There should be only one outlet on the circuit, and you should connect only your telephone system (and associated peripherals) to it. The wire (including the ground) should be run unbroken directly back to the electrical panel. The grounding conductor should be insulated and isolated. There are far too many stories of photocopiers, air conditioners, and vacuum cleaners wreaking havoc with sensitive electronics to ignore this rule of thumb.

The electrical regulations in your area must always take precedence over any ideas presented here. If in doubt, consult a power quality expert in your area on how to ensure that you adhere to electrical regulations. Remember, electrical regulations take into account the fact that human safety is far more important than the safety of the equipment.

The Equipment Room

Environmental conditions can wreak havoc on systems, yet it is quite common to see critical systems deployed with little or no attention given to these matters. When the system is installed, everything works well, but after as little as six months, components begin to fail. Talk to anyone with experience in maintaining servers and systems, and it becomes obvious that attention to environmental factors can play a significant role in the stability and reliability of systems.

Humidity

Simply put, humidity is water in the air. Water is a disaster for electronics for two main reasons: 1) water is a catalyst for corrosion, and 2) water is conductive enough that it can cause short circuits. Do not install any electronic equipment in areas of high humidity without providing a means to remove the moisture.

Temperature

Heat is the enemy of electronics. The cooler you keep your system, the more reliably it will perform, and the longer it will last. If you cannot provide a properly cooled room for your system, at a minimum ensure that it is placed in a location that ensures a steady supply of clean, cool air. Also, keep the temperature steady. Changes in temperature can lead to condensation and other damaging changes.

Dust

An old adage in the computer industry holds that dust bunnies inside of a computer are lucky. Let's consider some of the realities of dust bunnies:

- Significant buildup of dust can restrict airflow inside the system, leading to increased levels of heat.
- Dust can contain metal particles, which, in sufficient quantities, can contribute to signal degradation or shorts on circuit boards.

Put critical servers in a filtered environment, and clean out dust bunnies regularly.

Security

Server security naturally involves protecting against network-originated intrusions, but the environment also plays a part in the security of a system. Telephone equipment should always be locked away, and only persons who have a need to access the equipment should be allowed near it.

Telephony Hardware

If you are going to connect Asterisk to any traditional telecommunications equipment, you will need the correct hardware. The hardware you require will be determined by what it is you want to achieve.

Connecting to the PSTN

Asterisk allows you to seamlessly bridge circuit-switched telecommunications networks# with packet-switched data networks.*

#Often called *TDM networks*, due to the time division multiplexing used to carry traffic through the PSTN.

* Popularly called VoIP networks, although Voice over IP is not the only method of transmitting voice over packet networks (Voice over Frame Relay was very popular in the late 1990s).

Because of Asterisk's open architecture (and open source code), it is ultimately possible to connect any standards-compliant interface hardware. The selection of open source telephony interface boards is currently limited, but as interest in Asterisk grows, that will rapidly change.[†] At the moment, one of the most popular and cost-effective ways to connect to the PSTN is to use the interface cards that evolved from the work of the Zapata Telephony Project (*http://www.zapatatelephony.org*), which has evolved into DAHDI.

Analog interface cards

Unless you need a lot of channels (or a have lot of money to spend each month on telecommunications facilities), chances are that your PSTN interface will consist of one or more analog circuits, each of which will require a Foreign eXchange Office (FXO) port.

Digium, the company that sponsors Asterisk development, produces analog interface cards for Asterisk. Check out its website (*http://www.digium.com*) for details on its extensive line of analog cards, including the venerable TDM400P, the latest TDM800P, and the high-density TDM2400P. As an example, the TDM800P is an eight-port base card that allows for the insertion of up to two daughter cards, which each deliver either four FXO or four FXS ports.[‡] The TDM800P can be purchased with these modules preinstalled, and a hardware echo-canceller can be added as well.

Other companies that produce Asterisk-compatible analog cards include:

- Rhino (*http://www.rhinoequipment.com*)
- Sangoma (*http://www.sangoma.com*)
- Voicetronix (*http://www.voicetronix.com*)
- Pika Technologies (*http://www.pikatechnologies.com*)

Digital interface cards

If you require more than 10 circuits, or require digital connectivity, chances are you're going to be in the market for a T1 or E1 card.[§] Bear in mind, though, that the monthly charges for a digital PSTN circuit vary widely. In some places, as few as five circuits can justify a digital circuit; in others, the technology may never be cost-justifiable. The more competition there is in your area, the better chance you have of finding a good deal. Be sure to shop around.

[†] The evolution of inexpensive, commodity-based telephony hardware is only slightly behind the telephony software revolution. New companies spring up on a weekly basis, each one bringing new and inexpensive standards-based devices into the market.

[‡] FXS and FXO refer to the opposing ends of an analog circuit. Which one you need will be determined by what you want to connect to. Appendix A discusses these in more detail.

[§] T1 and E1 are digital telephony circuits. We discuss them further in Appendix A.

The Zapata Telephony Project originally produced a T1 card, the Tormenta, that is the ancestor of most Asterisk-compatible T1 cards. The original Tormenta cards are now considered obsolete, but they do still work with Asterisk.

Digium makes several different digital circuit interface cards. The features on the cards are the same; the primary differences are whether they provide T1 or E1 interfaces, and how many spans each card provides. Digium has been producing DAHDI cards for Linux longer than anyone else; it was deeply involved with the development of DAHDI (formerly Zaptel) on Linux, and has been the driving force behind DAHDI development over the years.

Sangoma, which has been producing open source WAN cards for many years, added Asterisk support for its T1/E1 cards a few years ago.‖ Rhino has had T1 hardware for Asterisk for a while now, and there are many other companies that offer digital interface cards for Asterisk as well.

Channel banks

A *channel bank* is loosely defined as a device that allows a digital circuit to be de-multiplexed into several analog circuits (and vice versa). More specifically, a channel bank lets you connect analog telephones and lines into a system across a T1 line. Figure C-2 shows how a channel bank fits into a typical office phone system.

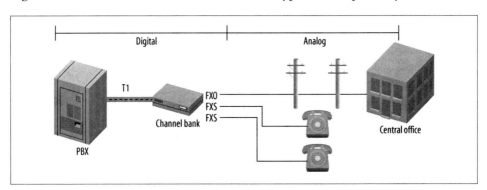

Figure C-2. One way you might connect a channel bank

Although they can be expensive to purchase, many people feel very strongly that the only proper way to integrate analog circuits and devices into Asterisk is through a channel bank. Whether that is true or not depends on a lot of factors, but if you have the budget, they can be very useful.# You can often pick up used channel banks on

---

‖ It should be noted that a Sangoma Frame Relay card played a role in the original development of Asterisk (see *http://linuxdevices.com/articles/AT8678310302.html*); Sangoma has a long history of supporting open source WAN interfaces with Linux.

#We use channel banks to simulate a central office. One 24-port channel bank off an Asterisk system can provide up to 24 analog lines—perfect for a classroom or lab.

eBay. Look for units from Adtran and Carrier Access Corp. (Rhino makes great channel banks, and they are very competitively priced, but they may be hard to find used.) Don't forget that you will need a T1 card in order to connect a channel bank to Asterisk.

Other types of PSTN interfaces

Many VoIP gateways exist that can be configured to provide access to PSTN circuits. Generally speaking, these will be of most use in a smaller system (one or two lines). They can also be very complicated to configure, as grasping the interaction between the various networks and devices requires a solid understanding of both telephony and VoIP fundamentals. For that reason, we will not discuss these devices in detail in this book. They are worth looking into, however; popular units are made by Sipura, Grandstream, Digium, and many other companies.

Another way to connect to the PSTN is through the use of Basic Rate Interface (BRI) ISDN circuits. BRI is a digital telecom standard that specifies a two-channel circuit that can carry up to 144 Kbps of traffic.* Due to the variety of ways this technology has been implemented, and a lack of testing equipment, we will not be discussing BRI in very much detail in this book.

Connecting Exclusively to a Packet-Based Telephone Network

If you do not need to connect to the PSTN, Asterisk requires no hardware other than a server with a Network Interface Card. However, you still may need to install the DAHDI kernel modules, as DAHDI is required for using the MeetMe() application for conferencing.

Echo Cancellation

One of the issues that can arise if you use analog interfaces on a VoIP system is echo. *Echo* is simply what you say being reflected back to you a short time later. The echo is caused by the far end, but you are the one that hears it. It is a little-known fact that echo would be a massive problem in the PSTN were it not for the fact that the carriers employ complex (and expensive) strategies to eliminate it. We suggest that you consider adding echo-cancellation hardware to any card you purchase for use as a PSTN interface. While Asterisk can do some work with echo in software, it does not provide nearly enough power to deal with the problem. Also, echo cancellation in software imposes a load on the processor; hardware echo cancellers built into the PSTN card take this burden away from the CPU.

Hardware echo cancellation can add several hundred dollars to your equipment cost, but if you are serious about having a quality system, invest the extra money now instead

---

* BRI is very rarely used in North America but is very popular in Europe, and Digium has produced the B410P card to address this need.

of suffering later. Echo problems are not pleasant at all, and your users will hate the system if they experience it.

Several software echo cancellers have recently become available. We have not had a chance to evaluate any of them, but we know that they employ the same algorithms the hardware echo cancellers do. If you have a recently purchased Digium analog card, you can call Digium sales for a keycode to allow its latest software echo canceller to work with your system.† There are other software options available for other types of cards, but you will have to look into whether you have to purchase a license to use them.‡ Keep in mind that there is a performance cost to using software echo cancellers. They will place a measurable load on the CPU that needs to be taken into account when you design a system using these technologies.

 For more on the topic of echo cancellation, see Appendix B.

Types of Phones

We all know what a telephone is—but will it be the same five years from now? Part of the revolution that Asterisk is contributing to is the evolution of the telephone, from a simple audio communications device into a multimedia communications terminal providing all kinds of yet-to-be-imagined functions.

As an introduction to this exciting concept, we will briefly discuss the various kinds of devices we currently call "telephones" (any of which can easily be integrated with Asterisk). We will also discuss some ideas about what these devices may evolve into in the future (devices that will also easily integrate with Asterisk).

Physical Telephones

Any physical device whose primary purpose is terminating an on-demand audio communications circuit between two points can be classified as a physical telephone. At a minimum, such a device has a handset and a dial pad; it may also have feature keys, a display screen, and various audio interfaces.

---

† This software is not part of a normal Asterisk download because Digium has to pay to license it separately. Nevertheless, it has grandfathered it into all of its cards, so it is available for free to anyone who has a Digium analog card that is still under warranty. If you are running a non-Digium analog card, you can purchase a keycode for this software echo canceller from Digium's website.

‡ Sangoma also offers free software echo cancellation on its analog cards (up to six channels).

This section takes a brief look at the various user (or endpoint) devices you might want to connect to your Asterisk system. We delve more deeply into the mechanics of analog and digital telephony in Appendix A.

Analog telephones

Analog phones have been around since the invention of the telephone. Up until about 20 years ago, all telephones were analog. Although analog phones have some technical differences in different countries, they all operate on similar principles.

When a human being speaks, the vocal cords, tongue, teeth, and lips create a complex variety of sounds. The purpose of the telephone is to capture these sounds and convert them into a format suitable for transmission over wires. In an analog telephone, the transmitted signal is *analogous* to the sound waves produced by the person speaking. If you could see the sound waves passing from the mouth to the microphone, they would be proportional to the electrical signal you could measure on the wire.

Analog telephones are the only kind of phones that are commonly available in any retail electronics store. In the next few years, that can be expected to change dramatically.

Proprietary digital telephones

As digital switching systems developed in the 1980s and 1990s, telecommunications companies developed digital private branch exchanges (PBXs) and key telephone systems (KTSs). The proprietary telephones developed for these systems were completely dependent on the systems to which they were connected and could not be used on any other systems. Even phones produced by the same manufacturer were not cross-compatible (for example, a Nortel Norstar set will not work on a Nortel Meridian 1 PBX). The proprietary nature of digital telephones limits their future. In this emerging era of standards-based communications, they will quickly be relegated to the dustbin of history.

The handset in a digital telephone is generally identical in function to the handset in an analog telephone, and they are often compatible with each other. Where the digital phone is different is that inside the telephone, the analog signal is sampled and converted into a digital signal—that is, a numerical representation of the analog waveform. We discuss digital signals in more detail in Appendix A; for now, suffice it to say that the primary advantage of a digital signal is that it can be transmitted over limitless distances with no loss of signal quality.

The chances of anyone ever making a proprietary digital phone directly compatible with Asterisk are slim, but companies such as Citel (*http://www.citel.com*)§ have created gateways that convert the proprietary signals to Session Initiation Protocol (SIP).‖

ISDN telephones

Prior to VoIP, the closest thing to a standards-based digital telephone was an ISDN-BRI terminal. Developed in the early 1980s, ISDN was expected to revolutionize the telecommunications industry in exactly the same way that VoIP promises to finally achieve today.

> There are two types of ISDN: *Primary Rate Interface* (PRI) and *Basic Rate Interface* (BRI). PRI is commonly used to provide trunking facilities between PBXs and the PSTN, and is widely deployed all over the world. BRI is not at all popular in North America, but is common in Europe.

While ISDN was widely deployed by the telephone companies, many consider the standard to have been a flop, as it generally failed to live up to its promises. The high costs of implementation, recurring charges, and lack of cooperation among the major industry players contributed to an environment that caused more problems than it solved.

BRI was intended to service terminal devices and smaller sites (a BRI loop provides two digital circuits). A wealth of BRI devices have been developed, but BRI has largely been deprecated in favor of faster, less expensive technologies such as ADSL, cable modems, and VoIP.

BRI is still very popular for use in videoconferencing equipment, as it provides a fixed-bandwidth link. Also, BRI does not have the type of quality of service issues a VoIP connection might, as it is circuit-switched.

BRI is still sometimes used in place of analog circuits to provide trunking to a PBX. Whether or not this is a good idea depends mostly on how your local phone company prices the service, and what features it is willing to provide.#

---

§ Citel has produced a fantastic product, but it is limited by the fact that it is too expensive. If you have old proprietary PBX telephones, and you want to use them with your Asterisk system, Citel's technology can do the job, but make sure you understand how the per-port cost of these units stacks up against replacing the old sets with pure VoIP telephones.

‖ SIP is currently the most well-known and popular protocol for VoIP. We discuss it further in Appendix B.

If you are in North America, give up on this idea, unless you have a lot of patience and money and are a bit of a masochist.

IP telephones

IP telephones are heralds of the most exciting change in the telecommunications industry. Already, standards-based IP telephones are available in retail stores. The wealth of possibilities inherent in these devices will cause an explosion of interesting applications, from video phones to high-fidelity broadcasting devices to wireless mobility solutions to purpose-built sets for particular industries to flexible all-in-one multimedia systems.

The revolution that IP telephones will spawn has nothing to do with a new type of wire to connect your phone to, and everything to do with giving you the power to communicate the way you want.

The early-model IP phones that have been available for several years now do not represent the future of these exciting appliances. They are merely a stepping-stone, a familiar package in which to wrap a fantastic new way of thinking.

The future is far more promising.

Softphones

A *softphone* is a software program that provides telephone functionality on a non-telephone device, such as a PC or PDA. So how do we recognize such a beast? What might at first glance seem a simple question actually raises many. A softphone should probably have some sort of dial pad, and it should provide an interface that reminds users of a telephone. But will this always be the case?

The term *softphone* can be expected to evolve rapidly, as our concept of what exactly a telephone is undergoes a revolutionary metamorphosis.* As an example of this evolution, consider the following: would we correctly define popular communication programs such as Instant Messenger as softphones? IM provides the ability to initiate and receive standards-based VoIP connections. Does this not qualify it as a softphone? Answering that question requires knowledge of the future that we do not yet possess. Suffice it to say that, while at this point in time softphones are expected to look and sound like traditional phones, that conception is likely to change in the very near future.

As standards evolve and we move away from the traditional telephone and toward a multimedia communications culture, the line between softphones and physical telephones will become blurred indeed. For example, we might purchase a communications terminal to serve as a telephone and install a softphone program onto it to provide the functions we desire.

---

* Ever heard of Skype?

Having thus muddied the waters, the best we can do at this point is to define what the term *softphone* will refer to in relation to this book, with the understanding that the meaning of the term can be expected to undergo a massive change over the next few years. For our purposes, we will define a softphone as any device that runs on a personal computer, presents the look and feel of a telephone, and provides as its primary function the ability to make and receive full-duplex audio communications (formerly known as "phone calls")[†] through E.164 addressing.[‡]

Telephony Adaptors

A *telephony adaptor* (usually referred to as an ATA, or Analog Terminal Adaptor) can loosely be described as an end-user device that converts communications circuits from one protocol to another. Most commonly, these devices are used to convert from some digital (IP or proprietary) signal to an analog connection that you can plug a standard telephone or fax machine into.

These adaptors could be described as gateways, for that is their function. However, popular usage of the term *telephony gateway* would probably best describe a multiport telephony adaptor, generally with more complicated routing functions.

Telephony adaptors will be with us for as long as there is a need to connect incompatible standards and old devices to new networks. Eventually, our reliance on these devices will disappear, as did our reliance on the modem—obsolescence through irrelevance.

Communications Terminals

Communications terminal is an old term that disappeared for a decade or two and is being reintroduced here, very possibly for no other reason than that it needs to be discussed so that it can eventually disappear again—once it becomes ubiquitous.

First, a little history. When digital PBX systems were first released, manufacturers of these machines realized that they could not refer to their endpoints as telephones—their proprietary nature prevented them from connecting to the PSTN. They were therefore called *terminals*, or *stations*. Users, of course, weren't having any of it. It looked like a telephone and acted like a telephone, and therefore it *was* a telephone. You will still occasionally find PBX sets referred to as terminals, but for the most part they are called telephones.

The renewed relevance of the term *communications terminal* has nothing to do with anything proprietary—rather, it's the opposite. As we develop more creative ways of communicating with each other, we gain access to many different devices that will allow us to connect. Consider the following scenarios:

---

† OK, so you think you know what a phone call is? So did we. Let's just wait a few years, shall we?

‡ E.164 is the ITU standard that defines how phone numbers are assigned. If you've used a telephone, you've used E.164 addressing.

- If I use my PDA to connect to my voicemail and retrieve my voice messages (converted to text), does my PDA become a phone?
- If I attach a video camera to my PC, connect to a company's website, and request a live chat with a customer service rep, is my PC now a telephone?
- If I use the IP phone in my kitchen to surf for recipes, is that a phone call?

The point is simply this: we'll probably always be "phoning" each other, but will we always use "telephones" to do so?

Linux Considerations

If you ask anyone at the Free Software Foundation, they will tell you that what we know as Linux is in fact GNU/Linux. All etymological arguments aside, there is some valuable truth to this statement. While the kernel of the operating system is indeed Linux, the vast majority of the utilities installed on a Linux system and used regularly are in fact GNU utilities. "Linux" is probably only 5 percent Linux, possibly 75 percent GNU, and perhaps 20 percent everything else.

Why does this matter? Well, the flexibility of Linux is both a blessing and a curse. It is a blessing because with Linux you can truly craft your very own operating system from scratch. Since very few people ever do this, the curse is in large part due to the responsibility you must bear in determining which of the GNU utilities to install, and how to configure the system.

Conclusion

In this appendix, we've discussed all manner of issues that can contribute to the stability and quality of an Asterisk installation. How much time and effort you should devote to following the best practices and engineering tips in this appendix all depends on how much work you expect the Asterisk server to perform, and how much quality and reliability your system must provide. If you are experimenting with Asterisk, don't worry too much; just be aware that any problems you have may not be the fault of the Asterisk system.

What we have attempted to do in this appendix is give you a feel for the kinds of best practices that will help to ensure that your Asterisk system will be built on a reliable, stable platform. Asterisk is quite willing to operate under far worse conditions, but the amount of effort and consideration you decide to give these matters will play a part in the stability of your PBX. Your decision should depend on how critical your Asterisk system will be.

Index

Symbols

! (exclamation mark), in section name, 92
(hash symbol)
    comment, 188
    delimiter between map names, 226
$[] (dollar sign square brackets)
    Asterisk expressions, 195
${DIALSTATUS} variable, 170
${eventextra} CEL variable, 544
${eventtime} CEL variable, 544
${eventtype} CEL variable, 544
${EXTEN} channel variable, 128, 240
${IPADDR} option (dundi.conf), 506
${NUMBER} option (dundi.conf), 506
${SECRET} option (dundi.conf), 506
${SECRET} variable, 514
* (asterisk), Asterisk character separator, 251
* logger.conf type, 525
, (comma), voicemail.conf, 418
/var mount point, 36
3WAY_END event, 538
3WAY_START event, 538
9, accessing external lines, 132
=> (same) operator, 112
[] (square brackets) contexts, 108
_ (underscore), pattern matching, 125
| (pipe character)
    delimiter, 167
    support for, 113
    voicemail.conf, 418

A

a(folder) (VoiceMailMain() application), 171
A(x) (Page() application), 230
A2Billing, 564
AA (Automated Attendant), 331–340
    building, 336–340
        dialplan, 338
        incoming calls, 339
        recording prompts, 336
    compared to an IVR, 331
    designing for you, 332–336
        dial by extension, 336
        greeting, 333
        invalid handler, 335
        main menu, 334
        timeout, 335
Aastra, SIP-based paging, 232
ABANDON event, 298
accent of prompts, internationalization, 190
acceptdtmf option (agents.conf), 281
accountcode CDR field, 527
accountcode CEL event field, 539
accountlogs option (cdr.conf), 532
accounts
    connecting to the AMI over HTTP, 459
    scanning for valid accounts, 565
    XMPP accounts, 316
ACD queues, 261–299
    agents.conf, 281
    announcement control, 287–291
    changing penalties dynamically, 285
    local channels, 293–296
    overflow, 291
    priority queue, 283
    queue member priority, 284
    queue members, 266–274
    queues.conf, 275–280
    simple ACD queue, 262–266

We'd like to hear your suggestions for improving our indexes. Send email to *index@oreilly.com*.

announce-position option (queues.conf), 278, 288

announce-position-limit option (queues.conf), 278, 288

announce-round-seconds option (queues.conf), 278, 288

announcement control, 287–291

ANSWER AGI command, 483

answer CDR field, 528

ANSWER CEL event type, 537

Answer() application, 113

anti-monopolistic practices, 593

aoc option (manager.conf), 464

APIs, securing Asterisk network APIs, 572

app dialplan applications, 12–15

appdata CEL event field, 539

application map grouping, 227

[applicationmap] section, features.conf, 225

applications
    AddQueueMember() application, 268, 285, 295
    AGI() application, 477
    dahdi_genconf application, 137
    Dial() application, 207, 211, 434
    dialplan applications, 529, 540
    dialplan syntax, 113
    Directory() application, 171
    DISA() application, 324, 325
    Festival application, 434, 440
    GoSub() dialplan application, 207–211
    GotoIf() application, 199, 201
    GotoIfTime() application, 202
    Hangup() application, 200
    JabberSend() dialplan application, 421
    Macro() application, 205
    MeetMe() application, 218, 319
    MeetMeCount() application, 219
    NoOp() dialplan application, 516
    Page() application, 229, 231, 235
    PauseQueueMember() application, 268
    Playback() application, 337
    prompt-recording application, 394
    Queue() application, 291, 294, 499
    Read() application, 390
    Record() application, 337
    RemoveQueueMember() application, 268, 295
    SendFAX() dialplan application, 450
    Set() application, 198

SIPAddHeader() voicemail application, 176

SLA applications, 318

SLATrunk() application, 321, 326

text2wave application, 442

UnpauseQueueMember() application, 268

VoiceMail() application, 292

Zapateller() application, 217

appname CEL event field, 539

apps option (cel.conf), 540

APP_END CEL event type, 538

app_mysql addon module, 23

app_saycountpl addon module, 23

app_set (asterisk.conf), 75

APP_START CEL event type, 538

app_voicemail.so module, 162

ARA (Asterisk Realtime Architecture), 368–375
    dynamic realtime, 371–375
    static realtime, 368

architecture, 9–28
    dialplan, 25
    file structure, 24–25
        configuration files, 24
        logging, 25
        modules, 24
        resource library, 25
        spool, 25
    hardware requirements, 26
    modules, 10–24
        addon modules, 23
        bridging modules, 15
        CDR modules, 15
        channel drivers, 17
        channel event logging modules, 16
        codec translators, 18
        dialplan applications, 12–15
        dialplan functions, 19
        format interpreters, 18
        PBX modules, 21
        resource modules, 21
        test modules, 24
    versioning, 26

arguments
    using in GoSub() subroutines, 209
    using in macros, 206

ARRAY() function, 362

astagidir (asterisk.conf), 72

astctl (asterisk.conf), 75

astctlgroup (asterisk.conf), 75

G

membermacro option (queues.conf), 278
members
    penalizing, 284
menuselect, 59–64
    about, 59
    interfaces, 60
    scripting, 63
    using, 61
menuselect command, 76
menuselect system, 59
menuselect.makeopts, 64
message encoding, AMI, 466
messages, voice messages, 172
messagewrap option (voicemail.conf), 165
messaging, unified messaging, 590
metrics, static realtime, 370
MFC/R2 protocol, configuring, 141
MG2, 101
MGCP (Media Gateway Control Protocol),
        624
Microsoft SQL, configuring ODBC for, 350
military option (voicemail.conf), 166
min-announce-frequency option
        (queues.conf), 278, 288
minmemfree (asterisk.conf), 74
minpassword option (voicemail.conf), 165
minsecs option (voicemail.conf), 159
mobile option (dundi.conf), 506
model option (dundi.conf), 507
module reload app_queue.so command, 283,
        287
module reload cdr_adaptive_odbc.so function,
        376
modules, 10–24
    addon modules, 23
    bridging modules, 15
    CDR modules, 15
    channel drivers, 17
    channel event logging modules, 16
    codec translators, 18
    dialplan applications, 12–15
    dialplan functions, 19
    file structure, 24
    format interpreters, 18
    PBX modules, 21
    resource modules, 21
    test modules, 24
[modules] section (modules.conf), 76
modules.conf, 56, 75

moh show classes command, 371
monitor-format option (queues.conf), 279
monitor-type option (queues.conf), 275, 279
months (GotoIfTime() application), 203
motherboards, requirements, 646
moveheard option (voicemail.conf), 159
MP3 codec, 629
MP3 format, 79
MPLS (Multiprotocol Label Switching), 631
multicast paging
    Cisco SPA Telephones, 234
    MulticastRTP channel, 233
multiplelogin option (agents.conf), 281
Multiprotocol Label Switching (MPLS), 631
multirow Functionality with func_odbc, 360
music on hold, licensing, 79
music, format, 79
musicclass option (queues.conf), 264, 275,
        288
musiconhold option (agents.conf), 282
musiconhold.conf, 58, 79
mxml encoding type, 470
MySQL
    configuration, 345
    configuring ODBC for, 349
    installing CentOS, 343
    installing for Ubuntu, 343
    verifying binary data, 386

N

n (Page() application), 230
name-to-extension mapping, 244
naming
    contexts, 144
    extensions, 110
    macros, 205
    phones, 85
    variables, 123
NANP (North American Number Plan)
    about, 248
    ENUM, 249
    toll fraud, 127
NAPTR records, 250, 253
NAT (Network Address Translation)
    H.323, 624
    IAX, 620
    SIP, 622
nat (sip.conf), 93
network APIs, security, 572

P

P (VoiceMail() application), 170
p (VoiceMailMain() application), 171
packages, Asterisk packages, 30
packet-switched networks
    about, 615
    hardware requirements, 656
Page() application, 231, 235
Page() command, 234
PAGELIST variable, 235
pagerbody option (voicemail.conf), 161
pagerdateformat option (voicemail.conf), 161
pagerfromstring option (voicemail.conf), 161
pagersubject option (voicemail.conf), 161
paging, 229–236
    places to send your pages, 230–235
        combination paging, 234
        external paging, 230
        multicast paging via the MulticastRTP
            channel, 233
        set-based paging, 231
        VoIP paging adaptors, 234
    zone paging, 235
paradigm shift, telephony, 580
parameters, configuring Asterisk, 88
Park:<exten@context> virtual device, 302
parkcall (features.conf), 225
parked calls, timed-out parked calls, 224
parkedcallhangup (features.conf), 223
parkedcallrecording (features.conf), 223
parkedcallreparking (features.conf), 222
parkedcalltransfers (features.conf), 222
parkeddynamic (features.conf), 223
parkedmusicclass (features.conf), 223
parkedplay (features.conf), 222
parkext (features.conf), 217, 222
parking
    call parking with dialplan, 217
    features.conf, 221–229
        application map grouping, 227
        [applicationmap] section, 225
        [featuremap] section, 225
        [general] section, 222–224
        parking lots, 228
parkinghints (features.conf), 222
parkingtime (features.conf), 217, 222
parkpos (features.conf), 217, 222
PARK_END CEL event type, 538
PARK_START CEL event type, 538

parsing files, 242
passwordlocation option (voicemail.conf),
    165
passwords
    generating, 97
    secure passwords, 94
    setting, 257
    strong passwords, 567
    validating, 162
pattern matching
    dialplan, 125–129
        ${EXTEN} channel variable, 128
        examples, 127
        syntax, 125
    DUNDi and Asterisk configuration, 516
PAUSE event, 298
PAUSEALL event, 298
PauseQueueMember() application, 268
PBX modules, 21
PBXs, and IVR systems, 389
pbxskip option (voicemail.conf), 160
pbx_ael module, 21
pbx_config module, 21
pbx_dundi module, 21
pbx_loopback module, 21
pbx_lua module, 21
pbx_realtime (asterisk.conf), 75
pbx_realtime module, 21
pbx_spool module, 21
PDFs, fax to PDF, 448
peer CEL event field, 539
peer definitions, dundi.conf, 506
peering, 590
peers, authentication, 634
penalizing queue members, 284
penalties, changing dynamically, 285
PENALTY event, 299
penaltymemberslimit option (queues.conf),
    276
performance
    paging multiple sets, 232
    server requirements, 641
    transcoding, 18
periodic-announce option (queues.conf), 279,
    289
periodic-announce-frequency option
    (queues.conf), 278, 288
permit option (dundi.conf), 507
permit option (manager.conf), 462

About the Authors

Leif Madsen first got involved with the Asterisk community when he was looking for a voice conferencing solution. Once he learned that there was no official Asterisk documentation, he co-founded the Asterisk Documentation Project. Leif is currently working as a consultant, specializing in Asterisk clustering and call-center integration. You can find out more about him at *http://www.leifmadsen.com.*

Jim Van Meggelen is a founding partner and CTO of Core Telecom Innovations, Inc., and iConverged LLC, providers of open source telephony solutions for the enterprise. He has more than 20 years of enterprise telecom experience, and has been working with VoIP for most of his career.

Russell Bryant is the Engineering Manager for the Open Source Software team at Digium, Inc. He has been a core member of the Asterisk development team since the Fall of 2004. At the first AstriCon in 2004, he was named the release maintainer for Asterisk's first major release series, Asterisk 1.0. He has since contributed to almost all areas of Asterisk development, from project management to core architectural design and development.

Colophon

The animals on the cover of *Asterisk: The Definitive Guide* are starfish (*Asteroidea*), a group of echinoderms (spiny-skinned invertebrates found only in the sea). Most starfish have fivefold radial symmetry (arms or rays branching from a central body disc in multiples of five), though some species have four or nine arms. There are over 1,500 species of starfish.

Starfish live on the floor of the sea and in tidal pools, clinging to rocks and moving (slowly) using a water-based vascular system to manipulate hundreds of tiny, tube-like legs, called *podia*. A small bulb or *ampulla* at the top of the tube contracts, expelling water and expanding the starfish's leg. The ampulla relaxes, and the leg retracts. At the tip of each leg is a suction cup that allows the starfish to pry open clam, oyster, or mussel shells. Starfish are carnivores; they eat coral, fish, and snails, as well as bivalves.

Starfish can flex and manipulate their arms to fit into small places. At the end of each arm is an eyespot, a primitive sensor that detects light and helps the starfish determine direction. Starfish also have the ability to regenerate a missing limb. Some species can even regrow a complete, new starfish from a severed arm.

The cover image is from the Dover Pictorial Archive. The cover font is Adobe ITC Garamond. The text font is Linotype Birka; the heading font is Adobe Myriad Condensed; and the code font is LucasFont's TheSans Mono Condensed.

Related Titles from O'Reilly

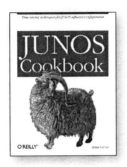

Networking

802.11 Wireless Networks: The Definitive Guide,
  2nd Edition

Asterisk: The Future of Telephony, *2nd Edition*

Cisco IOS Cookbook, *2nd Edition*

Cisco IOS Access Lists

Cisco IOS in a Nutshell, *2nd Edition*

DNS & BIND Cookbook

DNS and BIND, *5th Edition*

Essential SNMP, *2nd Edition*

Exchange Server Cookbook

IP Routing

IPv6 Essentials, *2nd Edition*

IPv6 Network Administration

JUNOS Cookbook

JUNOS Enterprise Routing

LDAP System Administration

Managing NFS and NIS, *2nd Edition*

Network Troubleshooting Tools

Network Warrior

RADIUS

ScreenOS Cookbook

sendmail, *4th Edition*

sendmail Cookbook

SpamAssassin

Switching to VoIP

TCP/IP Network Administration, *3rd Edition*

Time Management for System Administrators

Using Samba, *3rd Edition*

Using SANs and NAS

VoIP Hacks

Windows Server 2003 Network Administration

Wireless Hacks, *2nd Edition*

Zero Configuration Networking: The Definitive Guide

Our books are available at most retail and online bookstores.
To order direct: 1-800-998-9938 • *order@oreilly.com* • *www.oreilly.com*
Online editions of most O'Reilly titles are available by subscription at *safari.oreilly.com*

Get even more for your money.

Join the O'Reilly Community, and register the O'Reilly books you own. It's free, and you'll get:

- $4.99 ebook upgrade offer
- 40% upgrade offer on O'Reilly print books
- Membership discounts on books and events
- Free lifetime updates to ebooks and videos
- Multiple ebook formats, DRM FREE
- Participation in the O'Reilly community
- Newsletters
- Account management
- 100% Satisfaction Guarantee

Signing up is easy:

1. **Go to: oreilly.com/go/register**
2. **Create an O'Reilly login.**
3. **Provide your address.**
4. **Register your books.**

Note: English-language books only

To order books online:
oreilly.com/store

For questions about products or an order:
orders@oreilly.com

To sign up to get topic-specific email announcements and/or news about upcoming books, conferences, special offers, and new technologies:
elists@oreilly.com

For technical questions about book content:
booktech@oreilly.com

To submit new book proposals to our editors:
proposals@oreilly.com

O'Reilly books are available in multiple DRM-free ebook formats. For more information:
oreilly.com/ebooks

Spreading the knowledge of innovators   oreilly.com

Buy this book and get access to the online edition for 45 days—for free!

Programming Python, 4th Edition

By Mark Lutz
December 2010, $64.99
ISBN 9780596158101

With Safari Books Online, you can:

Access the contents of thousands of technology and business books

- Quickly search over 7000 books and certification guides
- Download whole books or chapters in PDF format, at no extra cost, to print or read on the go
- Copy and paste code
- Save up to 35% on O'Reilly print books
- **New!** Access mobile-friendly books directly from cell phones and mobile devices

Stay up-to-date on emerging topics before the books are published

- Get on-demand access to evolving manuscripts.
- Interact directly with authors of upcoming books

Explore thousands of hours of video on technology and design topics

- Learn from expert video tutorials
- Watch and replay recorded conference sessions

To try out Safari and the online edition of this book FREE for 45 days,
go to *www.oreilly.com/go/safarienabled* and enter the coupon code GUPHAZG.
To see the complete Safari Library, visit safari.oreilly.com.

CPSIA information can be obtained at www.ICGtesting.com
Printed in the USA
LVOW021103031011

248859LV00003B/64/P

9 780596 517342

5345 144

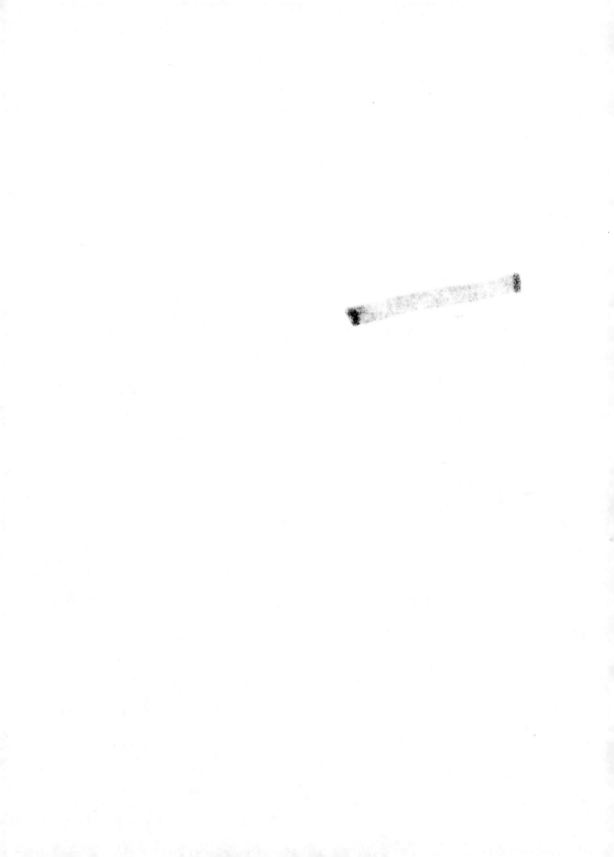